7th European Biosolids and Organic Residuals Conference and Exhibition

Proceedings of the 7th European Biosolids and Organic Residuals Conference

Monday 18th November 2002
To Wednesday 20th November 2002

Edited by
Paul Lowe
and
John A Hudson

Volume 2

© Aqua Enviro Technology Transfer

Volume 2
ISBN 1-903958-06-7

Published by:
Aqua Enviro Technology Transfer

Writing this forward for the 7th European Biosolids and Organic Residuals Conference Exhibition and Workshop makes one realise how much activity has taken place in the last 12 months both in legislation, regulation, research, process development, design, commissioning and operation of biosolids and organic residuals techniques and technologies.

Public concern over the release of pathogens into the environment from biosolids and the organic waste industry has been matched by a response from the industry. Looking through the papers for this conference the term "log-reduction" has become a standard reporting determinand. So whatever technology you are interested in, be it farm waste, compost, dried sewage sludge, lime treated sludge etc., you will find in these proceedings a wealth of data and information on how these processes perform in terms of pathogen reduction. Of course that is not the only reoccurring theme. The long-term sustainability of biosolids and organic residuals recycling is addressed. Information on metal content of sludges and soils can be found in many papers including the effect they have on growing plants.

While these data enable Regulators and Politicians to make value judgements on the sustainability of recycling operations it should not be forgotten that there are other important benefits both to the community, farmer and producers. Again important conclusions can be found in these proceedings and as such these papers form an up-to-date source of information to support the advantages to be gained from the utilisation of biosolids and organic residuals in agriculture and land restoration.

The industry continues to address the issues of process effectiveness and efficiencies. The Anaerobic Digestion process has formed the backbone of sewage sludge treatment for many years. Now we see Anaerobic Digestion coming of age in the organic waste industry. Papers to this conference not only address the problems associated with the anaerobic digestion of organic wastes but they also describe the advances which have been in pre-treatment and in process techniques for the anaerobic treatment of sewage sludge. No doubt some of these benefits will eventually be transferred across into the organic residuals treatment techniques. Equally the sewage sludge designers have a lot to learn from the organic residuals sector, who has found that high dry solids mixtures can be successfully stabilised by the anaerobic process.

Since our last conference, only a year ago, the value of lime as a means of effectively treating sewage sludge to achieve "advanced status" has grown with new techniques appearing in this compendium of papers together with more substantial operating data. In the last few years composting technology has taken a back seat, however, this year it is observed how interest in the various techniques has grown both for sewage sludge but in particular for organic residuals.

Of course the thermal processes continue to be explored. The proceedings contains case histories of how safety issues associated with sludge drying plants, which 12 months ago was a hot topic, has been addressed. High temperature systems both for pasteurisation, sterilising, and organic matter destruction all have a place in this year's conference proceedings offering one stop solutions. Analytical techniques and the interpretation of data is a going concern especially in the areas of pathogen resuscitation and enumeration techniques and in the meaning of the results for trace organic compounds.

New rules, regulations and Directives continue to drive the industry forward. The landfill Directive is seen to have an impact on the waste industry while the biosolids industry awaits the revision to the use of sewage sludge in agriculture directive and the impact of a European Soil Protection Strategy. Concerns of stakeholders have to be expressed. This year the final conference closing session is devoted to the impact of recently published (October 2002) draft regulations where such concerns will no doubt be expressed.

It is pleasing to see a number of papers produced by young people who have just entered the industry or are about to enter the industry. For some the presentation of a paper and review by more senior people is a new experience, however, the quality of some of these papers is very good even though there may be a few more figures and tables that the more experienced authors would offer. So I congratulate the young authors for making the effort to prepare their papers for such an enthusiastic and informed fraternity of sludge and organic waste audience. I would also like to express my thanks to the many authors who have taken time out to prepare their papers and presentations. Without your efforts there would be no conference. It is particularly encouraging to see an ever-increasing number of papers being presented at this conference from EU member States as well as USA and other countries. These proceedings do give an international perspective on the biosolids and organic residuals industries. John and I hope our editing has not been too severe.

I wish to express my thanks to Dr John Hudson, Dr Nigel Horan and Sarah Hickinson for the behind-the-scenes work that contributes so much to the success of the conference. My thanks go to Wastewater Technical Panel of the Chartered Institution of Water and Environmental Management (CIWEM) in particular Dr Tim Evans, the Panel Chair, and to the branch secretaries and branch chairs who help promote the event at local level.

Finally my thanks goes to CIWEM, CIWM, The Compost Association, The European Water Association and UK Water Industry Research for their endorsement of this conference and to the exhibitors who bring a focal point to the many breaks in the conference deliberations.

Paul Lowe
Conference Director
Aqua Enviro Consultancy Services
3 Shirley Court, South Marine Drive, Bridlington, Easy Yorkshire, UK YO15 3JJ
Tel: (44) 1 262 677 835 Fax: (44) 1 262 677 835 E-mail paul-lowe@aqua-enviro.co.uk
Web Site: www.european-biosolids.com
17 November 2002

THE EDITORS

Paul Lowe is the Senior Partner at Aqua Enviro Consultancy Services. He is a Chartered Chemist, a Fellow of the Royal Society of Chemistry. He is a Chartered Water and Environmental Manager a Fellow of the Chartered Institution of Water and Environmental Management, and a Member of the Chartered Management Institute. He is also a member of WEF. He is a Member of the Select Society of Sanitary Sludge Shovellers. For many years he was the secretary of the IWQ Large Wastewater Treatment Plant Group and is currently a member of that group and the Sludge Management Group. He has 40 years of experience in the water industry having worked for the Cities of Bradford, Leeds and Wakefield before joining Yorkshire Water on it's inception in April 1974 as Area Manger. In 1982 he moved to the Head Office of that organisation as Chief Technical Adviser leaving in 1989 to set up Aqua Enviro. He is a visiting lecturer at the University of Leeds, School of Civil Engineering were he shares his experience with post-graduate students. He has published over 30 papers on sludge and related topics and is often engaged as an "expert witness" on sludge and wastewater contractual disputes. After visiting the WEF Biosolids conference for the first time some nine years ago he became convinced that Europe needed an annual conference of similar status to provide a forum where managers, engineers and scientists could meet to exchange information and data on biosolids and organic residuals. He therefore established such a conference which in 1995, now in it 7th year and gaining influence throughout Europe.

John Hudson is an Independent Consultant working in the wastewater field. He holds a PhD from the University of London and a MSc in Chemistry. He is a Chartered Water and Environmental Manager and Fellow of the Chartered Institution of Water and Environmental Management. He is a Member of the Select Society of Sanitary Sludge Shovellers. He was awarded by WEF the Hatfield Prize in 1984. After working in the research department of British Rail he joined the City of Leeds Water Pollution Control Department in 1973 as Chief Chemist. On the formation of Yorkshire Water he became the Chief Water Pollution Control Chemist. He then served as Area Manager in the Harrogate, the Yorkshire Dales and in the West Yorkshire Conurbation of Halifax, Huddersfield and Dewsbury. He has been at the forefront of research into sewage sludge and biosolids throughout his career with notable projects on the disinfection of sewage sludge and the comparison of aerobic and anaerobic treatment back in the 1980's. He has published over 25 papers on sewage sludge and related topics and jointly received the Institutions Publication award for a paper on the Operating Experiences of Sludge Stabilisation and Disinfection at Colburn STW. John became involved in sewage sludge incineration back in 1973 on the commissioning of the City of Leeds plant later riving his interest in 1987 when he took over responsibility for operation of the Calder Valley Sludge Incinerator. He has also has an active interest in the treatment and disposal of sewage sludge from small works. He is a visiting lecturer to the UK Water Training, the University of Salford and the Aqua Enviro Sludge Master Class.

PROCEEDINGS 7TH EUROPEAN BIOSOLIDS AND ORGANIC RESIDUALS WORKSHOP, CONFERENCE AND EXHIBITION

VOLUME 2 INDEX

8. DEVELOPMENTS IN ANALYTICAL TECHNIQUES FOR SLUDGE AND ORGANIC WASTES

Session	Paper No	Authors	Paper Title
Session 8	Paper 056	Sanin, F D., Middle East Technical University, Turkey	Enzymatic extraction of activated sludge extracellular polymers
Session 8	Paper 057	Yuncu, B., Sanin, F.D., Yeli, U., Middle East Technical University, Turkey	An investigation of Pb(II) biosorption in relation to surface characteristics of activated sludge
Session 8	Paper 058	[1]Lang, N., [1]Smith,S.R., Pike,E.B., [2]Rowlands, C., [1] Imperial College, UK, [1]Severn Trent Water, UK	Methods for the enumeration of salmonella species in sewage sludge and soil
Session 8	Paper 059	[1]Patria, L., [1]Ducray, F*, [2]Huyard A., [1]Anjou Recherche (Vivendi), France, [2]CIRSEE (Ondeo), France	Biosolids of agronomic quality
Session 8	Paper 060	Thompson,C.K, Alcontrol Laboratories, UK	Issues related to biosolids and organic residuals chemical and microbiological analysis

9. APPROPRIATE TECHNOLOGIES

Session	Paper No	Authors	Paper Title
Session 9	Paper 061	Hertle, A., GHD Pty Ltd, Australia	An investigation into the causes of deteriorated sludge dewatering performance
Session 9	Paper 062	[1]Thomas, G., [2]Smith, J., [3] Haywood D [1]Carl Bro, [2]Wessex Water, [3]Ashact Ltd, UK	Treatment of sludge from chemically-dosed wastewater – a case study
Session 9	Paper 063	Mountford, L., Simon-Hartley Ashbrook, UK	Sludge thickening for small sewage treatment works
Session 9	Paper 064	[1]Whitlock, L., [2] Sellgren, A., [1]GIW Industries Inc, USA [2]Lulea University of Technology, Sweden,	Pumping a simulated paste-like sludge with a modified centrifugal pump
Session 9	Paper 065	Mormede, S., Atkins Consultants Ltd, UK	The use of vermistabilisation for the treatment of liquid sewage sludges
Session 9	Paper 066	Cooper P, Willoughby, Cooper DN. ARM, UK	Advances in the use of reed bed for sludge drying
Session 9	Paper 067	Vella, P., Carus Chemical Company, USA	Permanganate applications at public works, industrial and agricultural sites
Session 9	Paper 068	Hobbs,P.J.[1], Misselbrook,T.H.[1], Noble R.[2], and Persaud K.C.[3], [1]Institute of Grassland and Environmental Research, [2]Horticultural Research Institute,. UMIST.	Chemical indicators that determine the olfactory response from organic sources
Session 9	Paper 069	Findlay, E., COPA Ltd, UK	The significance of speed of rotation and wholelife costs in RBC tender evaluation.

063 did not make printing deadline

10. THERMAL PROCESSES

Session	Paper No	Authors	Paper Title
Session 10	Paper 070	Reed G.P., Zhuo,Y., Paterson, N., Dugwell, D.R., Kandiyoti,R., Imperial College of Science Technology and Medicine, UK	Sewage sludge gasification: trace element depletion and enrichment in the output solid stream.
Session 10	Paper 071	Paterson, N., Zhuo, Y., Reed, G.P., Dugwell, D.R., Kandiyoto, R., Imperial College of Science, Technology and Medicine, UK	Processing of sewage sludge: pyrolysis and gasification in a laboratory scale, spouted bed reactor.
Session 10	Paper 072	Svanstrom, M., Fröling M Chambers University of Technology, Sweden	Environmental assessment of energy recovery from sewage sludge through supercritical water oxidation.
Session 10	Paper 073	Gilbert, A.B., Bigot, B.,Cretenot D OTVB, UK	Wet oxidation using ATHOS a WAO alternative; update
Session 10	Paper 074	[1]Ludwig, P., [2]Gross, G. [1] Infraserv GmbH & Co Hoechst KG, [2]Messer Griesheim GmbH, Germany	First oxygen debottlenecking of large scale fluidised bed furnace for sewage sludge incineration
Session 10	Paper 075	Chauzy, J., Cretenot D, Fernades, P., Patria, L., Anjou Recherche (Vivendi) Vivendi Water Systems, France	BIO THELYS®: a new sludge process for sludge minimisation and sanitation
Session 10	Paper 076	Casey T, Bridle T, Ashford P,Environmental Solutions International Ltd,	THE ENERSLUDGE™ PROCESS – Operational Experience from commercial facility.
Session 10	Paper 077	[1]Wilson S, [2]Panter K, [1]Scottish Water, [2] EBCOR Ltd,	Operating experience of Aberdeen Cambi thermal hydrolysis plant 2002

11. NOVEL PROCESSES

Session	Paper No	Authors	Paper Title
Session 11	Paper 078	Stephenson, R., Laliberte, S., Elson P,. Paradigm Environmental Technologies Inc, Canada	The *MicroSludge*™ process to destroy biosolids and pathogens from combined primary and secondary solids from a municipal WWTP
Session 11	Paper 079	[1]Shaw, J.,[2] Nemeth, L.,[3] Henderson, G., [1]Paradigm Environmental Technologies Inc, Canada, [2]EarthTech Canada Inc, Canada, [3]EarthTech Australia Inc, Australia	Technical and economical benchmarking of the MICROSLUDGE™ process with biological and non-biological disposal options
Session 11	Paper 080	Gilbert, A.B[1]. Fernandes, P[2]. [2].Baratto, G [1]OTVB Ltd [2]Anjou Recherche	SAPHYR™: a stabilising biosolids treatment process with flexible disposal routes
Session 11	Paper 081	[1]Walker, J, [2] Glendinning, S [1]Jim Walker Consultant, [2]University of Newcastle, UK	*In-situ* dewatering of lagooned sewage sludge using electrokinetic geosynthetics (EKG)
Session 11	Paper 082	[1]Mene, R., [2]Lebrun, T., [1] Ondeo Degremont Ltd, UK,[2]Ondeo Degremont SA, France	BIOLYSIS®: cutting edge technology for the reduction of sludge quantities in activated sludge plants
Session 11	Paper 083	J.E.Taylorand [2]T Donnely [1]Environmental Performance Technologies [2] University of Newcastle	The Removal of Nitrogen and Phosphorus from the oxidation of Biosolids using the Advanced Fluidised Composting (AFC) Process™

079 did not make printing deadline

12. STRATEGY AND SUSTAINABILITY

Session	Paper No	Authors	Paper Title
Session 12	Paper 084	Toogood, S., Effluvium Ltd, UK	Having your sludge cake and eating it? A Discussion of the Psychology of Biosolids Application
Session 12	Paper 085	[1]de la Fuentes, [2] L., Sauders, B., [3]Klemes, J., [1]GAIKER, Spain, [2]ADAS Consulting, [3]UMIST	AWARENET: Agro-food wastes minimisation and reduction network
Session 12	Paper 086	Navarro, R.R., Perez, J.M.R., Valencia City Council, Spain	Life Ecobus: collecting used cooking oils to their recycling as biofuel for diesel engines
Session 12	Paper 087	Reekie, C., Burness, UK	PPP Waste Management Projects in Scotland
Session 12	Paper 088	McQueen, A ,Williams M Scottish Water, UK	Use of biosolids to protect the Machair, an internationally important habitat
Session 12	Paper 089	[1]Crowe, N., [2] Delaney, N., [1] ENTEC Ltd, [2]Nicholas O'Dwyers, Ireland	Sludge management in the Republic of Ireland
Session 12	Paper 090	Evans, T.D. [1] Lowe, N.R. [2] [1] Tim Evans Environment, [2] Enviro Consulting,	Partnership for sustainable of organic resources on land

086 and 089 did not make printing deadline

13. PATHOGEN REDUCTION

Session	Paper No	Authors	Paper Title
Session 13	Paper 091	[1]Siljehag, P., [2]Low, E., [1]University of Edinburgh, UK, [2]Mott MacDonald, UK	Controlling pathogens in sludge using Hazard Analysis and Critical Control Point methodology
Session 13	Paper 092	[1]Davies, G.,[1] Vuong, J., [2]Griffiths, A. ,[2]Millns N. ,[1]Pell Frischmann, UK [2]South West Water Ltd, UK	HACCP implementation
Session 13	Paper 093	Mormede, S., Clark P. Atkins Water	Secondary storage of digested sludge improves bacteria kill
Session 13	Paper 094	(1) Liz Bowman, [2] Patrick Coleman [1]Target Alliance (Thames Water) [2]Target Alliance (Black and Veatch)	Selection of a pre-pasteurisation technology for the new Reading sewage treatment works
Session 13	Paper 095	[1]Mayhew, M, [2]Davies, W.J., [1]United Utilities, UK, [2]Alpha Environmental Technology, UK	Experiences with the Alpha Biotherm biosolids pasteurisation process at United Utilities Ellesmere Port Sludge Processing Centre
Session 13	Paper 096	Brenton, M., Simon-Hartley Ashbrook , UK	Log Reduction – Two processes to meet US EPA Class A and European advanced criteria standards

096 did not make printing deadline

14. THERMAL DRYING

Session	Paper No	Authors	Paper Title
Session 14	Paper 097	[1] Brown M, [2] Thompson K [3] Bentley D E, [1] Damar Engineering Ltd. [2] TVD (UK) Ltd. [3] DryVac Environmental	Combined filter press/dryer technology -an option for enhanced treatment -
Session 14	Paper 098	Wild, R., Clay, S., Lilly, B., Severn Trent Water	Sludge drying in the East Midlands region of the UK
Session 14	Paper 099	[1] Groenewegen PJ, [1] [2] Wilson A, [3] Callander C, [4] Christison M, [1] Flo-Dry Engineering Limited, [2] New Plymouth District Council [3] Beca Carter Hollings and Ferner Ltd [4] Hutt Valley Water Services Ltd	Thermal Sewage Sludge Drying – The New Zealand Experience
Session 14	Paper 100	[1] Gary Meades, [2] Marcel Andrews [1] Anglian Water [2] Halcrow	Fast track construction of a sludge dryer
Session 14	Paper 101	Dehing, F., Englebert, B., SEGHERSBetter technology, Belgium	Drying-Pelletising of different types of sludge and biosolids
Session 14	Paper 102	Birkinshaw I, Duggleby J Portosilo Ltd	Safe storage of sewage sludge
Session 14	Paper 103	Manchester S Building Research Establishment	Case Study: Fire in a thermal sewage sludge dryer
Session 14	Paper 104	[1] Whipps A, [2] Lloyd Robin, [1] Pell Frischmann Consultants, [2] South West Water.	The Impact of the HSE and DSEAR on the UK Water Industry.

101 did not make printing deadline

15. THE NEW UK SLUDGE REGULATIONS

Session	Paper No	Authors	Paper Title
Session 15	Paper 105	McDonnell, E., DEFRA, UK	Potential impact of the EU Soil Strategy on the revision of the sludge directive.
Session 15	Paper 106	Rowlands, C., Severn Trent Water, UK	The impact of the revised sludge to land regulations upon practices in England and Wales
Session 15	Paper 107	[1] Crathorne B, [2] Rowlands C, [3] Bryson P, [4] Cochrane J, [4] Sweet N. [1] Thames Water and UK Water, [2] Severn Trent and UK Water, [4] Environment Agency	Implementation of HACCP Controls under the new Sludge (Use in Agriculture) Regulations.

105, 106 and 107 did not make printing deadline

POSTERS

Session	Paper No	Authors	Paper Title
Exhibition area	Poster 1	Briancesco, R., Chiaretti, G., Coccia,A.M., Della, L.S., Marini, R., Bonadonna, L., Instituto Superiore di Sanita di Roma, Italy	Microbiological quality of composting products
Exhibition area	Poster 2	Beccaloni, E., Coccia, A.M., Gucci P M B., Musmeci, L., Stacul,E., Ziemacki,G., Italian National Institute of Health, Italy	Presence of indoor microbial and inorganic pollutants with relation to stabilised biowaste production.
Exhibition area	Poster 3	[1] Mancini L., [1] De Lucia B, [2] Romano V., [2] Portincasa F. [1] Dipartimento di Scienze delle Produzioni Vegetali. [2] Acquedotto pugliese s.p.a. Via Cognetti, 36. 70100, Bari, Italia.	Bioagronomic Evaluation Of Soil Organic Or Mineral Fertilisation: Assessment Of Gladiolus Productivity.

ENZYMATIC EXTRACTION OF ACTIVATED SLUDGE EXTRACELLULAR POLYMERS

F. Dilek Sanin

Department of Environmental Engineering, Middle East Technical University, 06531 Ankara, Turkey
Phone: + 90 312 210 2642, Fax: + 90 312 210 1260, e-mail: dsanin@metu.edu.tr

ABSTRACT

This study examines the enzyme hydrolysis as a possible method of extracellular polymer extraction from mixed culture activated sludge. Knowing that the major components of extracellular polymers are carbohydrates and proteins, these components are targeted during the developed extraction procedure. Towards this end a carbohydrate specific enzyme (α-amylase) and a protein specific enzyme (proteinase) are used during the study. The extraction kinetics is investigated first for each enzyme studied, using mixed culture activated sludge samples grown in replica reactors. Following the determination of extraction time, enzyme dose optimization is investigated. Results indicate that the extraction of extracellular polymers by enzymes (both α-amylase and proteinase) is a quick process reaching equilibrium within two to three hours. As the doses of enzymes are increased, the extracted polymer quantities increase up to a certain dose, beyond which not much extraction was observed. Carbohydrate hydrolyzing enzyme (α-amylase) extracted some small amount of proteins along with the carbohydrates and protein-hydrolyzing enzyme (proteinase) extracted some carbohydrates together with the proteins indicating that proteins and carbohydrates exist bound to each other in the extracellular polymer network of sludge.

KEYWORDS

Activated sludge, α-amylase, carbohydrates, extracellular polymers, proteinase, proteins

INTRODUCTION

The success of gravity separation of biomass in activated sludge systems is dependent on the bioflocculation of the microbial slurry entering the secondary clarifier. Despite its importance in the overall effectiveness of the activated sludge process, bioflocculation remains still poorly understood. The inability of the wastewater treatment profession to adequately predict the rate and extent of bioflocculation is a handicap in terms of both economics and pollution control[1].

Flocculation is a state of a system such that it affects the subsequent behavior of that system. The recent interest in bioflocculation of activated sludge organisms originates from this fact. Flocculated state of activated sludge determines the physical properties of sludge such as dewatering, rheology, and settleability. Since all of these are important in the operation of wastewater treatment plants, the identification of mechanisms of bioflocculation becomes important.

For the activated sludge flocs to form, individual microorganisms, at least a couple of them have to get together forming the core, a colony. Then they further aggregate among themselves, with some dead bacteria, extracellular materials, and larger microorganisms like protozoa and filamentous bacteria. At the end of this process, flocs with various sizes and shapes and inhomogeneous composition are formed. It is now accepted that the extracellular polymers are functional in bioflocculation. These polymers can be contributed from the microbial system itself or by the wastewater. The first group includes the extracellular and intracellular polymers produced by bacteria during growth. Presence of intracellular polymers in the floc structure can be explained by bacterial lysis and death by which intracellular molecules remain in floc attached to it. The second group includes polymers like cellulose, humic acids, etc. which are carried by the wastewater and brought into the system and then they interact within the floc structure. They are believed to be loosely adsorbed and weakly bound to the rest of the floc[2]. Most bacterial ECPs are believed to be made up of polysaccharides, proteins, nucleic acids and lipids.

For further understanding of the bioflocculation phenomena as well as the composition of the extracellular polymers, a variety of extraction techniques have been developed to remove the polymer from the floc structure. These cover some physical and chemical techniques which may range from mild to very harsh methods. The critical issue in polymer extraction has been the selection of technique. If the technique is harsh it may end up lysing the bacterial cells causing intracellular contamination of the extracellular medium. On the other hand, if the technique is a mild technique it would not extract much of the polymers and would be useless.

One of the methods to extract these polymers is the use of certain enzymes that degrade the extracellular polymers or their specific components. In the literature, studies related to enzyme extraction especially for ECP analysis are very limited.

Lysozyme and celluase enzymes were applied to investigate the composition of extracellular polysaccharides[3]. Then, a floc forming bacterium, *K. cryocrescens*, was isolated from activated sludge using a proteolytic enzyme, Actinase E, as a deflocculating agent[4]. Deoxyribonuclease, ribonuclease and protease were applied to sludge to obtain purified activated sludge polysaccharides[5]. Cellulase was used to see its effect on viscosity of sludge[6]. Cellulase, lipase, proteinase enzymes were used to modify the surfaces of activated and digested sludge and it was seen that for activated sludge together with polysaccharides, proteins were important surface polymers[7]. Since ECP is believed to be largely polysaccharide in nature, polysaccharide enzyme preparation was used in mitigating the effects of extracellular polysaccharide deposition on soil hydraulic conductivity[8]. The enzymes were also used for better understanding the role of extracellular proteins and carbohydrates [9]. Pronase was used to hydrolyze extracellular protein and cellulase was used for hydrolyzing polysaccharides.

The most general reaction when proteinaceous materials are subjected to protein degrading enzyme is the hydrolysis of proteins to amino acids. This frees the $-NH_2$ and $-COOH$ groups of the peptide bond. In general proteinases appear to be less specific in their action than other enzymes; they do not cleave bonds only at certain sites. The hydrolysis of peptide bonds occurs at random and the protein is degraded to individual amino acids and short peptide chains[10]. Amino acid links with the next via a peptide bond, thus suggesting that a single enzyme can break all the subunits[11].

On the other hand, enzymes that hydrolyze polysaccharides are specific for the type of bond to be broken and the monomers involved. For example, most of the glucose units are liked head to tail in α-1,4 glycosidic bonds; whereas some are linked α-1,6 by providing branching points in the molecule[10, 11]. Cellulose is a polymer of glucose units with β-1,4 linkages and requires enzyme cellulase. Starch and glycogen are polymers of glucose having α-1,4 linkages, glycogen and one form of starch, amylopectin, are branched with α-1,6 linkages at the branched points. While the linear starch molecule formed by 1,4 glycosidic bonds, amylose, and the linear portions of amylopectin and glycogen can be degraded by amylase, complete degradation of branched molecules requires debranching enzyme α-1,6 glucosidase. Hydrolysis may convert polysaccharide to monosaccharides, disaccharides, or oligosaccharides[10]. Since enzymes are specific in their action, they can be used in a controlled manner in the extraction of ECPs and may end up being an effective method of polymer extraction. With this insight and knowing that the predominant components of ECP are polysaccharides and proteins, effectivenesses of a polysaccharide and a protein-hydrolyzing enzyme are investigated in this study as potential methods of ECP extraction. Therefore, the purpose of this study is to investigate the possibility of using enzymes in extracting ECPs; specifically α-amylase (a polysaccahride degrading enzyme) and proteinase (a protein degrading enzyme) are used in the study.

MATERIALS AND METHODS

Mixed culture bacteria grown in semi-continuous reactors were used during the experiments. The microbial seed was obtained from the primary settling tank effluent from the city of Ankara wastewater treatment plant.

Two replica reactors were operated with a working volume of 2 liters and at a mean cell residence time (MCRT) of 8 days. The reactors were located into the water bath, temperature of the system was kept at 20°C, and pH was 7.0±0.2. Required oxygen was supplied by air pumps which were also providing completely mixed conditions in the reactors. Dissolved oxygen concentration in reactors was kept at a minimum of 3 mg/L. The reactors were fed with synthetic medium, composition of which in terms of mg/L was: Glucose: 935; Peptone: 200; K_2HPO_4: 600; KH_2PO_4: 300; NH_4Cl: 225; $MgSO_4.7H_2O$: 112.5; $FeSO_4.7H_2O$: 3.75; $ZnSO_4.7H_2O$: 3.75; $MnSO_4.7H_2O$: 3.75; $CaCl_2$: 15; $NaHCO_3$: 180.

The average ammonia nitrogen in the feed was 59mg/L with the total nitrogen (TKN) being equal to 62mg/L.

Reactors operated under the above listed conditions were first brought to steady state, which was followed by measuring the effluent COD, MLSS, and MLVSS concentrations as described by Standard Methods[12]. Levelling of these parameters measured during several successive days indicated that the steady state has been reached. Once the steady state was reached, polymer extraction experiments were conducted on 250 mL of daily wasted sludge from each reactor.

As the polymer extraction method, a carbohydrate and a protein-extracting enzyme were used. With this purpose α-amylase enzyme was used to extract the carbohydrates in extracellular polymers of activated sludge. α-Amylase from *Bacillus subtilis* was supplied from Fluka (10070) with an activity of 50 U/g.

A protein-extracting enzyme is also employed in extracting extracellular proteins. Proteinase enzyme was used to extract the proteins in extracellular polymers of activated sludge. Proteinase from *Aspergillus sp.* was provided from Fluka (82539) having an activity of 3 U/g.

In developing a technique, extraction was investigated in terms of its time and dose dependent behavior. The time dependent extraction with each enzyme was tested separately at a fixed dosage. Next, enzyme dose was varied and the extracted polymers were quantified. In this part, an enzyme stock solution was prepared and according to the amount of microorganisms present in reactors, different dosages were applied for polymer extraction. Prior to extraction process, sludge samples (200mL) removed from the reactors were washed by centrifuging the samples at 3500 rpm for 15 minutes. The centrates were discarded and the pellets were resuspended to 200mL in Phosphate Buffer Saline (PBS) solution (NaCl: 4 g/L, KCl: 0.1 g/L, KCl: 0.1 g/L, Na_2HPO_4: 0.455 g/L). Polymer extraction tests were applied immediately after the washing step of the sludge. Resuspended pellets (200 mL) were transferred to an extraction beaker and then enzymes at predetermined amounts (based on the volatile solids (VS) measurements in the reactor) were added. Using a standard jar test apparatus operated at a stirring speed of 120 rpm performed polymer extractions. In each extraction test, sludge was accompanied by two controls. When the sample being tested was a sludge sample from a reactor with enzyme addition, the first control was a sludge sample from the same reactor without enzyme (amylase or proteinase) addition to see the effect of mixing on the release of polymers. Second control was the "enzyme only" added to 200 mL of PBS solution in an amount equal to that of enzyme (amylase or proteinase) added to sludge sample for extraction of polymers. "Enzyme only" was mixed at the same extraction conditions without any sludge in it. This control was especially important to know the amount of proteins originating from the addition of enzyme itself. In calculations of the extracted polymers control measurements were taken into account by subtracting their quantities.

In order to optimize the method, time dependent experiments were carried out at a fixed enzyme dose (amylase at 0.4 g/gVS and proteinase at 0.5 g/gVS) for a time period of 8 hours. Hourly samples were collected from the liquid phase of each extraction experiment including the controls starting from time zero.

Duplicate samples of supernatants removed from the settled sludge following extractions were centrifuged for 15 minutes at 3500 rpm. Centrates were analyzed for carbohydrate and protein content. Carbohydrates content of the extracted polymers were measured in duplicate samples by using phenol sulphuric acid method[13] using alginate as the standard. Protein content of the extracellular polymers was analyzed by using Lowry method[14] and Bradford method[15] immediately after the extraction of extracellular polymers. Bovine Albumin was used as the standard for calibration of the protein analysis methods.

After optimum time of extraction was determined, the effect of enzyme dose on the extraction of polymers was tested at this fixed time. Enzyme doses applied were 0.2, 0.4, 0.85, and 1.6 g/gVS for amylase and 0.2, 0.5, 0.8, 1.0 and 1.5 g/gVS for proteinase enzymes. Following the proteinase addition proteins are hydrolyzed into amino acids. Methods used after extraction detects only the proteins not the amino acids. To be able to determine how much protein is extracted at the end of each enzyme addition, total protein is measured and the extracted amount is determined by subtraction using the following technique. A 5 mL of sludge was taken from the well-mixed extraction vessel and centrifuged at 3500 rpm for 15 minutes. After discarding the supernatant, 25 mL 1 N NaOH was added to pellet in the test tube. Following 5 minutes boiling the contents of the test tube in a water bath, microorganisms were centrifuged at 3500 rpm for 15 minutes and samples removed from supernatant were analyzed for protein content. The same procedure for the solubilization of cell content was also applied to the control used to test the shear effect on polymer extraction. By subtracting proteins measured after proteinase extraction and that measured in the shear control, from the total protein content of the sludge, extracted proteins were calculated. The reliability of this method was verified before it is used in the extraction tests. A typical activated sludge sample with a mixed culture population was subjected to this analysis and the total protein was determined as described above. The total cell protein content, which was theoretically indicated to be between 40-60%[10], was determined to be 500 mg/g sludge (50%) by this method. For comparison purposes, same measurements were also conducted with Bradford protein measurement method. With this method only 10 % of the cell content was found to be protein. It was obvious that Bradford method underestimated the protein quantities. With this information, it was concluded that Lowry method is a reliable protein measurement method and that the cell disruption method applied in this study to expose all of the cellular protein for measurement worked successfully.

Oxygen uptake rate (OUR) of microorganisms were measured after each extraction to check the possibility of cell lysis originating from the extraction procedure. The method described in Standard Methods was applied[12]. A significant drop in OUR would indicate cell lysis, under these circumstances the experiments were discontinued due to a possible contribution of intracellular materials to extracellular polymers.

RESULTS AND DISCUSSIONS
AMYLASE EXTRACTION
TIME DEPENDENT EXTRACTION

Results of the 8-hour extraction test with amylase for reactor 1 are given in Figure 2. As seen in this Figure, measured carbohydrate concentration in the solution increases from the time of addition of amylase enzyme up to the end of 3 hours. Proteins measured by Bradford method increased until the end of 2-hour period. Beyond the 3-hour period, the extracted carbohydrates or the proteins do not change indicating that all extractable polymers are solubilized from the floc and contributed to the liquid phase within this time period.

Sludge from reactor 2 showed similar time dependent extraction patterns (results not shown) reaching equilibrium at about two hours, with polymer concentration values within 10% of the results presented for reactor 1. These results indicate that the enzyme extraction method is a relatively quick method. With this preliminary information, the time of extraction is fixed at 3 hours in the following dose-dependent experiments.

ENZYME DOSE OPTIMIZATION

Following time dependent experiments, the effect of different amylase doses on the extraction of polymers is tested. Figure 2 shows average values of polymers extracted from replica reactors (reactors 1 and 2). There was less that 3% variability between the results obtained from reactor 1 and reactor 2. In addition, the measured OUR values for reactor 1 under different amylase dosages are presented in Table 1. It is obvious from the results that, as the amylase dose increases, extracted protein and carbohydrate concentrations increase. Although amylase enzyme is carbohydrate specific, together with the carbohydrates some proteins (measured by Lowry method) are also extracted, even though their concentrations are low. These proteins can be the proteins released during the hydrolization of carbohydrates in ECP structure since carbohydrates and proteins are believed to be connected in ECP structure. Moreover, as one would expect, the increase in proteins extracted is smaller compared to increase in carbohydrates extracted with the increase in amylase dose. The increase in carbohydrates is sharp when amylase added is 0.2 g/gVS and when it is increased to 0.4 g/gVS. Carbohydrates still increased sharply when amylase was increased from 0.4 to 0.85 g/gVS. However, the increase in carbohydrates stabilized when the amylase dose was increased from 0.85 to 1.6 g/gVS, indicating that most carbohydrates in ECP structure that can be hydrolyzed by amylase enzyme are already solubilized. At an amylase dose of 0.85 g/gVS, extracted protein and carbohydrates are 3.7 and 11.9 mg/gVS, respectively. These values only increase slightly to 3.8 and 12.9 mg/gVS protein and carbohydrate, respectively at 1.6 g amylase/gVS. On the other hand, the OUR of microorganisms (Table 1) which indicate the viability, decreases relatively sharply when the amylase dose is increased from 0.85 to 1.6 g/gVS. These results altogether indicate that a limit on the extraction is reached at 0.85 g/gVS amylase dose in terms of extractable polymers. OUR results indicate that the viability is negatively affected after the dose of 0.85 g/gVS with the OUR value dropping from 4.6 to 3.9 mg/gVSS.h. From these results it is obvious that doses above 1.6 g/gVS would not make any difference in terms of polymers extracted, and would negatively affect the bacterial cells. So higher amylase doses were not tested thinking that the limit of ultimate polymer extraction has been reached with this enzyme.

PROTEINASE EXTRACTION
TIME DEPENDENT EXTRACTION

Results of the 8-hour extraction test with proteinase are given in Figure 3 for the first reactor. As seen in this Figure, the measured polymer concentration in the solution increases from the time of addition of proteinase enzyme up to the end of 1 to 2 hours. Results from the second reactor gave the same time dependent behavior with less than a few percent deviation of data from Reactor 1. Beyond this period, the extracted proteins and carbohydrates do not change indicating that all extractable polymers by proteinase are removed from the floc and passed into the liquid phase within this time period. With this preliminary information, the time of extraction is fixed at 3 hours in the following dose-dependent experiments with proteinase similar to those conducted with amylase.

ENZYME DOSE OPTIMIZATION

Proteinase enzyme is used to extract proteins in ECP structure. Figure 4 shows proteins and carbohydrates extracted as average values from replica reactors 1 and 2 at different proteinase doses. Table 2 lists the OUR values measured at each dose of proteinase enzyme. As can be seen from Figure 4, extracted protein and carbohydrate concentrations increase as the proteinase dose applied is increased.

Proteinase hydrolyze peptide bonds, therefore with the application of this enzyme on sludge samples, proteins in ECP structure are expected to be hydrolyzed into their amino acids and released into the solution. Similar to the extractions with amylase, some carbohydrates are extracted along with the proteins when the proteinase enzyme is used. The reason for this might be the ECP structure, as explained previously. It is expected that the proteins and carbohydrates to be bonded to each other in ECP structure. Extraction of one of them, to some extent, causes the hydrolysis of the other. As expected, the quantity of proteins extracted is overwhelmingly higher than the carbohydrate extracted especially at higher proteinase dosages. The quantity of polymers extracted showed an increase with the increasing dose of proteinase. The increase was sharper especially when the proteinase dose was increased from 0.2 to 0.5 g/gVS and then to 0.8 g/gVS as shown in Figure 4. The increase in concentration of extracted polymers become much less pronounced at a dose of 1 and then 1.5 g proteinase/gVS. The OUR results (Table 2) showed a more consistent decrease in the case of proteinase extraction than in the case of amylase extraction. Considering that the polymer quantity stabilizes at 1.5 g/gVS proteinase dose, the extractions were discontinued at this dose and higher doses were not tested.

Comparing the amount of carbohydrates and proteins extractable by these two enzymes, one can see that the quantity of proteins extracted by proteinase is higher than the quantity of carbohydrates extracted by amylase. The highest dose of proteinase (1.5 g/gVS) extracted 33.3 mg proteins/gVS, whereas, the highest dose of amylase (1.6 g/gVS) extracted 12.9 mg carbohydrates/gVS. This does not necessarily mean that there are more proteins in the sludge ECP network than carbohydrates. As discussed in the introduction, proteinase is a non-specific protein hydrolyzing enzyme, whereas, amylase is an enzyme specifically attacking the 1,4 glycosidic bonds. Results show that there are 1,4 glycosidic bonds present in sludge ECP network. However, comparing with the results in literature, non-specific methods of polymer extraction (like extraction with a cation exchange resin) extracts more carbohydrates from a similarly lab grown activated sludge (Durmaz and Sanin, 2001). Altogether these results indicate that 1,4 glycosidic bonds are abundant but not the only type of sugar bonds in the sludge ECP network. Extracted carbohydrate concentration with the application of proteinase is generally small compared to carbohydrates extracted from the reactors with the application of amylase.

CONCLUSIONS

Results of this study indicate that the extraction of extracellular polymers by enzymes is a mild process that successfully extracts the extracellular polymers. Enzyme extraction (by both α-amylase and proteinase) is a quick process reaching equilibrium within two to three hours. As the dose of enzymes is increased, the extracted polymer quantities increase up to a certain dose, beyond which not much extraction was observed. Comparing the amount of carbohydrates and proteins extractable by these two enzymes, one can see that the quantity of proteins extracted by proteinase is higher than the quantity of carbohydrates extracted by amylase. The highest dose of proteinase (1.5 g/gVS) extracted 33.3 mg proteins/gVS, whereas, the highest dose of amylase (1.6 g/gVS) extracted 12.9 mg carbohydrates/gVS. This does not necessarily mean that there are more proteins in the sludge ECP network than carbohydrates. Since proteinase is a non-specific protein hydrolyzing enzyme, it may find more bonds to attack, whereas, amylase is an enzyme specifically attacking the 1,4 glycosidic bonds. Results show that 1,4 glycosidic bonds are present but not the only type of sugar bonds in the sludge ECP network. α-amylase extracted some small amount of proteins along with the carbohydrates and proteinase extracted some carbohydrates together with the proteins indicating that proteins and carbohydrates exist bound to each other in the sludge extracellular polymer network.

REFERENCES

1. PARKER, D.S, MERRILL, M.S. AND TETREAULT, M.J. Wastewater treatment process theory and practice: the emerging convergence. *Water Sci. Tech.*, 1992, 25, (6), 301-315

2. KEIDING, K. AND NIELSEN. P. H. Desorption of organic macromolecules from activated sludge: effect of ionic composition. *Water Res.*, 1997??? 31, (7) 1665-1672

3. STEINER, A. E., MCLAREN, D. A. AND FORSTER, C.F. The nature of activated sludge flocs. *Water Res.*, 1976, 10, 25-30

4. KAKII, K., SHIRAKASHI. T. AND KURIYAMA. H. Isolation and characterization of a Ca^{+2} dependent floc forming bacterium. *J. Ferment Ttechnol.*, 1986, 64, 57-62

5. Horan, N. J. And Eccles, C. R. Purification and characterization of extracellular polysaccharide from activated sludges. *Water Res.*, 1986, 20, (11) 1427-1432

6. FORSTER C.F Sludge surfaces and their relation to the rheology of sewage sludge suspensions. *J. Chem.*

7. *Tech. Biotech.*, 1982, 32, 799-807

8. FORSTER C.F. Bound water in sewage sludges and its relationship to sludge surfaces and sludge viscosities. *J. Chem. Tech. Biotech.*, 1983, 33, 76-84

9. MAGESAN, G. N., WILLIAMSON, J. C., YEATES, G. W. AND LLOYD-JONES, A. RH. Wastewater C:N ratio effects on soil hydraulic conductivity and potential mechanisms for recovery. *Biores. Tech.*, 2000, 71, 21-27.

10. HIGGINS, M. J. AND NOVAK, J. T. Characterization of exocellular protein and its role in bioflocculation. *J. Env. Eng.*,1997, 123, (5) 479-485

11. GAUDY, A. AND GAUDY E. *Microbiology for environmental scientists and engineers.* 1980, Mc-Graw-Hill, New York, NY, 736 pages

12. BAILEY, J. E. AND OLLIS, D. F. *Biochemical Engineering Fundamentals.* 1986, Second Edition. Mc-Graw-Hill, New York. 1-85

13. APHA, AWWA, WPCF *Standard Methods for the Examination of Water and Wastewater.* 1998, 20th Edition. American Public Health Association, American Water Work Association, Water Environment Federation, Washington, D.C

14. DUBOIS, M., GILLES, K. A., HAMILTON, J. K., REBERS, P. A., AND SMITH, F. Calorimetric method for determination of sugar and related substances. *Anal. Chem.*, 1956, 28, 350-356

15. LOWRY, O. H., ROSEBROUGH, N. J., FARR, A. L.AND RANDALL, R. J. Protein measurement with the folin phenol reagent. *J. Biol. Chem.*, 1951, 193, 265-275

16. BRADFORD, M. M. A rapid and sensitive method for the quantification of microgram quantities of protein utilizing the principle of protein-dye binding. *Anal. Biochem.*, 1976, 72, 248-254

17. DURMAZ, B. AND SANIN F. D. Effect of Carbon to Nitrogen Ratio on the Composition of Microbial Extracellular Polymers in Activated Sludge *Wat. Sci. and Technol.*,2001, 44, (10), 221-229

TABLES

Table 1. Amylase dose versus OUR of microorganisms

Amylase Dose (g/gVS)	OUR (mg/gVSS.h)
0.2	5.0
0.4	4.8
0.85	4.6
1.6	3.9

Table 2. Summary of OUR values at various proteinase doses

Proteinase Dose (g/gVS)	OUR (mg/gVSS.h)
0.2	4.9
0.5	4.7
0.8	4.5
1.0	4.2
1.5	3.7

FIGURES

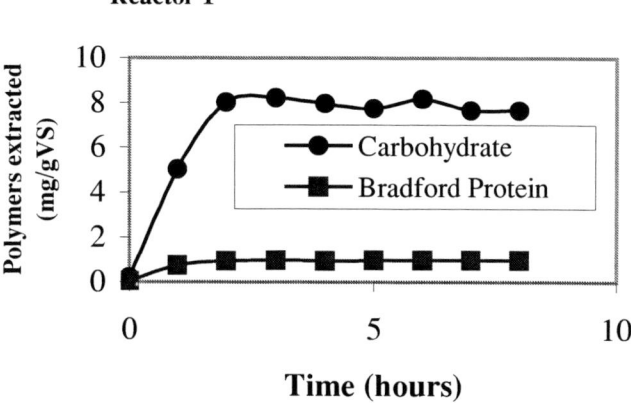

Figure 1. Time dependent amylase extraction results (dose 0.4g/gVS)

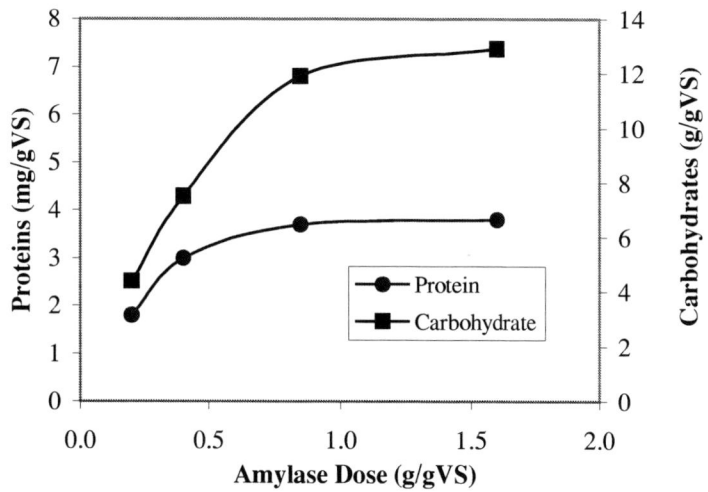

Figure 2. Effect of α-amylase dose on the extraction of extracellular proteins and carbohydrates

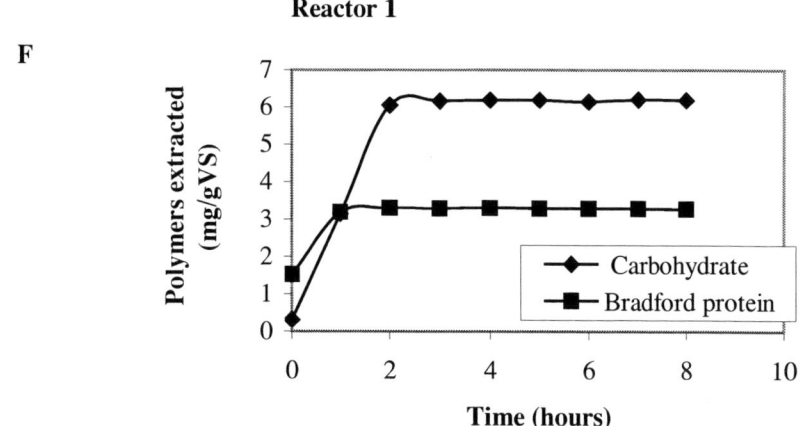

Figure 3. Time dependent proteinase extraction results (dose 0.5g/gVS)

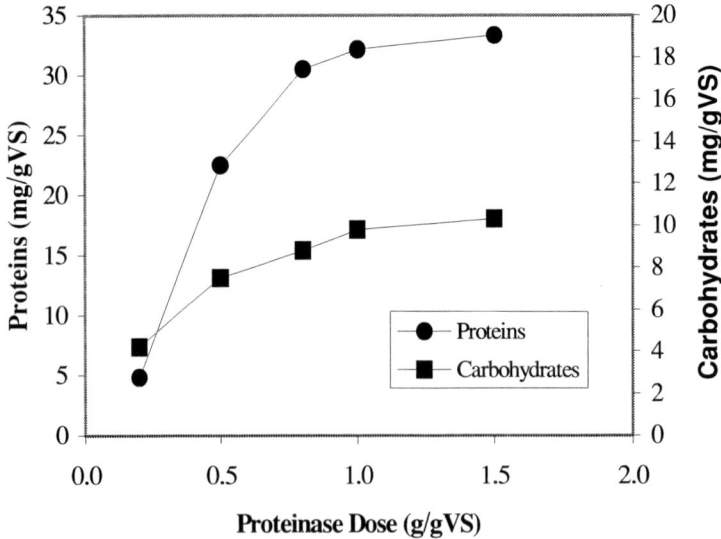

Figure 4. Effect of proteinase dose on the extraction of extracellular proteins and carbohydrates

AN INVESTIGATION OF PB (II) BIOSORPTION IN RELATION TO SURFACE CHARACTERISTICS OF ACTIVATED SLUDGE

Bilgen Yuncu, F. Dilek Sanin, Ulku Yetis*
Middle East Technical University, Department of Environmental Engineering, 06531 Ankara, Turkey
To whom all correspondence should be addressed
Tel:00 90 312 210 5868 Fax:00 900 312 210 1260
E-mail: uyetis@metu.edu.tr

ABSTRACT

The purpose of this study is to identify the mechanisms and the effect of extracellular polymer (ECP) composition of activated sludge on Pb(II) biosorption. Microorganisms cultured under different growth conditions are expected to have different compositions of extracellular polymers and hence, different biosorption capacities. For this purpose, three sets of reactors with different C/N ratios were set up. Isotherm and kinetic experiments were carried out with three different C/N ratios, and biosorption capacities for lead were examined. As C/N ratio increased, and the carbohydrate amount in ECP increased, the biosorption capacity of the system decreased. Also, with the purpose of understanding the mechanism of the process, released ion concentrations were measured. It was observed that during bisorption of Pb(II), Ca^{2+}, Mg^{2+} and H^+ ions were released into the solution. This shows that an ion exchange process was partly involved but it is not the major mechanism for biosorption.

KEYWORDS

Activated sludge, biosorption, C/N ratio, extracellular polymers, mechanism, Pb(II) removal.

INTRODUCTION

The increase in industrial activities has intensified environmental pollution with the accumulation of heavy metals, and other pollutants. The main concern with the heavy metals is their toxicity and tendency to accumulate throughout the food chain, which is a serious threat to the environment, animals and mankind. Therefore, the need for cost effective methods for the removal of metals is vital. Commonly used methods are precipitation, ion exchange, electrochemical processes and/or membrane processes. However, the application of such processes is sometimes restricted because of technical or economic constraints. The search for new technologies, involving the removal of toxic metals from wastewaters has directed attention to biosorption, based on the metal binding capacities of various biological materials[1].

The sequestering of metal ions by solid materials of biological origin is known under the general term "biosorption". It is well known that pure and mixed cultures of bacteria, algae, yeast, and other microorganisms can remove heavy metal ions from solution in significant quantities; microbial metal uptake varies from few microorganisms per gram to several percent of the dry cell weight[2]. Shumate and Strandberg[3] reported that pure cultures of microorganisms exhibited significant heavy metal uptakes ranging from 8% to 35% of the dry cell weight. They also suggested that mixed microbial cultures were more efficient than pure cultures in removing heavy metals. This means that biological pollution control processes could be effective in removing heavy metals from polluted waters. The capability of activated sludge cultures to remove heavy metals was observed and reported by Ruchoft[4]. Also, it has been shown by many other researchers that especially lead, copper, and zinc are removed relatively more efficiently by activated sludge organisms [5-8].

Factors, which may influence the biosorption process, include pH, the concentration of metal present in the wastewater and the composition and characteristics of wastewater and sludge. Heavy metal removal or uptake by sludge is a consequence of interaction between metals in the aqueous solution phase and the bacterial cell surface. Biosorption of metals is not based on only one mechanism. It consists of several mechanisms that quantitatively and qualitatively differ according to the species used, the origin of the biomass, and its processing. Metal sequestration follows complex mechanisms, mainly ion exchange, chelation, adsorption by physical forces; and ion entrapment in inter- and intrafibrillar capillaries and spaces of the structural polysaccaharide network as a result of the concentration gradient and diffusion through the cell walls and membranes[9].

Binding of heavy metals on the sludge surface can be attributed to the formation of surface complexes between metals and surface functional groups such as carboxyl, hydroxyl, and phenolic groups of the extracellular polymer (ECP). Many different species of bacteria isolated from the activated sludge have been shown to be able to produce ECP as capsules or a gelatinous matrix and such polymers have been shown to be involved in the adsorption of exogenous metal ions from solution[10,11]. Heavy metals are complexed by anionic legend found in polymers (e.g., proteins, lipids, polysaccharides, nucleic acids, lipoproteins, glycocalyxes, and others) of the cell membrane and cell wall[12].

It has been suggested that the composition of the growth medium is important in the production of metal-complexing proteins and exopolymers[2]. Feeding microorganisms with substrates at various carbon to nitrogen (C/N) ratios was shown to affect the composition of extracelullar polymers in terms of protein and polysaccharide content [13]. Microorganisms cultured under different growth conditions have different composition of ECP and hence it is expected that they will have different biosorption capacities.

The purpose of this study is to identify the mechanisms of Pb(II) bisorption and the effect of extracellular polymer (ECP) composition of activated sludge on its biosorption process. There are many studies in literature about the biosorption of heavy metals by activated sludge; nevertheless there is not much specific information on the bisorption mechanism(s) involved and especially the role of ECP in biosorption.

MATERIALS AND METHODS
REACTOR OPERATION

Mixed culture bacteria grown in semi-continuous reactors were used during the experiments. The microbial seed was obtained from the primary settling tank effluent from the city of Ankara wastewater treatment plant. Reactors had a working volume of 2 litres and were operated at a mean cell residence time (MCRT) of 8 days. The reactors were located into water bath and the temperature of the system was kept at 20°C, pH was 7.0 ± 0.2. Required oxygen was supplied by air pumps which were also providing completely mixed conditions in the reactors. Dissolved oxygen concentration in reactors was kept at a minimum of 3 mg/L.

Three sets of replica reactors were operated under three different C/N ratios. Each reactor set had two replica reactors. The first C/N ratio was selected as 21 (in terms of the ratio of COD to TKN as mg/L per mg/L). This set of reactors was operated to represent the typical operational conditions in activated sludge systems treating municipal wastewaters. These reactors were fed with the synthetic medium given in Table 1. The second set of reactors was operated at a C/N ratio of 9 to represent a carbon-limited situation. To adjust the amount of carbon, the synthetic medium given in Table 1 was modified by decreasing the glucose amount in the feed. The third set of reactors was operated at a C/N ratio of 43, represented nitrogen limited situation. The carbon content of these reactors was again adjusted by modifying the glucose amount in the feed given in Table 1.

Reactors operated under the above listed conditions were brought to steady state which was demonstrated by measuring mixed liquor suspended solids (MLSS) and mixed liquor volatile suspended solids (MLVSS) concentrations daily. Once the steady state was reached, biosorption kinetic and equilibrium (isotherm) tests were conducted using sludge wasted daily from each reactor. The wasted sludge was first centrifuged for 15 min and the supernatant was discarded. This was repeated for twice; and finally, the concentration of biomass suspension to be used in biosorption tests was measured.

BIOSORPTION TESTS

Biosorption studies that are mainly composed of sorption kinetic and sorption isotherm tests were performed as batch experiments using biomass suspension with a predetermined biomass concentration. Tests were held for each C/N ratio. Since the sorption performance may vary with temperature and pH, in all experiments the temperature was kept constant at 25°C, the initial pH was set to 4 using 0.01 M HNO_3. Lead(II) was added as $Pb(NO_3)_2$. All the biosorption tests were run in duplicate. Biosorptive capacity was calculated by using the equation;

$$[q = V(C-C_o)/ m] \qquad (1)$$

where q is the capacity as mg Pb(II) per g dry biomass, V is the volume of the sample, C and C_o are the final and initial Pb(II) concentrations in aqueous phase as mg/L and m is the amount of dry biomass as g.

SORPTION KINETIC TESTS

Kinetic studies were conducted by mixing Pb(II) solutions with the known amount of biomass and taking samples (10 ml) at different time intervals for 6 h beyond which there is no more sorption or desorption Three different initial Pb(II) concentrations were used for three C/N ratios in order to keep the initial Pb(II) to biomass concentration ratio constant (~ 0,6). The initial Pb(II) concentrations were 2000 mg/L, 3000 mg/L and 4000 mg/L for the C/N ratios of 9, 21 and 43, respectively. Such high

Pb(II) concentrations were deliberately selected in order to end up with the maximum Pb(II) sorptive capacity that will provide a fair comparison of the results. The biomass in the samples was removed by filtration through a 0.45 µm membrane filter and filtrates were analyzed by using flame atomic absorption spectrophotometer (AAS) (ATI Unicam 929), following pH adjustment to 2.0 using HNO_3. In all these measuremenst, the methods described in Standard Methods[14] were followed. Beside the Pb (II) ions absorbed, Ca^+ and Mg^+ ions released during the experiment were also analyzed by using AAS in order to investigate the mechanism of the process. At each sampling, the solution pH was also measured in order to follow the change of pH during sorption.

SORPTION EQUILIBRIUM TESTS

Batch isotherm experiments were held using flasks of net volume 50 mL which are placed in a shaking incubator operating at 200 rpm and 25°C. The biomass was added into each flask at a concentration of 2450, 4190 and 6660 mg/L for C/N ratios of 9, 21 and 43, respectively. Pb(II) concentration varied from 50 to 4000 mg/L for each C/N ratio at an initial pH 4. Kinetic tests have shown that almost 100% metal removal was achieved within the first 4 h of biomass-Pb(II) contact. So, 4 h was employed as the equilibrium time. The sample taken at the end of this time, was analysed for Pb (II) following the procedure described above.

RESULTS AND DISCUSSION
ADSORPTION KINETICS

Figures 1-3 present the time course of Pb(II) removal by activated sludge grown under three different C/N ratio conditions. In general, it can be seen from the results that Pb(II) uptake was very rapid for all C/N ratios. In all cases, there was a sudden decrease in Pb(II) concentration of the solution during the first 5 min. After 5 min, the Pb(II) uptake rate gradually decreased and it remained nearly constant; after 60, 60 and 90 min for the C/N ratios of 9, 21 and 43, respectively. This behaviour implicates the fact that sorption occurs at two stages: the first is the rapid surface binding and the second is the slow intracellular diffusion; independent of the metabolic state of the cells[15]. Similar results were reported by other researchers[15-18], while in some other studies single-step uptake was suggested for different biosorbents[19].

As it can be seen from Figures 1a, 2a and 3a, the biosorption capacity of reactors increases with the decrease of C/N ratio. The maximum capacities were approximately found as 950, 500 and 300 mg Pb(II)/g biomass for C/N ratios 9, 21 and 43, respectively. In all kinetic experiments, a pH decrease with biosorption was observed. This implies that H^+ ions were released into the solution during biosorption. On the other side, Figures 1b, 2b and 3b show that in parallel with H^+ ions release, there is also release of Ca^{+2} and Mg^{+2} ions with the uptake of the Pb(II). But, a comparison of the amount of Pb(II) ions adsorbed with the total ions (sum of Ca^{+2}, Mg^{+2} and H^+) released indicate that Pb(II) adsorbed is always higher than the amount of total ions released at all C/N ratios. This indicates that an ion exchange mechanism is involved in biosorption but the whole Pb(II) removal can not be explained by ion exchange.

ISOTHERM STUDIES

The adsorptive capacity of the activated sludge grown under different C/N ratios was determined by examining the relationship between the amount of Pb(II) adsorbed by biomass (q_e) and the equilibrium concentration at equilibrium (C_e). The results show that the maximum biosorptive capacities for the activated sludge grown under different C/N ratios were very high. It was found as 1000, 600 and 300 mg Pb(II)/g biomass, for the C/N ratios of 9, 21, and 43, respectively. It can clearly be seen that with the decrease of C/N ratio there is a remarkable increase in the Pb(II) sorption capacity of activated sludge. A two-fold increase in C/N ratio from 21 to 43 resulted in a two-fold decrease in the maximum capacity. Similarly, with an increase in C/N ratio from 9 to 21, there was about 40 % reduction in the maximum biosorptive capacity. On the other side, for the highest C/N ratio studied, the adsorptive capacity was found to increase gradually with an increase in equilibrium Pb(II) concentration indicating a low affinity of the activated sludge grown at a high C/N ratio. These results show that biosorption is directly affected by the C/N ratio that is directly related with ECP structure of the sludge. It is known that C/N ratio of the feed is a major determinant of the composition of ECP in activated sludge. As the C/N ratio increases, the carbohydrate content of ECP increases, whereas the proteins decrease[13].

It is a well known fact that binding of heavy metals on the sludge surface can be attributed to the formation of surface complexes between metals and surface functional groups such as carboxyl, hydroxyl, phosphate, phenolic and amino groups of the ECP[15, 21-24]. Ashkenazy et al.[20] reported that carboxyl groups were the dominant species in the Pb(II) biosorption mechanism but also groups containing nitrogen interacted with Pb(II) cations. The results of the present study disagree with this research. As it is stated above, with the increase of C/N ratio carbohydrate content of ECP –which means the carboxyl groups in the ECP structure – increases. On the other hand, biosorption capacity of activated sludge enhances with the decrease of C/N ratio so it can be said that Pb(II) uptake can not be attributed only to the carboxyl groups. Biosorption capacity is higher when the amine groups are dominant in the ECP structure. Nierboer and Richardson[25] proposed a classification based on the atomic properties and the solution chemistry of the metal ions. They discriminated metal ions into three classes: oxygen-seeking, nitrogen- and sulfur seeking, and borderline or intermediate class. Pb(II) ion is included partly in nitrogen- and sulfur seeking and partly in borderline class which means Pb(II) exhibits higher affinity for the $-NH_2$ and $-NH$ groups than carboxyl groups. Since it was demonstrated earlier that at lower C/N ratios, ECP contained more proteins than carbohydrates[13], higher adsorption of Pb(II) at lower C/N ratio can be easily explained. Higher protein content means that there are more $-NH_2$ and $-NH$ groups in ECP for Pb(II) to bind yielding much higher biosorption capacities at lower C/N ratios.

Besides these investigations, the effect of initial concentration of Pb(II) solution on biosorption can be seen from Figure 5. In all cases as the initial concentration increases, the equilibrium pH of the solution decreases. Especially for the C/N ratio of 9 and 21, there is a sharp decrease in equilibrium pH between the initial Pb(II) concentrations of 1000 mg/L and 1500 mg/L. This may be due to a change in biosorption mechanism with the initial concentration of the metal.

In attempt to determine the biosorption mechanism, total amount of ions released during biosorption were also analysed in isotherm experiments. It can be seen from Figure 5 that like the results from adsorption kinetic tests, in all cases the amount of Pb(II) adsorbed was higher than the amount of total ions released (Ca^{+2}, Mg^{+2}, H^+). Also, it is observed that both the adsorbed Pb(II) ions and total ion released during biosorption decreased with the increase of C/N ratio but the ratio of adsorbed Pb(II) ions to the total ions released remained nearly the same for all C/N ratios.

The results of isotherm studies were further studied in terms of conformity to the well known Langmuir and Freundlich isotherm models. The Freundlich expression is an empirical equation based on sorption on a heterogeneous surface and it is given by the Eq. 2.

$$q_e = K_f C_e^{1/n} \qquad (2)$$

where K_f and n are the Freundlich constants of the system.

The Langmuir expression is valid for monolayer sorption onto a surface and is given by the Eq. 3.

$$\frac{C_e}{q_e} = \frac{Q^0 b C_e}{1+bC_e} \qquad (3)$$

where Q^0 is the maximum possible amount of metallic ion adsorbed per unit weight of adsorbent to form a complete monolayer on the surface and b is an equilibrium constant related to the affinity of the binding sites.

The experimental data (q_e and C_e) obtained at different C/N ratios were fitted by linear regression method to evaluate the Freundlich and Langmuir model parameters by minimizing the sum of the squared deviations from experimental and predicted values. In Table 2, the estimated values of the model parameters along with the regression coefficient (R^2) which, permit to evaluate the model goodness[26] are given. The adsorption data obtained for activated sludge grown in reactors with C/N 9 and 21 were fitted better to Langmuir model, whereas Freundlich model was more suitable for the biosorption with the activated sludge of C/N 43. Langmuir or Freundlich sorption models are the most frequently used models for biosorption but they do not take into account electrostatic interactions or the fact that metal ion biosorption is largely an ion exchange phenomenon so these models are not very suitable for the prediction of the biosorption performance[27].

K_f and n, are indicators of adsorption capacity and adsorption affinity, respectively. Higher the value of n, lower will be the affinity. The values presented in Table 2 indicated that activated sludge grown at higher C/N rations represent a lower affinity for Pb(II). The magnitude of K_f at the C/N ratio of 9 also showed the higher capacity of activated sludge at low C/N values. A much higher b value obtained with activated sludge grown under the C/N ratio of 9 also confirmed the higher affinity of activated sludge at lower C/N ratios. Larger values of b for the C/N ratios of 9 and 21, also implied the strong binding of Pb(II) to the activated sludge cells at low C/N ratios.

CONCLUSIONS

In this study it is demonstrated that activated sludge is an efficient biosorbent for the removal of Pb(II). Also, the extracellular polymer composition directly affects the Pb(II) removal capacity of activated sludge. The kinetic and isotherm experiments conducted at constant temperature (25°C) with three different C/N ratios (9, 21 and 43) showed that with the increase of C/N ratio, there is a remarkable decrease in the biosorption capacity of activated sludge. This can be attributed to the presence of the higher amount of carbohydrates and lower amount of proteins at higher C/N ratios.

The measurement of the Ca^+, Mg^{+2} and H^+ ions which are released into the aqueous phase implied that ion exchange plays an important role in biosorption of Pb(II) but the whole mechanism can not be explained by only ion exchange phenomenon.

REFERENCES

1. VEGLIO, F. BEOLCHINI, F. Removal of heavy metals by biosorption : a review. *Hydrometallurgy*. 1997, 44, (3), 301-316.

2. CHANG, D. FUKUSHI, K. GHOSH, S. Stimulation of activated sludge cultures for enhanced heavy metal removal. *Wat. Env. Res.* 1995, 67, (5), 822-827.

3. SHUMATE, S E. II. STRANDBERG, G W. Accumulation of metals by microbial cells. *Comprehensive Biotech.* M. Moo-Young et.al .(Eds). Pergamon Press. New York. 1985, 235.

4. RUCHOFT, C C. The possibilities of disposal of radioactive wastes by biological treatment methods. *Sew. Works. J.* 1949, 21, 877.

5. ROSSIN, A C. STERRITT, R M. LESTER, J N. The influence of process parameters on the removal of heavy metals in activated sludge. *Water. Air. Soil Pollut.* 1982, 17, (2), 185-198.

6. ATKINSON, B W. BUX, B. KASAN, H C. Bioremediation of metal-contaminated industrial effluents using waste sludges. *Wat. Sci. Tech.* 1996, 34, (9), 9-15.

7. BUX, B. NAIDOO, D. KASAN, H C. Laboratory-scale biosorption and desorption of metal ions using waste sludges and selected acids. *South African Journal of Science.* 1996, 92, 527-529.

8. OLIVER, B G. COSGROVE, E G. The efficiency of heavy metal removal by a conventional activated treatment plant. *Wat. Res.* 1974, 8, 869-874.

9. VOLESKY, B. HOLAN Z R. Biosorption of heavy metals. *Biotechnol. Prog.* 1995, 11, (3), 235-250.

10. RUDD, T. STERRITT, R M. LESTER J N. Formation and conditional stability constants of complexes formed between heavy metals and bacterial extracellular polymers. *Wat. Res.* 1984, 18, (3), 379-384.

11. TIEN, C T. HUANG, C P. Adsorption behavior of Cu(II) onto sludge particulate surfaces. *Journal of Environ. Eng.* 1987, 113, (2), 285-299.

12. FUKUSHI, K. CHANG, D. GHOSH, S. Enhanced heavy metal uptake by activated sludge cultures grown in the presence of biopolymer stimulators. *Wat. Sci. Tech.* 1996, 34, (5-6), 267-272.

13. DURMAZ, B. SANÝN F.D. Effect of carbon to nitrogen ratio on the composition of microbial extracellular polymers in activated sludge. *Wat. Sci. Tech.* 2001, 44, (10), 221-229.

14. APHA, AWWA, WEP, Standard methods for examination of water and wastewater. 19th edition. United Book Press. Baltimore. 1995.

15. YETIS, U. DOLEK, A. DILEK, F B. OZCENGIZ, G. The removal of Pb(II) by *Phanerochaete Chrysosporium*. *Wat. Res.* 2000, 34, (16), 4090-4100.

16. CHANG, J S. LAW, R. CHANG, C C. Biosorption of lead, copper and cadmium by biomass of *Pseudomonas Aeruginosa* PU21. *Wat. Res.* 1997, 31, (7), 1651-1658.

17. LEUNG, W C. CHUA, H. LO, W. Biosorption of heavy metals by bacteria isolated from activated sludge. *Applied Biochemistry and Biotechnology.* 2001, 91-93, 171-184.

18. MATHEICKAL, J T. YU, Q. Biosorption of lead from aqueous solution by marine algae *Ecklonia Radiata*. *Wat. Sci. Tech.* 1996, 34, (9), 1-7.

19. GUIBAL, E. ROULPH, C. LECLOUREE, P. Uranium bisorption by the filamentous fungus *Mucor Miehei* pH effect on mechanisms and performance of uptake. *Wat. Res,* 1992, 26, (8), 1139-1145.

20. ASHKENAZY, R. GOTTLIEB, L. YANNAI, S. Characterization of acetone –washed yeast biomass functional groups involved in lead biosorption. *Biotech. and Bioeng.*1997, 55, (1), 1-10.

21. GONZALEZ, M E R. WILLIAMS, C J. GARDINER, P H E. Study of the mechanism of cadmium biosorption by dealginated seaweed waste. *Environ. Sci. Tech.* 2001, 35, (14), 3025-3030.

22. SENTHILKUMAAR, S. BHARATHI, S. NITHYANANDHI, D. SUBBURAM, V. Biosorption of toxic heavy metals from aqueous solutions. *Bioresource Tech.* 2000, 75, (2), 163-165.

23. SCHNEIDER, I A H. RUBIO, J. Sorption of heavy metal ions by the nonliving biomass of freshwater macrophytes. *Environ. Sci Tech.* 1999, 33, (13), 2213-2217.

24. MAMERI, N. BOUDRIES, N. ADDOUR, L. BELHOCINE, D. LOUNICI, H. GRIB, H. PAUSS, A. Batch zinc biosorption by a bacterial nonliving *Streptomyces Rimous* biomass. *Wat. Res.* 1999, 33, (6), 1347-1354.

25. NIEBOER, E. RICHARDSON, D H S. The replacement of the non-descript term heavy-metals by a biologically and chemically significant classification of metal- ions. *Env. Poll. Ser. B.* 1980, 1, (1), 3-26.

26. HO, Y.S. McKAY, G. Correlative biosorption equilibria model for a binary batch system. *Chem. Eng. Sci.* 2000, 55, (4), 817-825.

27. SCHIEWER, S. VOLESKY, B. Modelling of the proton-metal ion exchange in biosorption. *Environ. Sci Tech.* 1995, 29, (12), 3049-3058.

ACKNOWLEDGEMENTS

The authors gratefully acknowledge financial support from the Turkish Scientific and Technical Research Council (TUBITAK), under Grant No. ICTAG-C002.

TABLES

Table 1. Synthetic feed medium composition

Constituent	Concentration (mg/L)
Glucose	935
Peptone	200
K_2HPO_4	600
KH_2PO_4	300
NH_4Cl	225
$MgSO_4 \cdot 7H_2O$	112.5
$FeSO_4 \cdot 7H_2O$	3.75
$ZnSO_4 \cdot 7H_2O$	3.75
$MnSO_4 \cdot 7H_2O$	3.75
$CaCl_2$	15
$NaHCO_3$	180

Table 2. Freundlich and Langmuir isotherm constants.

	Freundlich[*]			Langmuir[**]		
	n	K_f	R^2	Q^0	b	R^2
C/N=9	1,3883	22,3254	**0.8941**	1111,111	0,0138	**0.9533**
C/N=21	1,2857	13,3045	**0.8546**	714,2857	0,0145	**0.9106**
C/N=43	1,2453	0,9030	**0.9501**	526,3158	0,000751	**0.8176**

[*]K values are given in $(mg/g)(L/mg)^{1/n}$ and 1/n is unitless
[**]Q^0 and b values are given in metal/g biomass and L/g, respectively.

FIGURES

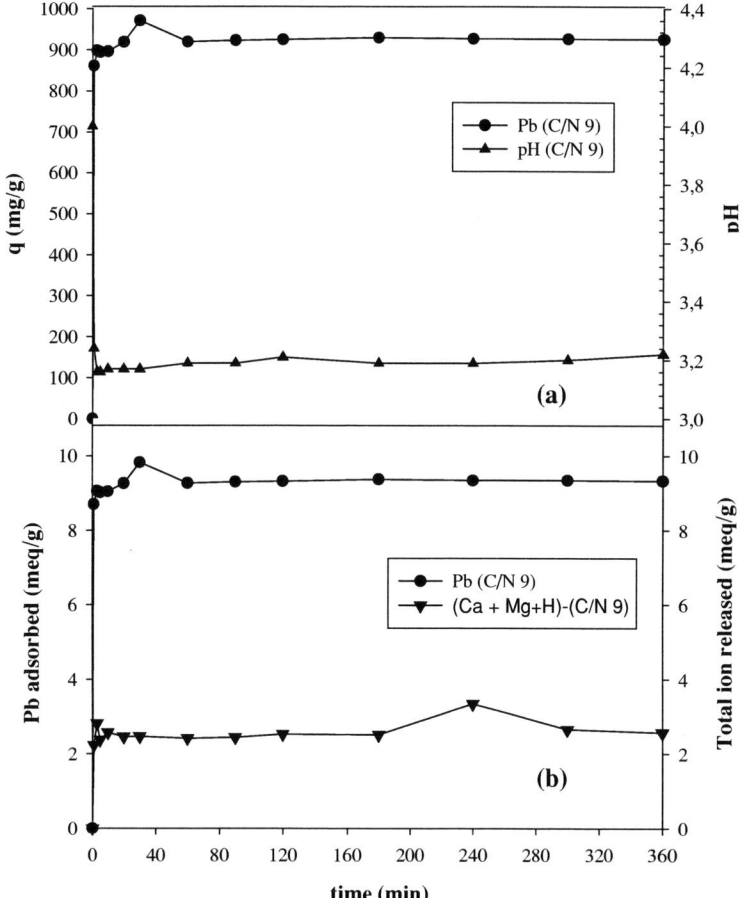

Figure 1. Biosorption kinetics at C/N 9.

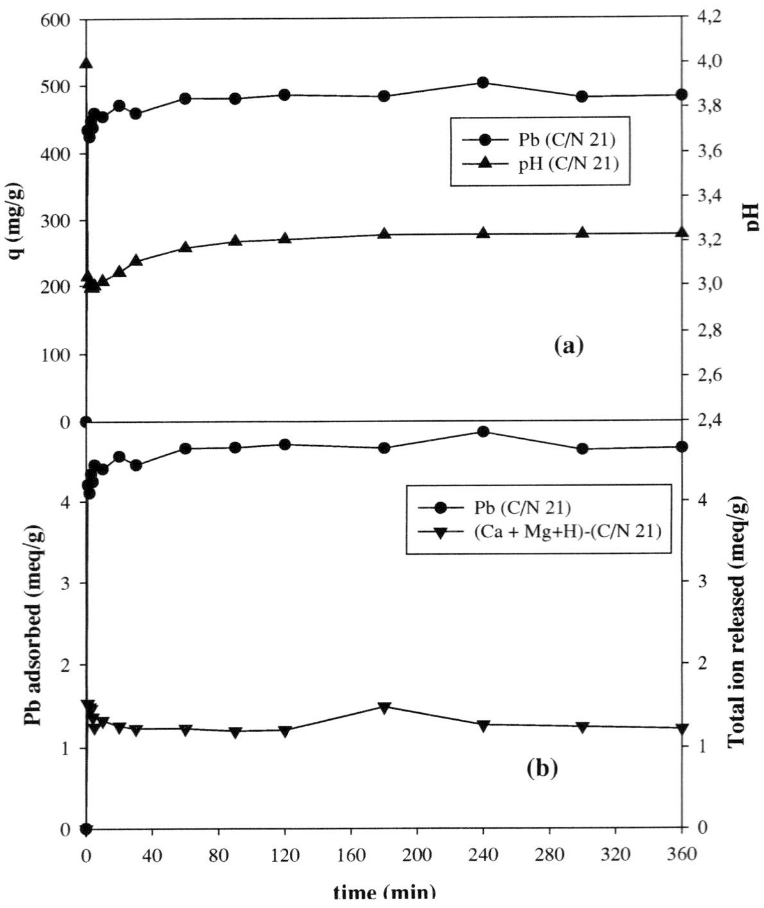

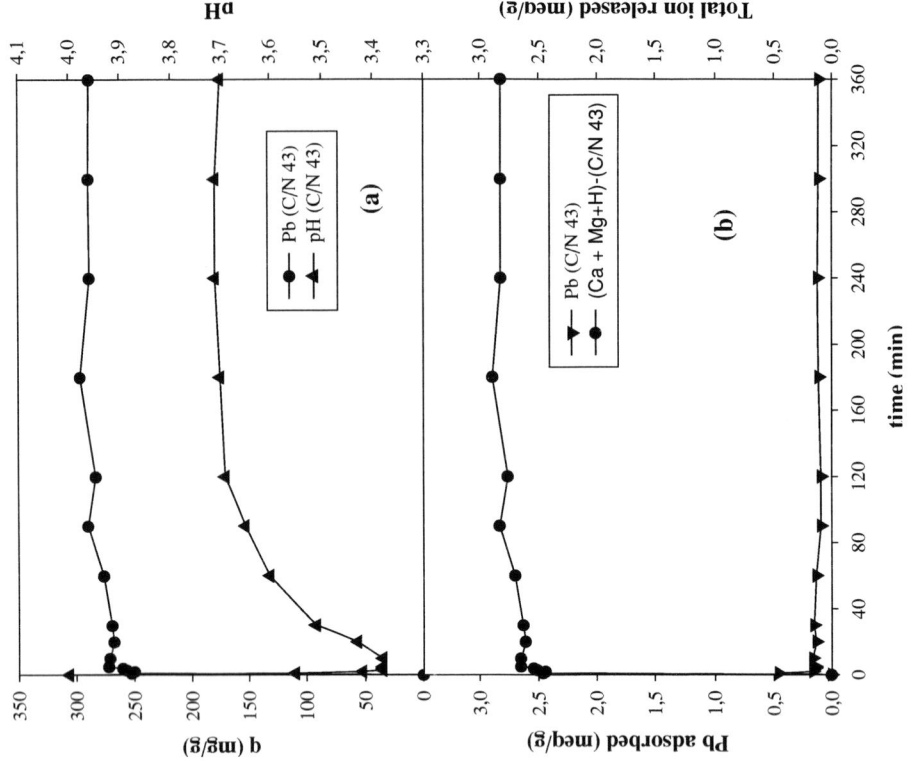

Figure 3. Biosorption kinetics at C/N 43.

Figure 2. Biosorption kinetics at C/N 21.

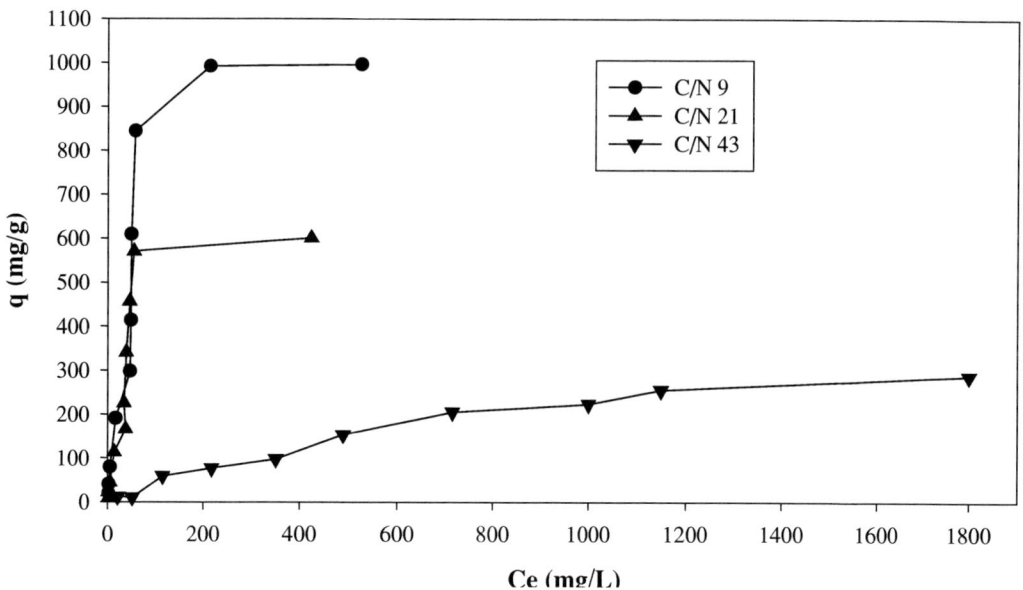

Figure 4. Equilibrium isotherms of Pb(II) sorption by activated sludge grown at different C/N ratios.

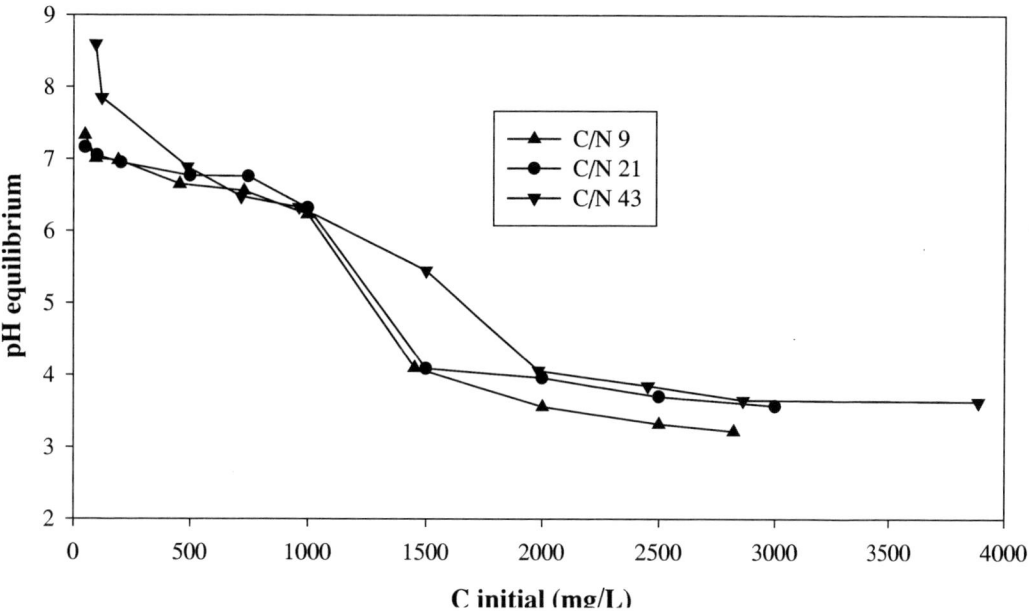

Figure 5. Effect of initial Pb(II) concentration on the equilibrium pH after biosorption by activated sludge.

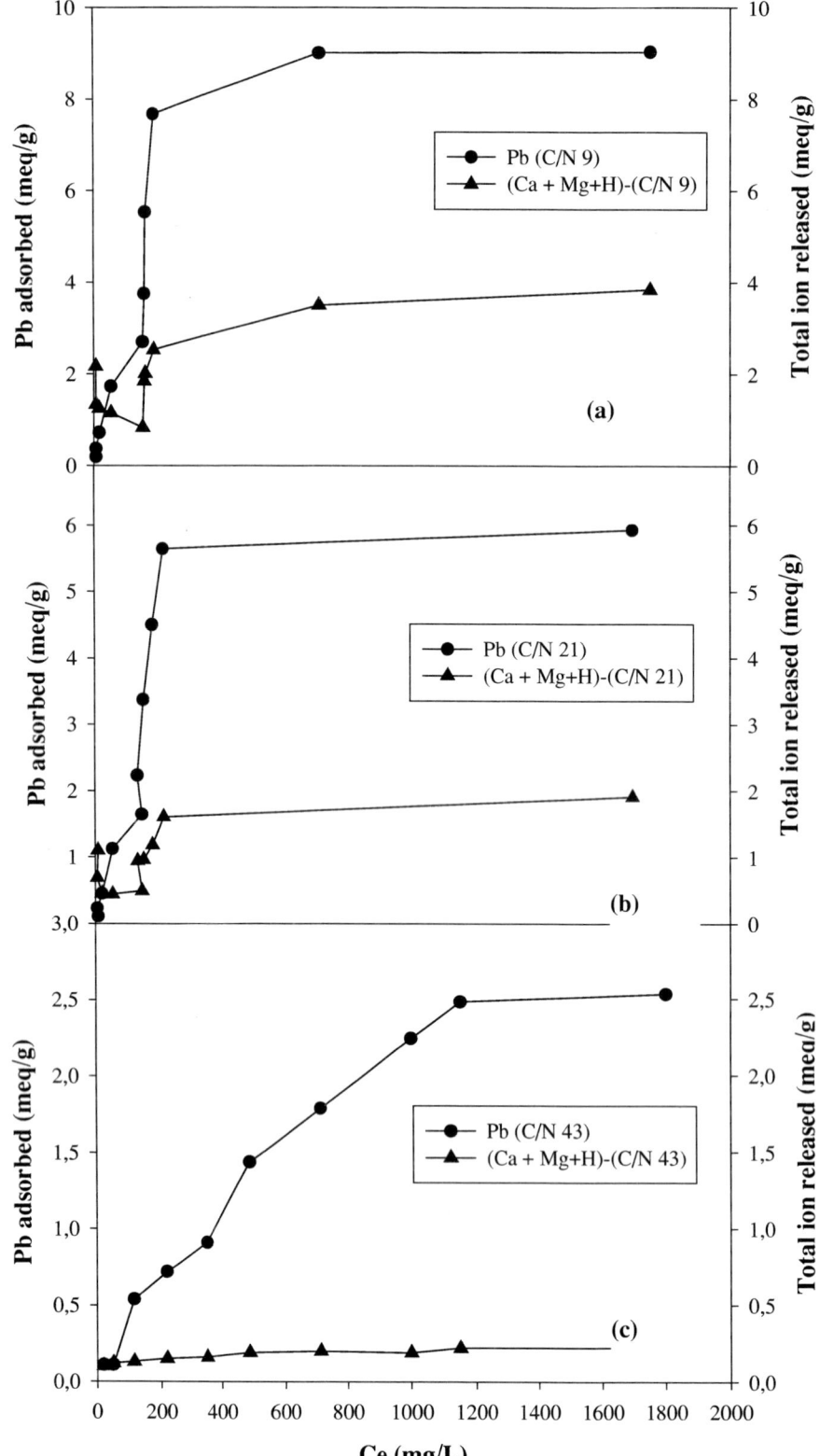

Figure 6. Amount of Pb(II) ion adsorbed and total ion released (Ca^{+2}, Mg^{+2} and H^+) versus equilibrium concentrations a) C/N 9 b) C/N 21 c) C/N 43.

METHODS FOR THE ENUMERATION OF SALMONELLA SPECIES IN SEWAGE SLUDGE AND SOIL

Nicola L Lang[1], Stephen R Smith[1], Edmund B Pike[2] and Chris L Rowlands[3]

[1] Centre for Environmental Control and Waste Management, Department of Civil and Environmental Engineering, Imperial College, London, [2] Sonning Common, Reading, [3] Severn Trent Water Ltd, Birmingham.

Tel: 020 75946051; Fax: 020 78239401; E-mail: nicola.lang@ic.ac.uk, s.r.smith@ic.ac.uk

ABSTRACT

A new rapid technique proposed for the enumeration of *Salmonella* species in sewage sludge, based on membrane filtration (MF) and chromogenic agar, was applied to conventional mesophilic anaerobically digested biosolids. However, overgrowth of the indigenous background flora on the agar plates prevented enumeration and identification of *Salmonella* colonies. Three alternative methods for *Salmonella* enumeration, based on the Most Probable Number (MPN) technique, were tested with samples of raw and digested sludge and a control broth spiked with *Salmonella typhimurium* and *Escherichia coli*. The presence of *Salmonella* was determined by streaking onto four different types of agar including: two conventional agars, xylose lysine desoxycholate (XLD) and brilliant green agar (BGA), and two chromogenic agars, Rambach (RAM) and Oxoid Salmonella Chromogenic Agar (OSCA). RAM was superior to OSCA, which was susceptible to interfering organisms expressing a similar chromogenic reaction to *Salmonella* spp., and XLD was comparable to RAM. BGA was the least suitable agar for presumptive identification of *Salmonella* based on colour reaction and colony morphology. The enumeration of *Salmonella* in tetrathionate broth supplemented with novobiocin and iodine-iodide solution was identified as the most suitable MPN method for sludge and was compared to MF by the further examination of raw and digested sludge, and soil spiked with *S. typhimurium*, and unspiked samples of these matrices.

KEY WORDS

Chromogenic agar, membrane filtration, MPN, *Salmonella* enumeration, sewage sludge

INTRODUCTION

Salmonella is one of the principal causes of gastro-intestinal disease in humans and infections are widespread in farm animals. The potential occurrence of this bacterium in sewage sludge is therefore regarded as a key indicator of the microbiological quality of sludge.

Future legislation supporting the agricultural use of sewage sludge will establish new microbiological standards for conventional and enhanced-treated biosolids products. This development will require the Water Industry to routinely measure indicator and pathogenic microorganisms in sewage sludge. The enumeration of these organisms in sludge-treated soil is also important for research purposes to establish the microbiological risks and environmental decay of enteric organisms in soil, and to assess the suitability of harvesting and cropping restrictions that apply to sludge-treated farmland.

Until recently, there has been relatively little development of isolation/enumeration techniques for *Salmonella* in sewage sludge, and the validity and suitability of the available methods of detection remains uncertain (Yanko *et al.*, 1995; Smith and Brobst, 2001). New plating agars for chromogenic bacterial identification have been recently applied to the examination of sewage sludge in the UK (UKWIR, 2000). These methods are designed to optimise the release of pathogenic organisms from biosolids matter and quantitatively resuscitate potentially sub-lethally damaged bacteria, followed by enumeration on chromogenic media (UKWIR, 2000).

Salmonella spp. were selected as a target organism for enumeration in ongoing research at Imperial College of pathogen survival in sludge-treated soil (Smith, 2001). This required the determination of background numbers of *Salmonella* in sludge being applied to experimental field plots and the chromogenic procedure developed for biosolids (UKWIR, 2000) was followed. However, overgrowth of the indigenous microbial flora in digested sludge was observed on the membranes preventing enumeration of the bacteria in this principal conventionally treated sludge type (Figure 1). Here, we present an investigation of this phenomenon and the suitability of the assay for the enumeration of *Salmonella* spp. in sludge and soil, and alternative detection methods are also examined and compared.

MATERIALS AND METHODS

The isolation of salmonellae requires a number of successive stages: concentration or pre-enrichment, enrichment, isolation and identification on selective media, and confirmation. Enrichment either suppresses growth of indigenous flora or enhances growth of *Salmonella* allowing them to predominate in mixed culture. Identification on differential agar is by colonial characteristic appearance and morphology.

MF TECHNIQUE FOR THE ENUMERATION OF *SALMONELLA* SPP.

SAMPLE PREPARATION

A 25 g sample of sludge or soil was aseptically transferred into 225 ml of phosphate buffered saline (PBS) to achieve a 10-fold dilution. Samples of biosolids were homogenised by stomaching (Seward) for 2 minutes in sterile stomacher bags and soil samples were transferred to Duran bottles containing approximately 10 g of sterile glass beads (Dudley *et al.*, 1980) and were agitated on a shaker (Gerhardt) at 200 rpm for 4 minutes. Sample solutions were clarified by centrifugation (Gallenkamp) at $200 - 300\ g$ for 1 minute and the supernatants were decanted into sterile containers.

Resuscitation and enumeration

Serial decimal dilutions were prepared in PBS and a 10 ml aliquot of each dilution was filtered under vacuum through a 0.45 µm, gridded sterile membrane (Pall Gelman) for recovery of cells. The membrane was transferred to an absorbent cellulose fibre filtrate pad (Pall Gelman), soaked in tetrathionate broth (TB) containing novobiocin (antibiotic) and an iodine-iodide solution, and incubated at $37\ °C \pm 1.0\ °C$ for 24 h (UKWIR, 2000). Filters were aseptically transferred to Rambach agar plates (RAM) and incubated $37\ °C$ for 24 h and 48 h for selective enumeration by typical reddish pink colonies (UKWIR, 2000).

Confirmation

Verification of *Salmonella* colonies against potential false positives was undertaken by spraying a 1% (v/v) solution of 4-methylumbelliferyl caprylate (MUC) in ethanol onto the filters and examination under UV light at 344 nm. Under these conditions, colonies fluorescing after 1 minute were recorded as positive.

MPN TECHNIQUES FOR THE ENUMERATION OF *SALMONELLA* SPP.

Available MPN techniques for *Salmonella* (APHA, 1998 and HMSO, 1982; 1994) were reviewed and tested using liquid raw sludge (LRS), dewatered mesophilic anaerobically digested sludge (DMAD) and a broth spiked with known numbers of *E. coli*, to represent the background flora of raw and conventionally treated sludge types, and *Salmonella,* the target organism.

Preparation of cell suspension for spike inoculum

Purity plates of *E. coli* (NCTC 9001) and *S. typhimurium* (NCTC 12416) on tryptone soya agar (TSA) were incubated at $37\ °C$ overnight, sub-cultured into 10 ml of tryptone soya broth (TSB) and incubated again at $37\ °C$. The TSB cultures were diluted to a concentration of approximately 10 cfu ml^{-1} and a 1 ml aliquot was transferred to 100 ml of TSB and incubated at $37\ °C$ for 16 - 24 h. A 50 ml fraction was centrifuged at 12,500 rpm at $20\ °C$ for 20 minutes. The supernatants were discarded and the pellets re-suspended in 20 ml of TSB. The inoculum culture was further diluted to contain approximately 3 $\log_{10}$ cfu ml^{-1} of *E. coli* and 1 $\log_{10}$ cfu ml^{-1} of *S. typhimurium*.

Sample preparation

Samples were prepared following the MF method described above.

Resuscitation and enumeration

The combinations of enrichment and plating media of three MPN methods selected for evaluation are presented in Table 1. A 100 ml aliquot of the clarified supernatant cell suspension was filtered through a filter aid (Hyflo supercell) on to an absorbent pad. The pad was removed using sterile forceps and aseptically transferred to 100 ml of 0.1 % peptone for Method 1 and Method 2 and 100 ml of buffered peptone water (BPW) for Method 3. The filter funnel was rinsed to recover all of the filter aid into the sample suspension and the samples were shaken vigorously to disperse the filter aid and cells into the solution.

A 3 x 5 MPN matrix was prepared from 100 ml of the cell suspension using TB modified with brilliant green and *l*-cystine in Method 1, or novobiocin and iodine-iodide solution in Method 2. All tubes were incubated at $37°C \pm 1.0°C$ for 18 - 24 h. For Method 3, BPW was decanted into five 10 ml aliquots and one 50 ml aliquot and incubated at $37°C \pm 1.0°C$ for 18 - 24 h. Following incubation, each BPW tube was sub-cultured into two selective enrichment broths. A 0.1 ml aliquot was transferred into 10 ml of Rappaport-Vassiliadis (RV) medium and incubated at $42°C \pm 1.0°C$ for 18 – 24 h and 1 ml was added to 9 ml of tetrathionate broth (TB), supplemented with novobiocin and iodine-iodide solution, and incubated at 37°C ± 1.0°C for 18 - 24 h.

A streak plate was performed from each MPN tube (Method 1 and 2) and from each selective enrichment broth (Method 3) on four different types of laboratory media including: two chromogenic agars (Rambach (RAM) and Oxoid Salmonella Chromogenic Agar (OSCA)) and two conventional agars (brilliant green agar (BGA) and xylose lysine desoxycholate agar (XLD)). All plates were incubated at 37°C ± 1.0°C for 24 h and 48 h for selective isolation.

CONFIRMATION
A selection of colonies (between 3 – 5) displaying typical *Salmonella* characteristics were streaked for purity onto tryptone soya agar (TSA), incubated overnight at 37 °C ± 1.0 °C, and confirmed using a *Salmonella* latex agglutination test (Oxoid).

RESULTS AND DISCUSSION
EXAMINATION OF DIGESTED AND RAW SEWAGE SLUDGE FOR *SALMONELLA* BY THREE MMP TECHNIQUES

Existing methods for detection of *Salmonella* are generally based on a MPN or presence/absence test. The MPN technique is a dilution-counting method and depends on the statistical probability of the presence of bacteria assumed to be randomly distributed in a sample. It is highly sensitive although relatively imprecise, but is inherently of more value than a presence/absence test, as it is quantitative and can detect small numbers of *Salmonella* (Pike and Fernandes, 1981). A potential disadvantage of MPN methodologies, however, is that they are relatively time consuming and require several incubation and confirmation steps. Sorber and Moore (1987) found that most studies reporting *Salmonella* recoveries from sewage sludge used MPN enumerations, comprising of a variety of pre-enrichment and enrichment broths, and plating agars. A wide range of different media has been used for the isolation of *Salmonella* and there are currently at least eight commercially available agars for *Salmonella* growth and differentiation of Enterobacteriaceae (Rambach, 1990). The selection of enrichment medium usually depends on the type of sample to be examined (Rhodes and Quesnel, 1986) and Dutka and Bell (1973) advocated using a variety of enrichment broth-plating agar combinations to determine the highest numbers of *Salmonella*. Three MPN methods, with combinations of enrichment and plating media, were adapted from published sources for the examination of biosolids in this investigation (Table 1).

Kenner and Clark (1974) have commented on the degree of work involved in obtaining accurate results from the isolation of *Salmonella*. Enumeration times may vary in the range of 3 – 5 days and, with serotyping, may be longer than 7 days. Thus, a disadvantage of Method 3 is the additional 24 h required for the pre-enrichment step, which increased the testing time from 3 to 4 days (without confirmation) by this technique. However, Cherry et al. (1972) found that two-step enrichment was unnecessary because most samples gave positive results from only a single enrichment stage. One-step enrichment reduces testing time, which is an important factor dictating analysis cost and rate. In addition, a single enrichment step is comparable to the MF technique developed by UKWIR (2000) for sludge and provides a relatively rapid and simple assay procedure.

Indigenous *Salmonella* spp. were not detected in either LRS or DMAD in the comparison of the different MPN procedures and agar types. However, all three techniques gave comparable recoveries of *Salmonella* spp. from the spiked inoculum broth, although Method 2 and 3 produced slightly higher recoveries of approximately $0.2 \log_{10} 100 \text{ g}^{-1}$ compared to Method 1 (Table 2).

E. coli in spiked samples of DMAD and LRS did not interfere with the detection of *Salmonella* colonies on any of the different agars tested indicating the effective suppression of other enteric microorganisms likely to be present in sewage sludge. RAM, OSCA and XLD were superior to BGA in terms of presumptive identification based on colour reaction and colony morphology. These results are in accordance with Carrington (1980), who reported that XLD was superior to BGA, and Yanko et al. (1995) who also showed that BGA was the least suitable medium for recovery of salmonellae. Consequently, BGA was not used in further method development work.

The laboratory study with pure bacterial cultures identified Method 2 as potentially the most suitable technique for enumerating *Salmonella* in sewage sludge. The method was not susceptible to interference from other enteric flora (*E. coli*) and recoveries of inoculated concentrations of *Salmonella* were determined within a 3-day testing schedule, which is a comparable time frame to MF (UKWIR, 2000). Further evaluation and comparison of MPN Method 2 was therefore undertaken with inoculated and unspiked samples of LRS, DMAD and soil.

EXAMINATION OF SEWAGE SLUDGE AND SOIL FOR *SALMONELLA* USING MPN
METHOD 2 AND MF

The first stage in the microbiological examination of sewage sludge or soil is designed to physically disrupt the matrix to release pathogens and provide a homogenous liquid sample for further examination. The recommended procedure for biosolids is to pummel the sample in a stomacher (UKWIR, 2000). However, this was not suitable for soil samples containing stones and solid debris, which rupture the stomacher bags used for homogenisation. Therefore, an alternative method for soil agitation was developed using glass beads and a laboratory shaker. Validation of the method for soil was reported in a previous paper (Lang *et al.*, 2002), which showed identical recoveries of *E. coli* from soils amended with DMAD by both homogenisation techniques. Dudley *et al.* (1980) also found that, for processing sludge, agitation with glass beads increased the recovery of organisms compared to homogenisation in a blender.

Salmonella was enumerated in low numbers in control (unspiked) samples of LRS by MF on OSCA and MPN on XLD, but was not detected on RAM or in the control (unspiked) soil (Table 3). The *Salmonella* strains isolated from raw sludge were serotyped as *S. Mbandaka* and *S. Oranienburg* at the Laboratory of Enteric Pathogens (Public Health Laboratory Service, Colindale, London). There are more than 2300 different *Salmonella* serotypes and between 5 and 10 cases of *S. Mbandaka* and *S. Oranienburg* infections are typically reported to the PHLS each quarter compared to 500 – 1000 *S. Enteritidis* infections (CDSC, 2002,b). Similar numbers of *Salmonella* were recovered for spiked samples of LRS and soil by both methods (Table 3). However, enumeration of both control and inoculated DMAD by MF on all agars was compromised to such an extent by confluent growth of indigenous organisms on the membrane surface that a result for *Salmonella* spp. could not be obtained despite a large dilution series (Figure 1). *Salmonella* was enumerated by MPN from spiked DMAD on RAM and XLD, although the value was 2 $\log_{10}$ 100 g^{-1} lower than the inoculum culture. The identification of *Salmonella* colonies was more successful by the MPN method than by MF because confluent growth is avoided by streaking only a small volume (10 µl) of the resuscitation/enrichment broth on to the agar. Yanko *et al.* (1995) state that detecting *Salmonella* spp. in biosolids is inherently difficult due to the high solids content and large numbers of competing bacteria. Furthermore, Rhodes and Quesnel (1986) observed disparities in the enumeration of *Salmonella* in sewage sludge between two enrichment broths. They concluded that high numbers of competitors in the sludge obscured salmonellas on some positive plates that were, as a consequence, erroneously scored negative. The results presented here demonstrate that the biosolids products examined contained low numbers of *Salmonella* spp. and isolation was dependant on the method and type of selective agar used and also on the sample type and nature.

The data on *Salmonella* enumeration in LRS and DMAD presented in Table 2 and 3 are entirely consistent with the highly variable numbers of *Salmonella* reported in sewage sludge. For example, Kenner and Clark (1974) isolated between 1 - 4 $\log_{10}$ 100 ml^{-1} in various sludge samples and Hess and Breer (1975) stated that 90 % of sludges they examined contained *Salmonella* spp. In contrast, Rhodes and Quesnel (1986) reported that sewage sludge contained a high background flora, but few salmonellas, and Pike and Fernandes (1981) found the numbers of salmonellae in raw sludge were low and variable. As *Salmonella* spp. occur only in very low numbers or are absent altogether, even in raw sludge, it could be argued from a technical standpoint that salmonellae are of relatively limited value as an indicator of sludge microbiological quality, and the effectiveness of sludge treatment processes at pathogen destruction, because this requires their presence in the incoming wastewater (Farrell, 1992).

A wide range of isolation media and procedures has been used for the enumeration of *Salmonella* in sludge and the sensitivity of detection of different microbiological procedures may vary considerably. For example, Jones *et al.* (1983) reported a 10 – 100 fold variation in the enumeration of *Salmonella* in sludged pasture by two laboratories using two different media. Isolation methods also vary in sensitivity and selectivity towards certain serotypes (Pike and Fernandes, 1981). For example, *S. dublin* isolated from sewage sludge required 48 h incubation on RAM compared to 24 h for *S. typhimurium* and *S. enteritidis* (UKWIR, 2000). Enrichment temperature also influences the sensitivity of detection and an incubation temperature of 42°C is used in many investigations of *Salmonella* in sewage sludge (Spino, 1966; Harvey and Price, 1968; Cheng *et al.*, 1971; Yoshpe-Purere *et al.* 1971; Cherry *et al.*, 1972; Kenner and Clarke,

1974 and Carrington, 1980). However, McCoy (1962) and Edgar and Soar (1979) observed improved isolation at 37 °C and Dudley et al. (1980) recommended an incubation temperature of 37°C was necessary to detect *Salmonella*. Dudley et al. (1980) considered that low recoveries of *Salmonella* indigenous to sludge were potentially caused by environmental damage to stressed organisms, particularly at high incubation temperatures, as well as the inhibitory nature of enrichment media. In this laboratory comparison of MF and MPN techniques, an incubation temperature of 37°C was used for the enrichment procedure for both techniques. Thus, *Salmonella* enumeration may vary depending on the sensitivity of the method, the selective medium used, the variability of *Salmonella* numbers in sludge and the type of sludge analysed. Therefore, implementing regulatory limits on the numbers of microbial pathogens in sewage sludge will be practically difficult without standardisation of microbial methods for laboratory examination of a diverse range of biosolids types. Yanko et al. (1995) evaluated specific *Salmonella* methods documented in the US EPA Part 503 regulations on pathogens in sludge (US EPA, 1993) and concluded that standard testing protocols had essentially been established without documenting the validity or suitability for intended purpose. A thorough evaluation of methods including a multilaboratory round robin testing programme was recommended. The work reported here has also emphasised the need for further evaluation and inter-laboratory comparison of the various microbiological methods available for enumeration of *Salmonella* in sewage sludge and soil.

Colour development by colonies of *Salmonella* on RAM from the examination of soil by MF was less distinctive compared to those obtained by MPN (Figure 2a). Furthermore, colonies of background flora expressed a chromogenic colour reaction similar to *Salmonella* on OSCA (Figure 2b). The turbid nature of supernatants of DMAD also affected colony colour when membranes were transferred to the agar for selective isolation (Figure 3). Therefore, proficient enumeration and identification of *Salmonella* in DMAD by MF would depend on laboratory competence and confirmation procedures. However, it is emphasised that these practical and operational difficulties with MF are in addition to the problems of microbial overgrowth observed with DMAD samples, which prevented the measurement of background numbers of *Salmonella* spp. in this sludge type by MF. The advantage of chromogenic agars over conventional bacteriological agar is their ability to identify bacterial species without resorting to extensive confirmation, shortening analysis times. However, the occurrence of false positives and ambiguous chromogenic colour reactions requires the additional confirmation of *Salmonella*, which undermines the rationale of using the more expensive chromogenic agars (£31 per l for RAM and £20 per l for OSCA compared to £3.50 per l of XLD).

The commercially available MUCAP test (Biolife) potentially offers a rapid approach to confirming *Salmonella* based on the detection of caprylate esterase using fluorogenic 4-methylumbelliferyl caprylate (MUC) under UV light at 366 nm for 1-5 min. UKWIR (2000) recommended the direct atomisation of MUC in ethanol onto the RAM membrane and observation under UV at 344 nm for 1 min for confirmation of *Salmonella*. A bluish fluorescence under UV indicates a positive *Salmonella* colony. However, visual inspection of membranes under UV was affected by supernatant residue and indigenous flora in DMAD samples, resulting in total fluorescence of the membrane. Similarly, Manafi (2000) reported that the MUCAP test was rapid and simple to perform, but was not specific to *Salmonella*, giving many false positives. *Salmonella* colonies were easily identified on sample membranes with low numbers of background flora and clearer supernatants, as was the case with spiked soil samples. An alternative confirmation method evaluated in this study was a latex agglutination kit (Oxoid), based on *Salmonella* antibodies. However, the test is not appropriate for non-motile strains (>0.1% of strains), requires an additional 24 h for sub-culturing on non-selective agar and was designed for use with the Oxoid Salmonella Rapid Test only. Despite the apparent disadvantages, all selected presumptive colonies were confirmed as *Salmonella* using this assay.

The controlled laboratory experiments reported here, evaluating *Salmonella* methodologies, have shown that MF is unsuitable for enumerating salmonellae in conventionally treated digested biosolids due to interference by the large background flora in the sludge. Under these circumstances, MF can potentially underestimate numbers of salmonellae as the identification of indigenous *Salmonella* colonies is ambiguous. Analysis time and cost may also be increased by the requirement for additional confirmation tests. Membrane filtration may be appropriate, however, for sample types (e.g. soil) with smaller background bacterial populations. Recoveries of *Salmonella* in spiked sludge samples by MPN were smaller than the numbers introduced in the inoculum, but in comparison to MF, MPN was less susceptible to interference from indigenous flora.

The data presented here emphasise the need for further development and validation work to produce standardised procedures for enumeration of *Salmonella* in the broad spectrum of different biosolids types currently available for application to agricultural land. This concurs with the view expressed over 20 years ago that, despite recent developments in laboratory media and techniques, there is no single method available that is suitable for all sample types and conditions (Carrington, 1980). Further work is recommended to establish specific methods applicable to different categories of biosolids that take account of their intrinsic microbial nature.

CONCLUSIONS

- The occurrence of *Salmonella* spp. in sludge is highly variable; they were not detected in dewatered digested sludge and were present in low numbers or absent in liquid raw sludge.
- Chromogenic and conventional agars recommended for *Salmonella* enumeration are susceptible to interferences from indigenous microorganisms in sewage sludge. Consequently, confirmation tests are required for bacterial identification, thus removing a principal advantage of more expensive chromogenic agar over conventional methods of enumerating *Salmonella* in sludge.
- Enumeration of *Salmonella* can be influenced by the sensitivity of the detection method and selective medium used. Therefore, a combination of enrichment broths and agar types may be necessary for reliable determination of *Salmonella* in sludge following the approach recommended by HMSO (1994) for the examination of *Salmonella* in water and associated materials.
- There are limitations to membrane filtration techniques for measuring small numbers of indigenous *Salmonella* in conventional digested sludge due to overgrowth on the agar plates by the large background flora in the sludge.
- An MPN method was adapted to enumerate *Salmonella* in digested biosolids with large indigenous microbial populations.
- Further development and validation work is needed to produce standardised techniques for the enumeration of *Salmonella* applicable to specific types of biosolids products.

ACKNOWLEDGEMENTS

The work described here forms part of a programme of research, Predicting Agricultural Benefit of Novel Biosolids Products, funded by the Environment Agency, Severn Trent Water, Thames Water, Yorkshire Water, Scottish Water and the EPSRC.

REFERENCES

1. APHA; AMERICAN PUBLIC HEALTH ASSOCIATION (1998) *Standard Methods for the Examination of Water and Wastewater*. 20th edition. United Book Press, Maryland.

2. CARRINGTON, E. G. (1980) *The Isolation and Identification of Salmonella spp. in Sewage Sludges: A Comparison of Methods and Recommendations for a Standard Technique*. Technical Report TR129. WRc Medmenham, Marlow.

3. CDSC; COMMUNICABLE DISEASE SURVEILLANCE CENTRE (2002a) Communicable Disease Report Weekly, 13 June.

4. CDSC; COMMUNICABLE DISEASE SURVEILLANCE CENTRE (2002b) Communicable Disease Report Weekly, 19 September.

5. CHENG, C. M., BOYLE, W. C. AND GOEPFERT, J. M. (1971) Rapid quantitative method for Salmonella detection in polluted water. *Applied Microbiology* 21, 662 – 667.

6. CHERRY, W. B., HANKS, J. B., THOMASON, B. M., MURLIN, A. M., BIDDLE, J. W. AND CROOM J. M. (1972) Salmonellae as an index of pollution of surface waters. *Applied Microbiology* 24, 334 – 340.

7. DUDLEY, D. J., GUENTZEL, N. M., IBARRA, M. J., MOORE, B. E., SAGIK, B. P. (1980) Enumeration of potentially pathogenic bacteria from sewage sludge. *Applied and Environmental Microbiology* 39, 118 – 126.

8. DUTKA, B. J. AND BELL, J. B. (1973) Isolation of Salmonellae from moderately polluted waters. *Journal of Water Pollution Control Federation* 45, 316 – 344.

9. EDGAR, D. AND SOAR, M. R. (1979) Evaluation of culture media for the isolation of salmonellas from sewage sludge. *Journal of Applied Bacteriology* 47, 237 – 241.

10. FARRELL, J. B. (1992) *Technical Support Document for Reduction of Pathogens and Vector Attraction Reduction in Sewage Sludge*. NTIS No. PB93 – 110609. NTIS, Springfield, VA.

11. HARVEY, R. W. S. AND PRICE, T. W. (1968) Elevated temperature incubation of enrichment media for the isolation of Salmonellae from heavily contaminated materials. *Journal of Hygiene* 66, 377 – 381.

12. HESS, E. AND BREER, C. (1975) Epidemiology of Salmonellae and fertilizing of grassland with sewage sludge. *Zentralblatt Fur Bakteriologie, Parasitenkunde, Infektionskrankheiten Und Hygiene. Erste Abteilung Originale. Reihe B: Hygiene, Praventive Medizin* 161, 54 – 60.

13. HMSO (1982) *Methods for the Isolation and Identification of Salmonellae (other than Salmonella typhi) from Water and Associated Materials.* HMSO, London.

14. HMSO (1994) *The Microbiology of Water Part 1 – Drinking Water. Methods for the Examination of Waters and Associated Materials.* Report on Public Health and Medical Subjects No. 71. HMSO, London

15. JONES, F., GODFREE, A. F., RHODES, P. AND WATSON, D. C. (1983) Salmonella and sewage sludge – microbiological monitoring, standards and control in disposing of sludge to agricultural land. In: *Biological Health Risks of Sludge Disposal to Land in Cold Climates.* University of Calgary Press, Alberta.

16. KENNER, B. A. AND CLARK, H. P. (1974) Detection and enumeration of Salmonella and Pseudomonas aeruginosa. *Journal of Water Pollution Control Federation* 46, 2163 – 2171.

17. LANG, N. L., SPANOUDAKI, K., SMITH, S. R., PIKE, E. B. (2002) The kinetics of microbial inactivation during treatment and after application of biosolids to land. *Biosolids: The Risks and Benefits – An Update on the Latest Research.* CIWEM Conference, 9 January, London.

18. MANAFI, M. (2000) New developments in chromogenic and fluorogenic culture media. *International Journal of Food Microbiology* 60, 205 – 218.

19. McCOY, J. H. (1967) The isolation of salmonellae. *Journal of Applied Bacteriology* 25, 213 – 224.

20. PIKE, E. B. AND FERNANDES, X. (1981) *Salmonellae in Sewage Sludges. An Analysis of Counts from Surveys of Sewage Works in England and Wales in 1978 and 1980.* WRc Report No. 71-S. WRc, Stevenage.

21. RAMBACH, A. (1990) New plate medium for facilitated differentiation of Salmonella spp. from Proteus spp. and other enteric bacteria. *Applied and Environmental Microbiology* 56, 301-303.

22. RHODES, P. AND QUESNEL, L. B. (1986) Comparison of Muller-Kauffmann tetrathionate broth with Rappaport-Vassiliadis (RV) medium for the isolation of salmonellas from sewage sludge. *Journal of Applied Bacteriology* 60, 161 – 167.

23. SMITH, J. E. AND BROBST, R. B. (2001) The US position on fate of pathogens in sewage sludge and the US 40CFR503 Regulation. *Pathogens in Biosolids and Their significance in Beneficial Use Programmes. Proceedings UKWIR/Aqua Enviro Pre-Conference Workshop, 6th European Biosolids and Organic Residuals Conference*, Aqua Enviro, 11 November, Wakefield.

24. SMITH, S. R. (2001) Overview of research in the Department of Civil and Environmental Engineering at Imperial College on microbiological aspects of biosolids treatment and agricultural use. *Proceedings UKWIR/Aqua Enviro Pre-Conference Workshop, 6th European Biosolids and Organic Residuals Conference*, Aqua Enviro, 11 November, Wakefield.

25. SORBER, C. A AND MOORE, B. E. (1987) *Survival and Transport of Pathogens in Sludge-amended Soil; A Critical Literature Review.* Report No. EPA/600/2-87/028. NTIS, Springfield, VA.

26. SPINO, D. F. (1966) Elevated temperature technique for the isolation of Salmonella from streams. *Applied Microbiology* 14, 591 – 595.

27. UKWIR; UK WATER INDUSTRY RESEARCH (2000) *Methods for the Detection and Enumeration of Pathogens in Biosolids.* Report No. 00/SL/06/5. UKWIR, London.

28. US EPA; US ENVIRONMENTAL PROTECTION AGENCY (1993) Part 503 Standards for the Use or Disposal of Sewage Sludge. *Federal Register* 58, 9387-9404.

29. YANKO, W. A., WALKER, A. S., JACKSON, J. L., LIBAO, L. AND GARCIA A. L. (1995) Enumerating Salmonella in biosolids for compliance with pathogen regulations. *Water Environment Research* 67, 364 – 370.

30. YOSHPE-PURERE, Y., RICKLIS, S. AND PAIST, M. (1971) A convenient method for isolation of salmonellae from sewage and contaminated sea water. *Water Research* 5, 113 – 120.

TABLES
Table 1 MPN techniques evaluated for the enumeration of *Salmonella* in sewage sludge

Method 1 (APHA, 1998)	[1]Method 2 (APHA, 1998; UKWIR, 2000)	Method 3 (HMSO, 1982; 1994)
25 g sludge ↓	25 g sludge ↓	25 g sludge ↓
Stomach 1 minute in 225 ml phosphate buffered saline ↓	Stomach 1 minute in 225 ml phosphate buffered saline ↓	Stomach 1 minute in 225 ml phosphate buffered saline ↓
Centrifuge 200 - 300 *g* 1 minute ↓	Centrifuge 200 - 300 *g* 1 minute ↓	Centrifuge 200 - 300 *g* 1 minute ↓
Filter 100 ml of supernatant through filter aid on an absorbent pad ↓	Filter 100 ml of supernatant through filter aid on an absorbent pad ↓	Filter 100 ml of supernatant through filter aid on an absorbent pad ↓
Transfer pad to 100 ml 0.1 % peptone ↓	Transfer pad to 100 ml 0.1 % peptone ↓	Transfer pad to 100 ml buffered peptone water ↓
Prepare MPN in tetrathionate broth with 1:50 000 brilliant green & 3 mg l^{-1} *l*-cystine ↓	Prepare MPN in tetrathionate broth with novobiocin & iodine solution ↓	Prepare MPN in BPW 5 x 10 ml and 1 x 50 ml ↓
Incubate 35 °C for 24 h ↓	Incubate 35 °C for 24 h ↓	Incubate 35 °C for 24 h ↓
Streak onto conventional and chromogenic agars (OSCA/RAM/XLD/BGA) ↓	Streak onto conventional and chromogenic agars (OSCA/RAM/XLD/BGA) ↓	Sub-culture into selective enrichment broths (RV/TB) ↓
Incubate 35 °C for 24 h	Incubate 35 °C for 24 h	Incubate 24 h at 35 °C for TB Incubate 24 h at 42 °C for RV & RV ↓
↓ Confirm *Salmonella* colonies and calculate MPN using standard probability tables	↓ Confirm *Salmonella* colonies and calculate MPN using standard probability tables	Streak onto conventional and chromogenic agars (OSCA/RAM/XLD/BGA) ↓ Incubate 35 °C for 24 h ↓ Confirm *Salmonella* colonies and calculate MPN using standard probability tables

[1] Method 2 was also used in the examination of soil samples

Abbreviations:

BGA, brilliant green agar
BPW, buffered peptone water
OSCA, Oxoid Chromogenic Salmonella Agar
RAM, Rambach chromogenic agar
RV, Rappaport-Vassiliadis broth
TB, tetrathionate broth
XLD, xylose lysine desoxycholate agar

Table 2 Enumeration of *Salmonella* in sewage sludge and inoculated broth by three MPN methods using two chromogenic (RAM, OSCA) and two conventional (BGA, XLD) agars

	Salmonella MPN 100 g^{-1} (log$_{10}$)		
	Inoculated broth	LRS	DMAD
Method 1			
RAM	2.90	<2.30	<2.30
OSCA	2.90	<2.30	<2.30
BGA	2.90	<2.30	<2.30
XLD	2.90	<2.30	<2.30
Method 2			
RAM	3.23	<2.30	<2.30
OSCA	3.23	<2.30	<2.30
BGA	3.04	<2.30	<2.30
XLD	3.23	<2.30	<2.30
Method 3 - RV			
RAM	3.18	<2.30	<2.30
OSCA	2.95	<2.30	<2.30
BGA	3.18	<2.30	<2.30
XLD	2.95	<2.30	<2.30
Method 3 - TB			
RAM	3.18	<2.30	<2.30
OSCA	2.95	<2.30	<2.30
BGA	3.18	<2.30	<2.30
XLD	3.18	<2.30	<2.30

LRS, liquid raw sludge
DMAD, dewatered mesophilic anaerobically digested sludge

Table 3 Comparison of *Salmonella* enumeration techniques in control (unspiked) and inoculated sewage sludge and soil samples

	Salmonella (MF) cfu 10 g^{-1} (log$_{10}$)	*Salmonella* (MPN) 100 g^{-1} (log$_{10}$)
CONTROL		
LRS		
RAM	<2.00	<2.30
OSCA	2.00	<2.30
XLD	<2.00	2.30
DMAD		
RAM	*	<2.30
OSCA	*	<2.30
XLD	*	<2.30
Soil		
RAM	<2.00	<2.30
OSCA	<2.00	<2.30
XLD	<2.00	<2.30
INOCULATED		
LRS		
RAM	3.58	5.20
OSCA	3.30	4.95
XLD	3.67	5.20
DMAD		
RAM	*	2.60
OSCA	*	<2.30
XLD	*	2.60
Soil		
RAM	3.48	>5.20
OSCA	3.56	>5.20
XLD	3.76	>5.20

*unable to determine owing to high background flora

LRS, liquid raw sludge
DMAD, dewatered mesophilic anaerobically digested sludge

FIGURES

◄─────────────── **Increasing dilution**

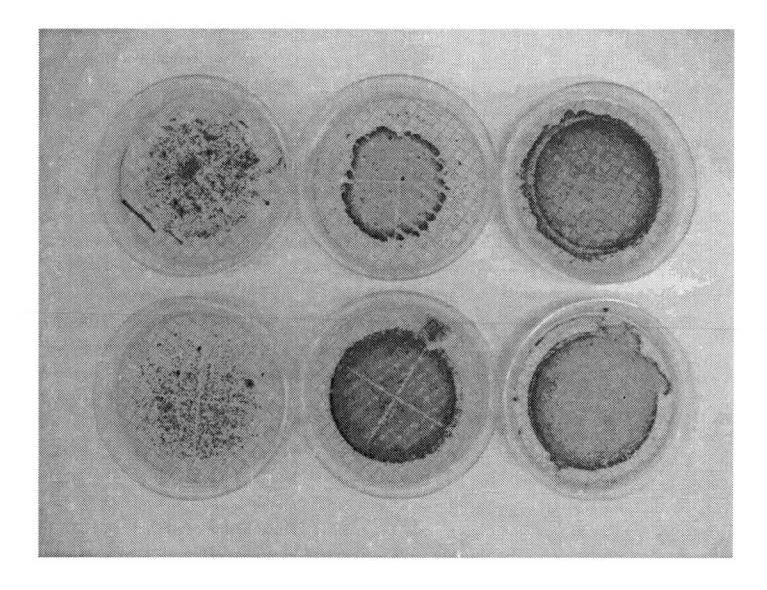

Figure 1 Interference and growth of background flora in digested sludge on duplicate membranes of RAM chromogenic agar plates

(a) *Salmonella* colonies by MPN and MF   (b) Non-salmonellae 'magenta' colonies

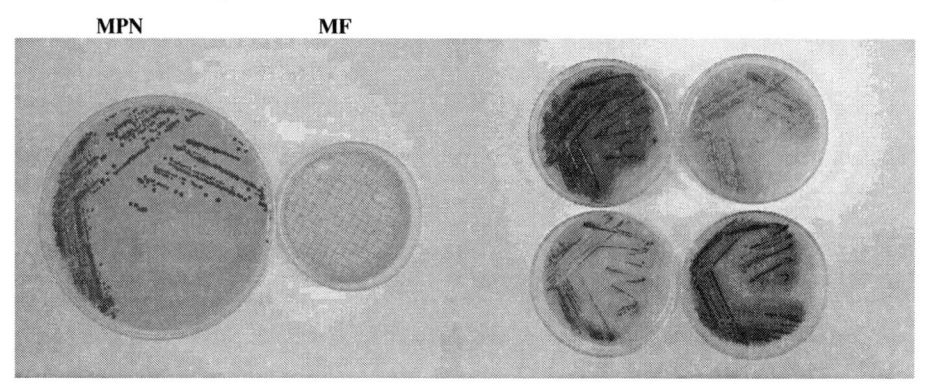

MPN   MF

Figure 2 (a) Comparison of *Salmonella* colony development on RAM from the examination of inoculated soil by MPN and MF techniques and (b) streak plates of colonies expressing a similar chromogenic colour reaction to *Salmonella* on OSCA

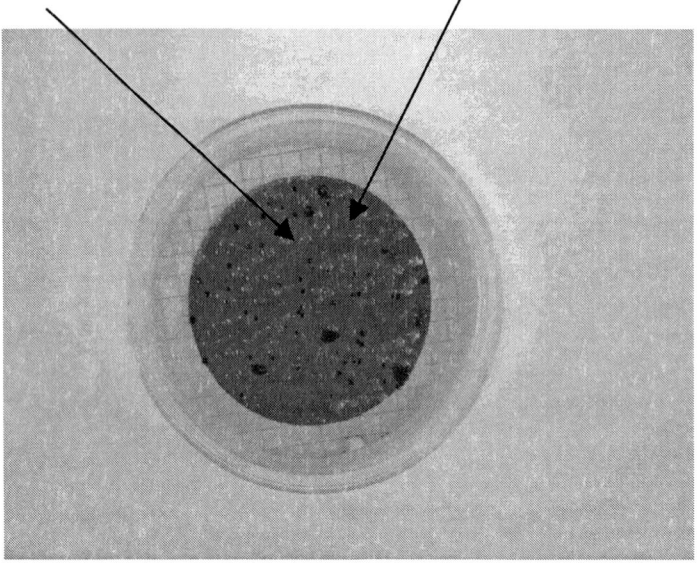

Figure 3 Effect of supernatant turbidity from digested sludge on membrane surface and chromogenic reaction colour of *Salmonella* colonies on RAM

BIOSOLIDS OF AGRONOMIC QUALITY

F. Ducray [2], A. Huyard [1], L. Patria [2]

[1]ONDEO Services-CIRSEE, 38 rue du Président Wilson, 78230 LE PECQ
Tél. 01.34.80.23.91. – Fax 01.30.53.62.11., E mail : alain.huyard@ondeo.com

[2]ANJOU RECHERCHE (Vivendi Water), Chemin de la Digue, BP 76, 78603 MAISONS-LAFFITTE CEDEX
Tél. 01.34.93.31.31. – Fax 01.34.93.31.10., E mail : florence.ducray@generale-des-eaux.net

ABSTRACT

Agronomic quality of biosolids is measured by analysis on metal and organic trace elements and evaluated by the compliance with the regulation limits. Today, most laboratories have adopted principles of good laboratory practices and have gained an accreditation for the analyses they propose.

In parallel, WWTPs operators have integrated a strong will of quality and good practices in their ways of operation in the past years. The lack of details (limit of quantitation, measurement error…) given by the laboratories with their analytical results impact strongly on the interpretation that can be made from these results. Some differences between laboratories are pointed out, showing the difficulty for operators to demonstrate a constant or even an improved agronomic quality of biosolids over time and for the Regulator to evaluate the pertinency of new standards

KEY WORDS

Analysis, Biosolids, EC Regulation, , sludge, Metal, Trace Organic Compound.

INTRODUCTION

Biosolids and compost from sewage sludge are either regarded as waste or as fertilisers, depending on European countries. In each country, biosolids quality must meet the EU Directive (86/278/EEC) requirements in term of metal and organic trace elements to be applied on land. At present this Directive is under revision. Quality procedures have been undertaken in the latest years in order to improve the global quality of biosolids. Procedures, such as 'sewage survey', focused on a close control of industrial influents discharges in sewers and their contents in metal and/or organic trace elements. They also included the construction and operation of combined sewage system overflows, dedicated storm water sewers and treatment processes, mainly to reduce the metal pollution issued from roads storm leaching. Those procedures were very effective at improving greatly the global quality of biosolids from wastewater treatment plants (WWTPs).

In parallel, analytical laboratories took actions to get official accreditations, in particular to supply the information of a respect or not of urban biosolids in term of trace elements contents.

Today, ISO 9001 and ISO 14001 normalisation procedures are applied to WWTPs. They tend to a continuous improvement of quality procedures and the setting of good practices procedures and quality indicators monitoring. In such approach it is expected more from an analytical result than to indicate a respect or not of a specific regulation. An actual demand exists for getting more information from the laboratories in order to establish closer quality comparisons, even below the regulation threshold.

This demand for more detailed information from biosolids analytical results is the object of the present paper. The main focus is a wider use of analytical results to respond to an approach of continuous quality improvement in plant operation and also to a more effective comparison of results from different laboratories.

The data commented on in the present paper will be based on a research program commissioned by the French government and the Ademe Agency to AGHTM and the 3 French utilities: Vivendi Water, Ondeo and Saur. Its main objective was to measure the sludge quality in France and to evaluate the impact of the proposal of European Sludge Directive on the French policy for sludge final routes.

REGULATION AND BIOSOLIDS FINAL ROUTES IN EUROPE

REGULATION

Today, to be used in agriculture, Biosolids must meet with metal and organic trace element limits set by the 86/278/EEC EU Directive. Those limits were initially set for metal trace elements such as lead, mercury, zinc, copper well known for their toxicity. Pathogens dissemination in the environment became then a concern. However several studies revealed that a chemical (such as lime addition), biological (digestion) stabilisation or sanitation process on one hand and the environmental competition on the other were sufficient to avoid contamination risks from spreading of sewage biosolids on land (Webber, 1984). Limits were set on some microorganisms considered as indicators of a general microbiological quality of biosolids.

In the 1960's, following several industrial incidents, the environmental pathway of organic compounds became a main concern. Webber and Lesage (1989) indicated than in 1974 more than 12,000 different organic compounds were used. Among those compounds some appeared to exhibit a long persistence in the environment and to be present in biosolids. Those ecotoxic compounds can travel long distances in the environment and accumulate in the food chain: fishes, mammals, humans. In 1979, EPA set a list of 129 priority substances that could eventually be found in biosolids. Since then, many studies focused on PCBs and PAHs.

In December 2000, a new worldwide convention on persistent organic pollutants forbade or limited the use of 12 compound families, in particular 8 pesticides, PCBs, dioxins and furans. At this point, the French regulation (arrêté 8/01/98) set threshold limits for biosolids for metals and organic compounds, and for pathogens when sanitation is requested. The only organic pollutants taken into account are PAHs and PCBs. Those compounds were selected due to their resistance to biodegradation and their persistence in soils. The threshold limits are presented in Table 1.

BIOSOLIDS FINAL ROUTES IN EUROPE

The national policies in term of biosolids final routes are directly linked to the national regulations based on the 86/278/EEC EU Directive. Some countries are well known for a political choice oriented towards incineration, others towards agriculture use. From the data presented in Figure 1, it is interesting to realise that the amount of biosolids spreaded on land include spreading normally called "agriculture use" but also other practices such as landscaping and compost application. The total amount of biosolids applied on land can in this way double in certain countries (as Finland). For the Netherlands or Austria, the amounts of sludges valorised to land or incinerated are in the same range.

The transformation of sludges to product is a good solution for their valorisation to land.

ORIGIN OF AGRONOMIC QUALITY BIOSOLIDS

The respect of regulatory thresholds on metal and organic trace elements is directly linked to the quality of wastewaters sent to WWTPs. It emphasises the importance of the politics of sewage survey and of the implementation of dedicated storm water sewers. The fact that biosolids are considered of agronomic value does not depend on their conditioning. Their texture might vary depending on the level of sludge volume reduction (liquid, pasty, dried, composted). They can be stabilised or sanitised. Table 2 presents an overview of the conditioning treatments that can be applied to sludge before spreading on land.

All the above conditioning treatments will not impact on biosolids contents in trace elements, but essentially on:
- Mode of biosolids spreading on land depending on texture
- Sanitation: definition of larger spreadable areas

END USERS OF BIOSOLIDS QUALITY MONITORING RESULTS

Initially, only 3 groups were concerned by the biosolids quality monitoring: biosolids producers (WWTPs operators), utilities dedicated to spreading on land, and farmers. Their main interest lies in the demonstration that biosolids spreaded on land meet the regulation requirements. Water Agencies and Government representatives supply technical and administrative support to confirm the compliance with the agronomic quality required of biosolids used on land.

Today, biosolids producers are responsible for the ultimate fate of biosolids. Therefore, the very first biosolids producers, that are the collectivities themselves, take a strong interest in the implications of current and pending regulations. Delegation contracts are revised in order to emphasise the constraints and the state of biosolids conditioning and final route options. The object is to develop perennial final routes for biosolids. One possible approach is first to implement good practices and quality procedures such as ISO 9001 and ISO 14001.

In the context, Government organisations (Ministry of Environment, Health, Agriculture, Water Agencies, etc,.) have a consultant role. They need to access very detailed and precise comparisons on conditioning and spreading practices, as well as on biosolids quality at different geographical scales (regions, states, Europe,). The European Commission and the peripheral organisations also require the same level of information and the need of detailed comparisons is always emphasised.

The circle of biosolids quality monitoring is extended even more following the different sanitary and health scares related, for example to BSE, foot and mouth disease. Those events questioned agricultural practices including sewage biosolids spreading on land. New players in the arena include consumer associations, food industry, supermarket industry and land-owners. Those new entries to sludge recycling require a quality guarantee which is both immediate and in time.

Today quality requirements are set on:
- Intrinsic biosolids quality: to be controlled by regulatory threshold limits,
- Immediate confirmation of quality at the time of biosolids conditioning and spreading practices: to be demonstrate: first regulatory threshold limits have not been exceeded, second to demonstrate the biosolids product production quality is consistent or improving, and an extended traceability supports spreading practices.

Those requirements imply different needs in the reading and detail of biosolids analyses.

ANALYSIS OF BIOSOLIDS QUALITY AND COMPARISON OF RESULTS FROM SEVERAL LABORATORIES

ANALYSIS OF BIOSOLIDS QUALITY

The analysis of biosolids quality takes into account 4 major steps as follows:
- Sampling on site and transport to the laboratory,
- Sampling in the laboratory on the biosolids sample received,
- Extraction, purification of elements to be analysed in the sample matrix,
- Actual analysis of the extract from the biosolids sample.

Each step, as well as the delay between them, will impact on the quality and the error of the analytical results. Each laboratory uses standard method or has developed its own analytical method. The objective of the analysis is to qualify a sludge in respect of limits set in the specific country's regulation.

All laboratories using a standard method have their own quantitation limit (LQ) for each compound. This LQ and the error on the analysis depend of the extraction methods, the analytical and detection techniques used. Information on analytical reports vary from laboratories. Few data (as quantitation limit, measure quality) are given with the stated result. Comparisons of quality of different sludges or results from different laboratories are difficult with the variation of the information reported on analytical report.

ROUND ROBIN TEST BETWEEN 3 LABORATORIES

The French government and the ADEME agency have commissioned a research program to measure the quality of sludges and to evaluate the impact of the proposed European sludge directive revision on sludge valorisation. A round robin test was performed between 3 laboratories on the metals, the organic compounds and pathogens analysis. Samples from 5 wastewater treatment plants (chosen with a size distribution) were analysed. For each parameter, the 3 laboratories have defined a quantitation limit.

The results for the metals Cd, Cr, Cu, Hg, Ni, Pb, Zn and the sum (Cr + Cd + Ni + Zn) are reported on Table 3. Mean and ratio standard deviation/mean (SD/M) are reported in the table for each metals in the 5 samples and for the 3 laboratories. The quantitation limit is indicated for each parameter. The ratio SD/M point the variability of results from several laboratories on the same sample. For a majority of the metals compounds, results agree between the 3 laboratories.

For 2 cases, variations are too large:
- Firstly for the special analysis of Hg. For 2 samples, the ratio SD/M is too high, reaching 101 %. The values obtained are below French standard
- Secondly, when values are lower, below the quantitation limit, we observed an amplification of the ratio SD/M.

FOLLOWING OF THE SLUDGE QUALITY

STATISTICAL ANALYSIS OF DATA

A statistical analysis was performed on values for metals content of sludge analysis (1999 and/or 2000). Data from 2500 analysis of sludges from 636 wastewater treatment plant were used. General statistics were calculated for each parameter. The distributions of the values for each parameters were drawn, taking in account the quantitation limits (LQ) from each laboratory. For example, the results for cadmium are presented in figure 2.

The analysis of cadmium data in wastewater sludges shows the different quantitation limit used for this parameter by the laboratories. For all the data, 4 quantitation limit were identified, respectively 1, 2, 3 et 5 mg Cd/kg MS. A large majority of these data for cadmium concentration in sludge are below 10 mg Cd/kg MS. But, regarding this distribution of data, 35 % have a value less or equal to the quantitation limit 5 mg Cd/kg MS.

In the case of cadmium, when values are less than the quantitation limits and these limits approach a regulation standard, to appreciate the quality of sludges became difficult. The existence of these different quantitation limits leads to underestimate mean concentration of cadmium in sludges. For each standardised method for analysis of parameter, it will be more convenient to specify these quantitation limits during the normalisation procedure.

LONG TERM BEHAVIOUR OF SLUDGE QUALITY

The concern over making progress in sludge quality is that indicators are needed such as regulation standards for each parameter. The observation of sludge quality improvements can be done with the results given by the laboratories if, at the beginning, the concentration of the parameter exceed the regulation standard and decreases under it. Table 4 give an example with data on the metal content of sludge from the Acheres treatment plant (near Paris).

But, when the value of quantitation limits are near the regulation standard or the accumulated information on measurement error during analysis, no improvement of the quality of sludges could be observe if it complied previously with the regulation atnadard.

For the same reasons, it is difficult to compare quality of sludges at the various geographic scales (region, country, EC). Without information on measurement error, it is only possible to calculate the percentage of sludge complying with the regulation standard.

In table 5, for chromium, it is impossible to estimate if the differences of values are significant between 1997 and 1999-2000. In the case of cadmium, the same query appears on measurement error from different laboratories and on their choices of quantitation limits. How is it possible to compare a value of 3mg Cd/kg MS obtained in Europe and in France to a value estimated to be below 3mg Cd/kg MS in 1999-2000? The only possible observation from these data is that all those sludge complied with the French regulation.

Such concern about analytical comparability is already a European concern, which is extended to any matrix from biosolids, to biowaste, compost and soil. A PCRD project called HORIZONTAL was set up in order to develop horizontal and harmonised European standards in the field of sludge, soil, contaminated soil and treated biowaste to facilitate regulation of these major streams in the multiple decisions related to different uses and disposal governed by EU Directives. The HORIZONTAL project stresses the necessity for horizontal standardisation to go beyond the environmental media and legal areas. This means that independent of legal regulations (e.g. Sewage Sludge Directive, Biological Treatment of Biowaste proposal and the Landfill Directive) and of the type of material (treated biowaste, soil, sludge), an examination method will be used which leads to comparable and reliable results.

CONCLUSION

The compliance to the biosolids regulation in term of agriculture use is essential. It requires a check that biosolids are of agronomic quality, before spreading on land.

Laboratories today have implemented good laboratory practices and are usually accredited for the analyses they propose. The results they report are set to evaluate the compliance, or not, of a sample of biosolids to the regulation.

In parallel, WWTPs operators must develop quality and good practices procedures. For operators, the first reading of analytical results is for confirming the agronomic quality of the biosolids they produce. However, once the agronomic quality of biosolids is demonstrated, their concern is to have indicators of consistency or improved biosolids quality. In that respect, a second reading of analytical results would be necessary and could only come from more detailed analytical results.

Getting a systematic knowledge of quantitation limits and measurement errors on each individual analytical result would help in closer biosolids quality comparisons:

- In time, for the same biosolids (indicator of constant or improved quality),
- In space, between different biosolids (regional, national, European statistics)

ACKNOWLEDGEMENTS

The French Ministry of Environment and the Ademe Agency commissioned and gave their financial support to a research program used for the present paper. Acknowledgements go also to AGHTM and Saur for their support and participation in the project.

REFERENCES

1 WEBBER M.D., LESAGE S. "Organic contaminants in canadian municipal sludges" Waste management and Research (1989) 7, pp 63-82

2 AGHTM "Impact du futur projet européen sur la valorisation agricole des boues en agriculture – Campagne d'analyse sur 60 boues de STEP" Report (2002a) 140 pages

3 AGHTM "Impact du futur projet européen sur la valorisation agricole des boues en agriculture – Analyse statistique des données ETM et CTO de boues de stations d'épuration" Report (2002b) 70 pages

4 ADEME "Les micropolluants métalliques dans les boues résiduaires des stations d'épuration urbaines" (1995)

5 SEDE/Arthur Andersen "Disposal and recycling routes for disposal"
Report for the European communities (2001)

Table 1 : Proposed revision of EEC regulation (2001 - 3rd draft)

Heavy Metals (mg/kg MS)				Trace Organic Compounds (mg/kg MS)	
	2005	2015	2025		
Cadmium	10	5	2	PAH (11 compounds)	6
Chromium	1000	800	600	PCB (7 compounds)	0.8
Copper	1000	800	600	LAS	2600
Mercury	10	5	2	NPE	50
Nickel	300	200	100	DEHP	100
Lead	750	500	200	AOX	500
Zinc	2500	2000	1500	PCDD/F (in ITEQ ng.kg MS)	100
Pathogens for hygienised sludges or biosolids					
Salmonella			Abs / 50 g MS		
E. coli			< 500 / g MS		

Table 2 : Diversity of the conditioning treatments that can be applied on agronomic quality

Concentration	Stabilisation	Dewatering	Post treatment
Gravity thickener Flotation Centrifuge	Anaerobic Digestion (1 or 2 phases) Thermophilic aerobic Digestion Thermal conditioning (Porteus) Saphyr	Sludge drying bed Belt filter Press Filter Centrifuge	Lime addition Thermocomposting or Compostong Drying

Table 3 : Mean and SD/M ratio for metal analysis of 5 sludges

PARAMETER	Limit of Quantitation (mg / kg MS)	Mean et standard deviation/mean ratio ( ) (in mg / kg MS et %)				
		A	B	C	D	E
Cadmium	4	2,67 **(43)**	14.33 (4)	2 (0)	2 (0)	2 (0)
Chrome	10	31 (20)	50.33 (10)	19 (9)	29.33 (15)	28 **(27)**
Copper	10	331.67 (6)	518 (6)	162.67 (7)	488.67 **(10)**	526.67 (3)
Mercury	1	2,2 (14)	4,77 (24)	2,4 **(101)**	2.93 (55)	3,93 (1)
Nickel	10	23.33 (16)	121.33 (8)	12 (17)	17.33 **(29)**	24.33 (13)
Lead	10	163.67 **(30)**	243.33 (5)	42.33 (13)	193.33 (6)	36.33 (13)
Zinc	10	982.67 **(17)**	1379.33 (5)	460 (15)	1326.67 (8)	726.67 (6)

* the maximum ratio standard deviation/mean (SD/M) is indicated for each parameter in bold

Table 4 : Decrease of metal concentration in sludge from Acheres plant

Parameter	Acheres's plant sludge Mean concentration in metals (in mg/kg MS)		
	1978	1987	1992-93
Cadmium	161	69	14
Chromium	636	383	186
Copper	1538	1111	764
Mercury	18	12	7
Nickel	216	138	68
Lead	1638	1060	576
Zinc	5670	4087	2845

Source ADEME (1995)

Table 5 : Metal concentrations in sludges : French and CEE data

Parameter	French Regulation (1998) (mg/kg MS)	Mean concentration in sludges (mg/kg MS)		
		CEE states members*	French data 1997*	AGHTM** 1999-2000
Cadmium	*15*	0,4 - 3,8	2,9	< 3
Chromium	*1000*	16 - 275	58,8	65,9
Copper	*1000*	39 - 641	309	299,3
Mercury	*10*	0,3 - 3	3	2,3
Nickel	*200*	9 - 90	31,9	29,4
Lead	*800*	13 - 221	106,7	94
Zinc	*3000*	142 - 2000	754,2	773,3

- Source : SEDE/Arthur Andersen study (2001) **AGHTM study (2002b)

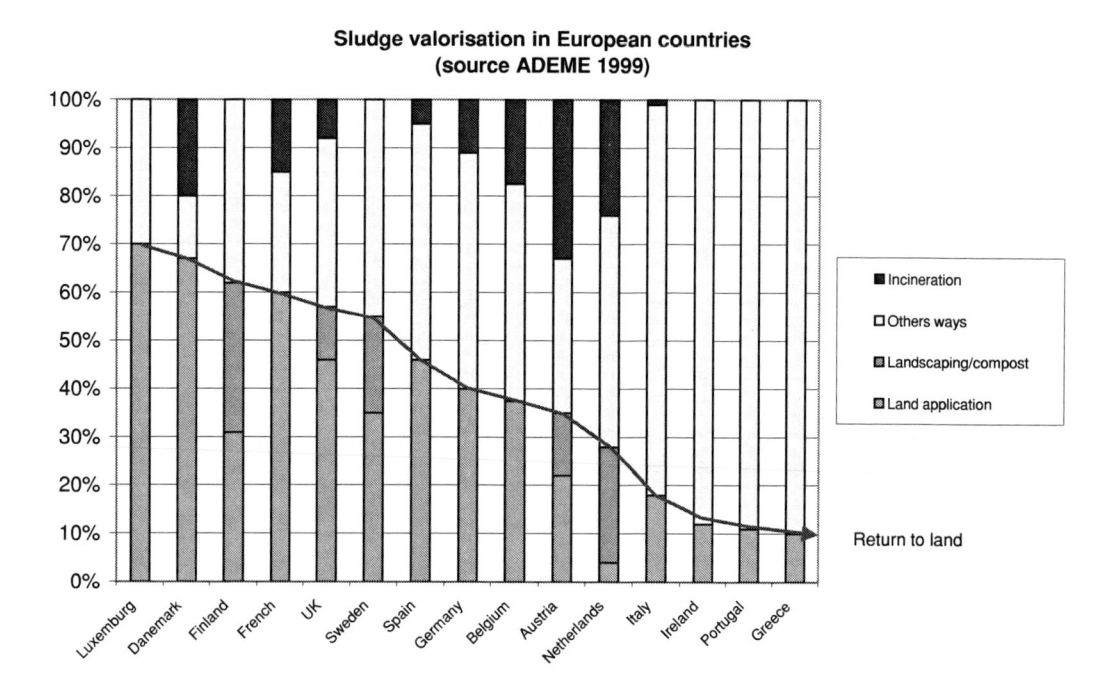

Figure 1: Biosolids final routes in Europe

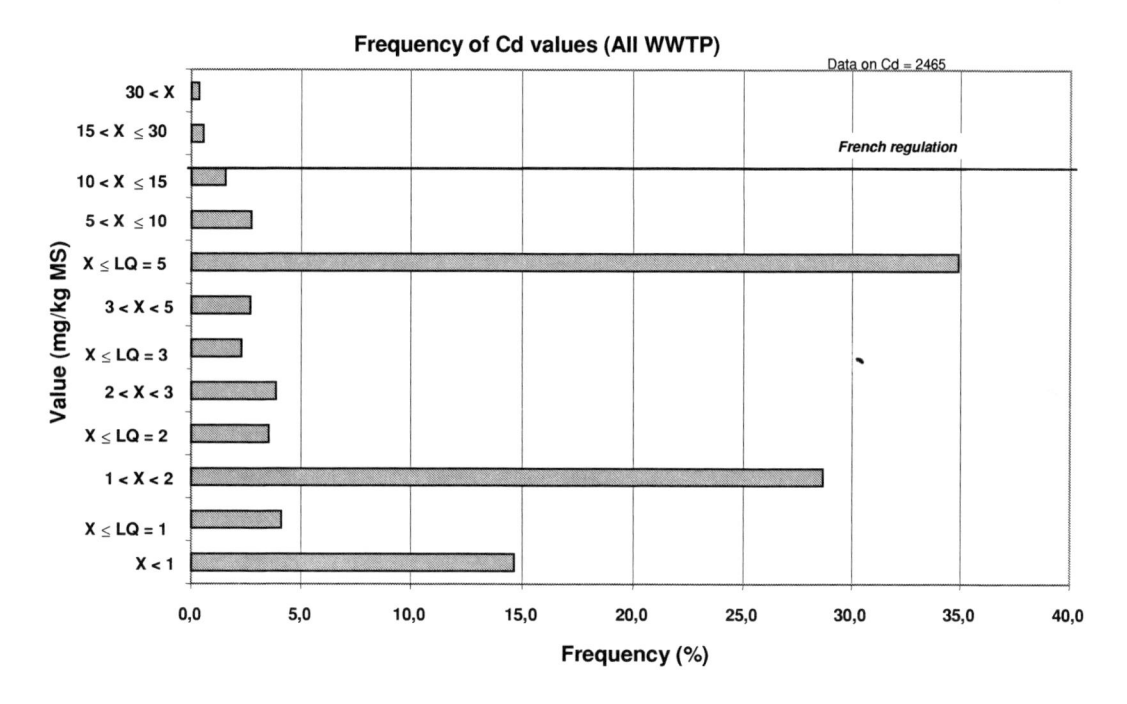

Figure 2: Distribution of Cadmium concentration in French Sludges (1999-2000)

ISSUES RELATED TO SLUDGE CHEMICAL AND MICROBIOLOGICAL ANALYSIS

K Clive Thompson, ALcontrol Laboratories, Mill Close Rotherham, S60 1BZ
Tel: +44 (0)1709 841078; Fax: +44 (0) 1709 841011;
E-mail clive.thompson@alcontrol.co.uk

ABSTRACT

In order to ensure a level playing field in Europe, it is essential that all sludge analysis standards are properly validated. This is not occurring at present. With the huge potential capital expenditure that could result from compliance failure on chemical and microbiological sludge parameters, it is considered essential that all CEN sludge analysis standards should be fit for purpose. Often chemical and microbiological sludge analysis compliance measurements underpin decisions costing many orders of magnitude more than the cost of the relevant sludge analysis. False negative or positive compliance failures have a very large cost implication. The cost of upgrading a sludge treatment process can be prohibitive. Thus it is essential that users of results have confidence in their laboratories and are satisfied that there is a level playing field across Europe with respect to compliance analysis measurements in relation to the sludge directive[1]. This paper will attempt to discuss validation of sludge analysis standards; microbiological analysis issues; the issue of prescribed methods or performance tested methods; sludge proficiency testing issues; the significance of AOX measurements and other relevant issues.

KEYWORDS

AOX test significance, Certified reference sludge materials, Helminth analysis of sludge, Microbiological analysis issues, Prescribed sludge analysis methods versus performance tested methods, Sludge analysis issues, Sludge analysis proficiency testing, Sludge interlaboratory analysis trials, Validation of sludge analysis methods.

VALIDATION OF SLUDGE ANALYSIS STANDARDS

It is now proposed that prior to acceptance and publication of all future CEN sludge analysis standards, the methods should be validated using an interlaboratory trial involving a minimum of ten laboratories in at least three different countries.

A documented protocol[2] is being considered for publication as a CEN Technical Report. This will involve organising a large international interlaboratory trial for each method. This would be achieved by submitting a range of different types of sludge samples and spiked samples with some suitable control samples to all participating laboratories by the trial organiser. Only experienced laboratories routinely carrying out sludge analysis will be allowed participate. It is recommended that each trial includes a range of sludges that is representative of the proposed scope of the method, as a minimum this should include an untreated (raw) sludge, a treated (an aerobically digested) sludge, a presscake, a thermally dried sludge and samples of any other sludge types relevant to the method being tested. In addition a spiking solution and a calibration check standard will generally be distributed to ensure absence of significant calibration standard preparation errors.

The main exercise variables to be agreed are: -
- Organising laboratory
- The documented method to be tested
- The number and nature of the samples
- Statistical calculations including rejection of outliers
- Payment arrangements. This is a very contentious area (see 2.9)
- Deadlines for sending out samples, receipt of results from participants and final exercise report from the organizer to WG1.
- Spiking protocol to calculate percentage recovery
- Analysis of a calibration check standard will also allow a check to be made on the comparative accuracy of the calibration standards used by all the participants. This can sometimes explain differences between laboratories.

Documentation to be sent with the samples will be: -
- A copy of the agreed method
- The spiking solution/material if relevant
- The instructions for the exercise
- The form for the submission of result
- The analysis start date (Four days after dispatch.)
- The analysis completion date (not more than two weeks after the dispatch date)
- The deadline for return of results. (Within three weeks of sample dispatch)

- Full contact details of the organizer to allow for queries and for the return of results

The interlaboratory trial organizer will provide a full report of the exercise to CEN TC308/WG1 and all participants by the agreed deadline. This will include:
- A listing of all of the results
- The mean value for each tested parameter
- The calibration check solution result
- The reproducibility (absolute and relative)
- The repeatability (absolute and relative)
- The percentage recovery (if relevant)
- The names and countries of laboratories that returned valid results
- The total number of results returned
- The number of different countries that returned valid results
- The temperature of the samples upon receipt at the participating laboratories.
- The number of outliers from non-rejected laboratories
- Any exceptional circumstances/problems reported by any of the laboratories on an anonymous basis
- Summary of performance: - i.e. interpretation of figures and data
- A detailed description of the samples and of the interlaboratory trial.

The organizer will obtain a written undertaking from each laboratory agreeing to participate, agreeing to follow the provided method exactly and meet the deadlines for commencement and completion of analysis and for submission of the results on the appropriate form.

If possible, samples shall be in the same form as would normally be submitted to a laboratory for analysis.

The transporting of sludge samples across trans-national boundaries can create logistical difficulties in that sludge samples are regarded as a significant biological hazard that can generate inflammable gases. Also ensuring stability of non-conservative liable parameters in sludges such as bacteria, organic substances mainly in organic substances (e.g. ammonia, alkalinity pH etc) is notoriously difficult. However, for international sludge analysis standards it is important that all of these standards are thoroughly tested and the results found to be fit for the intended purpose prior to publication. For conservative parameters such as toxic metals, this is not too difficult as it is possible to pasteurise the sludges. It is important that all major types of sludges are covered.

In order to organise and run these proposed interlaboratory trials, some funding is required. To date it has been found impossible to obtain any significant funding for this work. This is despite continuing lobbying over the last three years. If no funding is forthcoming then a number of key sludge analysis standards are likely to be published with inadequate method validation data. It will then not be possible to estimate the typical uncertainties of the results obtained using these methods.

All methods must be able to cope with sludges containing 1-2% m/m dry solids up to 95-100% m/m dry solids. This is not an easy task especially for microbiological methods. For labile organic parameters pasteurisation of the sludge may significantly reduce the analyte concentration, but this can be overcome by spiking the pasteurised sludge (contained in suitable containers). For very label substances it is possible to supply the participating laboratories with a spiking solution, where a small specified volume is added to the sludge sample bottle just prior to the analysis and the bottle contents thoroughly mixed.

Figures 1 – 3 illustrate the need for good precision and low bias. Figure 1 also shows the effect of uncertainty of results around a compliance limit. The less precise a result is, the bigger the grey area between a definite (95%) confidence fail and a definite 95% confidence pass. Figure 3 illustrates the problem of method bias. This can be significant for some sludge analysis parameters.

MICROBIOLOGICAL ANALYSIS ISSUES

For microbiological parameters, the situation is much more difficult than for chemical parameters. For instance, for *Salmonella spp.* testing where the third draft of the sludge directive[2] requires absence in 50g of 'as received' sludge. In this instance pasteurisation is not appropriate, as it will kill competitive microorganisms, which may well be present at a level many orders of magnitude greater than the number of *Salmonella spp.* present. For this case a sludge containing competitive organisms and no Salmonella would be sent out at 4°C with a guaranteed 24-hour delivery. Then each participating laboratory would add a supplied Lenticule® or Vitroid® containing a known low number of a relevant *Salmonella* species. If necessary, these Lenticules or Vitroids can be supplied in a stressed form. For *E.coli*, which is normally one of the major microbiological components of sludge, pasteurisation may be an appropriate option with addition of a suitable *E.coli*

Lenticule® or Vitroid® by the participating laboratory.

It is appreciated that this proposed protocol is less than ideal, but with the current limited resources available to organise these trials it would appear to be a pragmatic way forward.

For *Salmonella spp.* (and *E.coli*), unlike chemical parameters, two other aspects must be considered:
- Relevant *Salmonella* species to be covered by the standard
- The detection of sub-lethally damaged bacteria

There are a very large number of different Salmonella species. It is impossible to test the proposed standard against all of these. At least one (centre of excellence) laboratory should test the standard, using the main relevant strains. A good example with the current proposed Salmonella standards, is *S.dublin.* Plates for S.dublin plates should be enumerated at both 24 & 48 compared with species such as *S.enteriditis, S.typhimurium* and *S.seftenburg* where 24 hours only is satisfactory.

Also, the issue of sub-lethally damaged bacteria needs to be addressed. After most sludge treatment processes to reduce pathogen numbers, most remaining *Salmonellae* and *E.coli* will be in a sub-lethally (stressed) form. If a laboratory does not apply suitable resuscitation techniques, then the result can be orders of magnitude below a laboratory employing a fit for purpose resuscitation technique.

Thus, to ensure a level playing field across Europe, it is important to ensure that all laboratories are using suitably validated methods. If they do not use the CEN standard methods they must be able to demonstrate equivalence of results with the CEN standard using a range of typically analysed sludge samples.

Some European countries are lobbying for the inclusion of helminth (nematodes and cestodes) in the forthcoming sludge directive. There are a significant number of potential relevant pathogenic helminths in Europe in the area ranging from Southern Greece to Northern Finland. Thus the first problem will be agreeing the species to be covered. The second problem is accurately assessing the viability of the helminths detected. (Are they alive or dead?). Unlike bacterial analysis where the detection occurs after bacterial replication, (non-viable or killed bacteria are not detected) with helminths no replication occurs and any compliance limit will relate to 'viable' helminths. There is no doubt that some will be present in some will be present in some sludges but if the sludge treatment has been effective killed them they are on longer of concern.

Also is there any evidence that viable pathogenic helminths constitute a significant risk from sewage sludge spreading? There appears to be little formal control over the use of farm manure in agriculture the amount of which is significantly greater (~ 50 times) than sewage sludge used on agricultural land. This route is considered to constitute a more significant risk for this parameter.

The other problem will be obtaining reference supplies of all the relevant helminth species in both viable and non-viable form so that laboratories can validate their methods and run suitable quality control samples. Also, until a truly independent third party proficiency scheme is available, it will be difficult to ensure confidence in the results produced. The typical cost of a properly validated helminth test method with fit for purpose QC protocols for viable cestodes and nematodes could be in the range £200-400 per sample.

In the author's opinion, it would be worth carrying out development work to assess whether a much lower cost bacterial surrogate could be used. (For instance Clostridium perfringens spores).

PRESCRIBED METHODS OR PERFORMANCE TESTED METHODS

One danger of specifying that only CEN standards can be employed to carry out regulatory sludge analysis, is that it can take well over five years before a new improved method can be adopted as a CEN standard. This will effectively delay or prevent development of new improved methods.

The author firmly believes that the best way forward is performance-tested methods. This allows laboratories to use any analysis method as long as it meets stated minimum performance characteristics. This could be achieved by developing a range of certified sludge reference materials (CRMs) that should be used for validation of all CEN sludge methods. For labile parameters it may be necessary to add these parameters via a suitable spike just prior to analysis. These 'CRMs' would be produced by the EC (or an approved sub-contractor), then before a laboratory could submit a sludge result for any regulatory purpose, it would have to show fit for purpose method validation data. Using this range of CRMs an 11-day duplicate analysis of all relevant CRMs would be carried out[3]. Then the within batch, between batch and total standard deviation, would be calculated to assess the precision of the method. The bias of the method can be calculated using the value obtained versus the certified value expresses as a percentage. A decision can then be made based on the performance data whether the proposed method is fit for purpose and comparable to the official CEN method.

If all laboratories had to use the same set of reference CRMs then it would be very easy to compare laboratory performance for both method validation and routine QC analysis carried out with each batch of analysis. The argument against this is that it would cost too much to implicate. However, the cost of not implementing this is a non-lend playing field in an area where results from ethical laboratories could cost their clients considerable capital treatment improvement costs, whilst lower cost analysis at less ethical laboratory with poor analytical recovery could report lower levels of chemical species and pathogens and thus meet compliance targets.

SLUDGE ANALYSIS PROFICIENCY TESTING

The issue of relevant proficiency schemes is another key area. Currently there is no proficiency scheme covering all the necessary sludge parameters (especially microbiological parameters).

An EC wide approved sludge proficiency scheme(s) covering all listed sludge compliance parameters in the final directive should be set up. Laboratories which intend to carry out regulatory sludge analysis should have to join this scheme and submit results for more than 80% of proficiency samples received. The results from each lab would be assessed for fitness for purpose on an annual basis by the scheme organiser. Laboratories with poor results would be then barred from submitting regulatory sludge results.

A blind proficiency scheme for six months should be considered to assess current analysis performance in the EC for regulatory sewage sludge analysis in routine laboratories. This would give an indication of the consistency of sludge analysis results across the EC.

ADSORBABLE ORGANOHALOGEN (AOX) TEST

The AOX method has been routinely used in Germany for sludge analysis for many years. The author has tried (unsuccessfully) to highlight the futility of this method. It is difficult to determine what actual chlorinated or brominated pollutants the method will actually determine. It is unlikely to detect chlorophenols, medium chlorinated paraffins or other volatile or semi-volatile chlorinated compounds, as these will volatise during the prolonged drying at 105°C needed for liquid sludges. No work appears to have been carried out on this aspect. The QC used in the proposed CEN AOX standard (4-chlorophenol) will almost certainly volatilise and be unlikely to be detected if the QC were to be carried through the sample drying step (which it is not!).

The method is too insensitive to detect PCBs and dioxins at expected or even highly polluted levels for these parameters. The current proposed limit in the draft directive is 500 mg/kg DS[1]

The AOX test will also respond to inorganic chloride and there is no formal check to ensure that any significant AOX detected is not caused by carryover of inorganic chloride. In the author's opinion, all AOX results should be accompanied with a sludge total inorganic chloride result. A sludge from a works with saline intrusion or salt run-off from treated roads in winter can contain 1000 mg/l chloride (equivalent to 25,000 mg/kg DS AOX in a 4% DS sludge.)

The AOX result has no toxicological significance and cannot be related to the substances responsible for the AOX result. However, the Germans stressed that they had used it for years; it was a cheap test (£UK15 – 30)(23-46Euros) and the levels in sludges had fallen over the last ten years. (The washing technique to remove inorganic chloride may also have improved!)

The AOX method is considered unsuitable for soils as there contain "natural chlorinated humic material" which is not considered of any toxicological concern. It was pointed out that many uncontaminated soils could contain high "natural" AOX contents.

The inorganic chloride potential interference issue was not resolved at the meeting. Also the scope currently is only 'sludge'. It was agreed that until the round robin interlaboratory results were available, soils would not be included. (A suitable soil would be included in the proposed trial).

A 1989 German study gave an RSD of 14% at AOX levels close to the 500mg/kg proposed compliance limit for an undried unspecified type of sludge. This would give 95% uncertainty limits of 500 ± 140mg/kg (i.e. 360 to 640mg/kg. Can this be considered fit for purpose for the proposed 500 mg/kg compliance limit? The author feels sure that an interlaboratory trial with three or more countries with a range of sludges will show much worse agreement. Taking a reproducible homogeneous 10 –100mg method specified aliquot of a dried sludge also causes problems!

A very significant problem in the UK is that it would appear that no water company has a working AOX system (and the capital cost is over £30K.) So the UK will be unable to take part in the proposed AOX interlaboratory trial or more importantly carry out development work to demonstrate the futility of the AOX method for sludge analysis.

OTHER RELEVANT ISSUES

There is an increasing level of regulation in Europe in the environmental area mainly being driven by public concerns and various liability groups. There is also a move towards a risk-based approach to sampling and analysis programmes. This is exemplified in the food area with increasing concerns with respect to food safety.

Regulators are looking for lower and lower limits of detection and also increasing the number of substances of concern, e.g. for sludge, concern has been expressed in relation to: - dioxins; co-planar PCBs; alkyl phenol ethoxylates chlorinated paraffins; polybrominated fire retardants; organotin compounds; endocrine active substances etc. The list continues to grow and the expense of analyses of many of these organic substances in a sludge matrix down to the required level is very expensive[4].

The same problem is observed with respect to "hygienic parameters". For example trying to prove absence of *Salmonella spp*. In a 50g aliquot of a thermally dried sludge. In food analysis, absence in 25 g is the normal reporting limit. It will be difficult to achieve a limit of less than one Salmonella in 50g of all "as received" sludges.

Often chemical and microbiological analysis compliancy measurements underpin decisions many orders of magnitude greater than the cost of the relevant analysis. The cost of upgrading a sludge treatment process can be prohibitive. Thus it is essential that users of results have confidence in their laboratories and are satisfied that there is a level playing field across Europe.

The decrease in the number of students attracted to science courses is another concern, the cost for Universities to run science courses is prohibitive compared with many non-science courses where the only major capital resource to be supplied is a computer. A group of eminent Universities are considering introducing very significant tuition fees (~ £10.000 per annum) for students wishing to undertake science courses. This is likely to accelerate the drift away from science-based subjects. The author predicts that there will be an increasing shortage of skilled analysts in the next ten years.

REFERENCES

1   Working Document on Sludge (3rd Draft) Brussels 27th April 2000 (ENV.E.3/LM)

2   Protocol for Carrying out Chemical and Microbiological Analysis Interlaboratory Trials in CEN/TC 308/WG1 (Issue 5) Draft CEN TR Report

3   Cheeseman, R. V. and Wilson, A.L. (Revised by Gardner, M. J.), A Manual on Analytical Quality Control for the Water Industry. NS 30. Water Research Centre 1989

4   Jones K.C. and Northcott G. L. – Lancaster University "Organic Contaminants in Sewages Sludges: A Survey of UK Samples and a Consideration of their Significance"   Final Report to The Department of the Environment, Transport and the Regions Water Quality Division. December 2000

FIGURES

Importance of Result Confidence Limits

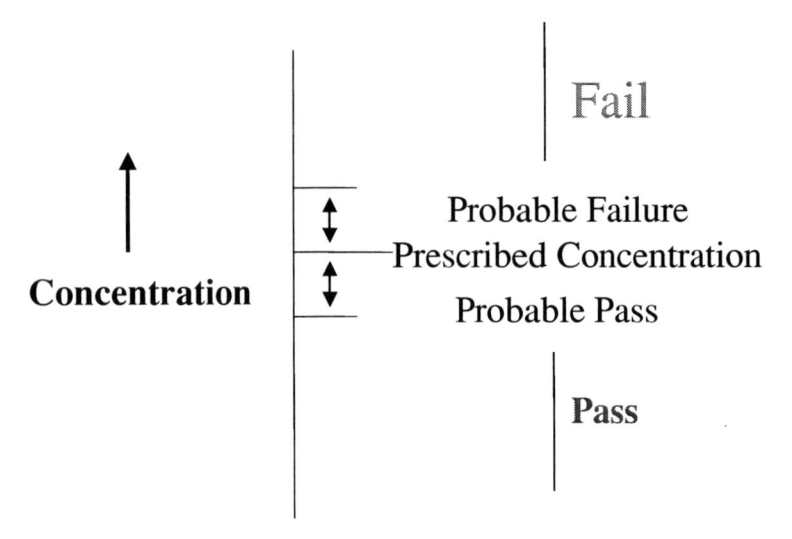

Concentration ↑

Fail

Probable Failure
Prescribed Concentration
Probable Pass

Pass

This depiction assumes no bias of the result

Figure 1: Confidence Limits

GOOD PRECISION, NEGLIGIBLE BIAS

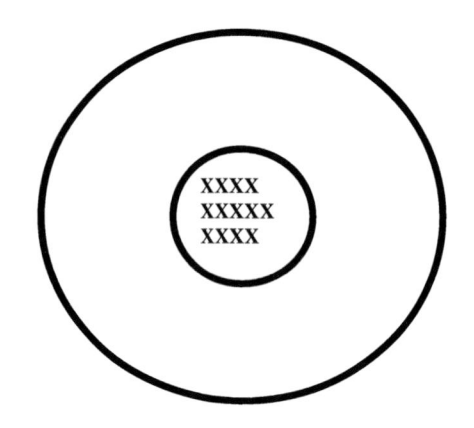

Desired Analysis Results

Figure 2: Precision and Bias – Analysis Results

GOOD PRECISION, SIGNIFICANT BIAS

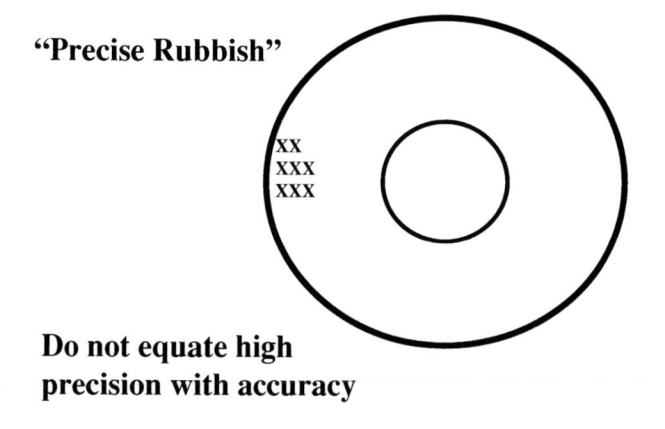

"Precise Rubbish"

xx
xxx
xxx

Do not equate high
precision with accuracy

Effect of Method Bias

Figure 3: Precision and Bias – Method bias

(This effect can be observed with some sludge methods)

AN INVESTIGATION INTO THE CAUSES OF DETERIORATED SLUDGE DEWATERING PERFORMANCE

Arnim Hertle
GHD Pty Ltd, TEL.: +61 8 9429 6926, FAX: +61 8 9323 9980,
EMAIL: ahertle@ghd.com.au

ABSTRACT

A major WWTP incurred a drop in the performance of its mechanical dewatering facilities that jeopardised the established transporting and reuse options for the dewatered sludge due to the excessive water content in the cake and the associated unfavourable cake handling properties. An investigation into the causes of this problem and possible remedies was carried out comprising an assessment of the dewatering equipment as well as an assessment of the dewatering process and the digestion process. Whilst a number of contributing factors was found in all three areas, the key problem is believed to be the problem of motivating operators for the proper maintenance and optimisation of aged open belt filter presses, which present to the operator as a very unpleasant work.

However, as this problems is very difficult to resolve, a fast relief and improvement of the situation was achieved by mixing the secondary digester. Together with some minor mechanical works the dewatering performance could be improved and 1% in cake DS was gained immediately without any capital cost or other exceptional expenditure.

KEY WORDS

Mesophilic anaerobic digestion, sludge digestion, sludge dewatering, maintenance

INTRODUCTION

The Beenyup wastewater treatment plant (WWTP) is the second largest of three WWTP that serve the Perth metropolitan area. Its catchment are the northern suburbs of Perth and is almost fully residential. The WWTP is owned and operated by the Water Corporation of Western Australia. It comprises primary and secondary wastewater treatment and is currently receiving an average dry weather flow of around 98,000 m^3/d and a BOD load equivalent to 383,000 PE.

Sludge treatment takes place in five digesters with shallow conical bottoms, floating roofs; the digesters were operated as four mixed and heated primary digesters and one secondary digester, which was neither mixed nor heated. The WWTP operates belt filter presses to dewater the anaerobically digested sludge prior to on-site storage and transportation to an external storage facility and reuse.

It was reported that the dewatering facilities never achieved the design performance, however, on top of an ongoing slow deterioration of the cake (= dewatered sludge) dryness a significantly steeper decline suddenly occurred. This saw the cake dry solids (DS) content drop down as low as 13% which in turn leads to transportation and handling problems which can jeopardise the removal of the sludge from the WWTP and its beneficial reuse. A secondary concern is the increased operating costs caused by the increased transport costs due to the extra volume of dewatered sludge cake.

A project was therefore initiated in order to establish whether and how the dewatering plant performance can be improved without necessarily replacing the core equipment. The possible causes for the observed under-performance can be complex and can comprise polymer or dewatering equipment related limitations as well as problems in the digestion process. The project hence took a broad approach to the problem and investigated the whole sludge processing train starting from raw primary sludge and excess activated sludge thickening through the digesters to the dewatering plant itself.

EXISTING SLUDGE PROCESSING FACILITIES

Raw primary sludge and mechanically thickened excess activated sludge (TEAS) is continuously fed to the digestion plant. The digestion plant comprises five digesters, of which up to two can be operated as secondary digesters. The digesters are flat bottom / floating roof type tanks of 4,200 m^3 maximum gross volume each with a "PEARTH" gas injection mixing system. The mixing system comprises 8 lances per tank, which are sequentially fed. At the time of this investigation, the digestion plant was operating with four mixed and heated primary digesters and one un-mixed and un-heated secondary digester.

From the secondary digester, sludge is withdrawn via eccentric screw pumps and pumped to the gravity drainage decks (GDD) mounted on top of the belt filter presses (BFP). Before the sludge is discharged onto the

GDDs, polymer solution is dosed into the sludge pipelines. A density meter is installed in the feed sludge line to the BFP feed pumps, however, it is not used for control purposes.

Each BFP has one feed pump and one polymer dosing pump allocated. Sludge flow is measured with magnetic flow meters on the discharge side of the feed pumps. Usual operation entails starting the pumps up at reduced speed and after about 1 hour ramp them up to the maximum possible flow (determined by the capacity of the BFPs). Reclaimed effluent can be added to the sludge on either side of the pumps; in the past this practice has reduced polymer consumption, however it had been discontinued.

Polymer is automatically made up from powder, using potable water, to a 0.4% solution; the ageing time is approximately 45 minutes before the solution is being used. The polymer dosing pumps draw aged polymer solution from the ageing tank; they are flow paced by the BFP feed pump speed. Before being dosed, the polymer solution is post-diluted with reclaimed effluent. Post-dilution is at a ratio of around 1:5.

Potassium permanganate solution is dosed to the digested sludge to oxidise dissolved sulphides in the sludge and thus control the H_2S emissions from the BFP units. This is solely an occupational health and safety related operation. The dosing point is just downstream of the sludge flow meters, i.e. upstream of the flocculant dosing points.

The flocculated sludge is thickened on the GDDs which are installed directly on top of the BFPs. The thickened sludge falls directly from the GDDs onto the BFP belts and is further dewatered there. Finally, the dewatered sludge is conveyed into the sludge storage hopper.

METHODOLOGY

The approach taken for this investigation comprised two main elements:
1. An in situ assessment of the layout, condition and performance of the dewatering belt filter presses; and
2. A theoretical analysis comprising a mass balance for the sludges of the WWTP and a process assessment based on the results from the mass balance.

These are described in more detail in the following sections.

ASSESSMENT OF THE SLUDGE DEWATERING FACILITIES

A site visit was carried out together with the WWTP superintendent. The visit focussed on the dewatering equipment. A number of issues was identified during this site visit that were likely to contribute to the poor dewatering performance of the BFPs. The main items that were found are listed below together with a short explanation why they were considered relevant.

PROCESS SET UP

A number of process related aspects were found that were considered critical in regard to the dewatering result. The item that was considered most important was that the feed to the BFPs is more or less constant in regard to volume, however, over time the DS content in the sludge withdrawn from the un-mixed secondary digester drops from ~4% to ~0.5%. As the polymer dosing is only flow paced, this will inevitably lead to over- or underdosing of polymer for most of the time. The effects are twofold: Underdosing leads to a lower cake dryness and lower capture rate. Overdosing may also have adverse effects on cake dryness and capture rate, but mainly it causes premature belt blinding.

The polymer dosing and sludge flocculation area were also found to be less then optimal. Firstly there is the risk that overdosed potassium permanganate may react with polymer and lead to an increased polymer consumption. This tied in with the general observation that the approximate polymer dose rate at the BFPs was 6.2 kg/t DS which is considered very high for BFPs, in particular for BFPs operating on digested sludge. However, the high polymer dose rate is more likely due to the fluctuation of the DS content in the feed and the attempt to find a polymer dose rate that allows robust operation of the presses.

Also, the dilution water flow could not be monitored due to range limitations of the flowmeters. This could have caused unnecessarily high water addition with associated unnecessarily high volumetric load on the GDDs.

The polymer dosing points themselves did not appear to be engineered for equal polymer distribution and maximum turbulence. In the extreme, bad mixing can lead to a stream of un-reacted polymer inside the sludge flow in the GDD feed pipe. A high turbulence at the polymer dosing point facilitates quick dispersion of the polymer in the sludge such that a maximum number of sludge particles gets into contact with the polymer.

The flock maturation time appeared to be too short. The polymer dosing point is only some 3 m upstream of the GDD feed chamber. The total residence time in the pipe and the feed chamber was estimated to be less than 10 seconds. Typical maturation times are 30 seconds or more where, after the initial intensive mixing of polymer with sludge, the mixture is usually given some time with only gentle movement to facilitate small flocks attaching to each other to form larger flocks.

MECHANICAL SET UP AND DESIGN

The installation also was constrained by various mechanical issues: The turbulence in the GDD entry appeared to be fairly high and this was considered a negative effect because an overly turbulent feed zone will destroy some of the flocks that have formed and hence be detrimental to the separation of water and the capture rate.

The belt speed also appeared to be fairly high. For a given feed flow there is an optimum belt speed where drainage is maximal. At lower speed the increasing thickness of the sludge layer will increasingly be an obstacle to drainage and the thickened sludge dryness will decrease. At higher speed the residence time of the sludge on the belt is too short to allow complete drainage to take place and the thickened sludge dryness will also decrease.

It was further found that the sieve area of the GDD was not particularly well used. Firstly, the feed weir width was restricted to about 1/3 of the belt width because in the past the lower flow velocities over the full weir width led to unequal flow distribution when solids built up on the beach. Due to the feed weir restrictions only about 2/3 of the GDD belt width was actually being used. This results in a thicker sludge layer than required which makes water drainage more difficult. Secondly, the sludge layer was constricted to a narrower width to suit the BFP belt width by two side baffle panels at about 4/5 of the GDD length. As the sludge was collected over the whole width of the belt and transferred to the BFP via a chute anyway, the side baffles were considered to waste GDD area.

In the press section of the dewatering units it was found that the drainage zone of the BFP was fairly short and that the wedge zones did not have any belt supports, but the belts were supposed to be kept in (wedge-shaped) position by their tension. Due to the weight of the sludge on it however, the lower belt was sagging, defying the principle and purpose of the wedge zone. A second effect was that a lot of water was still hanging on the underside of the lower belt and was then pressed back into the sludge when the belts rolled over the first roller.

It was also noted that all rollers of the presses were fully closed. Most BFP designs include a perforated first roller to facilitate water transport through both sides. However, as the investigated BFPs are lacking this feature, a lot of water is unnecessarily carried through to the smaller rollers. Even half way down the press free water could still be seen being caught between the belts and the rollers and jets of water squirting out through the belt, however, it is unusual to have that much water left so far down the press.

Lastly, it was found that the filtrate system could also contribute, probably to a minor extent, to the low cake dryness. Filtrate from higher parts of the units is discharged into trays in the middle of the units. It was found that the momentum of the flow was such that some filtrate topped over and fell onto the rollers and belts below. This unnecessarily added water to this section of the units.

WEAR

Significant wear was identified as a major contributing factor to the performance drop of the sludge dewatering facilities. The GDD / BFP units are 12 years old and have always been operated at their capacity limit because the small sludge hopper and the vicinity to residential housing restricted dewatering to normal day hours. This and the generally corrosive atmosphere in the press room has lead to a significant degree of general mechanical degeneration of the presses and the ancillary mechanical equipment. Further to that, individual points were noted that were considered important in regard to the dewatering performance.

It was found that the sludge ploughs on the GDDs did not touch the belt. The purpose of the ploughs is to break and turn the sludge layer on the belt to facilitate drainage. If the ploughs don't touch the belt, a thin layer of sludge will remain on the belt untouched and can thus form an obstacle to drainage.

Further to that, the belt support grids of the GDDs were worn to an extent that the belts sagged. Together with wear of the ploughs, the sagged support grids were the reason why the ploughs don't reach down to the belts any more. A sagged belt leads to a concentration of sludge in the centre of the deck, rather than using the full width.

The drainage properties of the belts could not be assessed. However, physical wear was clearly visible and typically belts have already lost significant drainage capacity due to blinding before physical wear becomes clearly evident. Typical belt lifetime (based on drainage properties / blinding) is 3,000 to 8,000 h with the average probably being around the 4,000 h mark.

OVERALL ASSESSMENT

It was found that some design features of the sludge dewatering area facilities were not optimal. Some process issues lead through to the review of the digestion process below. The main problems however, were found to be related to maintenance and optimisation. This is discussed further in the conclusions.

DIGESTION PROCESS MASS BALANCE AND REVIEW

As second step of the investigation, a mass balance for the sludge train of the WWTP was established. The mass balance was based on operational data from the plant's logbook for the last three years.

An evaluation of that data clearly showed that the dry solids content in the dewatered sludge was, despite significant scattering, on a continuous downward trend. For the same period, the residual content of organic matter (volatile solids, VS) was found to be increasing continuously. These two trends are shown in Figure 1. The increasing trend in residual VS in the digested sludge shown in Figure 1 is likely to contribute to significantly to the deterioration in the dewatering system's performance.

The cause for the increasing residual organic matter in the digested sludge is most likely a steady decrease in hydraulic retention time due to steadily increasing flows, again with significant scatter, as shown in Figure 2.

Averaging over longer periods to dampen the scatter of the data leads to a relatively good match of the curve describing the development of the digester feed flows and the curve describing the development of the content of residual organic matter in the digested sludge. This is shown in Figure 3.

Since the last digester clean out was 5 years ago, an additional to the impact of the increased hydraulic load is possibly the impact of rags and other un-digestable matter collected in the digesters due to general shortcomings of their shape and shortcomings of the mixing system and its operation. However, experience from previous digester clean outs was that the volume of rags was typically less than 5% of the total digester volume. An increase from 0% to 5% or even 10% dead volume in the digesters over 5 years is unlikely to lead to a significant deterioration in the VS destruction rate. The reason is that the theoretical total hydraulic retention time in the 5 digesters is around 25 days; after 25 days the VS destruction rate that is typically already quite low so that a 2.5 days decrease in digestion time has little to no impact on the total VS destruction in the system.

REVIEW OF DIGESTER OPERATION

It is noted that with 74 – 77%, the VS in the digested sludge is still higher than the organic content in many European sludges. The VS in the digester feed sludge is approximately 83.5% - 84%. This is due to:
- The sewerage system where sewage and stormwater are collected separately;
- A system with only little infiltration due to very long dry periods and sandy soils; and
- A low content of fine inorganics because the sandy soils in Perth can only lead to infiltration of easily settlable inorganics that are removed in the grit chambers rather then being carried further and being removed in the primary settling tanks.

A VS balance shows that the digesters, although operating robustly, only accomplish a VS destruction of 39%. This is quite low, even when the secondary digester is not considered as an active reactor because it is not mixed.

For a further analysis of this finding the last six months of operational data were therefore compared with theoretical data as shown in Figure 4.

The theoretical data for Figure 4 was calculated for a plant with 4 primary digesters only based on 1st order reaction kinetics with a rate factor of $0.25\ d^{-1}$ and assuming that 60% of the total organic matter is digestable, which is within the typical range for municipal sludge of 50% to 70%[1]. For the calculation of the existing HRT, the dead volume in the digesters was assumed to be 10%.

According to the model calculation a VS destruction of 49% of the total VS would have been expected; significantly more than the observed average VS destruction. Using the same calculus backwards shows that the low VS destruction rate of 39% is equivalent to the performance of a typical primary digester with a HRT of 7.5 days only. This is equivalent to only 37% active volume in the digesters. The same methodology was applied to the minimum and maximum values so that data ranges were obtained for the assessment rather than single points. It is noted that although the range of equivalent HRTs is quite broad, it is still significantly distant from the range of the existing HRTs.

Three main causes could be considered responsible for the discrepancy between expected and measured digester performance:

1. Lower content of digestable VS in the total VS of the digester feed sludge. As the sludge is normal, nearly 100% residential sludge, there is little reason to assume that it is less than average in terms of digestability. However, there is a possibility that the chlorination of the raw sewage, which has being carried out for the last several years as an interim measure of odour control, leads to toxic byproducts in the sludge which then impede the digestion process;
2. Insufficient mixing to minimise dead volume in the digesters and achieve good contact between sludge and bacteria. Mixing is via gas injection through 8 lances per tank which are fed sequentially one after the other. Digester operating practice was to operate the mixing system in each tank only 8 h/d, which is just one hour per day and lance; and
3. Build up of rags and sediment reducing the active volume of the tanks.

A conclusion as to which of those possible reasons is the main cause of the poor digester performance could not be reached at the time of the investigation.

RECOMMENDATIONS FOR SHORT TO MEDIUM TERM PERFORMANCE IMPROVEMENTS

The following recommendations were made in order to improve the performance of the sludge digestion and dewatering plants:
- Mixing of the secondary digester to make the tank available as an active reactor and in order to provide a constant feed DS concentration to the BFPs to enable optimisation of the polymer dose rate. No static thickening function would be lost because the secondary digester As the tank has no facility to withdraw sludge liquor and hence act as a thickener, there is no reason why it couldn't be mixed or even heated. An alternative arrangement to achieve better control over the feed specific polymer dose rate is to tie the density meter in into the control system.
- Increasing the mixing hours in all digesters to 24 h per day and operate with three primary digesters and two secondary digester in order to maximise VS destruction.
- Optimisation of the polymer dosing points and reduction of the dilution water rate as far as possible to minimise the hydraulic load on the GDDs and optimise flock maturation and optimisation of the belt speed.
- Alteration of all filtrate pipework such that all discharges are to the filtrate pits rather than to any other trays inside the GDD / BFP units.
- Replacement of the worn GDD belt supports and the worn GDD ploughs with new ones.
- Removal of the side baffles at the discharge end of the GDDs and the GDD feed weir restrictors, or at least moving them further out.
- Installation of an adjustable sludge weir at the discharge end of the GDDs to increase sludge retention time on the belt and facilitate drainage by applying some slight compression, installation of belt supports for both belts in the wedge zone and replacement of the first roller of the BFPs with a perforated roller.

STATUS QUO: FIRST IMPROVEMENTS AND THEIR RESULTS

Since the recommendations above were made in early August, the following steps were taken to improve the plant performance:
- Mixing of the secondary digester;
- Increased mixing hours;
- Replacement of worn belt supports on three out of four GDDs; and
- Trials to determine the optimum belt speed.

Unfortunately mixing could not be increased to 24 hours per day; it turned out that 10 hours per day was the practical limit to avoid excessive ragging problems in the dewatering equipment. In particular, mixing has to cease before dewatering commences. It appears that the amount of rags collected in the digesters is such that these problems cannot be overcome by a slow "bleed out" of the rags but only by a cleanout.

However, as per middle of September, despite the mixing limitations and despite large quantities of imported sludge of around 15% of the WWTP's own sludge production the VS content in the digested sludge dropped and the cake dryness increased from around 13% to around 14% as shown in Figure 5.

CONCLUSION

A thorough analysis of the dewatering facilities and of the dewatering process and the preceding digestion process showed that the dewatering performance of the BFPs has continuously fallen over the last years, very likely due to the continuously increasing load to the WWTP.

However, whilst the analysis showed that the digestion process was operating less efficient than it could, it had to be acknowledged that it was operated in that way for years and very robustly. In regard to the dewatering plant this implies that the performance of the BFPs could have been on a higher level all the time, but the performance would still be on a steady decline due to the increasing load. Further, the recent sudden drop cannot be attributed to process issues.

It became evident in the course of this investigation, that the BFPs could have performed better if ongoing maintenance and optimisation works would have been carried out regularly. However, due to the "dirty" working environment around open frame belt filter presses, those works had not been carried out as required. It is believed that the cumulative effect of delayed or neglected maintenance and optimisation works was the main reason for the very low cake dryness of only 13% although a definitive cause could not be determined conclusively. It is noted that although the client is fully aware of problems in regard to proper maintenance of the BFPs, practice shows that it is very difficult to motivate staff to carry them out.

The general conclusion is that the client was caught in a vicious circle where the unpleasant working environment at the open frame BFPs caused delayed or neglected maintenance which in turn led to deteriorations on the machines that made the working environment even more unpleasant and dirty. The example at the Beenyup WWTP has shown that these issues are factually uncontrollable for the client and that the effects can accumulate to an extent where a serious plant failure is inevitable.

The consequence is that, although process alterations together with some maintenance improved the performance of the digestion and the dewatering plants noticeably, failure events will continue to occur if the key problems in maintaining and operating the BFPs cannot be resolved, in particular because of the age of the units, their high sludge loads and the generally corrosive atmosphere in the press room.

The recent trend towards fully encapsulated dewatering machines that can be seen in many places can probably be taken as evidence that the problems experienced at the Beenyup WWTP are not a single event, but typical.

REFERENCES

1. ROEDIGER, H., ROEDIGER, M. AND KAPP H. *Anaerobe alkalische Schlammfaulung*. 4th Edition. Oldenbourg Verlag, Germany. 1990.

ACKNOWLEDGEMENTS

The author likes to thank Margaret Domurad, Geoff Kendall and Paul Jackson from the Water Corporation of Western Australia for their enthusiastic support of this investigation and their assistance in interpreting the enormous quantity of operational data.

FIGURES

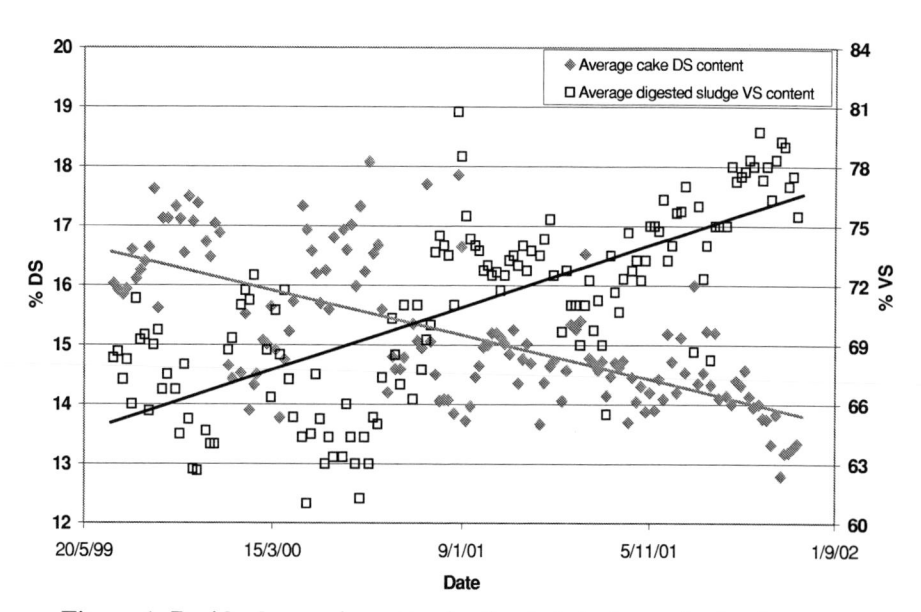

Figure 1: Residual organic matter in digested sludge and cake dryness

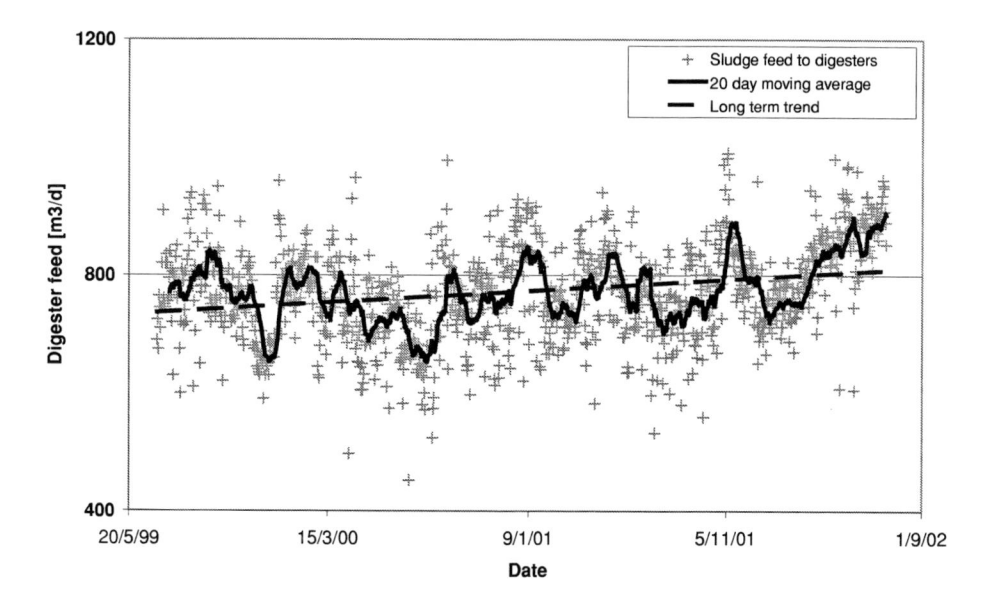

Figure 2: Development of digester feed flows

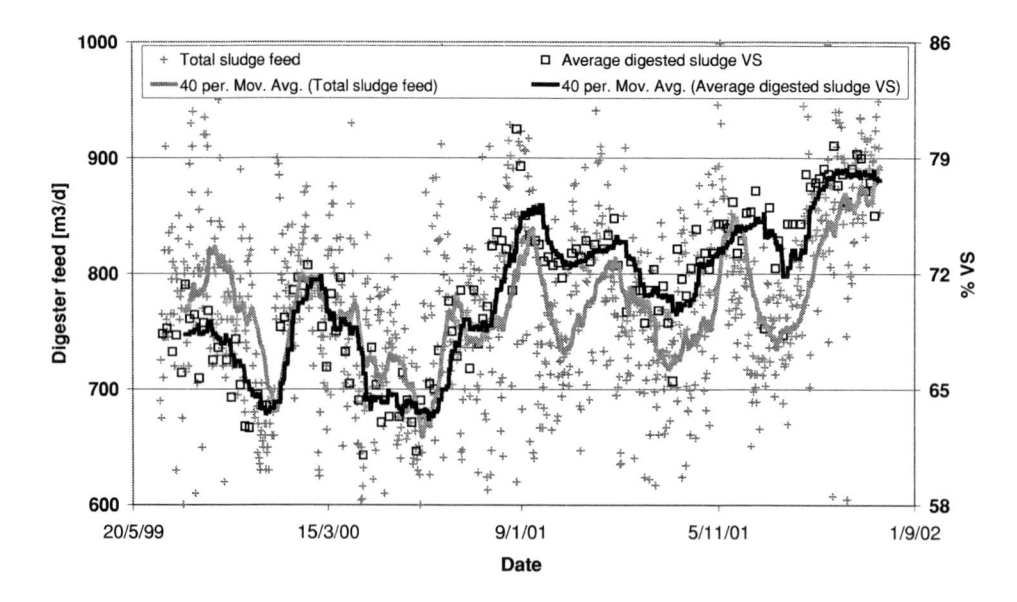

Figure 3: Trends in digester feed flows and residual organic matter in the digested sludge

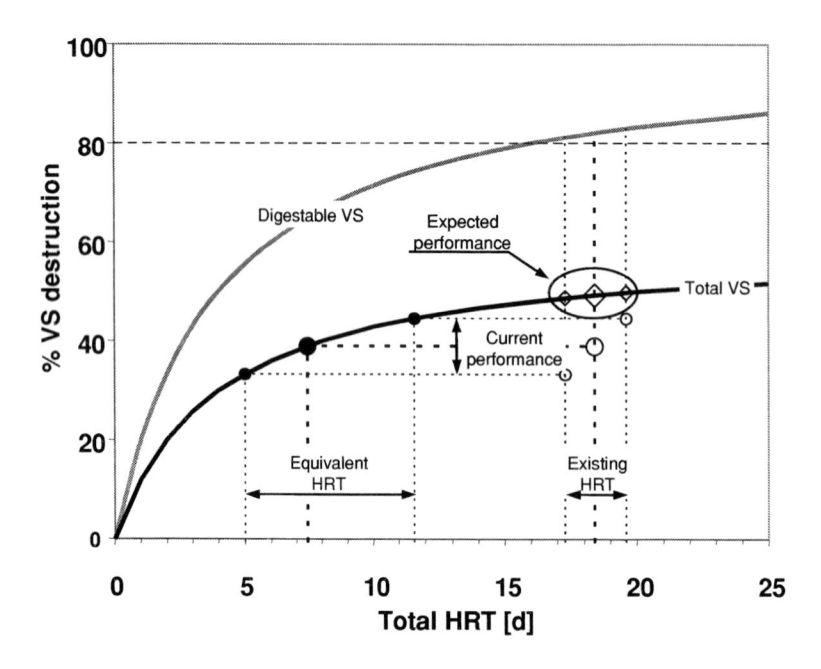

Figure 4: Assessment of digester performance

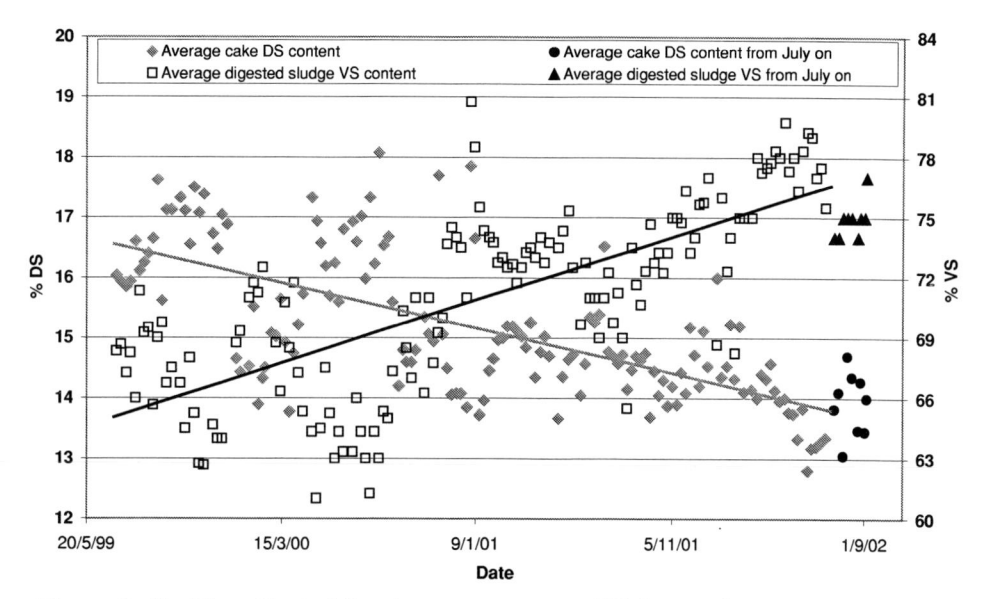

Figure 5: Positive effect of first improvements on VS destruction and cake dryness

TREATMENT OF SLUDGE FROM CHEMICALLY-DOSED WASTEWATER A CASE STUDY

Gordon Thomas[1], Jill Smith[2] and David Hayward[3]

[1] Carl Bro, 13/23 King Square Avenue, Bristol, BS2 8HU
Tel: 0117 923 2221, Fax 0117 924 7416, email: gordon.thomas@carlbro.com
[2] MWH Wessex joint venture, Taunton
Tel: 01823 225109, Fax 01823 274466, email: jill.smith@mwhwessex.co.uk
[3] Ashact Ltd., Bridge House, Station Approach, Great Missenden, Bucks, HP16 9AZ
Tel: 01494 891100, Fax: 01494 890320, email: david.hayward@ashact.com

ABSTRACT

The paper focuses on the treatment of sludge from chemically-dosed wastewater, in particular, the effect of dosing ferric compounds into raw sewage for phosphorus removal. The following aspects are considered:
- the increased sludge production in terms of solids and volume
- the changed characteristics of the sludge and their effect on the treatment selection

The wastewater treatment works considered is a works of population equivalent 30,000 serving the town of Dorchester and a brewery. The effluent treatment stream consists of inlet works, primary settlement tanks, biological trickling filters (some plastic media, others stone) and humus tanks. The humus tank sludge is returned to the primary tanks for co-settlement. The initial design for the sludge treatment consisted of refurbished existing picket fence thickeners, anaerobic digesters, digested sludge thickening with a belt thickener, and storage of thickened sludge followed by disposal to agricultural land. The paper considers the amendments to the initial design undertaken in order to deal with the increased sludge production and changed sludge characteristics mentioned above. Mention is made of the options evaluated during the design process.

Now that construction of the amended design has been completed and the new works is in operation, a comparison is made between theoretical and actual quantities of sludge production.

KEY WORDS

Ferric dosing, phosphorus, sludge production, sludge thickening.

INTRODUCTION

This case study relates to a completed project at the Dorchester WWTW for Wessex Water. The project was required to meet a new phosphorus effluent consent of 2 mg/l Total P (as an annual average) amongst other effluent quality improvements. It was planned to use dosing of ferric salts to raw sewage upstream of the primary tanks to achieve the phosphorus removal. The new tighter effluent consent had to be met by the end of March 2002. Dorchester is the ancient County Town of Dorset and as such is mainly residential but with one large brewery. The raw sewage is, therefore, domestic with a substantial brewery waste component. The population equivalent of the raw sewage BOD load is 30,000 of which approximately 30% is brewery load. The effluent treatment stream is conventional UK practice with inlet works, primary tanks, biological trickling filters and humus tanks. The humus tank sludge is returned to the primary tanks for co-settlement.

The project was procured for Wessex Water under the leadership of the Montgomery Watson Harza / Wessex joint venture using a design and construct contract. The main contractor was Morgan Est (formerly Miller Water) and the designers were Carl Bro Group. Process design assistance was provided to Carl Bro by Ashact Ltd.

EXISTING SLUDGE TREATMENT

The co-settled sludge was pumped to three picket fence thickener tanks and from these the thickened sludge was transferred to two anaerobic digesters. Digested sludge overflowing from the top of the digesters was stored in two digested sludge storage tanks prior to tankering away for disposal to land (Fig.1). The sizes of the main treatment components were:

Picket fence thickeners 3 x 90 m^3 = 270m^3 total
Digesters 2 x 250 m^3 = 500 m^3 total
Digested sludge storage 2 x 160 m^3 = 320m^3 total

Emergency sludge storage lagoons were also available if required.

The main perceived problems with the sludge treatment were considered to be:
- very poor solids separation in the picket fence thickeners (around 60% solids removal)
- lack of storage capacity for the digested sludge exacerbated by thin sludge from the digesters.

INITIAL DESIGN

A preliminary design had been carried out prior to the main contract start. This envisaged three main improvements (Fig.2):
1) detail modifications to the inlet and outlet of two picket fence thickeners to improve the solids separation and conversion of one thickener into an imported sludge reception tank
2) installation of a belt thickener after the digesters to thicken the digested sludge
3) increased sludge storage (additional tank)

SOLIDS PRODUCTION

It was acknowledged that the use of ferric salts would increase the sludge solids produced by the works. It was essential to quantify this increase in order to confirm the loadings on the sludge treatment stages and thus to size them correctly. The methods available to predict the increased sludge production on a dry weight basis seemed to produce widely varying results. On the one hand 'rule of thumb' predictions gave increases as low as 20% whilst on the other hand theoretical calculations gave increases as high as 100% depending on the assumptions made. These variations together with expected changes in the sludge characteristics gave rise to great concern regarding the correct design for the sludge treatment.

THEORETICAL FERRIC QUANTITIES

The chemical reactions taking place when ferric salts are added to wastewater are not simple. Phosphorus in wastewater is found in three principal forms: orthophosphate ion, polyphosphates and organic phosphorus. The predominant form changes with pH and with the degree of treatment. In practical terms phosphorus is measured as total P but the different forms make exact prediction of the sludge solids difficult to achieve.

The theoretical calculations of the sludge production were based upon published US EPA design information[1]. The precipitation of phosphorus with ferric ions can be shown simply as:

$$Fe^{3+} + PO_4^{3-} \rightarrow FePO_4$$

From the above chemical equation the mole ratio of Fe:PO$_4$ (and thus the mole ratio of Fe:P) is 1:1. However, it is known that this is an oversimplification and, because of competing reactions, more ferric compound than the stoichiometric quantities will be required. From jar tests it was anticipated that the mole ratio of Fe:P would be 1.5:1. That is 50% more than ferric compound than that estimated by the simple chemical equation. The US EPA Design Manual states that the mole ratio will vary with the degree of phosphorus removed. The following mole ratios are given for aluminium:phosphorus:

P reduction	Mole Ratio Al:P
75%	1.38:1
85%	1.72:1
95%	2.3:1

As the stoichiometric mole ratio for Al:P is the same as Fe:P, similar mole ratio increases might be expected for ferric compounds. Our target was 85% removal based upon a raw sewage concentration of 10 mg/l P and an effluent concentration of 1.5 mg/l P (consent standard of 2 mg/l). For peaks of phosphorus up to 15 mg/l a mole ratio Fe:P of 2:1 might be expected for 90% removal of phosphorus. The selection of the mole ratio has a major impact upon the consumption of chemical and thus upon the quantity of sludge estimated. Since the atomic weights of Fe and P are 56 and 31 respectively the following weight ratios of Fe:P can be tabulated as follows:

Mole Ratio Fe:P	Weight Ratio Fe:P
1:1	1.8:1
1.5:1	2.7:1
2:1	3.6:1

For the design of the sludge treatment a 2020 design horizon was used. However, in order to compare the actual quantities produced with the calculated quantities it is appropriate to use data for 2002. The basic parameters for the works used in the calculations are given in table 1 for 2002

The sewage solids removal in the primary tanks was assumed to be 65% of the input solids load. Therefore the sludge dry solids (DS) from settlement excluding ferric contributions was:

$$0.65 \times (2200 + 380) = 1680 \text{ kgDS/d}$$

For calculation of the ferric sludges a mole ratio Fe:P of 1.5 was used based on the jar tests. The weight ratio Fe:P was then 2.7:1 giving the following requirement for weight of Fe:

$$116 \times 2.7 = 314 \text{ kg/d}$$

For ferric sulphate (Mistrale 600) 14% w/w Fe, SG 1.6 the volume of chemical is:

$$(314/0.14)/1.6 = 1400 \text{ l/d}$$

The ferric ion is precipitated in two forms: ferric phosphate FePO$_4$ and ferric hydroxide

$Fe(OH)_3$. Using the molecular weights the weight of the ferric precipitates can be calculated:
The P load removed = 0.85 x 116 = 99 kg/d (equivalent to 99 x 56/31 = 179 kgFe); therefore the ferric phosphate weight is:
99 x [56 + 31 + (16x4)]/31 = 480 kgDS/d

The remaining ferric ion is wasted in competing reactions producing hydroxide:
(314-179) x [56 + ((16+1)x3)]/56 = 260 kgDS/d

ADDITIONAL SLUDGE QUANTITIES
In addition to these precipitates, soluble organic material is adsorbed on the precipitate and removed. This additional component has been estimated as follows:
It is assumed that 35% of the BOD load is soluble,
i.e. 0.35 x 1900 = 665 kg/d
30% of this soluble BOD fraction is then assumed to be adsorbed on the ferric precipitate,
i.e. 0.30 x 665 = 200 kg/d
Using a volatile matter solids (VM):soluble BOD ratio of 1.6 and taking a %VM of 85% the resulting increased sludge quantity is:
200 x 1.6/0.85 = 375 kgDS/d

In addition to these precipitates the US EPA Design Manual states that the ferric precipitate may incorporate other cations and anions into its complex ligand structure when it settles. This is because US data has suggested that higher chemical sludge yields are often experienced in practice compared with values obtained from simple stoichiometry. However, in the context of this project, it was not thought justifiable to make further allowances.

THEORETICAL SLUDGE QUANTITIES – SUMMARY
The resulting theoretical sludge solids production is therefore built up from several components:
Sludge solids from settlement = 1,680 kgDS/d
Ferric phosphate sludge = 480 kgDS/d
Ferric hydroxide sludge = 260 kgDS/d
Sludge from adsorption
 of soluble BOD= 375 kgDS/d
Total sludge solids (rounded) = 2,800 kgDS/d

SLUDGE CHARACTERISTICS
The sludge characteristics are modified by the addition of chemicals. The sludge from the primary tanks is normally thinner and the dewatering properties are different to conventional sludges. In parallel with the design being undertaken at Dorchester, Wessex Water had started to commission another works with ferric dosed sludge. This experience suggested an average percentage dry solids of 1.5% instead of the 3 or 4% normally obtained with co-settled primary/humus sludges. Clearly this has considerable impact on the volumes of sludge produced as can be see in table 2.

A further concern was the possibility of black ferric sulphide coloration caused by reaction of the ferric compound with septic sewage from the drainage network. At other works a black ferric sulphide precipitate had been produced. This passed through the works and caused the effluent to have a black coloration.

IMPACT ON DESIGN
There was considerable uncertainty in the volumes of sludge production dependent upon the assumptions used in the design. These are given in table 3.
The uncertainties were further complicated by the need to consider the 2020 design horizon and also possible large increases in the principal trade load of the brewery.

PICKET FENCE THICKENERS
In most scenarios the picket fence thickeners were overloaded. Although improvements to the picket fence thickeners were planned, the existing solids capture was only 60%. The settleability of the ferric sludge was expected to be poorer than the existing sludge. It was therefore considered unlikely that the solids capture could be sufficiently improved to prevent substantial quantities of solids returning to the effluent treatment plant in the return liquors.

DIGESTERS
At the higher sludge production levels the existing digesters would have hydraulic retention times that would be too low for satisfactory operation.

DIGESTED SLUDGE BELT THICKENER AND STORAGE
The capacity of the belt thickener (with polyelectrolyte dosing) was adequate to deal with the increases in sludge production. The operating hours would have to be greater than originally planned but this could be accommodated. In most cases the digested sludge storage capacity was adequate at least in 2002. Additional sludge lagoons were available if required.

OPTIONS CONSIDERED
It was believed at an early stage that it was too risky to retain the picket fence thickeners as originally proposed. It was decided to replace them with a mechanical thickening stage. There was not sufficient time to purchase and

install a new permanent mechanical thickener and so the options of hiring a mechanical thickener or operating the belt thickener in an alternating dual mode (digested and raw thickening) were considered. In order to deal with the digester overload, the possibility of abandoning the digestion process and exporting the untreated sludge to another works for further treatment was considered. An alternative was to tanker away any surplus overload to another works and retain the digesters operating at their maximum capacity.

FINAL DESIGN

In the final design the raw sludge thickening stage was carried out using a hired mechanical thickener. The hired thickener (with polyelectrolyte dosing facilities) was positioned next to the existing picket fence thickeners, which were converted to buffer storage tanks. Each picket fence thickener tank had a volume of $90m^3$ and so two were used as raw sludge storage ($180m^3$) and one as thickened sludge storage ($90m^3$) prior to the digesters. The tines were removed from each picket fence thickener and submersible mixers installed in their place (Fig. 3).

The available raw sludge storage of $180m^3$ was equivalent to only one day's peak production and therefore an operator was hired with the thickener so that it could be operated 7days/wk. A rotary drum thickener and operator was available at relatively short notice. This was obtained and installed without delaying the project programme.

In order to prevent the development of anaerobic sludge, with the consequent production of black ferric sulphide, Nutriox (calcium nitrate) was planned to be added to the raw sludge prior to the holding tanks. The existing digesters were retained with the facility to tanker away any excess thickened raw sludge if required. The remaining stages of treatment were left unchanged: belt thickener for digested sludge thickening, increased digested sludge storage.

COMMISSIONING AND OPERATION

Construction of the additions to the sludge treatment stream was completed in January 2002. This included temporary installation of the hired drum thickener, which was brought into operation on 18th January 2002. Dosing of ferric sulphate did not start until 25th February 2002 and so it was possible to directly see the effect on sludge production of the addition of the ferric compound (prior to 18th January the records of sludge production were not detailed because of a lack of measurement and because of effects on the raw sewage from the very low solids capture of the picket fence thickeners). Also, because it was required to establish parameters for a more permanent solution to the raw sludge thickening problem, the sludge treatment plant was the subject of a data-gathering exercise[2].

FERRIC DOSING

The ferric sulphate solution was dosed into the inlet channel to the primary tanks. The dosing system was controlled using a non-linear control based on incoming flow. The initial dose rate was set at a level below what was anticipated to meet the P reduction required so that the works would adjust to the coagulant and to limit the impact on the sludge production (25 February to 19 March 2002). Following three weeks at this dose rate it was then increased to a level at which it was expected that the effluent would comply with the new consent, which was then to come in force within two weeks. Four weeks of operating at the increased dose rate (19 March to 16 April 2002) showed that the works was comfortably meeting the P consent and the decision was made to reduce the dose rate slightly.

From Table 4 it can be seen that the average phosphorus load was slightly lower than the design load. Also, the dose rate of ferric was less than the design dose rate not only because of the lower phosphorus load but also because of the lower mole ratio Fe:P required.

NUTRIOX DOSING

Nutriox dosing started on the same day as the ferric dosing at an average daily dose of 20l/d. After 5 weeks there was no apparent formation of ferric sulphide and so the dose was reduced by 20%. Subsequently different dose rates were used including a period with no dose. Analysis of the liquors from the drum thickener showed that the Nutriox was capable of reducing the dissolved sulphide levels by approximately a third. In general, levels of sulphide in the raw sludge have been low but the Nutriox facility has been retained in case raw sludge has to be stored over a weekend for example.

SOLIDS PRODUCTION

Solids production from the primary tanks substantially increased after addition of the ferric sulphate (refer Fig 4). Day to day production of raw sludge solids was very variable because of various factors during the commissioning phase. Therefore, in order to try to show trends more clearly the values for each week have been averaged to give a figure for production in kgDS/d. The solids production was generally in the range 1800 to

2200kgDS/d but with a peak week average of 2791kgDS/d at the end of May 2002. This compares well with the design estimate of 2800kgDS/d. However, in practice the solids removal performance of the primary tanks was higher than that used in the design calculations. On the other hand, the actual ferric dose was lower than predicted. When the actual loadings and performance removal data were used in the design calculations the solids production was of the order of 3,000kgDS/d (Table 5).

SLUDGE PRODUCTION

Generally, the expected characteristics of the ferric-dosed sludge were found to prevail (Fig 5). The percentage dry solids of the ferric-dosed sludge was half that of the raw sludge without ferric dosing. The reduction from 3%DS to 1.5%DS coupled with the increased solids yield caused a threefold increase in the raw sludge volume prior to thickening (Fig. 6). It was necessary to operate the drum thickener 7 days per week (as expected) in order to cope with the primary sludge volume. The mean volume of raw sludge produced during the period May-June 2002 was 132.5 m^3/d. However, during the peak week 27 April to 3 May 2002 the volume produced was 176m^3/d. The corresponding design estimate was 188m^3/d.

During the commissioning period the automatic desludging of the primary tanks has been improved which has tended to produce a thicker sludge and therefore a reduced volume. In operation the drum thickener was able to thicken the raw sludge to 6%DS for the digestion process. In order to maintain a minimum retention time in the digesters small quantities of raw sludge were tankered off site. Similarly, the belt thickener was used to thicken the digested sludge from 4.5%DS to 7%DS for storage and tankering away. A most important point was that the tightened effluent consent was met by the end of March deadline and thus the major objective of the project was achieved.

CONCLUSION

In the normal design situation where major improvements are being made to several aspects of a sewage treatment works no very precise figure can be calculated for the sludge production after the addition of chemical dosing because of the number of variables involved. However, it can be said that:

The solids production will increase substantially when the ferric chemical is added on a regular basis to the primary settlement. In this case the average increase in solids production was 35% but with peaks of approximately 70%.

Thinner sludge will be produced leading to a substantial increase in volume.

The theoretical formulae used for sludge production tended to provide figures for the peak week rather than the average. It is then a matter of judgement whether the design should be based upon the values for the peak week or a figure closer to the average case.

REFERENCES

1. U.S. E.P.A., Process Design Manual for Phosphorus Removal, 1976.

2. SMITH, J. Dorchester P Removal with Mistrale 600 Commissioning Report, Internal MWHWessex document, 2002.

ACKNOWLEDGEMENTS

This paper is published with the kind permission of Wessex Water and The MWH Wessex joint venture.

TABLES

Table 1: Basic parameter for works design

Dry weather flow (summer)	8,600 m³/d
Average flow including return liquors	11,800 m³/d
Full flow to treatment (FFT)	21,000 m³/d
Domestic BOD load	1,100 kg/d
Brewery BOD load	550 kg/d
Total raw sewage BOD load (incl. all trade and return liquors)	1,900 kg/d
Total raw sewage SS load (incl. all trade and return liquors)	2,200 kg/d
Humus sludge SS load	380 kg/d
Total raw sewage P load (incl. all trade and return liquors)	116 kg/d

Table 2: Impact on the volume of sludge produced

Item	Without chemical dosing	With chemical
Solids production	1,600 kgDS/d approx	2,800 kgDS/d
Percentage dry solids	3%	1.5%
Sludge volume	53 m³/d	187 m³/d

Table 3: Degree of uncertainty in sludge volumes

	'Rule of Thumb' 20% increase	Theoretical Mole ratio Fe:P 1.5:1	Theoretical Mole ratio Fe:P 2:1
Sludge solids 2002	1900 kgDS/d	2800 kgDS/d	3000 kgDS/d
Sludge volume 3%DS	63 m³/d	93 m³/d	100 m³/d
Sludge volume 1.5%DS	126 m³/d	186 m³/d	200 m³/d

Table 4: Ferric Dosing: Actual and Design

Date	Flow m³/d	Dose l/d	Inlet P mg/l	Effluent P mg/l	P load kg/d	Effluent Fe mg/l	Mole Ratio Fe:P
25/2/02 to 19/3/02	11426	870	8.08	2.06	92	0.67	1:1.2
19/3/02 to 16/4/02	9870	940	8.16	1.21	80	0.81	1:1.5
16/4/02 to 9/5/02	9694	825	7.74	1.24	76	0.68	1:1.35
Design 2002	11600	1400	10	1.5	116	N/a	1:1.5

Table 5: Indicative Sludge Production Data

Item	DESIGN FIGURES (WITH FE)	04/02/02 to 24/02/02 Actual	04/02/02 to 24/02/02 Theoretical output	16/04/02 to 09/05/02 Actual	16/04/02 to 09/05/02 Theoretical output
Dry weather flow (summer) m^3/d	8,600	-	-	-	-
Average flow including return liquors m^3/d	11,800	12,400	-	9,911	-
Full flow to treatment (FFT) m^3/d	21,000	-	-	-	-
Total raw sewage BOD load (incl all trade and return liquors) kg/d	1,900	1,350	1,350	1,507	1,507
Total raw sewage SS load (incl all trade and return liquors) + humus sludge DS kg/d	2,580	3,040	3,040	2,795	2,795
Total raw sewage P load (incl all trade and return liquors) kg/d	116	120	120	76	76
Percentage BOD removal in primary tanks	55%	47%	45%	80%	79%
Percentage SS removal in primary tanks	65%	76%*	55%	91%	85%
Ferric dose as Fe kg/d	**314**	**Nil**	**Nil**	**185**	**185**
Average raw sludge dry solids production kg/d	2,800	1,548	1,672	2,226	3,100
Average raw sludge percentage dry solids	1.5%	2.2%	3%	1.3%	1.5%
Average sludge volume m^3/d	188	70	56	167	207

*Removal rate is high relative to sludge output but it is based on very few values

FIGURES

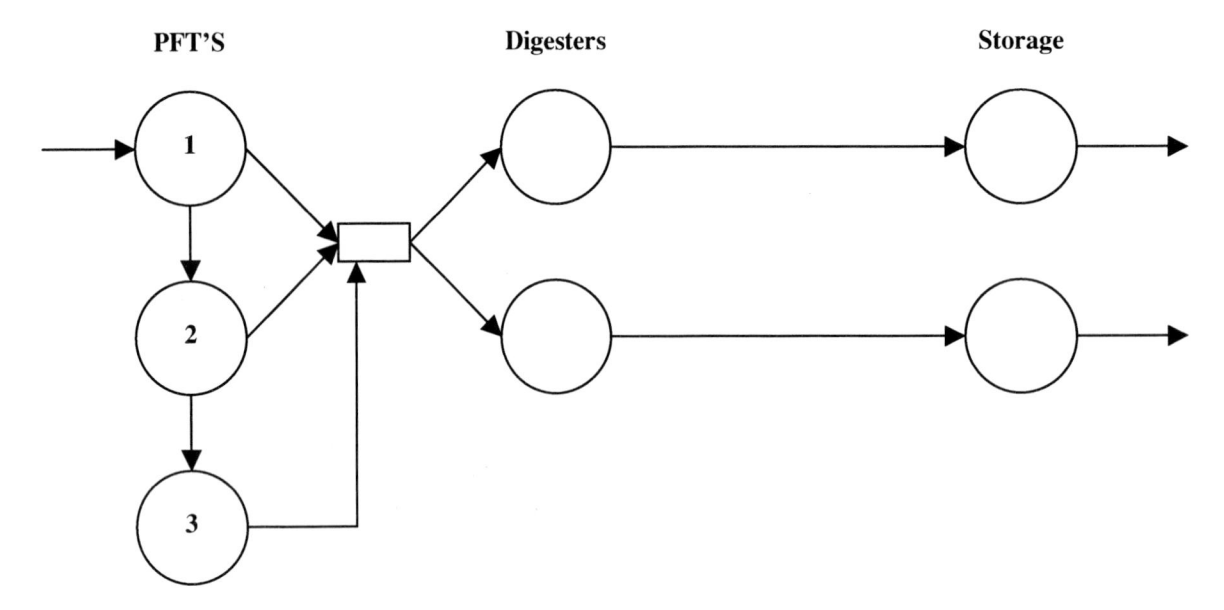

Figure 1 – Existing Treatment (before modification)

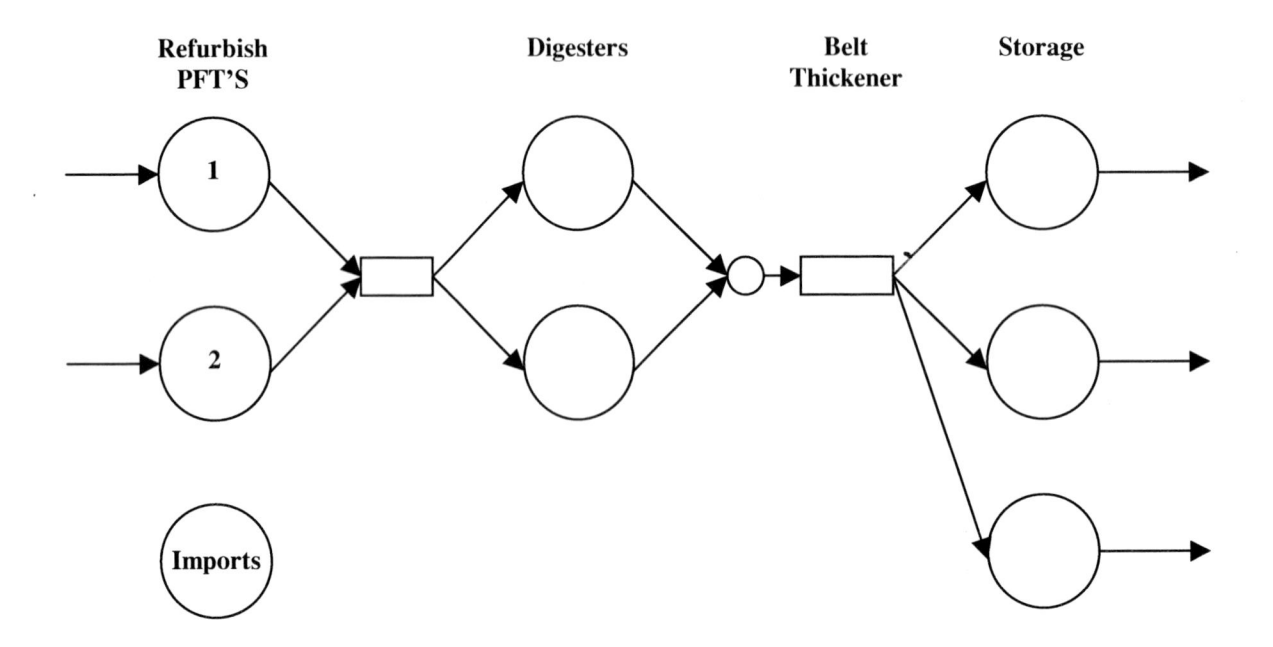

Figure 2 – Initial Design

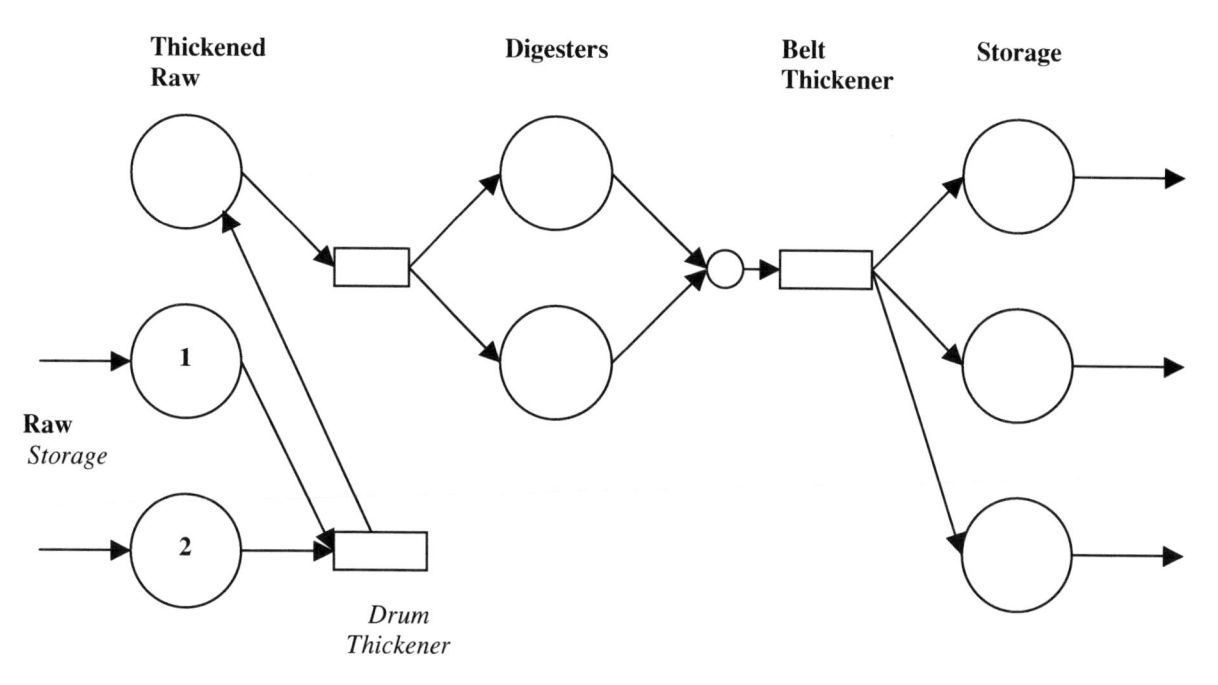

Figure 3 – Final Design

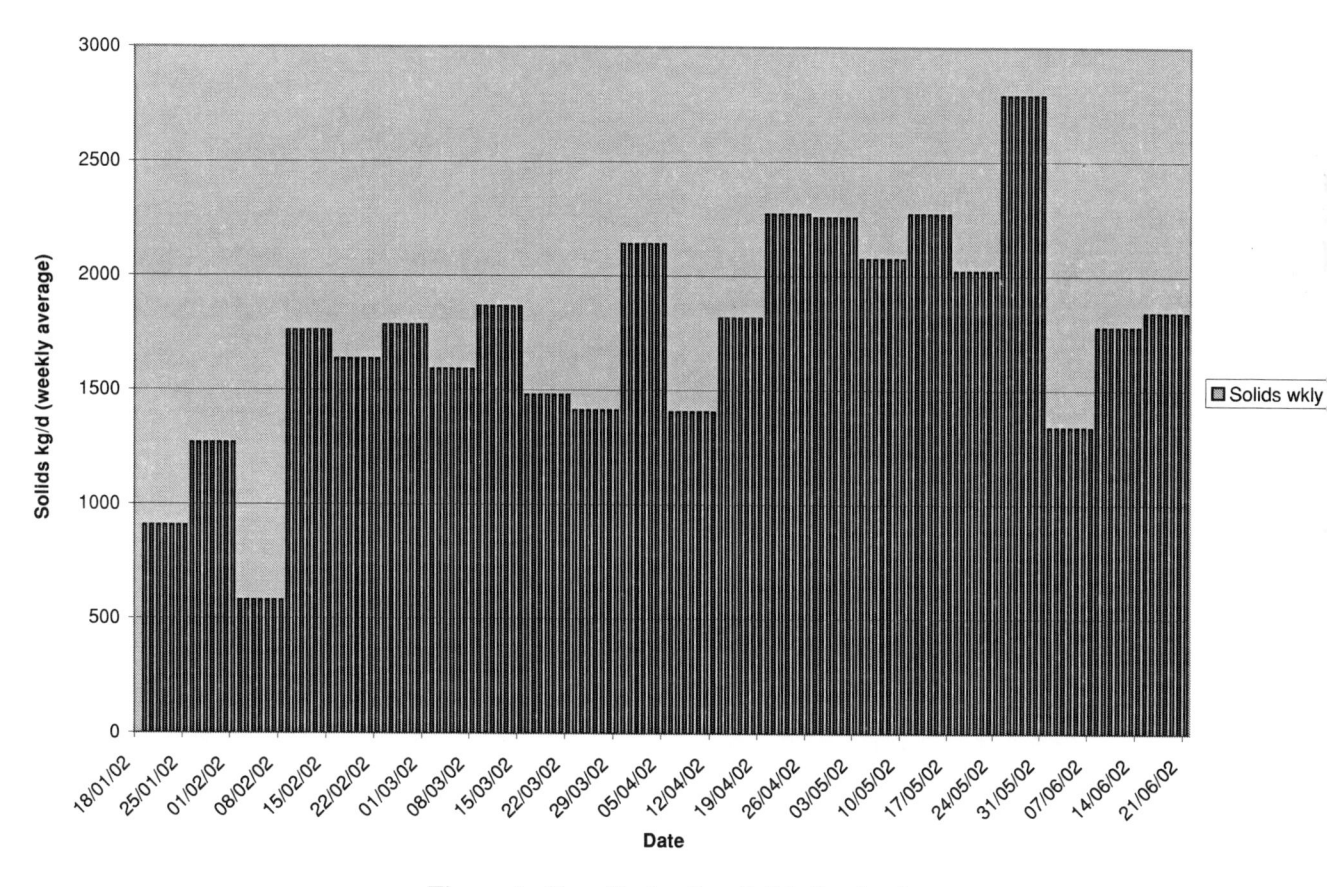

Figure 4 – Raw Sludge Dry Solids Production

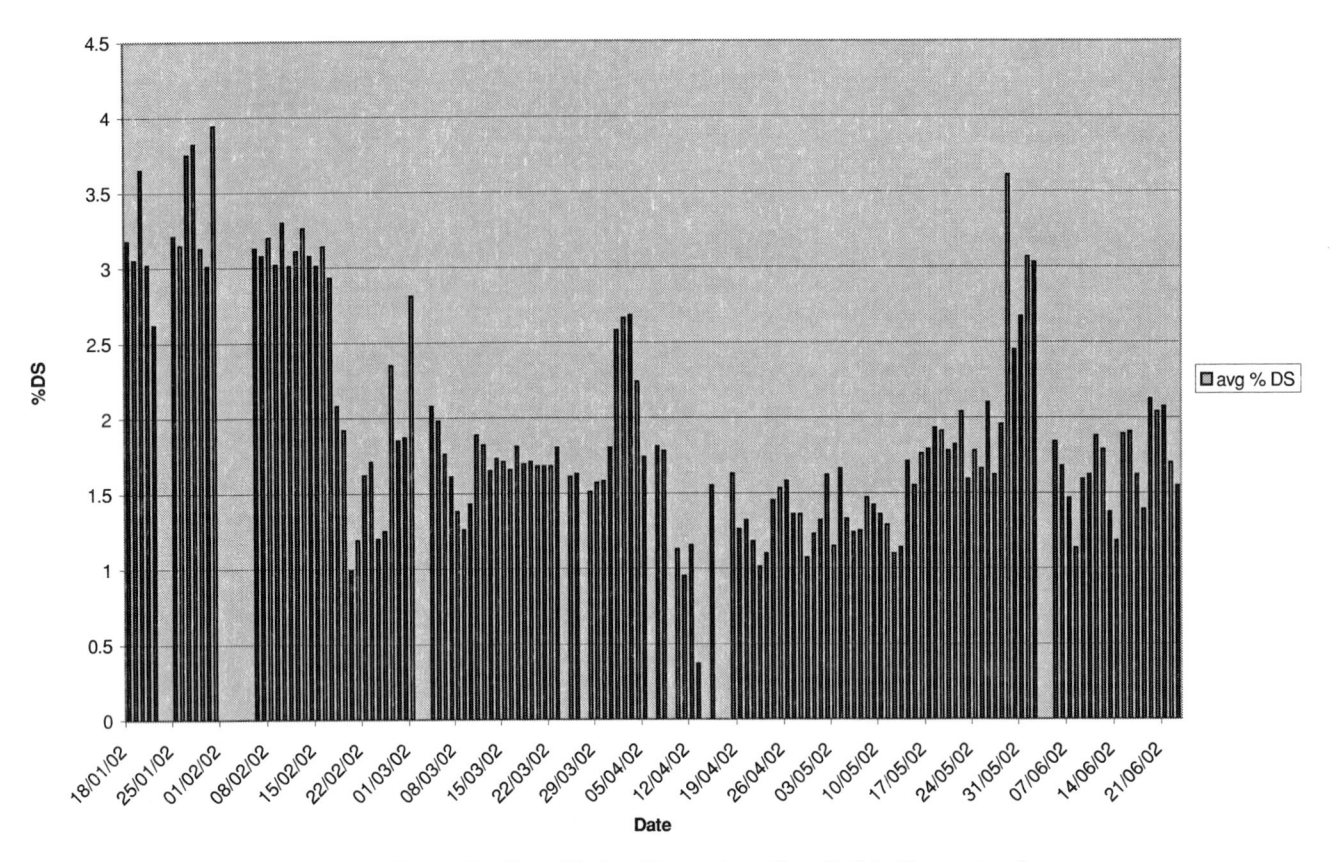

Figure 5 – Raw Sludge Percentage Dry Solids Concentration

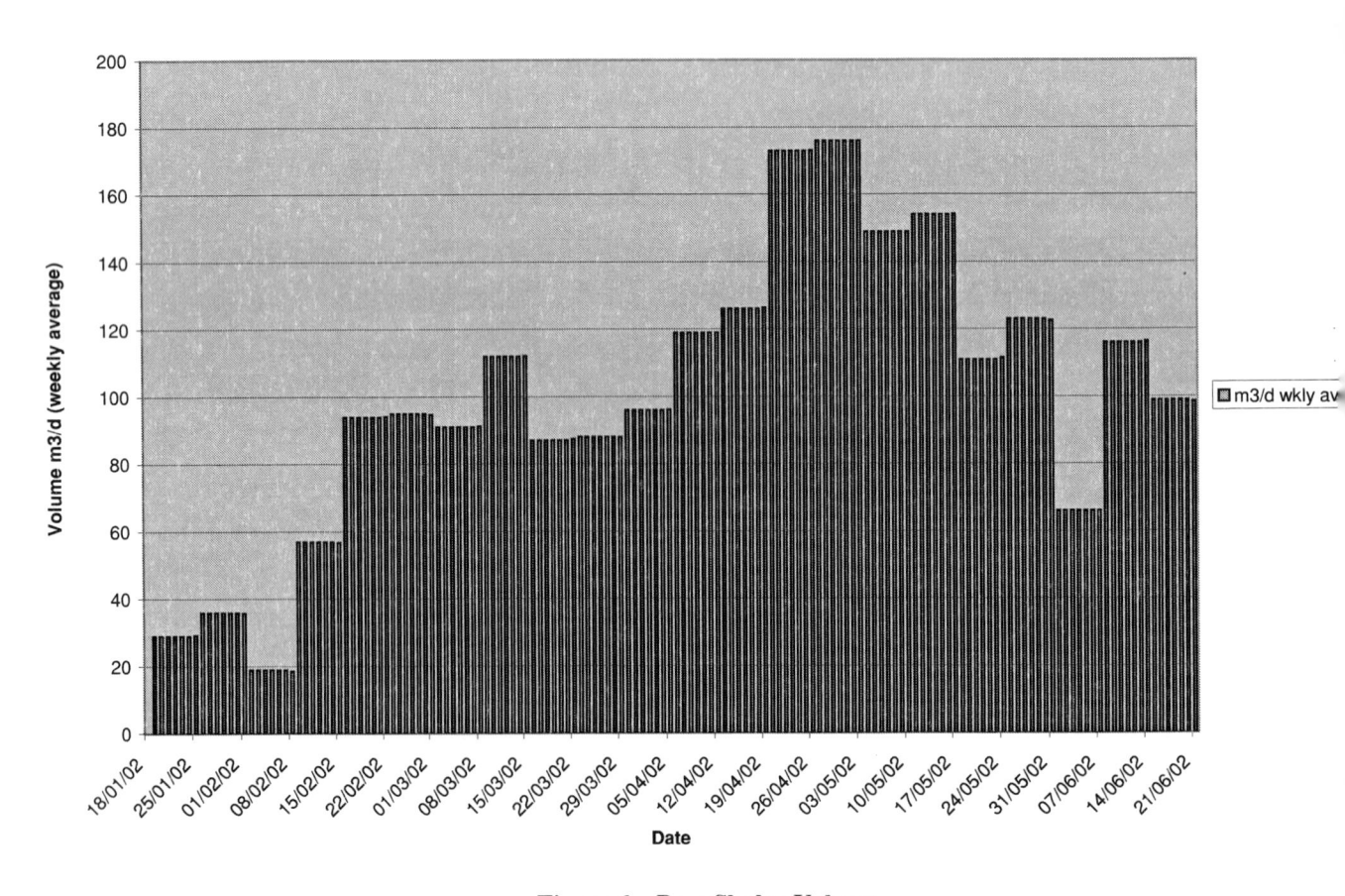

Figure 6 – Raw Sludge Volume

SLUDGE THICKENING FOR SMALL SEWAGE TREATMENT WORKS

Mountford, L.,
Simon-Hartley Ashbrook, UK

ABSTRACT

The effect of regulations and policies upon sewage sludge treatment and disposal options have been well documented. The more visible, large inner city treatment works have been identified for many years, and for the vast majority effective sludge treatment facilities now exist or are well into the design and construction phase. However, there are also many smaller capacity works that will require improved sludge treatment, many for the first time. Due to improved treatment standards, higher quantities of biological sludge is being produced for subsequent treatment, disposal or recycling. Selection of sludge thickening equipment for these smaller sites shall be discussed. We will also detail End User considerations when incorporating thickening units at strategically situated locations within a region, the impact on tanker fleet size, tanker movements, operating costs and operator manning levels.

This paper failed to make the printers deadline. It should be available on the conference CD

PUMPING A SIMULATED PASTE-LIKE SLUDGE WITH A MODIFIED CENTRIFUGAL PUMP

Lee Whitlock[1], Anders Sellgren[2]
[1]*GIW Industries Inc., U.S.A.,* [2]*Lulea University of Technology, Sweden*
[2]*Phone: +46920491497, Fax: +46920491697, E-mail: anders.sellgren@sb.luth.se*

ABSTRACT
A phosphate clay slurry was found to have similar rheological properties as sewage and fibrous paper mill sludges when handled at typical solids concentrations by mass of 6 and 3%, respectively. Dewatered sewage sludges are often pumped at paste-like conditions at concentrations in excess of 15%, corresponding to yield stresses larger than 200 Pa. Adding fine sand to a phosphate clay slurry has simulated high concentrations. At the GIW Hydraulic Testing Laboratory in the U.S. pipeline loop experiments with an open shrouded and augered centrifugal pump (impeller diameter 0.3 m) showed that a simulated sludge product could be pumped at yield stresses well over 200 Pa.

KEY WORDS
Centrifugal pump, non-Newtonian, paste, rheology, sludge pumping, yield stress.

INTRODUCTION
In practise the flow behaviour of sewage sludges varies considerably from the different processing steps and from plant to plant depending on local conditions, such as the amount of industrial effluents. Johnson (1981) reported extensive large-scale experimental work on pipeline friction losses and centrifugal pump performances with various sewage sludges was reported by An extension of this work resulted in the well-known and often referred to design guidelines, Frost (1983). The dilemma of formulating generalised results with sewage sludges is shown in Figure 1, with data from Johnson (1981).

Figure 1 shows that the primary sludge with the largest total solids content (TS) has the lowest friction losses. There is also a tendency of lower friction losses for primary sludge than for water flow at velocities exceeding about 2.5m/s. The sludge in Figure 1 with the lowest total solids has the largest loss for laminar flow below a velocity of about 2.5m/s, i.e. the transition region between laminar and turbulent flow. The corresponding transitional velocity for the intermediate TS-value in Figure 1 is about 1.5m/s.

The head and efficiency of centrifugal pumps are generally lowered to some extent when pumping sewage sludges for TS larger than 3 to 4%. The relative reduction of the water head and efficiency for a constant flow rate and rotary speed may be defined by the ratios and factors shown in Figure 2.

Performance experimental results with a centrifugal pump for the digested sludge in Figure 1 are shown in Figure 3. It follows from Figure 3 that the drop in head is 10% and 25% for efficiency and efficiency , i.e. the efficiency is influenced more than the head. The effect is more pronounced at high flow rates. As a result the flow rate at the best efficiency point tends to be smaller when pumping sludge, i.e. the pump behaves like a smaller pump.

The non-Newtonian behaviour of sewage sludges shown in Figure 1 which deviates strongly from the water loss curve for low velocities indicates a yield stress, i.e. the minimum stress required causing the sludge to flow. Dewatered sewage sludges with TS-values in the range of 15 to 25 % behave in a paste-like way with high yield stresses. Experimental results (Sellgren et al 1999a, 1999b) from regular centrifugal slurry pumps have shown sharp reductions in head at low flow rates (about 40 % lower than the best efficiency point flow rate) with highly non-Newtonian industrial slurries with a yield stress in the range of 100 Pa.

The objective here is to present experimental pumping performance results with a modified centrifugal slurry pump when pumping a simulated paste-like sludge.

EXPERIMENTAL RESULTS
The experimental work was conducted at the Hydraulic Testing Laboratory, GIW Industries Inc., U.S.A. Here, slurry pipeline hydraulics and pump performances can be investigated in loops with pipe diameters of up to 0.5 m and pipeline lengths of up to 200 m: a pipeline-loop system with pipe diameter of 0.075 m was used for these experiments. The pump was an open shrouded and augered GIW LCC-type (3-vane, all-metal) centrifugal pump with an impeller diameter of 0.3 m, see Figure 4.

Phosphate clay sludge with a S.G. of 1.11 was used to simulate paste. Adding fine sand with an average size of 135 microns increased the consistency. The rheological properties and the pumping characteristics of the slurries were determined from differential pressure drop and flow rate measurements.

Calculated shear stress versus the rheological scaling parameter 8V/D (V = velocity, D = diameter of pipe) are shown in Figure 5 for only phosphate sludge and a sand-mixture with S.G. = 1.60 together with sludge data from other sources. Sand was then added at V/D=400 up to a mixture S.G. of 1.67- corresponding to a shear stress value of about 360Pa. The slope of the curve for S.G.=1.60 is about 0.2 in Figure 5. Rheograms were then constructed for S.G-values of 1.60 and 1.67, following the procedures described in, for example Wilson et al (1997). It was assumed that the slope for S.G.=1.67 remained the same (0.2) as for 1.60. The rheograms with the results represented by a Bingham model are shown in Figure 6.

Figure 7 shows how the pump head and efficiency were lowered by the highly viscous clay-sand mixture. When the pump in Figure 7 was operating in the best efficiency region (0.015 – 0.020 m^3/s), the reductions in head was about 10% and about 15% for efficiency. However, the performance became very sensitive to small variations in the mixture S.G. for lower flow rates. It can be seen in Figure 7 that the pump can produce head fairly well for S.G. values of 1.62 to 1.65 corresponding to yield stress values larger than 200Pa. However, it cannot maintain the head at a S.G. of 1.67, when an unstable head curve is created.

DISCUSSION AND CONCLUSIONS

From tests at the GIW Hydraulic Laboratory experiences with regular closed impeller slurry pumps have sometimes indicated similar instability problems as shown in Figure 6, for slurries with yield stresses of about 100 Pa. The open shroud and augered impeller seems to have been the determining factor for pumping in excess of 200 Pa without the problem of an unstable head curve for lower flow rates.

The rheology of the simulated paste here may correspond to dewatered sewage sludge with a TS-value in the range of 15 to 18 % (Sellgren 2000), see Figure 5. However, this is not to say that the pumpability of true sewage sludge can be directly related to the rheological properties. Pump blockages, deposits and feeding are problems that may occur also for well suited centrifugal pump types when handling sewage sludges with TS-values above 4%. Guidelines to centrifugal pump selection in sewage sludge application were given in a recent paper by Lancett (2001).

Guidelines on pump curve reductions for viscous <u>Newtonian</u> fluids are available from, for example, The Hydraulic Institute (1983). Resulting reductions show principally the same trends as in Figure 3, i.e. a larger pump should be chosen in an application to pump the required flow rate with operation within the best efficiency region. The sludge from Figure 3 cannot be considered Newtonian. However, it can be considered moderately non-Newtonian compared to sludges with yield stresses larger than 100Pa.

The results reported here with highly non-Newtonian media with pronounced yield behaviour showed that a selection of a large pump might introduce a risk of operating at flow rates where an unstable head curve can occur.

REFERENCES

ABS-Scanpump AB report. Paper stock flow. The calculation of friction loss and pump performance, 1991

FROST R.C. How to design sewage sludge pumping systems, Technical report, TR185, Water Research Centre, January 1983

HYDRAULIC INSTITUTE 1983 Standards for centrifugal, rotary and reciprocating pumps, 14th Edn. Hydraulic Institute, Cleveland, OH, U.S.A.

JOHNSON M. First report on the WRC sewage sludge pumping project. Technical report, TR162, Water Research Centre, April 1981

LANCETT M. A practical guide to centrifugal pump selection for sewage sludge applications. Proceedings, 6th European Biosolids and Organic Residuals Conference, Wakefield, UK, November 2001

SELLGREN A. and ADDIE G. and YU W.C. Effects of non-Newtonian mineral suspensions on the performance of centrifugal pumps. *Journal of Mineral Processing and Extraction metallurgy Review. vol. 20. pp 239-249, 1999 a.*

SELLGREN A. and ADDIE G.R. and JUZWIAK J.H. Factors involved in the pumping on non-settling slurries with centrifugal pumps. Proceeding. *Rheology in the mineral industry II*. E.J. Wasp. Ed., Kahudu, Oahu, Hawaii, U.S.A., 1999 b.

SELLGREN A. Effective pumping of sewage sludge, Proceedings, The VA-2000 Conference, Jonkoping, Sweden, September 19-20, 2000 (In Swedish)

WILSON K.C., ADDIE G.R., SELLGREN A. and CLIFT R. Slurry transport using centrifugal pumps. Blackie Academic and Professional, London, U.K., 1997.

ACKNOWLEDGEMENTS

This investigation was financially supported by the VA-Forsk research foundation, Sweden and the GIW Industries Inc., U.S.A.

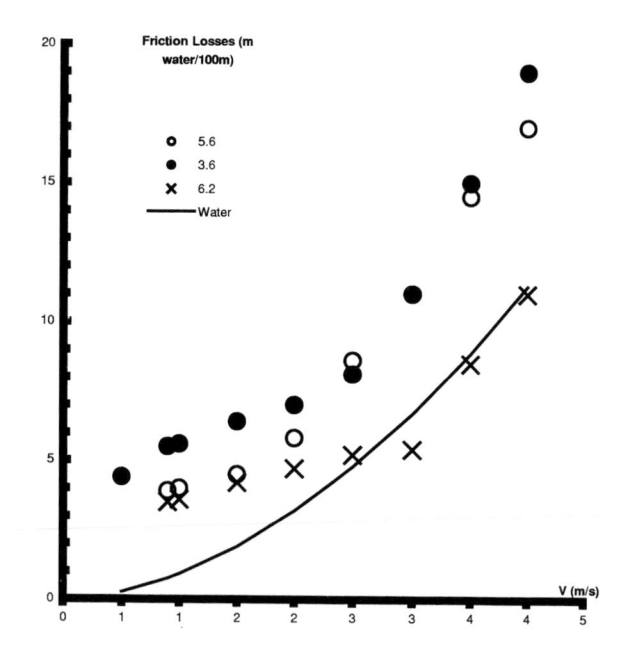

Figure 1. Friction losses versus velocity for three sludges in a 0.1 m in diameter pipe. The TS-contents of 5.6, 3.6 and 6.2% corresponded to digested, activated and primary sludges, respectively. Data from Johnson (1981).

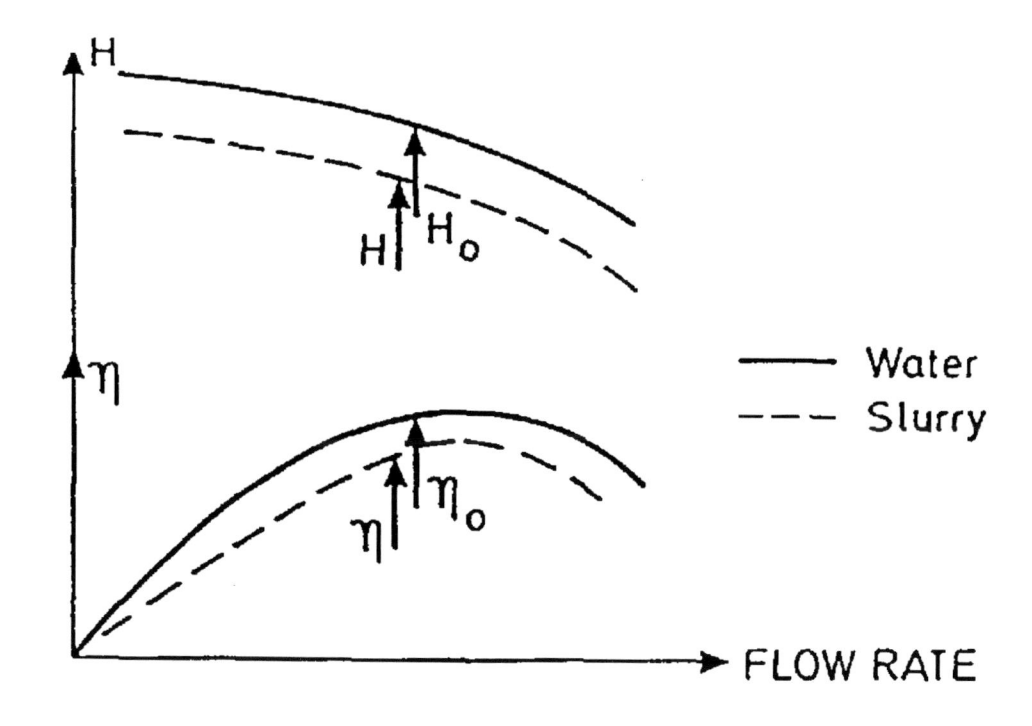

Head ratio: $HR = H/H_0$ Efficiency ratio: $ER = \eta/\eta_0$

Head reduction factor: $R_H = 1 - HR$ Efficiency reduction factor: $R_\eta = 1 - ER$

Figure 2. Sketch defining the reduction in head and efficiency of a centrifugal pump pumping a slurry

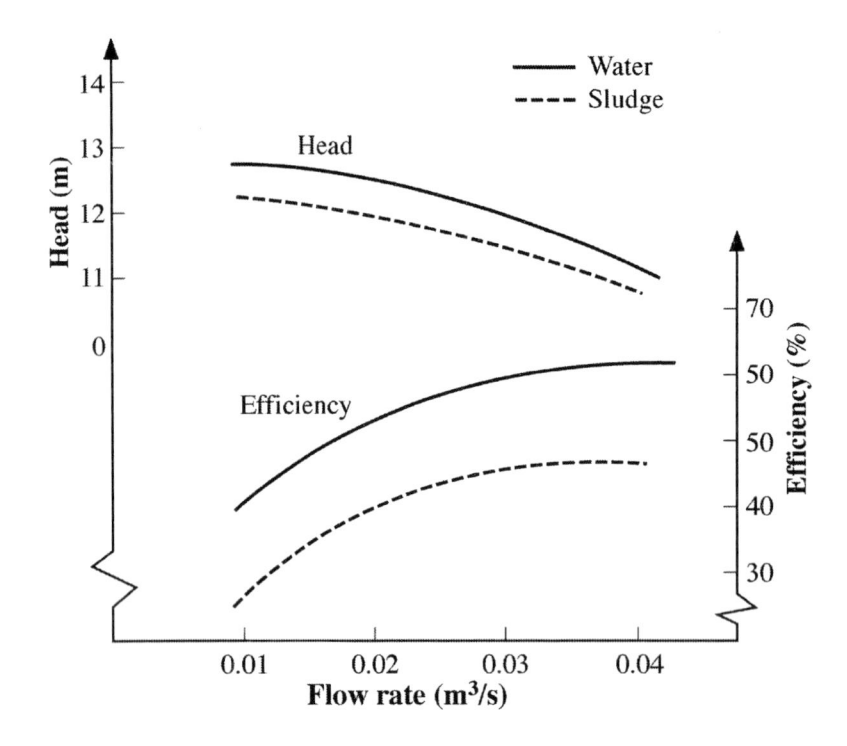

Figure 3. Reductions in water head and efficiency when pumping the digested sludge in Figure 1 at 6.7% with a 0.1 by 0.1 centrifugal pump with a 4-vane impeller (diameter=0.3m) running at 960 rpm.

Figure 4. Open shrouded and augered centrifugal pump (impeller diameter =0.3 m)

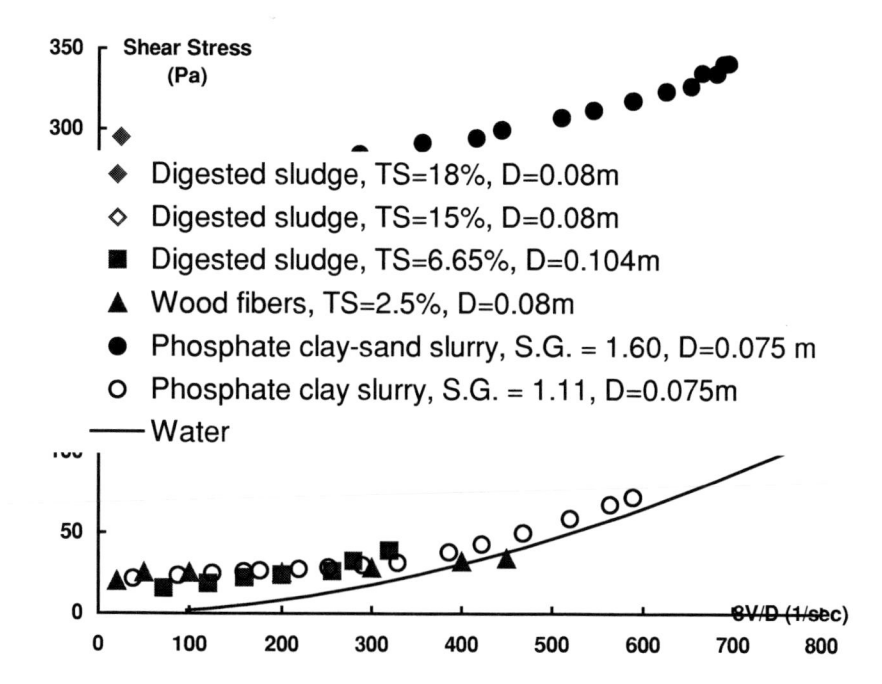

Figure 5. Evaluated shear stress versus 8V/D for phosphate clay (S.G.=1.11) and clay-sand mixture (S.G.=1.60). Digested sludge data (6.65%) from Johnson (1981) and from Sellgren (2000) for TS = 15% and 18%. Wood fibre data from ABS- Scanpump AB (1991)

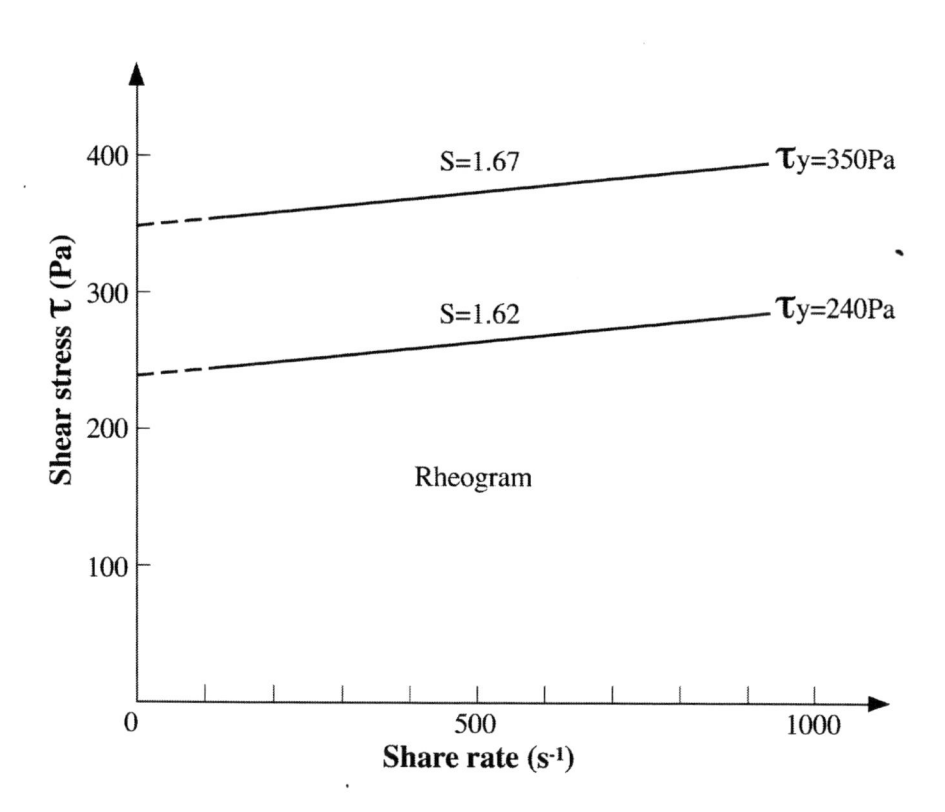

Figure 6. Rheogram with estimated yield stresses. S is mixture S.G.

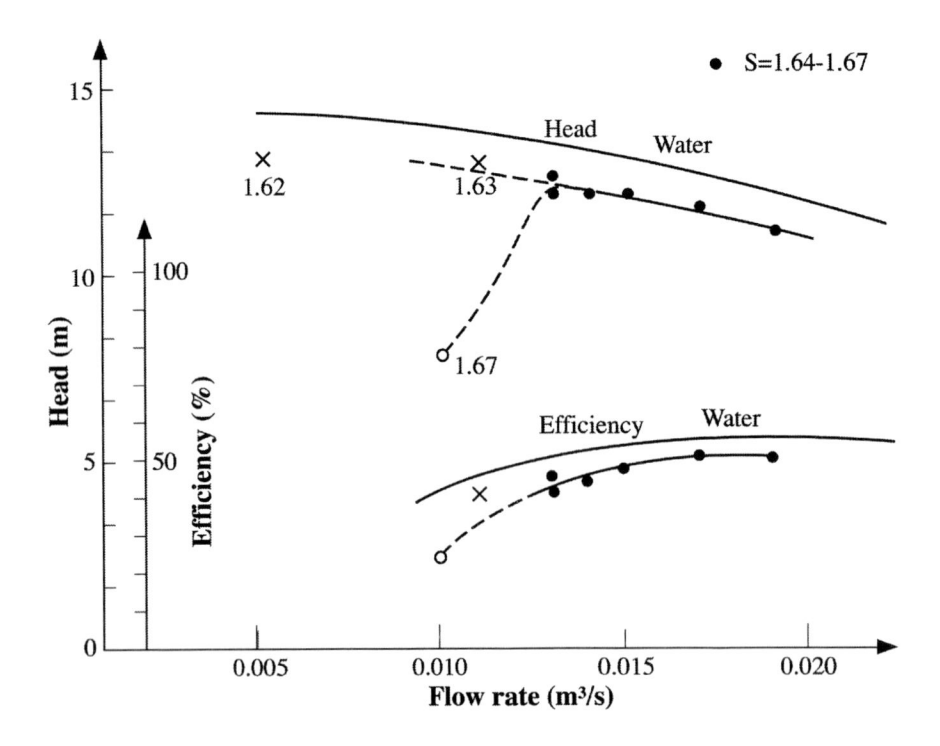

Figure 7. The effect of slurry on the pump head and efficiency at different mixture S.G.-values.

THE USE OF VERMISTABILISATION FOR THE TREATMENT OF LIQUID SEWAGE SLUDGES

Sophie Mormede
Atkins Water, Epson, Surry UK
Tel: 01372 726140; Fax: 01372 740055; e-mail: sophie.mormede@atkinsglobal.com

ABSTRACT

Vermisabilisation is the use of earthworms to process sewage sludges, converting them into fertiliser. Numerous studies have been completed to assess the viability of vermistabilisation systems for sludge treatment, most of which were undertaken with cake sludges.

This paper reviews the data collected to date from a 2 year on-going research programme to develop a large scale vermistabilisation unit for the continuous, fully automated treatment of liquid sludges. Vermistabilisation is proposed as a single-step treatment process for small rural works currently disposing of raw sludges (which is to be phased out under the new Sludge Regulations). This technology is also proposed as a cost-effective, environmentally acceptable alternative to conventional treatment processes for small rural works.

A prototype large-scale, fully automated unit was commissioned in April 1999 at Clayton West sewage treatment works. In its present configuration, this unit provides a 60% reduction of total suspended solids, and 50% of COD.

KEY WORDS

Sewage sludge, small works, vermistabilisation, vermiculture

INTRODUCTION

Approximately 50% of all sludge produced in the UK is recycled to agriculture. On large works most of this sludge is anaerobically digested and dewatered prior to disposal (in accordance with the UK Code of Practice). On small rural works however the quantities of sludge produced are relatively small and thus the cost of extensive sludge treatment prohibitive. Since the availability of suitable agricultural land around small rural works is usually high, raw sludges are often applied directly. It is estimated that approximately 25% of all the sludges disposed to agriculture in the UK are raw sludges derived from small rural works[1].

However, under the "Safe Sludge Matrix", all plants between 10,000 and 15,000pe and those between 2,000 and 10,000pe discharging to fresh water will be required to have secondary treatment by December 2005. Therefore, implications for sludge treatment on these rural sites are considerable. A variety of sludge management strategies are possible. For example treating the sludges in centralised processing facilities has been suggested. This solution offers the advantage of economies of scale but can result in considerable tankering costs being incurred and is thus not appropriate in all circumstances. Traditional rural works treatment technologies, such as lime stabilisation remain costly and capital intensive. In recognition of this a number of water companies are investigating and developing alternative treatment technologies. Vermistabilisation being a low technology with low capital cost and relatively simple operation offers potential with this respect.

Vermistabilisation is the use of earthworms to process sewage sludge, converting them into a valuable fertiliser. The benefits of vermistabilisation are well known and documented[3]. These include:

- Fragmentation of sludges, converting them into castings and increasing the surface area available for drying and microbial decomposition
- Improved aeration due to tunnelling action of worms
- Conversion of nitrogen to nitrite (improving the availability for plants)
- Increased concentration of available phosphorus, potassium and magnesium
- Reduction in Carbon : Nitrogen ratio
- Decrease in particle size and thus increase in moisture holding capacity of sludges (to a level comparable with peat)
- Reduction in odour characteristics of the waste
- Bioaccumulation of heavy metals

The present paper reviews results to date from a 2 year ongoing research program funded by Yorkshire Water and East of Scotland Water on the development of a large scale automated vermiculture unit treating liquid sludge. This project was carried out as the continuation of research concentrating on the development of vermistabilisation units treating cake sludges[2].

MATERIALS AND METHODS

The development work leading to the design of the Continuous Liquid Sludge Vermistabilisation (CLSV) unit used in the present trial as well as results obtained when treating thickened sludge have been reported in detail elsewhere[2]. The CLSV unit is composed of a circular tank with a 6m diameter (surface area of $28m^2$), divided into 8 test sections, and is fully covered. Sludges are delivered onto the bed via a rotating gantry. The feeding mechanism is designed to be continuous (i.e. 24 hours/day) and automated.

The circular design allows the bed to be split in to eight segments each with its own media and drainage port for effluent collection. Methods for heating the unit were considered bearing in mind the fact that the processing capabilities of worms are very temperature dependent. However since the unit has a certain amount of in-built insulation (being covered and containing considerable compost-type material), the inclusion of a heater unit was not deemed a requirement for the initial design. The capability to include a heating mechanism on the unit at a later date has however been included.

Due to the need for accurate data collection during the research stage and the Health and Safety issues associated with access, the research unit is 'over-designed'. In reality the final design could be as simple as a rectangular drying bed with an 'easy to remove' lid, low specification motor and standard drying bed drainage.

The unit was constructed in early 1999 at Clayton West Sewage Treatment Works in South Yorkshire. A first trial was conducted between 1999 and 2001, treating thickened sludge[2]. In April 2001, the media from the previous trial was removed and new media and worm biomass was installed in the unit. The worms used in the unit are *Eisenia foetida*, for which the typical generation time (i.e. from egg through to a reproductive adult producing eggs themselves) is between 10 – 15 weeks depending on temperature.

In order to monitor closely the efficiency of the CLSV unit, the inlet sludge and outlet sludge at each of the eight ports are sampled once or twice a week (depending on the analysis to be carried out). These samples were analysed for total and volatile suspended solids, COD and ammonia.

RESULTS AND DISCUSSION

Therefore the results presented here concentrate on the liquid sludge trial, started in June 2001. The results include all data obtained to June 2002.

MEDIA

The physical properties of the media employed for vermistabilisation determine the efficiency of the bed. For a most efficient vermistabilisation unit, the media has to be optimised for the growth of both worms and bacteria (on which the worms feed). Optimum conditions include good drainage, 70 to 90% humidity, constant food supply etc.

Various media are being tested in the 8 sections of the CLSV unit. These include green waste, forest mulch, straw and plastic media. The green waste and forest mulch media (sections 1 to 4) are the most popular with the worms, whereas the straw media (sections 5 and 6) showed signs of clogging and water logging (hence the migration of the worms towards the other medias). However, after almost a year continuous operation, all media are still of acceptable standards.

In the section covered with plastic media, part of the worms have made their way into the plastic media and live there or at least feed there. Further test have shown that, if the plastic blocks are moved to another area (of new media for example), up to 30% of the worms will migrate back from the plastic blocks into the media within a week, providing an easy way of recovering part of the worm population when changing the media.

SOLIDS LOADING RATES

The feeding regime of the vermiculture unit was maintained as such: feeding for 2 hours three times a day. The solids loading rate obtained during the period of interest is detailed in figure 1. The flow was controlled in order to achieve a weekly solids loading rate of approximately $1kg/m^2$. This target was exceeded early March 2002, when the solids loadings of the wastewater increased from a typical value of 300mg/l total solids to over 1200mg/l, due to the desludging of the humus tank onsite. In order to avoid such discrepancies in the feed stream, the inlet to the vermiculture unit was subsequently moved to avoid receiving sewage coming from the desludging of the humus tank.

SOLIDS REMOVAL EFFICIENCY

Vermistabilisation of wastewater relies on the ability of the media to capture solids whilst allowing the liquid to drain through. The extent of solids capture is quantified by comparing the content of the feed and effluent. Figures 2 and 3 represent the solids concentrations in the feed and effluents of the various beds. The results from beds of similar media were similar (usually less than 10% standard deviation) and therefore averaged for ease of comparison.

Figures 2 and 3 show that the solids content of the feed is erratic, and as a consequence, so is that of the various effluents. However, there is still no differentiation of the various media on the grounds of solids removal: none of the media appears to be failing to retain the solids. Figure 4 and 5 represent the percentage of solids removal achieved by each type of media. As for the previous figures, the results where the media is duplicated are averaged. Figures 4 and 5 confirm that the various media are performing in a similar way. However, it seems that the media made up of green waste covered with plastic blocks has been performing significantly worse than the others since early May. This might be due to the clogging of the plastic with worm casts.

The total and volatile solids removal efficiency varied between 40 and 90%, averaging about 65% throughout the trial to date (see Figure 6). Total solids removal is slightly higher than that of volatile solids, but this is not significant due to the variability of the results. Moreover, the total suspended solids in the effluents varied between 100 and 150mg/l whatever the value of the inlet sludge, showing the limit of efficiency of the bed.

COD AND AMMONIA REMOVAL

Figures 7 and 8 represent the COD and ammonia levels in the feed and effluents. As with the solids contents, results for similar media were averaged and the chemical oxygen demand (COD) and ammonia concentrations were highly variable in the feed. The COD concentration in the inlet feed varied between ca 200 and 2000mg/l. That of ammonia was between 8 and 55mg/l.

Figure 9 shows the average removal efficiency in each bed over the entire trial period to date. The COD removal efficiency varies between 50 and 55% for the various beds, the difference is not significant considering the wide variability of the data (see figures 7 and 8). The ammonia removal efficiency is notably lower, between 15 and 30% depending on the beds. The beds made up of forest mulch present the highest ammonia removal, at 30%; whereas the other beds were significantly less efficient at retaining ammonia.

GROWTH TRIALS

The by-products of the vermiculture treatment of wastewater are a cleaner effluent and the worm bed when replaced. The quality of the effluent obtained has been detailed in previous sections. The worm bed needs changed once a year or less depending on the loading of the bed. When it has been used, the worm bed is mainly made-up of worm casts, and therefore commonly known as vermicast. This product is stable and enriched in nutrients, and can be used as a fertiliser.

Throughout July and August 2001, growth trials were conducted to evaluate the value of the worm casts as a fertiliser. Grass was grown on either soil, soil mixed with 50% vermicast, or on 100% vermicast. The height of the grass was regularly measured and is reported in figure 10, which shows that the bigger the proportion of casts in the soil, the taller the grass. Therefore vermicast has definite value as a fertiliser.

The nutrient composition of the various soils was also analysed, and is expressed in table 1. Table 1 shows that the vermicast is highly enriched in nutrients, with at least twice as much N, P and K as traditional soil. Moreover, vermicast is a well aerated product, stable and of peaty odour. Therefore vermicast can be utilised as a valuable fertiliser.

CONCLUSIONS

The trial to date has showed that:
- Vermiculture can be applied FOR the treatment of wastewater
- All the types of bed tested (green waste, forest mulch and straw) have a life expectancy of at least a year, with straw maybe least suitable of all three
- Adding stones or plastic media did not seem to change the efficiency of the beds
- Plastic media can be used to recover part of the worm population prior to changing the beds
- The solids removal of the three media tested is about 65% for total and volatile solids
- The COD removal is at least 50% for all medias
- The ammonia removal is between 15 an 30%, with the best results for the forest mulch media
- The vermicast produced is of high fertilising value, with twice the NPK concentration of traditional soil

Therefore, vermistabilisation of sludges is an efficient way of treating liquid sludges. Because of its ease of operation and low capital cost, it is a process method particularly adapted to small remote treatment works.

ACKNOWLEDGEMENTS

Yorkshire Water and East of Scotland Water Authority have supported the research presented here financially. The work is part of an on-going two-year study to develop a fully automated liquid sludge vermistabilisation system.

REFERENCES

1. Bruce A., Biosolids- Dull? It certainly isn't, Water and Environment Manager pp8 (1996)

2. Clark, P., and Peebles, J., Advances in the development of an automated liquid sludge vermistabilisation system for the treatment

of sewage sludge (1999). In the proceeding of the November 1999 CIWEM conference.

3. Edwards C.A., Breakdown of animal, vegetable and industrial organic wastes by earthworms. In "Earthworms in waste and environmental management" (eds Edwards and Neuhauser), SPB Academic Publishing, The Hagues (1988)

TABLES

Table 1: Nutrient composition of the various soils, expressed in percentage dry weight

	100% soil	50% soil 50% vermicast	100% vermicast
N (% dry weight)	0.20	0.23	0.43
P (% dry weight)	0.09	0.10	0.20
K (% dry weight)	0.09	0.09	0.23
Total solids (%)	82.8	63.9	29.2

FIGURES

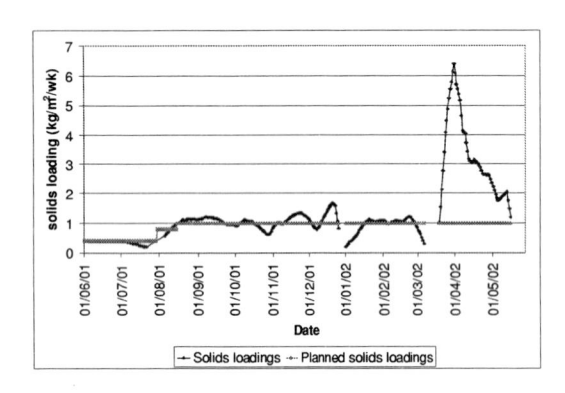

Figure 1: Solids loading rate, expressed in kg/m^2/week

Figure 2: Total suspended solids in the feed and effluents, expressed in mg/l

Figure 3: Volatile suspended solids in the feed and effluents, expressed in mg/l

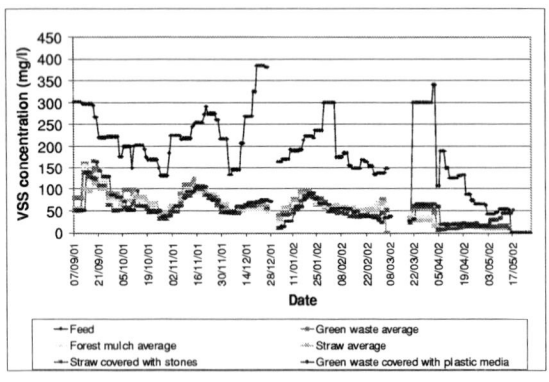

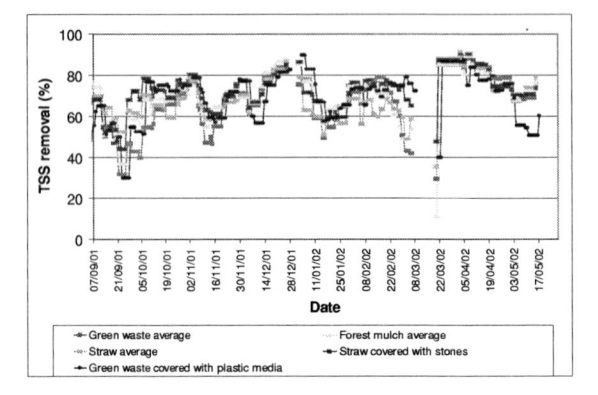

Figure 4: Removal of total suspended solids by the various media, expressed in %

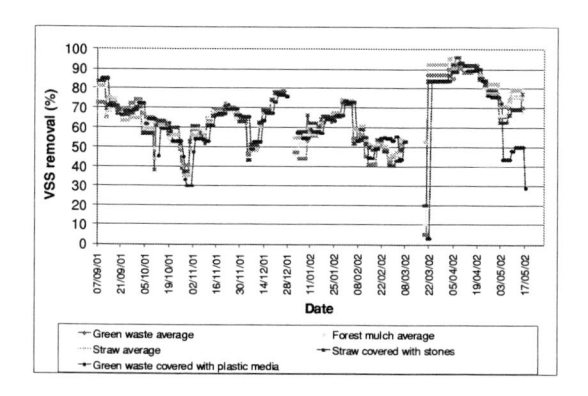

Figure 5: Removal of volatile solids by the various media, expressed in %

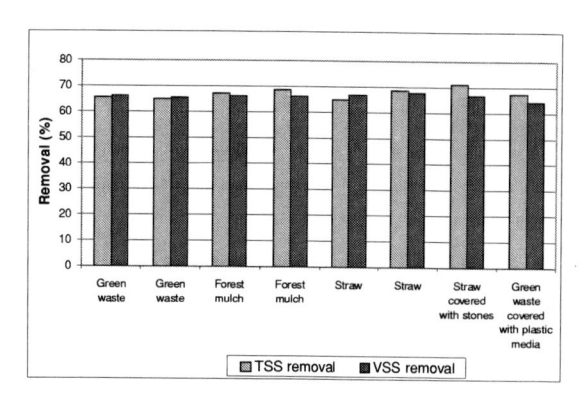

Figure 6: Average of solids removal throughout the last year of the trial, expressed in %

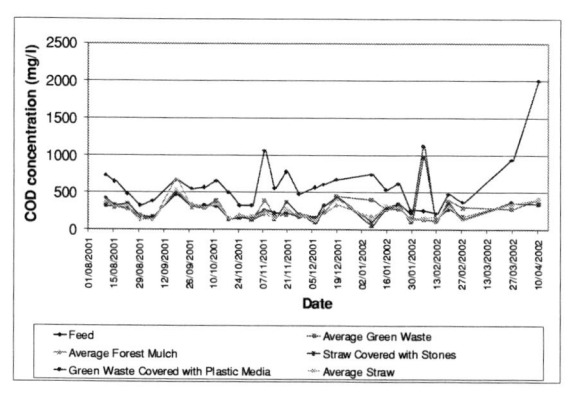

Figure 7: COD in the feed and effluents, expressed in mg/l

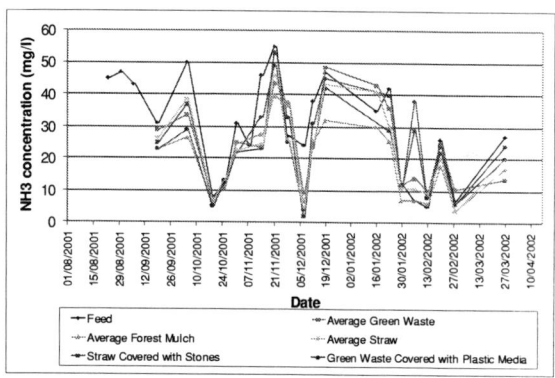

Figure 8: Ammonia levels in the feed and effluents, expressed in mg/l

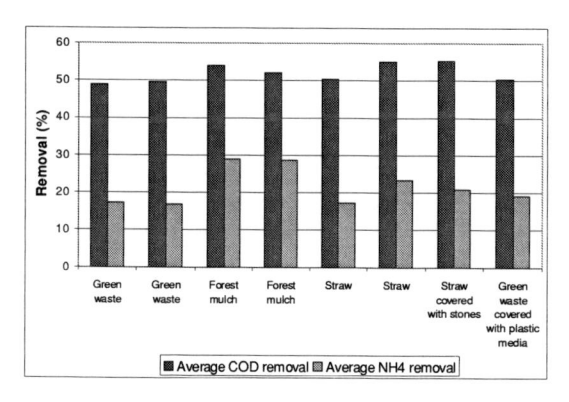

Figure 9: Average COD and ammonia removal over the entire trial to date

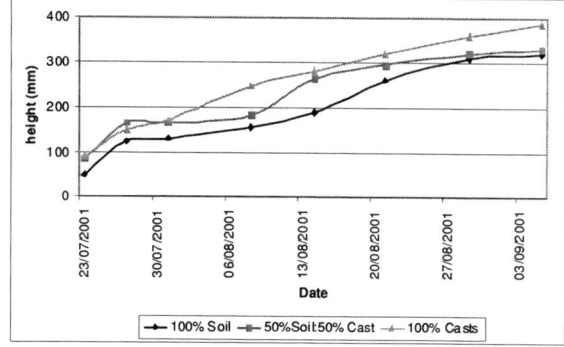

Figure 10: Height of grass, expressed in mm

ADVANCES IN THE USE OF REED BEDS FOR SLUDGE DRYING

Paul Cooper, Neil Willoughby and David Cooper
Arm Limited, Rydal House, Rugeley, Staffordshire, WS15 3HF, United Kingdom
Tel: (44) 1889 583811, Fax: (44) 1889 584998, E-Mail: Info@Armreedbeds.Co.Uk

ABSTRACT

The article reviews the design and performance experience with Sludge Drying Reed Beds over the past 14 years. There are very few in the UK but there is much more experience in Europe and in particular in Denmark. The design of and experience with 2 UK systems is described. The final solids concentration depends upon the concentration in the initial sludge dose. It is possible, when treating anaerobically digested sludges with 3 to 4%DS, to achieve about 90% volume reduction and a final solids content of up to 40% but with weaker unthickened activated sludges with 0.3 to 0.6% DS a reduction in volume of greater than 97% is possible but with a final solids concentration of 10 to 20% DS. Most of the information refers to domestic sewage sludge but there is also a report here of some work done on drying an agricultural sludge.

KEY WORDS

Agricultural sludge, constructed wetlands, design, reed beds, sewage sludge, sludge, sludge drying reed beds, Wastewater treatment.

INTRODUCTION

We have been using Reed Beds [also called Constructed Wetlands] in the UK since 1985 and there are now more than 600 systems in operation. There are now many variants of these systems for different purposes in use throughout the World and most of them have examples in the UK but one of them seems to have been largely ignored, the Sludge Drying Reed Bed (SDRB). We have only been able to find 6 full-scale systems and a few pilot tests of SDRBs in the UK. Why is this? We suspect that it is due to two reasons. There were many conventional Sludge Drying Beds in use at sewage treatment works [STWs] in the UK up to the 1960s but they were gradually abandoned or taken out of use during the 1960s and 1970s. Some of these were large mechanised systems but most were simple beds at small village works, which were manually emptied. They were gradually replaced because in the wet British climate they did not dry the sludge well enough or rapidly enough and because better alternative sludge drying techniques were being applied in the 1970s.

Sludge removal from rural sewage treatment works pose a problem to the operators in that it is rarely cost-effective or practical to provide mechanical dewatering to reduce the volume of sludge. The costs of transporting this sludge with a high water content are high and from the environmental viewpoint it is regarded as undesirable to have sludge tankers travelling along small country roads.

In the last 25 years interest has developed in using reeds (*Phragmites australis*) to enhance the performance of conventional drying beds. Lienard, 1990 [14] reports that Seidel and Kickuth, the early German researchers with reed beds, tried using reeds in sludge drying beds for physical-chemical sludges at the nuclear research centre in Karlsruhe in the late 1960s Bittaman and Seidel, 1967 [1] and Kickuth, 1969 [10]. The process was then used at a number of other German sewage works sites in the 1970s and 1980s. The results and some design details from these sites, have been reported by Lienard 1990 [14].

The first UK contact with Sludge Drying Reed Beds was at Windelsbleiche in Germany in 1985 during a water industry study tour Boon 1985 [2]. A Vertical Flow bed was being used to treat chemical sludges from a textile factory. Other SDRBs were seen in the USA for dewatering sewage sludges and chemical sludges from potable water treatment using iron salts Boon, 1986 [3].

The use of reed beds for sludge drying has been taken up very enthusiastically and successfully in Denmark, Nielsen 1990, 1993, 1994, 2002 [15, 16, 17, 18], USA, Kim 1990, 1994 [11, 12] and France, Lienard and Esser, 1990, 1995 [13, 14] but has not attracted as much enthusiasm in the UK. In Denmark there are very large SDRBs which are remotely dosed by computer control through the telephone network and monitored by video links. They are widely used in USA for village systems especially down the East Coast. SDRBs are now being taken up for small village systems in France and Belgium, Lienard, *et al* 1990, 1995 [13, 14] and De Maeseneer 1997 [8]. There are only about 10 examples in the UK. Early attempts at designing and operating SDRBs in the late 1980s failed in the UK because inadequate design or operation. One system was built in an old concrete tank that unfortunately had leaks and the *Phragmites* died. Another system was abandoned after the operators dosed far too much sludge and the reeds were submerged and died. These early mistakes may have set back the adoption of SDRBs in the UK but really these were elementary mistakes, which were easily rectified by good design and operation. Subsequent installations have proved to be far more successful.

The benefits claimed for application at small, remote sites are:
- Many years of operational security.
- Tanker access not required.
- Dewatering on sites where previously non-viable.
- Low maintenance.
- In some cases no power requirement.
- High quality return liquors.
- Can be self-built by operators in remote locations.

For larger systems the benefits are:
- Operational security for up to a decade.
- Low power usage.
- No chemical additives for dewatering.
- Low maintenance,
- No noise,
- High quality return liquors,
- Unmanned operation if done via remote sensors and control.

REVIEW OF SLUDGE DRYING REED BED DESIGN

The principles that allow reed planted sludge drying beds to dry the sludge more rapidly are as follows:
- Stem, rhizomes and roots enhance drainage of water by providing channels in depth.
- Windrock produces holes in the sludge surface.
- Evapotranspiration is enhanced by the presence of leaves.
- Mineralisation.

MEDIA
The beds are normally about 800mm deep and made up of layers of gravel topped off with sand. The WRc/Severn Trent Water manual recommends 700mm of 5-10mm gravel topped with 100mm of sharp sand Cooper et al, 1996 [6]. Nielsen 2002 [18] uses 350 to 450 mm of gravel topped with 150 mm of filter sand.

DESIGN LOADING RATES
Estimating the sludge volumes and mass to be treated is the most important factor in the correct design. Once this has been determined then the loading rate is the next most important factor. A range of different loading rates have been reported.

EXPERIENCE IN THE USA
Sludge drying reed beds have been used since the mid 1980s for villages and small towns in the Eastern USA. 30kgDS/m^2 year is more usual in the USA, Kim 1990 and 1994 [11, 12]. In their work in the Pacific North Western USA Burgoon and colleagues 1997 [4] used a method of gradual increase in the dose from 10kg DS/m^2 year at the start through to 65kgDS/m^2 year [the design loading rate]. The sludge loading was gradually increased in two weekly steps.

EXPERIENCE IN FRANCE
Lienard 1995 [13] reports that too high a sludge loading in the first growing year can lead to the death of the *Phragmites*. A similar sensitivity has been noted with agricultural sludges [see later]. A value of 70gDS/m^2 day is quoted. This is equivalent to 25 kgDS/m^2 year. The same author has recommended a dose of only 18kgDS/m^2 year in the first year but after that with the reeds fully established the loading can be increased to 60kgDS/m^2 year.

EXPERIENCE IN BELGIUM
De Maeseneer 1997 [8] recommends growing the reed for 1 year before loading the bed with sludge. 20 to 30kgDS/m^2 year for *aerobically* stabilised sludge De Maeseneer, 1997 [8] this is equivalent to 1.0 to 1.5 m^3 of 2%DS sludge/m^2 year or 1.5m^2 year for a sludge, which contains 45 to 65% organic matter. De Maeseneer et al, 1997 [8] describe the use of the system for use in drying black anaerobic sludge from dredging operations in Belgium and Holland containing metals.

EXPERIENCE IN DENMARK
60 kgDS/m^2.year is used in Denmark (Nielsen, 1993, 2002) for Surplus Activated Sludge from a plant with long sludge age [>20 days]. For Activated Sludge plants with a lower sludge age or for a mixture containing SAS and anaerobically digested sludge the loading rate is reduced to 50kgDS/m^2 year. Care has to be taken in allowing for beds being rested and for the planned emptying of beds.
Nielsen, 2002 [18]. There are two very large systems in Denmark,
  a] Kolding dealing with 125,000pe and 2,168 tonne/year in an area of 62,000 m^2,
  b] Skive dealing with 123,000 pe and 2,000 tonne/ year [but which may be extended later to 2,850 tonne/year] in an area of 99,000m^2.

The Kolding value works out at 35kgDS/m^2 year and that for Skive at 29kgDS/m^2 year. This shows the effect of allowance for beds being out of commission to allow for the resting and planned emptying upon the spot loading of 60 kg DS / m^2 year. The quality and type of the sludge can have a substantial affect. Nielsen, 2002 [18] reports treating mainly Surplus Activated Sludge [SAS] or mixtures of SAS and Anaerobically digested sludge in Denmark. In this paper he recommends the use of the Capillary Suction Time [CST] to assess the sludge quality and define the loading rate as shown in Table 1.

OVERALL % VOLUME REDUCTION

Nielsen 1993 [16] reports on the results from 6 early beds built in Denmark covering 3 beds receiving Activated Sludge [2 anaerobically digested], 2 beds taking chemical sludge mixed with Activated Sludge and 1 Raw [primary] sludge ranging size from 215m^2 to 11,6000m^2 and 1,000 to 30,000 pe. Table 2 shows the basic data on these systems and Tables 3 and 4 show the percentage reductions that were achieved in two of these beds were 91 to 97%.

In some studies in the arid North Western USA Burgoon *et al*, 1997 [4] report a volume reduction of 98%. In this arid region they estimated that evapotranspiration accounted for 10% of the water loss. They included temperature, and wind speed in their evaluation.

The degree of volume reduction is obviously related to the original solids concentration in the sludge. This is clearly seen from Tables 3 and 4 that the degree of volume reduction is 90 or less for the more concentrated sludges [3 to 4%TS] but approaching 97 or 98% for the weak activated sludges with about 0.4% TS. We suspect that they will both be approaching a similar final concentration but from the different starting points.

FINAL SOLIDS CONCENTRATION

Nielsen says that an average dry matter content of 35 to 40% can be expected on the basis of the early Danish beds at Allerslev and Regstrup Nielsen, 1993 [16]. He mentions a maximum theoretical value of 46% Dry Solids. Since this early work by Nielsen there have been very many systems built in Denmark. In a very recent paper given at the 8th International Conference on Wetland systems for Water pollution Control he has reported on the situation in Denmark after 14 years experience Nielsen, 2002 [18]. There are now in 2002 a total of 95 systems in operation and by next year there are expected to be 105 systems. Some of the systems are dosed remotely by use of video and computer links from a control centre using a detailed loading, resting and emptying plan.

Lienard *et al* 1995 [13] from CEMAGREF have built a number of systems in France. They report on the treatment of a weak activated sludge [3%DS, 3,000mg/l] from a bulking activated sludge plant with high SVI values [indicating a poorly settling sludge]. They operated 3 beds one of which was unplanted as a Control. All three beds produced sludge at 10 to 11% DS with the Volatile Solids at 50 to 62 % down from an average of 73% in the activated sludge. This indicated an aerobic mineralisation of the sludge in the SDRBs. Like other workers they found that it was possible to increase the loading rate in the summer over that in the winter. They concluded that the results show that it is possible to treat 55kgDS/m^2 year after the first growing season.

COMMISSIONING

Nielsen, 2002 [18] states that the optimal time for planting is May/June, which coincides with our experience with other Reed Bed systems. He states that loading of sludge may commence after weed control has been effected and 1 to 2 months after planting. Weed control may be affected by adding sludge to about 5cm. However if this is not successful in controlling the weeds then flooding with effluent to a depth of 10cm for up to a maximum of 14 days.

It is dangerous to load the system to its ultimate loading rate at the start of operation. It is essential to allow the reeds to grow horizontally and vertically and so sludge must not be added in quantities that inhibit this. The reeds must never be submerged. It is essential to get a dense growth of reeds before the planned loading rate is reached. The commissioning period can extend to 2 years Nielsen, 1994 [17].

Lienard 1995 [13] has recommended a loading of only 18 kg DS/m^2 year in the first year after which it can be raised to 60kgDS/m^2 year after the reeds are fully-established. Nielsen, 2002 [18] states that the loading rate can increase 10-fold over the 10 years of operation.

RESTING

Applications of sludge should be followed by a rest period. The rest period should not be too great during the early stages as the quick-drying thin layer of sludge may become too dry and harm the reeds. Conversely, on a well-established bed, the sludge should be left for an adequate length of time to dry out before the next sludge application otherwise the higher water content of the sludge layers can cause anaerobic conditions in the sludge mass. In his recent paper Nielsen, 2002 [18] describes the regime used at different sites and how they vary with operational time. He quotes resting times of 55 to 65 days for older beds.

Cooper, 1995 [5] [see later section] designed the SDRBs on the Isle of Islay as single beds. These beds were to be dosed at 5cm of surplus activated sludge once per week thus allowing for 6 days resting.

OPERATIONAL PERIOD AND EMPTYING

The Danish systems are designed to operate for an average of 10 years producing a dried sludge with 40% dry solids. Emptying of the first-used cells starts in year 8 and is completed by year 12 Nielsen, 2002 [18]. There should be no problem of disposing of the dried sludge on farmland provided it does not contain any toxic metals.

UK SYSTEMS

The details of the UK systems are not well recorded or published. The following are a list of the ones that the authors are aware of.

- The first system was that built at Oaklands Park [a Camphill Village Trust Community] in 1989 designed by Uwe Burka to treat septic tank sludge from 65 pe.
- The Centre for Alternative Technology, Machynlleth, Powys, West Wales have a SDRB system.
- Ballygrant STW, Isle of Islay, a West of Scotland Water site built in August, 1995 to a design by the first author whilst at WRc Swindon.
- Severn Trent Water have built 5 test beds at 3 sites but they have no full-scale systems yet.
- Birmingham University/ARM agricultural sludge trials at ARM Ltd site at Rugeley, Staffs.
- Wild Fowl and Wetland Trust, Gloucestershire, Slimbridge. Designed and built by Chris Weedon of Watercourse Systems Ltd for a population of 250pe. 2 beds each 50 m^2 with a freeboard of 400mm.
- Watch Oak Farm, another Camphill Village Community at Thornbury near Bristol designed by Chris Weedon to treat septic tank sludge for 15 pe.
- Kiells STW, Isle of Islay, West of Scotland Water, built in 1999 to design done by the first author whilst at WRc Swindon.

There are believed to be a few more systems in the UK but no details have been published and no performance data available.

UK PERFORMANCE FOR TREATMENT OF DOMESTIC SEWAGE SLUDGES

BALLYGRANT AND KIELLS STWS, WEST OF SCOTLAND WATER

These two systems were designed by the first author whilst at WRc Swindon UK (Cooper, 1995)[5] and Cooper et al, 1996 [6]. They were constructed by the local operations staff on the Isle of Islay in the Inner Hebrides. There are two small treatment works in the middle of the island, which have circular activated sludge aeration units with a central settlement tank. The aeration unit had long retention times and so the sludge was a well-stabilised aerobic activated sludge.

The village of Ballygrant had a population of about 100 people and there was organic matter contributed by a small abattoir. The village of Kiells was located about 10 miles away and had a similar circular activated sludge plant. The sludge treatment and disposal arrangement was to take the waste sludge to the mainland at Campbelltown, Argyll by road tanker. In order to do this it was necessary to travel via two car ferries from Islay to the next island and from there to the mainland.

The only data available regarding the sludge and its rate of production rate was the MLSS in the aeration units of the activated sludge plants. This data showed sharp peaks and troughs in MLSS concentration. Since the sites were close together and the peaks and rapid reductions of MLSS occurred at the same times for the two works it seemed likely that there had been a washout of MLSS from the settlement tanks caused by storm events. The gradual recovery in MLSS concentration allowed the rate of growth of biomass to be calculated for both works. This allowed both the population equivalents [pe] for both works and the annual rate of excess sludge production to be calculated. This information allowed the two SDRBs to be accurately sized and designed. Figure 1 shows a plan view of the system at Ballygrant STW with the SDRB fitted into the limited space in the site. Figure 2 is the side elevation showing the media layers and the drainage and aeration pipes [installed to allow extra oxygen ingress to the lower layers of the bed].

These 2 systems were designed for an Activated Sludge Plant with 6,000mg MLSS/litre or 0.6%TS with a freeboard of

0.9m above the surface of the media. The system was designed to handle a dose of sludge about 5 cm deep twice each week. This bed was calculated to last for 1.8 years if only 90% reduction was achieved but 3.6 years if 95% reduction in volume could be achieved. In the event the Water Company staff were very rightly cautious with this new technology and decided to see how the Ballygrant system performed before building the system at Kiells.

The West of Scotland Water operators transported sludge from the site at Kiells to Ballygrant from until 1999 when they decided that the system at Ballygrant had worked well and decided to implement the design for Kiells. The Ballygrant system was planted in August 1995 and took sludge from Ballygrant and Kiells until 1998 when it had to be emptied after 3 years operation. The new Kiells system started operation in 1999. It is estimated that the Ballygrant system achieved a reduction of about 97%. The Ballygrant system was built by the local treatment plant operating staff to a design provided by the first author whilst at WRc.

TREATMENT OF AGRICULTURAL SLUDGES

Edwards and colleagues from Birmingham University and ARM Ltd (Edwards *et al*, 2000)[9] and Cooper *et al*, 2001 [7] reported on the treatment of sludges from agricultural treatment. The sludge from a BAF [Biological Aerated Filter] treating piggery wastewater was treated in a set of three experimental SDRBs. Two of the beds were planted with *Phragmites australis* and a third bed was unplanted for use as a control.

Table 5 shows the characteristics of the sludge being treated and the performance of the system is shown in Tables 5, 6, 7 and 8. The feed sludge was similar in solids concentration to a digested sewage sludge Table 5 and Table 6 shows that the reduction in volume achieved was about 84 to 86%. This is similar that reported by Nielsen 1993 [16] for 3 and 4%ww sewage sludges. It is only a few percentage points better than the Control unplanted sludge but it was much more stable sludge. The unplanted sludge from Bed C slumped quite badly but the sludges from the planted beds did not. Table 8 shows greater oxidation [mineralisation] of the sludges at depth i.e. improving mineralisation with time.

CONCLUSIONS

i] Planted SDRBs can dewater sludges arising from domestic sewage and agriculture wastewaters more effectively than unplanted beds.

ii] SDRBs are very appropriate to small works and should find more future application in the UK especially in remote difficult to service works. They seem very applicable to island communities since they allow local treatment. They have already been shown to work in this situation in Scotland.

iii] Work in Denmark has shown their application at large works where there are already two sites treating the sludges from greater than 120,000 people.

iv] The reduction in volume will depend upon the initial sludge concentration. For weaker sludges such as activated sludge with solids of about 0.6% DS the degree of reduction can be greater than 97% but for thickened sludges, such as anaerobically digested sludges, with 3 to 4 % DS the degree of reduction will be about 90%.

v] The final sludge concentration is usually in the range 20 to 40% DS. The 40% DS value may be achieved with surplus activated sludge.

vi] Loading rates of 60kgDS/m^2 year is quite safe however this is the spot value for a fully commissioned plant and does not take into account the time required for resting and emptying at the end of the filling cycle. When these periods are taken into account the overall loading rate may drop to 30 to 35 kgDS/m^2 year.

vii] The use of multiple beds will be required for large systems. This makes the commissioning, resting and emptying easier. Small systems with only one bed, for populations less than 100, have been successfully operated in the UK.

viii] Care has to be take not overload the beds in the first year and in particular in the Spring when high loading of raw sludges can be deleterious to new shoot growth. Some designers recommend growing the bed for up to a year before adding any sludge.

ix] Higher loadings may be used in summer, up to double winter loadings.

x] There is an urgent need for data to be collected from existing UK plants so that the design of these systems can be refined for future use.

xi] These systems are very appropriate for remote villages and can be built by the treatment plant operators or by a local building contractor.

REFERENCES

1. BITTAMANN, M. SEIDEL, K. *Entwaesserung und aufbereitung von chemieschlamm mit hilfe von pflanzen.* GWF (1967) vol. 108, No 18. 488-491.

2. BOON, A G. *Report of a visit by members and staff of WRc to investigate the Root Zone Method for treatment of wastewaters.* WRc report 376- S 1, WRc Stevenage, UK, August 1985.

3. BOON, A G. *Report of a visit to Canada and the USA to investigate the use of wetlands for treatment of wastewater.* WRc report 425 – S. WRc Stevenage, UK, June 1986.

4. BURGOON, P S. KIRKBRIDE, K F. HENDERSON, M. LANDON, E. *Reed beds for biosolids drying in the arid North Western United States.* Water Science and Technology (1997), Vol.35, No. 5. 287-292.

5. COOPER, P F. *"Design of Sludge Drying Reed Beds for Ballygrant STW and Kiells STW, Isle of Islay, Argyle Division, Strathclyde Regional Council"* WRc report UC 2591, Water Research Centre, Swindon, Wilts. November 1995. Also reported in summary in the next reference.

6, COOPER, P F. JOB, G D. GREEN, M B. SHUTES, R B E. *Reed beds and constructed wetlands for wastewater treatment* (1996). WRc Medmenham, Marlow, Bucks. UK. 202.

7. COOPER, P F. COOPER, D J. EDWARDS, J. BIDDLESTONE, J. *Treatment of sludges from sewage and agricultural wastewaters using Sludge Drying Reed Beds.* Paper presented to the 4th Workshop on Transformations of Nutrients in Natural and Constructed Wetlands, Trebon, Czech Republic, September 2001.

8. DE MAESENEER, J L. *Constructed wetlands for sludge dewatering.* Water Science and Technology (1997), Vol.35, No. 5. 279-285.

9. EDWARDS, J K. GRAY, K R. COOPER, D J. BIDDLESTONE, A J. WILLOUGHBY, N. *Reed Bed dewatering of agricultural sludges and slurries,* paper presented at the 7th International conference on Wetland Systems for Water Pollution Control, Orlando, Florida, USA, September 2000.

10. KICKUTH, R *Hoehere wasserpflanzen und gewaesserreinhaltung.* Schriftenreihe Vereinigung Deutscher Gewaesserschutz. (1969) EV_VDG, vol. 19. 3-14.

11. KIM, B J. CARDENAS, R. *"Use of Reed Beds for dewatering sludge in the USA"* Cooper, P F. Findlater, B C. [Editors] *Constructed wetlands in water pollution control,* Pergamon Press, Oxford, September 1990. 563-566

12. KIM, B J. *" Field evaluations of reed bed sludge dewatering technology: summary of benefits and limitations based on 4 years experience at Fort Campbell, Kentucky"* paper presented at the 67th annual Conference of the water and Environment Federation, Chicago, October, 1994.

13. LIENARD, A, DUCHENE, P. GORINI, D. *A study of activated sludge dewatering in experimental reed-planted or unplanted sludge drying beds.* Water Science and Technology (1995), Vol. 32, No. 3, 252- 261.

14. LIENARD, A. ESSER, D. DEQUIN, A. VIRLOGET, F. *"Sludge dewatering and drying in Reed Beds: an interesting solution? General investigations and first trials in France"* Cooper, P F. Findlater, B C. [Editors] *Constructed wetlands in water pollution control,* Pergamon Press, Oxford, September 1990. 257-267

15. NIELSEN, S M. *"Sludge dewatering and mineralisation in Reed Bed systems"* Cooper, P F. Findlater, B C. [Editors] *Constructed wetlands in water pollution control,* Pergamon Press, Oxford, UK, September, 1990. 245-255

16. NIELSEN, S. *"Biological sludge drying in constructed wetlands".* In constructed wetlands for water quality improvement (1993). G A Moshiri (Editor.), Lewis Publishers, Boca Raton, Florida, USA. 549-558.

17. NIELSEN S M *"Biological sludge drying in reed bed systems – six years of operations experience"* paper presented to the 4th International Conference on Constructed Wetlands for Water Pollution Control, Guangzhou, China, November, 1994.

18. NIELSEN, S M (2002) *"Sludge Drying Reed Beds"* paper presented at the 8th International Conference on Wetland Systems for Water Pollution Control, Arusha, Tanzania, 16 to 19 September, 2002.

TABLES

Table 1: Correlation between CST, area loading rate and number of beds (Nielsen, 2002).

CST (seconds)	Area loading rate(kg DS/m^2.year)	Number of beds
30	60	8
30 to 100	60	10
100 to 500	50	10
500 to 1,000	50	12 to 14
1,000	40	12 to 14
2,000	?	?

Table 2: Details of 6 reference plants in Denmark (Nielsen, 1993).

Town	Date of opening	pe	Sludge Load (Tonne TS)	Type of Sludge	No of Beds	Total Area m^2
Regstrup	8/1988	2,000	25	AS;DAS	4	415
Allerslev	9/1988	1,000	14	AS;DAS	2	215
Nakskov	11/1990	30,000	870	RS	10	11,600
Galten	11/1990	10,000	152	CS;AS	6	2,377
Gislinge	10/1991	3,520	42	AS	3	700
Rudkobing	1992	15,000	232	CS;AS	8	4,000

AS; Activated Sludge: 0.4%TS at Regstrup; 1% at Gislinge
DAS; Anaerobically Digested Activated Sludge: 3.0%TS at Regstrup and 4% at Allerslev
CS; Chemical Sludge/ Activated Sludge: 1%TS at Galten; 0.5 –1%TS at Rudkobing
RS; Raw primary sludge: 1-2 %TS at Nakskov
All the systems shown in Table 1 were planted with Common Reeds, *Phragmites australis*.

Table 3: Sludge Loading and reduction at Regstrup, Denmark, August 1988 to March 1991 (Nielsen, 1993).

Bed No.	Sludge Loaded m^3	Sludge Residue m^3	Reduction (%)
1	2313[a]	58	97.5
2	2132[a]	51	97.6
3	685[b]	71	89.6
4	2154[a]	56	97.4

[a] Dry matter content of sludge approx. 0.4%
[b] Dry matter content of sludge approx. 3%

Table 4: Sludge loading and reduction at Allerslev, Denmark, September 1988 to March 1991 (Nielsen, 1993)

Bed No.	Sludge Loaded m^3	Sludge Residue m^3	Reduction (%)
1	263[a]	18	93.2
2	253[a]	23	90.9

[a] Dry matter content of sludge approx. 4.0%

Table 5: Characteristics of the BAF sludge in the SDRB tests at the ARM Ltd site at Rugeley (Edwards *et al*, 2000).

Parameter	Mean Value
TS, w/w%	**4.07**
TSS, mg/l	31,364.00
Dissolved Solids, mg/l	4,645.00
VS, %	74.20
pH value	7.81
Specific Gravity	1.02
COD, mg/l	22,600.00
BOD_5, mg/l	7,000.00
NH_4N, mg/l	659.00
orthoPO_4P, mg/l	5,650.00

Table 6: Overall sludge loadings and residue heights in the tests at Rugelely, May 1997 to February 1999 (Edwards *et al*, 2000).

	Bed A	Bed B	Bed C
Sludge added, m^3	**7.260**	**8.460**	6.870
Height of sludge added, m	2.200	2.560	2.080
Final residue, m^3	1.330	1.020	1.340
Height of final residue, m	0.345	0.352	0.393
Reduction in height, %	84.300	86.300	81.100

Beds A and B were planted but Bed C was an unplanted Control bed.

Table 7: Mean concentrations of the percolates from the 3 beds at ARM Ltd Rugeley, 5 March to 18 August 1998 (Edwards *et al*, 2000).

Parameter	Bed A	Bed B	Bed C
COD, mg/l	1,822.00	**2,220.00**	1,962.00
BOD_5, mg/l	1,123.00	1,468.00	935.00
TSS, mg/l	809.00	917.00	770.00
Dissolved Solids, mg/l	3,674.00	4,055.00	3,290.00
pH value	8.02	7.65	8.17
NH_4N, mg/l	471.00	589.00	474.00
orthoPO_4P, mg/l	560.00	590.00	583.00

Table 8: Analysis of final sludge cores at Rugeley taken 25 February 1999 (Edwards *et al*, 2000).

	Bed A		Bed B		Bed C	
Depth	TS%	VS%	TS%	VS%	TS%	VS%
0-100mm	20.77	54.38	23.46	52.11	19.03	54.55
100-200mm	20.82	53.41	19.37	53.59	17.47	51.96
200-300mm	18.24	51.21	21.09	51.79	17.25	52.15
300-400mm	22.09	50.99	22.11	51.90	20.85	51.46
Mean Value	20.98	52.50	21.51	52.35	18.65	52.53

FIGURES

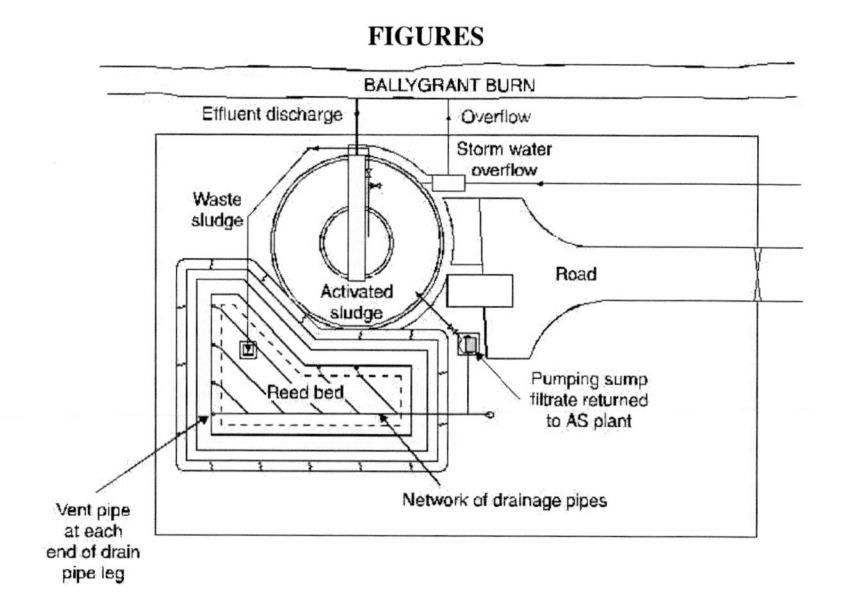

Figure 1. Layout of the SDRB at Ballygrant STW, West of Scotland Water, Isle of Islay (Cooper *et al*, 1996).

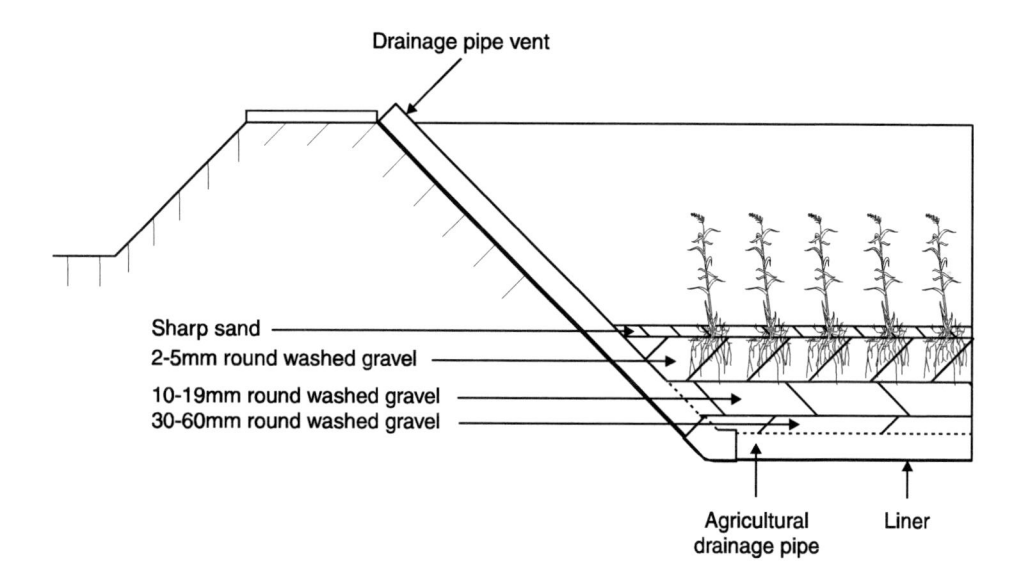

Figure 2. Side elevation of the SDRB at Ballygrant STW, Isle of Islay, West of Scotland Water showing the structure and the drainage and aeration pipework (Cooper *et al*, 1996).

PERMANGANATE APPLICATIONS AT PUBLIC WORKS, INDUSTRIAL AND AGRICULTURAL SITES

Philip Vella
Carus Chemical Company, LaSalle, IL USA
Tel: +1 815 224 6869, Fax+ +1 815 224 6841
Email phil.vella@caruschem.com

ABSTRACT

Hydrogen Sulphide (H_2S) is a common problem in both municipal and industrial wastewater collection and treatment systems. This colorless gas, known for its rotten egg smell, is produced by the reduction of sulfates and the decomposition of organic material under anaerobic conditions. It forms in all areas of a system from collection lines, force mains, lift stations, holding tanks, dewatering presses and drying beds. There are a number of chemical treatment options that can be used to control H_2S. These include oxidants, precipitants, and biological methods. Although potassium permanganate ($KMnO_4$) has been applied successfully for the treatment of H_2S and other odours in the United States for over 25 years, it is only recently that is has been gaining favor in the European community.

This paper will give an overview of sulphide and other related odour problems commonly found at public works, farm wastes, industrial sites, etc. and how potassium permanganate can solve these unwanted problems. In addition to the common crystalline form of potassium permanganate, a new controlled release permanganate, and a recently formulated liquid permanganate, and their potential in these applications, will be presented. To show the effectiveness of permanganate, a case study on odour control and dewatering improvement at a site in the UK will be shown.

KEY WORDS

Biosolids, corrosion, hydrogen sulphide, , odour, oxidation, potassium permanganate, scrubbing.

INTRODUCTION

The application of potassium permanganate for municipal wastewater treatment did not begin developing until the early 1980s. As the general public became more aware of the odours emanating from the collection, treatment, and disposal of sanitary waste, they applied pressure on utilities to control these odours. As utility employees became more aware of the dangers of being exposed to the toxic gas hydrogen sulphide (H_2S), they applied more pressure on the utilities to minimize their risk. As some of the technology of biosolids dewatering shifted from high pH or iron salts used with centrifuges to polymers and belt presses, more sulphide gas was being emitted. All of these factors, plus the residential growth around wastewater treatment plants, led to a greater need on the part of the utility to control hydrogen sulphide, the most prevalent odour source. One of the stronger reductants found in wastewater, hydrogen sulphide is usually generated in the collection and treatment processes when naturally occurring sulfates are reduced by bacteria in the system to hydrogen sulphide. This H_2S is not only odorous (rotten egg odour), but is also toxic and corrosive. At 10 ppm (V/V) in air, the odour of H_2S is extremely offensive. At 300 ppm it poses a definite health hazard and could be life threatening.[1] This problem can occur in the collection system and in the plant unit processes, but the worst area is the biosolids storage and dewatering process.

While not the only odorous chemical found in wastewater treatment plants, it is the most dangerous and yet easiest to control through chemical oxidation. Potassium permanganate has been proven to be one of the more effective oxidants that can be applied to quickly and cost-effectively control H_2S.[2] In 2001, 88% of the $KMnO_4$ used in municipal wastewater treatment systems was applied to control sulphides and other odorous chemicals generated and released from the biosolids dewatering process (Figure 1).[3]

Of the total, only 9% of the potassium permanganate sold to the municipal wastewater treatment market was used in the collection system and in the plant. About 3% was used for miscellaneous and unidentified applications.

PERMANGANATE CHEMISTRY

The chemistry of sulphide oxidation is relatively straight forward (Figure 2). Potassium permanganate reacts with sulphides to form a variety of end products ranging from elemental sulfur to sulfate. These possible reactions show that the $KMnO_4$:H_2S ratio can range from 3.1:1 for oxidation to elemental sulfur, to 12.4:1 if the oxidation end product is sulfate. In practice the ratio for odour control usually ends up with about 4-6 parts (w:w) of $KMnO_4$ for each part of dissolved H_2S. This ratio indicates a mixture of oxidation products ranging from sulfur to sulfate.[2]

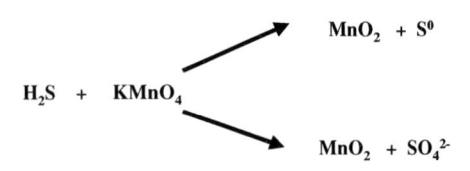

$$H_2S + KMnO_4 \nearrow MnO_2 + S^0$$
$$\searrow MnO_2 + SO_4^{2-}$$

Equation 1. Potassium permanganate –

$$3H_2S + 2KMnO_4 \longrightarrow$$
$$3S^0 + 2H_2O + 2KOH + 2MnO_2 \quad \text{3.1:1 w:w}$$

$$3H_2S + 8KMnO_4 \longrightarrow$$
$$3K_2SO_4 + 2H_2O + 2KOH + 8MnO_2 \quad \text{12.4:1 w:w}$$

hydrogen sulphide reaction

Equations 2-3. Potassium permanganate – hydrogen sulphide reactions

BIOSOILDS SULPHIDE OXIDATION

The solids that are separated from the sanitary sewage in the primary and secondary treatment processes are collected, concentrated, and dewatered before disposal. The dewatering process can consist of centrifuges, plate and frame presses or belt filter presses. Prior to 1980, the chemicals used in dewatering included lime, some polymers, and iron salts. During the treatment of this sludge or biosolids, the pH was raised high enough with lime so that the sulphides were not released. Newer higher rate belt filter presses were introduced to reduce the volume of lime sludge produced. These operate at a lower pH with no lime or ferric salts. However, the lower operating pH results in the release of sulphides.

It has also been theorized that high concentrations of sulphides can interfere with the dewatering efficiency and higher dosages of polymer may be required. The application of potassium permanganate for sulphide oxidation has also been proven to reduce the amount of polymer needed for good dewatering and to increase the through-put of sludge through the dewatering process equipment (Table 1).

The results shown in Table 1 indicate that the addition of $KMnO_4$ at a concentration to control the sulphides, reduced the polymer usage by almost 15% while increasing the through-put by almost 2%. The addition of hydrogen peroxide did not show the same improvement. In some cases this reduction in polymer cost will compensate for the cost of the permanganate treatment.

COLLECTION SYSTEMS AND IN-PLANT SULPHIDE CONTROL

Hydrogen sulphide can build up in collection systems due to long transmission lines where sulfate reducing bacteria react with the sulfates in the sewage. In addition to causing odours, the released H_2S can react with the moisture in the air and form corrosive sulfuric acid. This can lead to a deterioration of metal pipes causing failures in the collection system. This toxic gas can also be released in pumping stations and at the headworks of the treatment plant. Potassium permanganate can be applied to the system at convenient feed points to oxidize the H_2S to control odours and minimize corrosion.[4] An example of the effectiveness of potassium permanganate when compared to other commonly used treatment technologies is given in Figure 2. As shown, potassium permanganate can provide more effective odour control at significantly lower application rates.

LIVESTOCK ODOR CONTROL: CHEMICAL OPTIONS

With increasing population density, the odors resulting from livestock facilities is a growing concern. Of specific interest is the odors generated from swine production units. Recent research conducted by C. J. Clanton et al[5] compared the application of a wide variety of chemicals for the control of H_2S emissions. The chemical properties ranged in affect from pH adjustment to precipitation to oxidation. The chemicals that were tested included calcium hydroxide (lime $Ca(OH)_2$), ferric chloride ($FeCl_3$), ferrous chloride ($FeCl_2$), ferrous sulfate ($FeSO_4$), hydrogen peroxide (H_2O_2), potassium permanganate ($KMnO_4$), and sodium chlorite ($NaClO_2$).

The results of the study are presented in Table 2. As shown, permanganate was slightly more cost effective than peroxide under the conditions used and quantities of chemicals purchased.

A NEW PERMANGANATE PRODUCT: EXTRUDED POTASSIUM PERMANGANATE

There are a number of chemical treatment options that can be used to control H_2S. These include oxidants, precipitants, and biological methods. Most chemical applications to eliminate odours require the use of sophisticated mechanical feed equipment. This becomes a limiting factor when dealing with remote locations such as lift stations and manholes. The use of mechanical feed equipment, even a simple dosing pump, requires electricity, and in some cases, a source of clean water.

Although $KMnO_4$ has been applied successfully for the treatment of H_2S, it generally requires feed equipment. To overcome this need, a new material has been formulated that is a self-contained controlled release oxidant product. CAIROX-CR™ extruded potassium permanganate[6] is a new form of controlled release permanganate. Classified as an oxidizing solid, this product is specifically formulated for applications at municipal and industrial wastewater sites where the use of feed equipment may not be practical. It is used primarily for the control of odours and corrosion associated with hydrogen sulphide. This new product eliminates the need for feed equipment and its associated utilities. It can be used for remote locations and requires little or no routine maintenance.

CAIROX-CR™ extruded potassium permanganate is applied through a specially designed polyester mesh bag. The bag is filled with the extrusions and then lowered into the waste stream. Lowering more of the bag into the wastewater increases the amount of product fed into the system. As original material dissolves, the amount of product consumed can be measured by loss in weight and additional product can be added to the bag.

CASE STUDY: PERMANGANATE IMPROVES SCRUBBER OPERATION

A treatment works located in the southern UK currently uses a chemical scrubbing system as its primary odour control system. This facility processes both blended and works sludge. The scrubbing towers use sodium hypochlorite as the odour control chemical to remove H_2S from the exhaust air of the centrifuge and centrifuge room. Due to design limitations, the sodium hypochlorite dosing system cannot maintain a satisfactory chlorine residual in the towers when the centrifuge is in operation. The desired operating level of 1000 mg/L (as chlorine) can fall to 250 mg/L within 20 minutes during sludge processing. This results in a subsequent odour release in the centrifuge room. Depending on the sludge being processed, H_2S levels in the room can reach as high as 25 ppm that can result in serious health affects for the staff. In order to maintain the desired chlorine residual, excessive sodium hypochlorite has to be used resulting in chemical deliveries every other day. This was not acceptable to the works and alternatives to the current treatment system were investigated. Over the next 24 hours, H_2S readings ranged from 0-12 ppm (down from 82 ppm), resulting in a 74% drop in average H_2S levels compared to readings obtained in the spring (Figure 4).

Based on results obtained at other treatment works, it was decided to investigate the use of AQUOX® potassium permanganate to assist the towers in controlling odours. AQUOX® was injected as a 1% solution and dosing was controlled by linking the chlorine residual in the towers to the permanganate dosing pump. The results on hydrogen sulphide levels are shown in Figures 3 and 4. With the addition of the AQUOX® solution, the hypochlorite consumption was reduced resulting in significant hypochlorite cost savings as shown in Figure 5. The combined cost of both hypochlorite and permanganate is shown in Figure 6. As seen, the cost for treating the blended sludge with AQUOX® is the same but odours were significantly reduced. The works sludge, being more septic, required more permanganate. This increased the overall cost slightly but solved the odour problem resulting in >97% reduction in H_2S levels. In addition, higher sludge throughput was achieved with no increase in polymer usage.

From the data obtained, the application of AQUOX® potassium permanganate in this scrubber application:

1. Provided a cost-effective odour control solution to the Public Works
2. Reduced excessive sodium hypochlorite consumption
3. Allowed operation of centrifuges on works sludge with the existing odour control system
4. Lowered H_2S concentrations from the centrifuge, sludge cake and centrate
5. Increased centrifuge throughput with no increase in polymer usage, for additional cost savings
6. Reduced H_2S emissions during cleaning of plugged centrifuge hopper
7. Reduced corrosion of equipment and facilities
8. Improved workers' health and safety

CONCLUSIONS

Potassium permanganate has numerous applications in wastewater treatment from collection systems, in-plant applications to bio-solids improvement. It can be used for corrosion and odour control as well as destruction of hazardous and toxic compounds. In wastewater, the reactions of potassium permanganate are faster than other oxidants and control of hydrogen sulphide reduces odour complaints, minimizes corrosion, improves the dewatering operation, and provides for a safer environment for operators. Potassium permanganate is a truly versatile oxidant and the answer to many wastewater problems.

REFERENCES

[1] Odour and Corrosion Control in Sanitary Sewerage Systems and Plants. EPA/625/1-85/018 Design Manual 1985

[2] Ficek, K.J. Potassium Permanganate Controls Sewage Odours. Carus Chemical Company, Peru, IL Reprint. Form #310

[3] Carus Chemical Company Internal Report. 2000

[4] Vella, P. Potassium Permanganate Applications in Wastewater Treatment. Presented at the Workshop on Principles and Practices of Chemical Oxidation in Wastewater Treatment, sponsored by ICOA and Vanderbilt University, February 1995.

[5] Clanton, C. J. Nicolai, R. E. Schmidit, D. R. Presented at the ASAE Annual International Meeting, Toronto, Ontario, Canada, 1999.

[6] CAIROX-CR™ extruded potassium permanganate is a trademark of Carus Chemical Company

TABLES

Table 1. Test results comparing permanganate to peroxide

Chemical	Polymer Feed	Sludge Flow	Polymer Usage
None	51	112	12.3
H_2O_2	51	111	12.3
$KMnO_4$	42	114	10.5

Note: all values in the above Table are unit-less

Table 2. Chemical Cost Analysis for Swine Odor Control

Chemical	Cost/lb[1]	Amount Used[2] g/M^3	Cost Euro/M^3
Calcium hydroxide ($Ca(OH)_2$)	0.374	622	0.51
Ferric chloride ($FeCl_3$)	0.69	6260	23.81
Ferrous chloride ($FeCl_2$)	0.55	238	1.03
Ferrous sulfate ($FeSO_4$)	0.60	221	1.04
Hydrogen peroxide (H_2O_2)	1.09	30.7	0.074
Potassium permanganate ($KMnO_4$)	1.81	15.5	0.062
Sodium chlorite ($NaClO_2$)	3.90	150	1.29

1. Obtained from a local supplier in quantities expected to be used by livestock facilities
2. Amount needed for 50% reduction in H_2S emission.

FIGURES

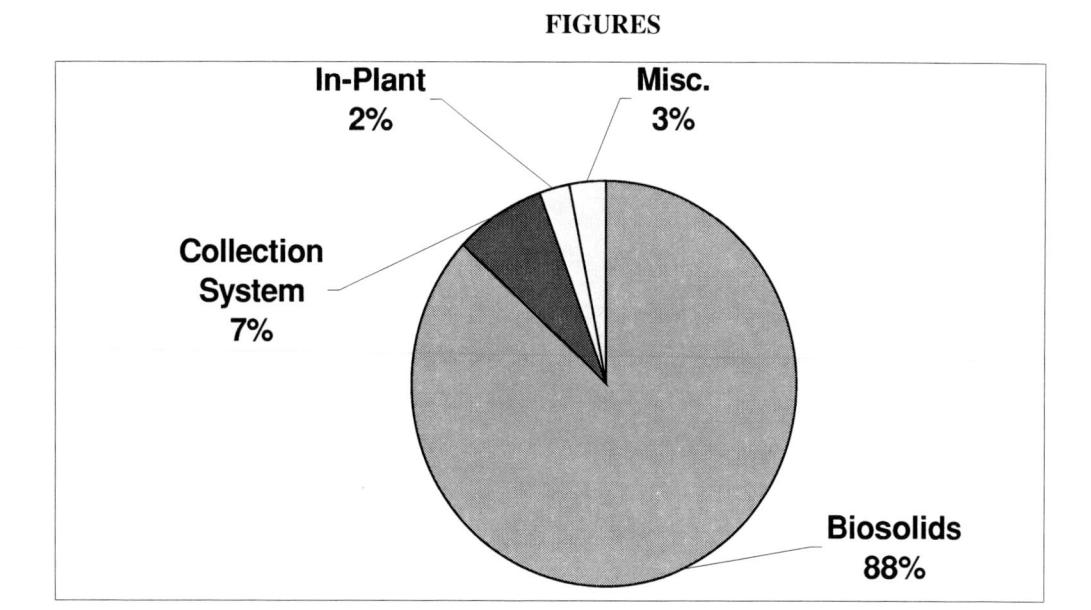

Figure 1. Primary 2001 applications of potassium permanganate in municipal wastewater

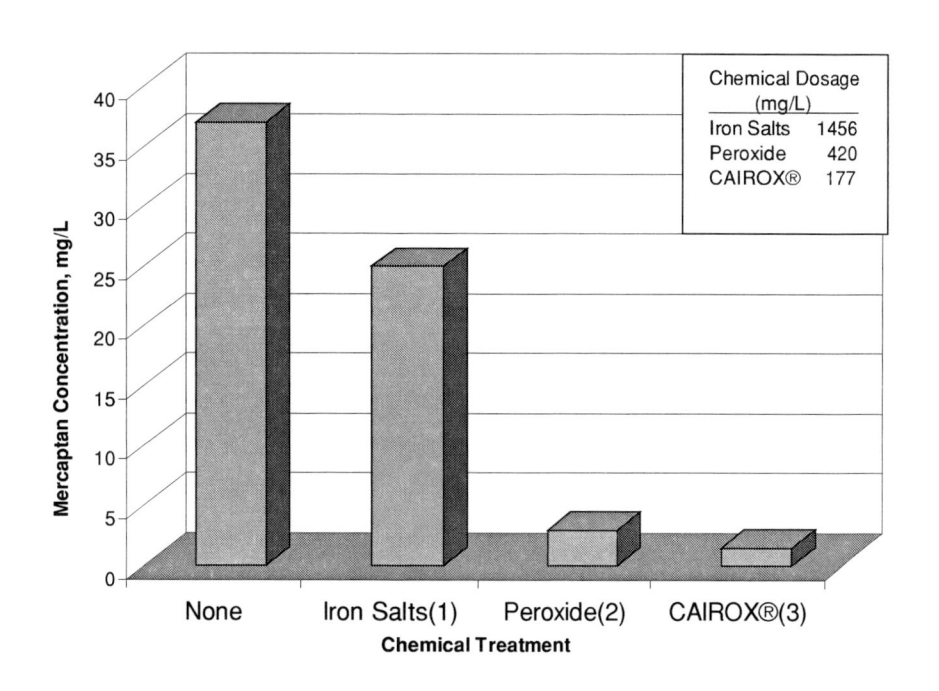

Figure 2. Technology comparison for odour control

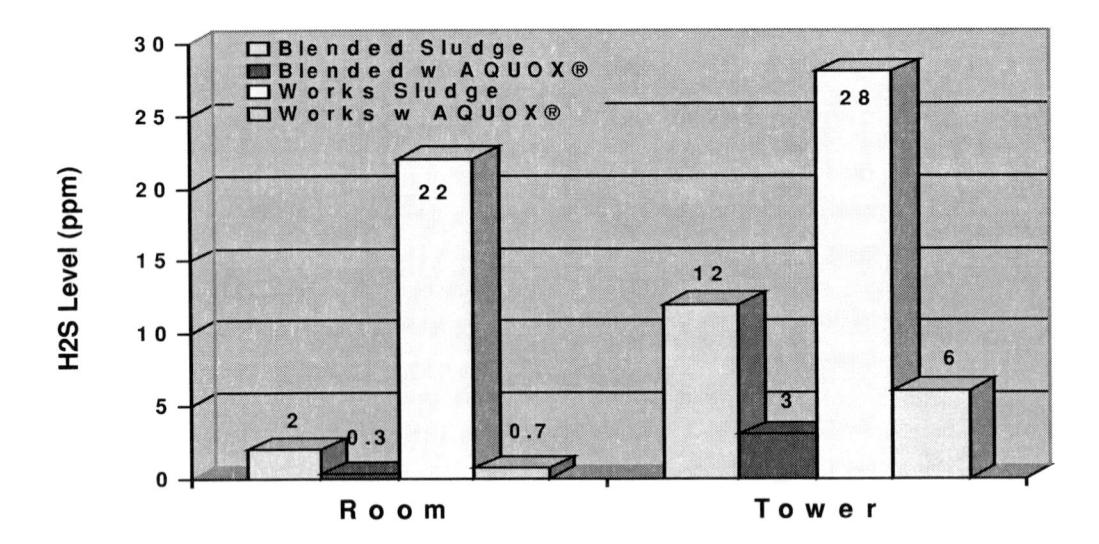

Figure 3. Effects of AQUOX® on H_2S levels: Centrifuge Room and Tower Inlet

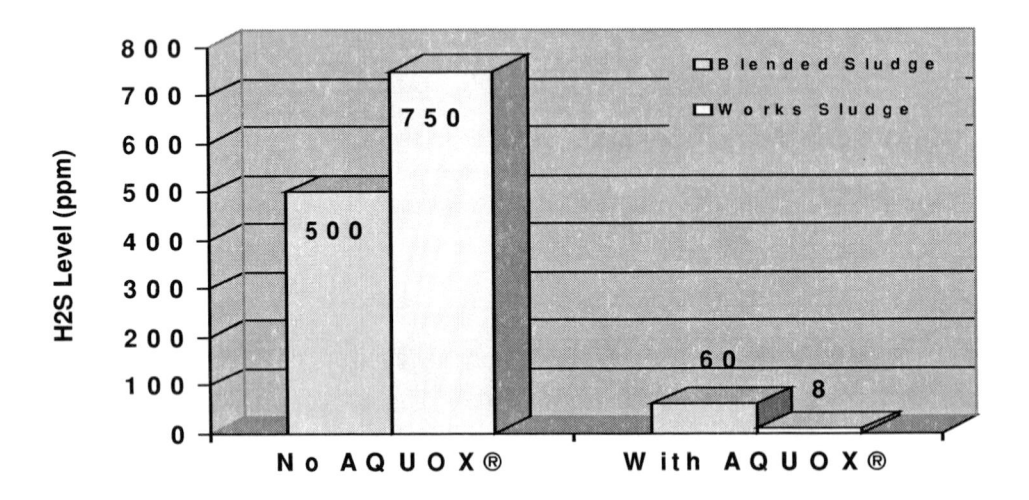

Figure 4. Effects of AQUOX® on H_2S levels: Centrifuge Extract Air

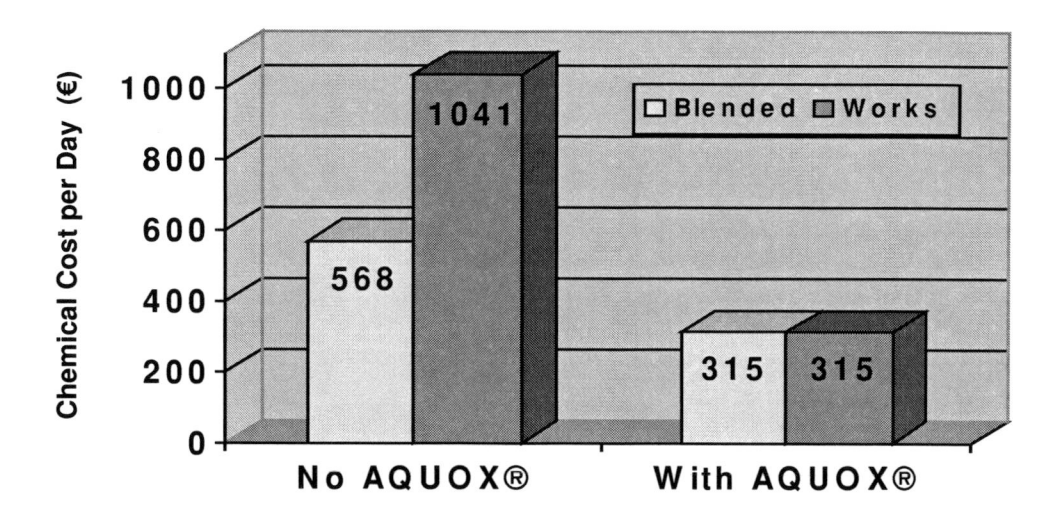

*AQUOX® solution dosed at 171 mg/L for Blended and 600 mg/L for Works Sludge

Figure 5. Chemical Cost of Hypochlorite Alone

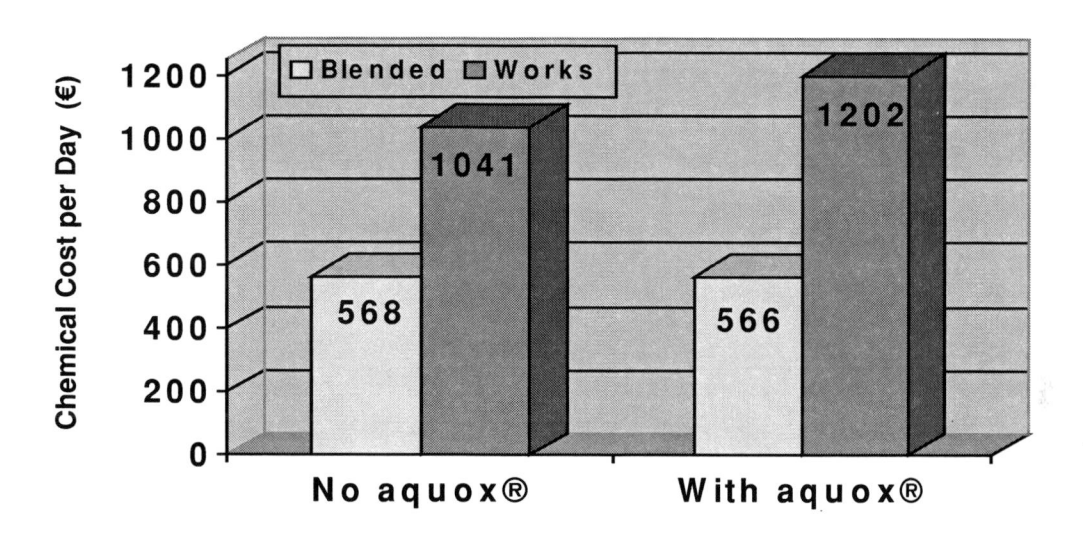

*AQUOX® solution dosed at 171 mg/L for Blended and 600 mg/L for Works Sludge

Figure 6. Chemical Cost of Hypochlorite & AQUOX®

CHEMICAL INDICATORS THAT DETERMINE OLFACTORY RESPONSE FROM ORGANIC SOURCES

Hobbs,P.J.[1], Misselbrook,T.H.[1], Noble R.[2], & Persaud K.C.[3],

[1]Institute of Grassland and Environmental Research, North Wyke, Okehampton, Devon, EX20 2SB.
[2]Horticultural Research Institute, Wellesbourne, Warwick, CV35 9EF.
[3]Department of Instrumentation and Analytical Science, UMIST, Manchester, M60 1QD.

ABSTRACT

Determination of a human olfactory response to livestock odours by instrumental means has been sought to simplify on-site measurement and reduce cost. A new approach has been developed using volatile components to identify the olfactory response from different manure types. The chemical concentration of the odorants was determined in the headspace of pig manure. The headspace of chicken manure composted with straw was determined using thermal desorption -gas chromatography- mass spectrometry with the intention of finding marker compounds indicative of the olfactory response. The major odorous compounds were identified as those significantly greater than their odour threshold values. An equation for each manure odour was determined and able to predict the odour concentration. For odour from pig manure the predictive compounds, that are also odorants, were H_2S, 4-methyl phenol and acetic acid. H_2S, dimethyl sulphide, butanoic acid, methanethiol and trimethylamine were the significant odorants emitted from chicken manure. For composting chicken manure the odour response was proportional to the sum of the combined concentrations of H_2S + DMS. Synthetic odour mixtures were produced in the concentration ranges of the selected odorants for each manure type. Models were developed to explain variations of olfactory response using a trained odour panel. Sensitivity tests surprisingly revealed that 4-methyl phenol could reduce the olfactory response in certain circumstances for pig manure odours. Multiple linear regression and interpolative neural network approaches were used and the merits of both are discussed. Direct and rapid measurement of the sulphides on the chicken manure composting site was possible using gas detector tubes. There was a close correlation between the odour concentration of the odour samples and, measured on-site with gas detector tubes.

KEYWORDS.

GC-MS, livestock manure, modeling, Odour, odour analysis olfactometry.

INTRODUCTION

Measurement of odours by either olfactometry or chemical analysis of volatile compounds has proved difficult because they are often near or at at their threshold of odour detection and these concentrations are in the parts per billion range $ppb(v)$ (10^{-9}). Interpretation of an instrumental measurement as a sensory response requires knowledge of the chemical composition of an odour that requires advanced analysis. Concentrating the odorants is necessary at these low concentrations followed by thermal desorption into a chromatography coupled with mass spectrometry system to identify and quantify emissions from manure[1]. They have been recognised as containing mostly sulphides, ammonia, volatile fatty acids, phenols and indoles. Fifteen odorants have been identified as major components contributing to the odour of livestock wastes[2], but at different concentrations for different livestock wastes. This was confirmed by the production of synthetic odour samples(unpublished).

In this experiment, we aimed to determine the olfactory response from mixtures of odorants commonly found in pig[3] and chicken manure[2]. Secondly synthetic mixtures were used to demonstrate that the correct approach to develop the models. The applicability of the models was evaluated using data obtained from both chicken and pig manure odours.

EXPERIMENTAL

The concentration of each of the major odorants were identified by GC-MS for pig and chicken manure sources. To build an interpretive model a mathamatically design was necessary. A second order uniform precision rotatable central composite design was used which essentially determines the most effective way to model the response as OC in terms of the best accuracy and precision[4].

A range of concentrations of hydrogen sulphide, 4-methyl phenol (fmp), ammonia and acetic acid were produced in tedlar sample bags to develop a model for the pig manure odour. Olfactory analysis was performed with an odour panel containing 8 people, selected according to recognised European criteria to determine the odour concentration. Odour concentration is the dilution factor necessary for a sample to achieve threshold odour concentration which is statistically percieved by 50% of the odour panel.

Multiple linear regression (MLR) was used to generate a model of the four odorant concentrations as explanatory variables for OC. A newly produced radial basis function neural network (RBFNN) was employed to allow complex functional forms to be modeled. Validation of the models was performed using odours from each of the manure types.

RESULTS
ODOUR FROM COMPOSTED CHICKEN MANURE

Odorants from the composted chicken manure that were found to exceed their detection thresholds by the greatest order of magnitude were, in decreasing order: H_2S, DMS, butanoic acid, methanethiol and trimethylamine[5]. Concentrations of NH_3 were generally above the detection threshold in most of the samples but did not affect OC. There was considerable variation in the relative difference between stages of composting in terms of OC and H_2S, DMS and NH_3 concentrations. In this case the model was derived directly from the samples and was best described as shown in Figure 1.

Measuring H_2S and DMS accounted for 45% (p=0.05) of the variance in OC from poultry manure compost after the exclusion of 2 outlier points. Neural network analysis identified 16 hidden nodes from the 24 data points and accounted for 75% of the variance for four odorant concentrations.

ODOUR FROM PIG MANURE

H_2S and FMP concentrations accounted for the majority of the 74.1% variance of the model. Acetic acid and ammonia concentrations had a marginal and insignificant influence respectively. The model with standard errors for each variable is expressed in the equation

$$OC = 950 + 2630.[H_2S] - 25617.[fmp] + 179.[acetic]$$

This compared to 77% for the multiple linear regression analysis using H_2S, fmp and acetic acid concentrations. Neural network analysis identified 16 hidden nodes from the 24 data points and accounted for 88% of the variance for four odorant concentrations.

Figure 2 shows the applicability of the model to samples where the odour concentrations and chemical concentrations were measured. 76% of the variance was accounted for, however the odour concentrations were 20% less than predicted indicating other odour components were contributing to the olfactory response[6].

DISCUSSION

There are some general phenomena that describe the olfactory response from pig and chicken manure. H_2S was a primary odorant and necessary to modelling the relationship to olfactory response. Ammonia was below its odour threshold concentration and had no effect on the OC of the odorant mixture nor in the description of the olfactory reponse.

For the pig manure odours 4-methyl phenol gave a negative OC effect with increasing concentration, which was not identified statistically as a crossover effect within the model. The multiple linear regression model utilises H2S, 4-methyl phenol and acetic acid. The concentration of the individual odorants did not follow an additive, geometric or average olfactory prediction. Radial basis function neural network software was able to improve upon multiple regression model to describe the OC with changing odorant concentrations. The one major limitation was that the the neural network model was only successful for OC values within the range of the model that is the range of the training database. However if we can limit the number of nodes used in the neural network we could approach a more normalised model that would be close to principal component analysis.

Only composted chicken manure was investigated where the major components were identified as hydrogen and dimethyl sulphides, methanethiol, butanoic acid and trimethylamine. However the model only included H_2S and dimethyl sulphide, where there was a close correlation between the OC of composting odour samples and the model adding H_2S + DMS.

CONCLUSION

The multiple linear regression approach was able to model the OC from pig manure odours and composted chicken manure. Our approach demonstrated that we have selected the appropriate volatile components that are also odorants to predict the OC for both types of manure.

The neural network approach operated well within the bounds of the training data presented to the process, however if the concentration of samples presented to the model to predict the odour concentration were outside of the range of the training data there was no correlation between the model and the olfactory concentration measured.

REFERENCES

1. Hobbs PJ. Odour analysis by gas chromatography. In: Stuetz RM, Frechen FB, eds. *Odours in Wastewater Treatment.* London: IWA publishing, 2001; First edn. 155-175.

2. Hobbs P J, Misselbrook T H and Pain B P, Assessment of odors from livestock wastes by a photoionization detector, an electronic nose, olfactometry and gas-chromatography mass-spectrometry. *Journal Of Agricultural Engineering Research* **60**: 137-144. (1995)

3. Hobbs P J, Misselbrook T H and Pain B F, Characterisation of odorous compounds and emissions from slurries produced from weaner pigs fed dry feed and liquid diets. *Journal Of The Science Of Food And Agriculture* **73**: 437-445. (1997)

4. Myers R.H. In: Response surface methodology, Allyn and Bacon Inc., Boston (1971)

5. Noble R, Hobbs P J, Dobrovin-Pennington A, Misselbrook T H and Mead A, Olfactory response to mushroom composting emissions as a function of chemical concentration. *Journal Of Environmental Quality* **30**: 760-767. (2001)

6. Hobbs P J, Misselbrook T H, Dhanoa M S and Persaud K C, Development of a relationship between olfactory response and major odorants from organic wastes. *Journal Of The Science Of Food And Agriculture* **81**: 188-193. (2001)

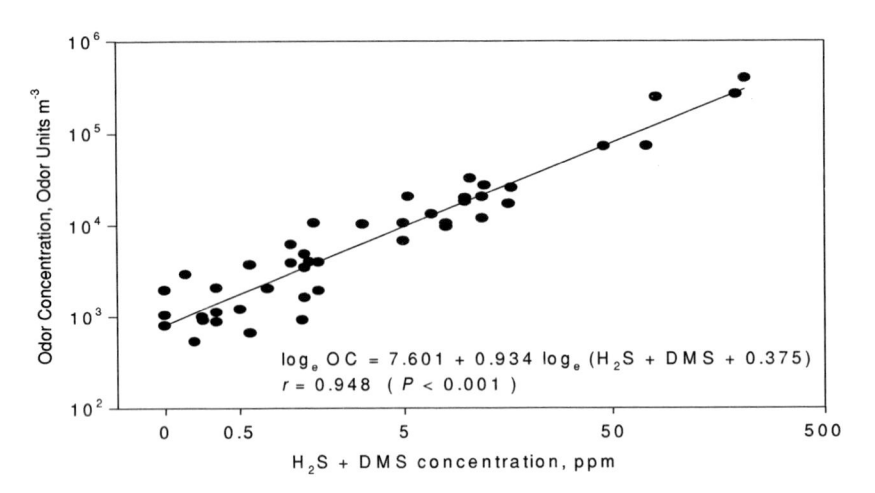

Fig. 1 Relationship between the combined on-site hydrogen sulphide and dimethyl sulphide concentrations and the 24-h odor concentration of mushroom composting odor samples. Each point is the mean of two sample determinations.

$\log_e OC = 7.601 + 0.934 \log_e (H_2S + DMS + 0.375)$
$r = 0.948 \ (P < 0.001)$

Figure 2 testing the pig manure odour

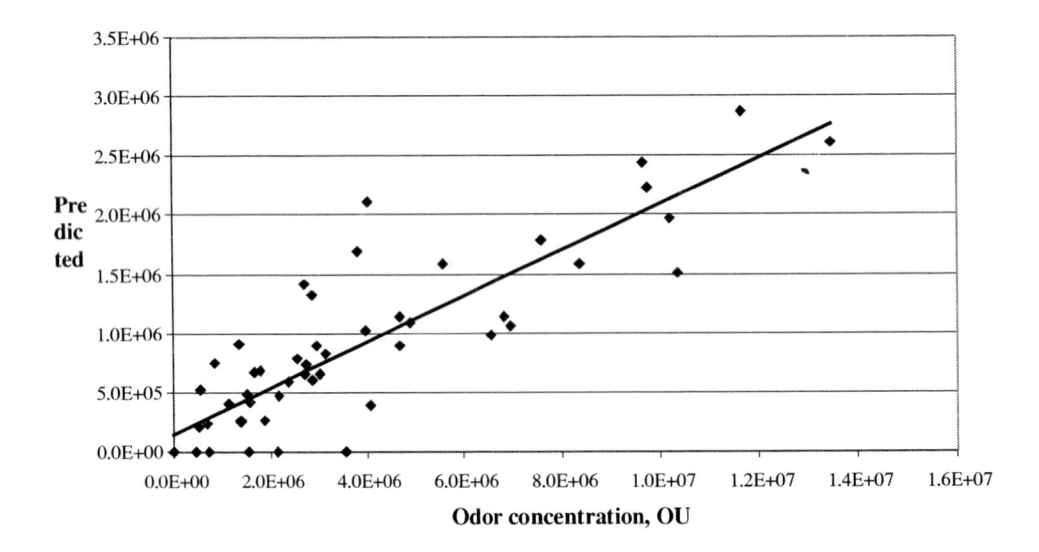

THE SIGNIFICANCE OF THE SPEED OF ROTATION AND WHOLE LIFE COSTS IN RBC TENDER EVALUATION

G E Findlay
COPA Ltd
Tel:+44 121 543 4800 Fax+44 121 543 4819. E-mail: eric.findlay@copawb.co.uk

ABSTRACT

Cranfield University sponsored by Severn Trent Water Ltd have researched the reasons for RBC failure.[1] Severn Trent Water Ltd has implemented those recommendations and as a result the performance of RBCs has improved. The Cranfield initiative has altered the Industry's view on RBCs [2] and has effectively resulted in a totally new approach to RBC design and selection.[3] Most Water Companies would agree they are fully conversant with the application of Whole Life Costs in option and possibly tender evaluation. The release of information following the Severn Trent / Cranfield Initiative has shed a new light particularly on the area of contract selection not previously considered as well as other non-process areas.

The paper considers the importance of the effect of speed of rotation and discounted cash flow techniques on the evaluation of tenders suggesting that such an evaluation is absolutely imperative if a robust process is to be achieved. It suggests that failure to carry out a proper analysis can lead to the wrongful selection resulting in short life failure and dissatisfaction with the process performance. The paper gives guidance on this matter.

KEYWORDS:

Fatigue, Beggiatoa, Microbiologically Induced Corrosion, Rotating Biological Contactors, Whole Life Costs.

INTRODUCTION

Whole-life Costing (WLC) is a technique developed for identifying and quantifying all costs, initial and ongoing, associated with a project or installation over a given period usually its life. Whole-life costs use the standard accountancy principle of discounted cash flow analysis (DCF), so that total costs incurred during a life cycle period are reduced to present day values. This allows a realistic comparison to be made of the options available.

A full life cycle analysis enables management to consider the full implication of total cost both in terms of both actual monetary value, and inconvenience of future maintenance and replacements. Experience has shown that future maintenance and the associated downtime costs can have a major effect on the capital decision.

The concept of discounting cash flow techniques is well used within the water industry. It enables the identification of the prefer option where several possibilities exist. A typical example might be the comparison of pumping to a sewage treatment works where adequate capacity exists as opposed to gravitating to an existing installation, which requires extension.

Once the preferred option is selected the onus is on the Engineer to ensure value for money is achieved by pursuing that self same principle. In many instances this is adopted but unfortunately not in every case and not with the full information available. The engineer is invariably faced with an uncertainty in terms of operational and maintenance costs to enable him to evaluate tenders properly due to lack of adequate operational experience. This can be particularly difficult where an RBC is being selected for this first time.

APPLICATION TO RBC'S

RBCs fail primarily as a result fatigue. [4] Failure occurs when the number of cycles on the appropriate S/N curve coincides with the design fatigue stress. If the Severn Trent Specification is adopted this should occur at 1.E+ 07 cycles at a biomass thickness of five mm and a maximum fatigue stress of 25 MPa.

Bending a paper clip back and forward is an interesting way of simulating failure. The clip will fail once its fatigue life has been exhausted. The period to failure is greater where the clip is more robust. RBCs are subjected to a similar stress reversal process. Regardless of the metal content because of fatigue they will all fail at some time. The period to failure is related to the amount of metal content, the quality of the design and the minimisation of stress raisers during manufacture and assembly. These requirements were carefully considered by Cranfield University when they designed and specified the requirements COPA have to achieve to ensure a 20 year life.(4)

In order to comply with EU regulations it was necessary to accept tenders from more than one supplier. Suppliers who wished to tender to Severn Trent during Asset Management Plans II investement period (AMP 2) were invited to submit details of their plant for audit by Severn Trent Engineering and Cranfield University, School of Engineering

The suppliers were asked to indicate the speed of rotation to which their plant was designed and the audit was conducted only at that speed. Not all plants were acceptable consequently the number of potential Tenderers was significantly reduced. One of those deemed acceptable chose not to tender. Only two companies regularly tendered during AMP 2 investment programme.

Unlike other plants currently available the rotors provided by COPA Ltd were specifically designed by Cranfield to meet both the Severn Trent Engineering and the strict Cranfield standard and are alone in that respect. [4,5] Where possible plants were modified "insitu" at the suppliers expense. Where a plant had failed and replacement by the same type was not possible the rotor was replaced by one from an approved supplier and costs recovered. Over £800,000 was recovered from suppliers. This took into consideration usage prior to failure and betterment.

Counsel advised that under the UK G/90 Standard Conditions of Contract (suitable for mechanical equipment supply) the supplier was only liable if failure occurred within 3 years of Take Over regardless of the specified time (or in Severn Trent's case six years). Consequently, several writs were issued to prevent any action becoming Statute Barred where there was reasonable evidence of possible failure whilst the supplier modified his plant to meet Cranfield's requirements.

Recently a plant provided by one of the approved suppliers and modified "insitu" has proven to be less than successful due to poor fabrication, which has led to the failure 10 years earlier than predicted. This gives rise to concern about the quality of other RBCs in the same category.

The Cranfield audit was at the speed of rotation normally employed by the particular suppliers and varied from 0.6 rpm to 0.75 rpm. The only RBC, which meets both the Severn Trent and Cranfield standards, is that provided by COPA and rotates at the gearbox speed of 1.1 rpm. [5]

SPEED OF ROTATION

It is possible to economise on metal content by reducing the speed of rotation. Clearly the slower the speed of rotation the longer it will take to reach the failure criteria. Clearly, other factors have an effect on life but failure as a result of fatigue is the prime reason for failure. Speed of rotation however, also has a major effect on the quality of the process and the life of the plant. It has been shown that reducing the speed of rotation below one rpm can result in the formation of the nuisance bacteria, Beggiatoa.

Beggiatoa is a sulphate reducing bacteria and anaerobic. Unlike conventional biomass it can grow quite thick imposing a load on the frame greater than design. This can result in a reduction in the life of the RBC and inhibit the performance of the process. It may also give rise to the formation of tubercles. The pH underneath the Beggiatoa is low and can give rise to the formation of Micro biologically Induced Corrosion (MIC). Even before the formation of MIC Hydrogen Embrittlement maybe accelerated due to increasing number of hydrogen ions accelerating crack propagation. Consequently, Severn Trent Water Ltd and other water companies are specifying a minimum speed of rotation of one rpm.

OPTIMUM LIFE

It is the fact of modern life that sewage regularly discharges to a treatment works for purification. The failure of an RBC due to fatigue, or any other reason, is therefore a major embarrassment and one that needs to be avoided. The cost of disruption, temporary treatment and replacement has been shown by Severn Trent to exceed twice the whole life cost of a rotor at five per cent over 20 years. Long life is therefore extremely important.

The quality of mild steel under fatigue conditions is such that it is difficult to exceed 20 years at the specified conditions consequently 20 years has been selected as the design life based on an S N curve operating in simulated sea water.

COMPARISON OF TENDERS

Lack of knowledge of the importance of speed of rotation on whole life costs, and the need to ensure a 20 year life has resulting in an inadequate evaluation of RBC tenders. Table 1 demonstrates the significance where the theoretical design life is 20 years (option 1) and 10 years (option 2). This clearly demonstrates that where the tenders are evaluated based on DCF techniques the longer life plant has a significantly lower net present value (NPV), by some 37%, and it is cheaper in capital costs in the longer term. This analysis takes no account of inflation. Table 1 does not into consideration the disruption caused when the RBC fails in the year 10 and has to be replaced

Table 2 demonstrates the effect of this on life cycle costs where the cost of disruption can be shown to amount typically to £136K. This results in an increase in NPV of some 162% and demonstrates that long life is in a significant factor in the economics of RBCs selection.

CONCLUSIONS

Rotating Biological Contractors are a very robust form of sewage treatment process and as a result of the success of Severn Trent Water Ltd they are becoming progressively more attractive.

In order to ensure is a higher quality robust process it the essential that the following is observe: -

1. Tenders should be invited strictly in accordance with Severn Trent Water Ltd's regional mechanical specification. This specifies a speed of rotation of not less than one rpm and not greater than two rpm, and a biomass thickness of five mm on the coarse packs and three on the fine. The designed fatigue stress is not less than 25 MPa.

2. Process and mechanical problems have been experienced where the speed of rotation is less than 1 rpm due to the development of Beggiatoa.

3. Tenders should be evaluated using discounted cash flow techniques and including all operational and revenue costs including disruption where a shorter life plant is considered and take into consideration the quality of manufacture and the speed of rotation.

REFERENCES

1. Brenner, R.C, Heidman, J.A, Opatken, E J, and Petrasker, A C (1984) "Design Information on Rotating Biological Contactors" EPA - 600/2 - 84 - 106 NN15 PB84 - 199561.

2. Weston, R.F (1985) (US National Technical Information Service). Report PB85 - 180545."Review of Current RBC Performance and Design Procedures"

3. MBA, D, Bannister, R and Findlay, G.E, (1999) "Mechanical Redesign of the Rotating Biological Contactor" - Wat Res V33 No 18 pp 3679 - 3688 (1999).

4. Findlay, G E (200) " The Production of High Quality RBCs for the UK Water Industry" Aquaenviro Waste Water Conference – Edinburgh

5. Mba .D., Bannister, R and Findlay G E (2002) "Manufacturing High Quality RBCs for the UK Water Industry" – Water Institute of Southern Africa-Biennial Conference Durban

ACKNOWLEDGEMENTS

The author wishes to acknowledge Severn Trent Water who provided the operational and financial data included in this paper.

Table 1: Comparison of Net Present Values for a 10 and 20-year life

Option	Description	Capital Cost	Year 1	Year 11'	Total Cost
No 1	10 year life	£90k	£90k	£90k	£180k
Discount Factor			1.0	0.527	
Net Present Value			£90k	£47.43	£137.43
No 2	20 year life	£100k	£100k	Nil	£100k
Net Present Value			£100	Nil	£100k

Table 2: Comparison of Whole Life Costs for a 10 and 20-year life including disruption costs

Option	Description	Capital Cost	Year 1	Year 11	Total Cost
No 1	10 year life	£90k	£90k	£90k	£180k
Cost of Disruption				£136k	
Total				£326k	
Discount Factor			1.0	0.527	
NPV			£90k	£172k	£262k
No 2	20 year life	£100k	£100k	Nil	£100k
NPV			£100k	£199k		£100k

SEWAGE SLUDGE GASIFICATION: TRACE ELEMENT DEPLETION AND ENRICHMENT IN THE OUTPUT SOLID STREAMS

G P Reed*, Y Zhuo, N Paterson, D R Dugwell and R Kandiyoti
Department of Chemical Engineering and Chemical Technology
Imperial College London
Prince Consort Road, London SW7 2BY, UK
* corresponding author: g.reed@ic.ac.uk

ABSTRACT

Gasification and pyrolysis processes have attracted considerable interest from water utilities as technologies for sewage sludge disposal, with the advantages of waste volume reduction, destruction of pathogenic bacteria and energy recovery. However, elements such as Ba, Cu, Hg, Pb and Zn are present in sewage sludges at levels significant to the disposal of the residual streams from a gasifier. The behaviour of these elements has been studied in an air blown laboratory-scale spouted bed gasifier, fuelled by crushed dried sewage sludge pellets. Measurements of trace element concentrations by ICP-AES have been used to determine their overall retention in the solid streams, as well as their relative depletion from the coarser bed residue and enrichment in the fines carried to the gas cleaning system. The effect of gasifier bed temperature and type of sewage sludge has been investigated. The depletion of Ba, Pb and Zn from the bed residue and their enrichment in the fines is enhanced by gasifier bed temperatures in excess of 900°C. The observed behaviour is discussed in relation to their speciation as predicted by thermodynamic equilibrium modelling. The potential implications of these findings for process design, operating conditions and residue disposal are discussed.

KEYWORDS:
Sewage Sludge, Gasification, Trace Elements

INTRODUCTION

The production of sewage sludge in the EU is forecast to grow, as population increases and becomes increasingly urbanised. In the past, the main disposal routes for sewage sludge have been marine disposal, application to agricultural land, landfill and incineration. Sewage sludge disposal has been the focus of much attention within the water industry in recent years, as some of these disposal routes have been banned or become subject to greater constraints. Disposal at sea was banned in 1998[1], because of concern about the marine environment. Any expansion of agricultural land application will be limited, because of concerns about the possibility of crop contamination by pathogenic bacteria and the build-up of toxic or phytotoxic trace elements. The shortage of landfill capacity, and increasing taxation, has combined to make this an increasingly expensive option. Incineration has met with increasing local resistance to planning applications, and is affected by government measures aimed at reducing CO_2 emissions by extending the use of wastes as fuel for energy production. Consequently, gasification has attracted considerable interest from water utilities as an alternative technology with the same advantages of destruction of pathogenic bacteria and volume reduction, and the additional benefits of energy recovery and lower cost atmospheric emissions control. However trace elements (defined as < 1000ppmwt) including heavy metals such as Ba, Cu, Hg, Pb and Zn are present in sewage sludges at levels significant to the disposal of the residual streams from the gasifier. The potential for some of these elements to be emitted during the pressurised co-gasification of sewage sludge with coal has been found[2], and there has been some qualitative success in explaining the observed behaviour by thermodynamic equilibrium modelling[3]. However the dispersed nature of sewage treatment operations appears to favour simple small scale plant operated at atmospheric pressure on sewage sludge alone, without the cost and infrastucture complexities of adding coal. The lower sulphur and chlorine and higher trace element contents of sewage sludge relative to coal make it likely that the trace elements will behave differently in this circumstance.

An EU 5th Framework project is in progress to investigate the use of sewage sludge gasification in a combined heat and power (CHP) system. As part of their contribution to this project, Imperial College London are studying the gasification of crushed, dried sewage sludge pellets in a laboratory scale gasifier; the details of tests on three sewage sludges (one of Danish origin, and two from the UK) and their performance as fuel are described in a paper to this conference by

Paterson[4]. In this paper the results of investigations into the behaviour of a selection of trace elements in the gasifier will be presented.

EXPERIMENTAL

A flow scheme for the equipment is shown in Figure 1. The gasifier is capable of operation at temperatures up to 1000 °C and pressures to 20 bar_a, with a continuous feed system that meters the fuel into the reactor at rates of up to 4 $gmin^{-1}$. Fluidising gases are either air/N_2 or air/steam/N_2 mixtures. With each type of mixture, the air:N_2 ratio is used to achieve the required bed temperature. Details of the fuel feed system, reactor construction, gas cleaning system and instrumentation are given elsewhere[5]. In this project, changes have been made to the mode of operation of the reactor to enable studies to be done under conditions relevant to the small scale CHP systems, using sewage sludge as the fuel. The main change has been to reduce the operating pressure to nominally 3 bara and to operate without added steam in the fluidising gas or sorbent addition to retain sulphur.

Each test started with an empty reactor. The wall temperature was set at the desired bed temperature and the O_2 concentration in the inlet gas set at nominally 5%. The sewage feed (size range 200-300 micron) was started and the air/N_2 input flows adjusted slightly to achieve the desired bed temperature. There was no solids offtake through the base, so char and ash built up in the reactor during the test. A typical test duration was 30 min, limited by the capacity of the feeding system. Full details of the tests are given by Paterson[4]. An analysis of the feedstocks and a summary of the test conditions is shown in Tables 1 and 2 respectively.

After cooling, the reactor was dismantled and the residual solids in the reactor were collected, weighed and sampled. In this work the solids collected in the bed and the finer fly ash elutriated from the bed were treated as separate samples. Samples of the feed, bed and flyash were then prepared for analysis. The sample preparation sequence commenced with digestion by concentrated H_2SO_4 at 150°C and Low Temperature Ashing at 520°C to remove carbon, followed by digestion of the mineral components in 40% HF and 70% $HClO_4$ at 250°C and finishing with dissolution in 10% HNO_3. The solutions prepared in this way were then analysed for trace and minor elements by ICP-AES. Hg was measured using a LECO AMA-254 dedicated Hg analyser. It should be noted that the reactor incorporates a quartz liner, which minimizes the risk of trace element contamination or losses by corrosion of the reactor. There was also no initial bed in the reactor at the start of these tests, to avoid any contamination or dilution of the trace element content of the final bed of residual solids.

RESULTS

An analysis of the minor and trace element contents of the sewage sludge feeds are shown in Table 3 and 4. The higher Fe content of the DK sludge and the higher Ca and Al contents of the UK sludges can be seen. The most significant trace elements can be seen to be Zn, Ba, Mn, Cu and Pb.

The repeatability limitations of the analysis imply that a mass balance closure on trace elements of between 80 and 120% should be regarded as acceptable. Although the measurement of concentrations in the fuel gas was impractical at this scale, the retention in the solids was within this range for all elements except Hg, Cr, Ni, Ba, Pb and Zn. The level of Hg in the solid residues was seen to be very low, in agreement with the complete release of this element seen in earlier work[6]. Cr and Ni retentions were higher than 100%, indicating contamination by corrosion of reactor components made from high temperature alloys. No further discussion of the Hg, Cr and Ni data is therefore given in this paper.

The depletion of trace elements from the bed char relative to the feed is a useful indicator of the volatility of individual elements; these may normalised relative to a non-volatile element

$$trace\ element\ depletion\ factor_{Al} = \frac{[TE]_{feed}/[Al]_{feed}}{[TE]_{bed}/[Al]_{bed}}$$

such as Al using the following expression

Depletion factors for all of the remaining elements measured are shown in Figures 2a-c; Al is also shown for clarity but by definition, the depletion factor for Al is always 1. The error in these values is estimated to be about 30%.

The extent to which trace elements released from the bed char may condense upon the fine flyash taken out by the gas cleaning system may be element specific. An enrichment factor relative to the bed char may be calculated and normalised relative to Al using the following expression:

For those tests where sufficient flyash for analysis was collected, enrichment factors for the same list of elements are shown in Figure 3a-c.

DISCUSSION

ENRICHMENT AND DEPLETION

Examination of the bed depletion data in Figures 2a-c shows that there is significant depletion of Pb for all three sludges in tests conducted at > 925°C, and in most tests above 900°C. There are also indications of Ba and Zn depletion in some of the higher temperature tests. Comparisons between the three sludges must take account of differences in the scale plotted, but the UK1 sludge shows stronger evidence of Ba depletion than any of the others.

The enrichment data in Figures 3a-c shows that Pb was especially enriched in the flyash, with enrichment factors in excess of 20 being observed in some tests. The enrichment of Pb is also seen to increase markedly with increasing temperature. The temperature plotted in Figure 3 is the temperature inside the bed, the temperature where the flyash sample was collected being lower than this but not measured. The strong dependency of enrichment on bed temperature suggests that a chemical reaction between Pb vapour in the gas and the flyash could have been controlling the rate of deposition on the flyash. Paterson[2] has shown that increasing the bed temperature increases the rate of flyash production, which is attributed to greater production of soot by cracking of tars and hydrocarbons released by the initial pyrolysis of the fuel. It may be that this additional soot has been involved in the capture of Pb by the flyash as well.

MODELLING

It is instructive to examine whether the trends in bed char depletion and flyash enrichment could have been predicted by thermodynamic equilibrium modelling. The typical system composition shown in Table 5 has therefore been modelled using the multiphase module of the MTDATA software from NPL, which applies a free energy minimisation approach and sources thermodynamic data from the SGTE database. An ideal gas phase and pure condensed phase models were assumed, together with global equilibrium. The system composition incorporates all of the major components (C,H,N,O) and the minor components S, Cl and Ca, which are known to be important to the speciation of trace elements in a reducing atmosphere[3]. Predictions of the gas phase speciation (mole fraction as a function of system temperature) for each of Ba, Pb and Zn in the system one at a time are shown in Figures 4a-c respectively.

Chlorine is seen to be important to the speciation of all three elements, which will compete to associate with it. Comparison of the predictions indicates that Ba should be the least volatile of the three, requiring temperatures in excess of 900°C before any gas phase species should form; this is in agreement with the observed bed depletion and flyash enrichment behaviour seen for the DK and UK2 sludges, but not that of the UK1 sludge; this may be due to the assumed Cl value being too low for the latter sludge. On the other hand Pb and Zn should both be mostly in the gas phase at temperatures above 650 °C, i.e. well below the range of the experimental parameters. It is likely that Zn in the sewage sludge is associated with some more complex/less volatile species that is not included in the thermodynamic database; similarly the strong enrichment and temperature dependency of Pb in the flyash suggests the formation by chemical reaction of another involatile species absent from the database (perhaps a Pb aluminosilicate or silicate).

IMPLICATIONS

Although the studies described here are ongoing, some of the implications are already apparent. It must be expected that none of the Hg will be retained in the gasifier bed char; also at gasifier temperatures of 900°C and above a substantial proportion of the Pb and some of the Ba and Zn will similarly not be retained. The extent to which this constitutes an emission problem will then depend on the effectiveness of the gas cleaning system at removing each volatile species from the fuel gas. The most important gas cleaning parameter is likely to be temperature. A wet scrubbing system is likely to be effective for removing all of these elements from the gas, with the possible exception of Hg. The experience of Hg removal by wet flue gas desulphurisation in coal-fired power stations is of variable performance dependent on coal chlorine content[7], poor removal generally being associated with low chlorine coals whose chlorine content is still higher than those typically found in sewage sludges. In a high chlorine system, much of the Hg will be oxidised and readily-soluble Hg^{2+}, whilst in the absence of chlorine the less soluble Hg^{o} will predominate. An activated carbon system for Hg^{o} removal from the fuel gas may therefore be needed. The wet scrubbing system will produce a contaminated aqueous effluent and a final sludge; whilst recycle of the aqueous effluent to the sewage treatment cycle is the obvious option, the possible microbiological effects within the sewage treatment works of other contaminants present such as cyanides (from HCN in the fuel gas) must be considered, and trace element levels in the sewage sludges fed to the gasifier will be increased. The trace element content of the final sludge will also be increased, and the speciation of captured elements such as Pb will then be important in determining their leachability on disposal to landfill.

CONCLUSIONS

The behaviour of trace elements in a sewage sludge gasification system has been investigated as a function of sewage type and gasifier temperature. All of the Hg, a substantial fraction of the Pb and some of the Ba and Zn are found to be released in the gasifier. The gasifier temperature is generally an important parameter, but this is not the case for Hg (which was always released), or for Ba with one of the sludges tested. High Pb enrichment in the flyash was found, with a strong dependency on gasifier temperature indicating capture by chemical reaction. Thermodynamic equilibrium modelling has been used to explain the volatility of Pb and the lower volatility of Ba, but the low volatility of Zn could not be explained within the limitations of existing thermodynamic data. Experimental and modelling activities aimed at extending our ability to predict the behaviour of trace elements in sewage sludge gasification are continuing, particularly to a wider range of sewage sludge compositions.

REFERENCES

1 Urban Waste Treatment Directive (91/271/EEC)

2 REED GP, DUGWELL DR, KANDIYOTI R. Control of Trace Element in a Gasifier Hot Gas Filter: Distribution to the Output Streams of a Pilot Scale Gasifier. *Energy and Fuels,* **2001**, 15, 794-800.

3 REED GP, DUGWELL DR, KANDIYOTI R. Control of Trace Elements in a Gasifier Hot Gas Filter: A Comparison with Predictions from a Thermodynamic Equilibrium Model. *Energy and Fuels,* **2001**, 15, 1480-1487.

4 PATERSON N, ZHUO Y, REED GP, DUGWELL DR, KANDIYOTI R. Processing of sewage Sludge: Pyrolysis and Gasification in a Laboratory Scale Spouted Bed Reactor. 7th *European Biosolids Organic Residuals Conference and Exhibition*, 17-20th November 2002, Wakefield, UK

5 PATERSON N, ZHUO Y, DUGWELL DR, KANDIYOTI R. *Energy and Fuels,* **2002**, 16, 127-135

6 RICHAUD R, LACHAS H, COLLOT A-G, MANNERINGS AG, HEROD AA, DUGWELL DR, KANDIYOTI R. Trace mercury concentrations in coals and coal-derived material determined by atomic absorption spectrophotometry. *Fuel,* **1998**; 77, (5), 359-368

7 BROWN TD, SMITH DN, HARGIS RA, O'DOWD WJ. Mercury measurement and its control: What we know, have learned, and need to further investigate. *J. Air & Waste Manage. Assoc.,* **1999**, 49, 628-640

ACKNOWLEDGEMENTS

The financial support of the European Union for this work under 5th Framework Contract No. ENK5-CT-2000-00050 is gratefully acknowledged.

TABLES

Table 1 Analysis of Sewage Sludges

Analysis	DK 1	UK 1 (Biogran)	UK 2
Proximate			
Total moisture, % ad	9.2	6.5	7.4
Volatile matter, % ad	50.5	50.2	52.6
Ash, % db	37.6	36.1	31.7
Ultimate, %, as analysed			
Carbon,	28.7	30.7	33.0
Hydrogen	3.9	4.1	4.5
Nitrogen	4.6	4.2	4.8
Sulphur	0.66	1.1	1.5
Oxygen (by difference)	15.4	17.3	17.1

ad: as determined in laboratory, db: dry basis,

Table 2 Test Conditions

Test No	Sewage sludge sample	Gasifier Temperature, °C	Fuel:Air mass ratio
10	DK	795	5.5
11		795	4.9
12		950	4.4
13		850	3.5
14		925	5.2
15		770	4.7
19		942	4.9
21	UK1	950	3.6
24		952	2.5
25		940	4.4
26		900	3.1
27	UK2	956	3.7
28		948	3.7
30		894	4.0

Table 3 Minor Element Analysis of Sewage Sludges

Sample	Minor Element analysis, ppm (wt)						
	Ca	Na	Mg	K	Fe	Al	Ti
DK	14500	2745	2740	4874	68679	9421	1305
UK1	38663	1058	4990	4684	9229	22426	1665
UK2	28732	2402	5120	3363	11751	19563	1776

Table 4 Trace Element Analysis of Sewage Sludges

Sample	Trace Element analysis, ppm (wt)									
	Ba	Co	Cr	Cu	Hg	Mn	Ni	Pb	V	Zn
DK	235	9	33	102	1.1	273	20	57	29	761
UK1	406	6	94	257	2.1	203	31	250	25	789
UK2	302	6	75	298	1.0	236	20	112	29	492

Table 5 System Composition and Constraints used in Modelling*

Pressure, atm	Content of Component, gmol									
	C	H	O	N	S	Cl	Ca	Ba	Pb	Zn
1	3.32	9.26	3.94	10.1	0.03	0.002#	0.144	0.0004	0.0001	0.0015

Notes: *taken from Test 28   #assumed value

FIGURES
Figure 1 Diagram of the Fluidised Bed Reactor

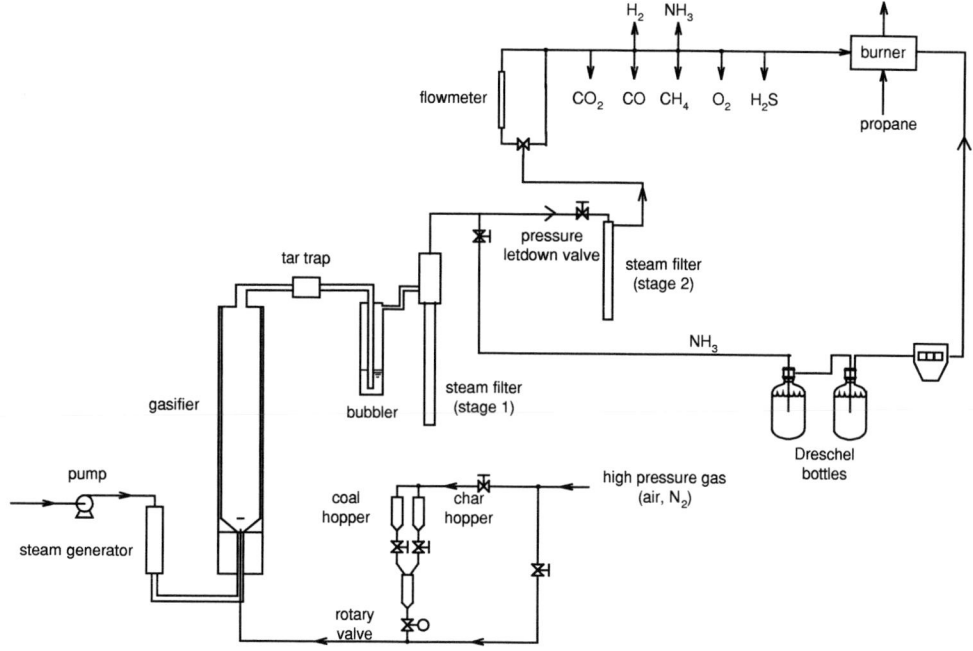

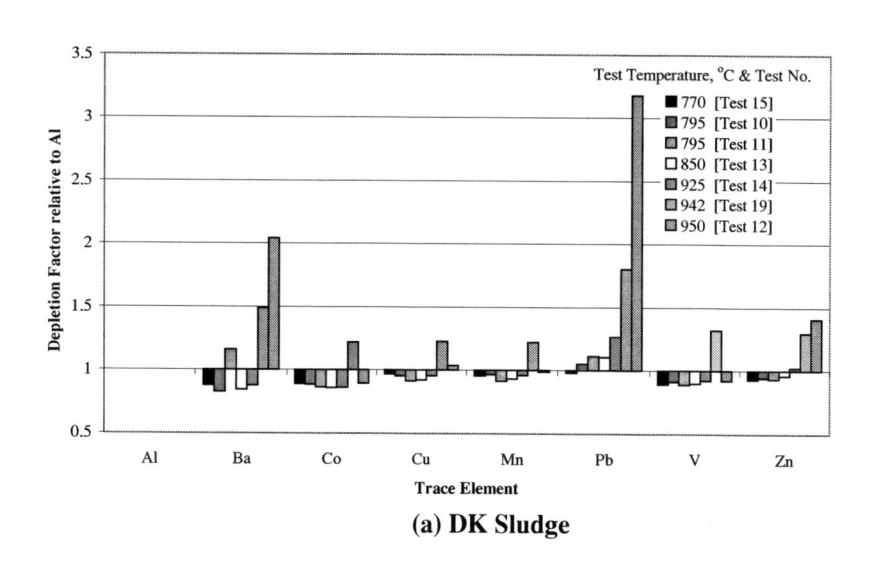

Figure 2 Bed Depletion for Trace Elements of Concern

(a) DK Sludge

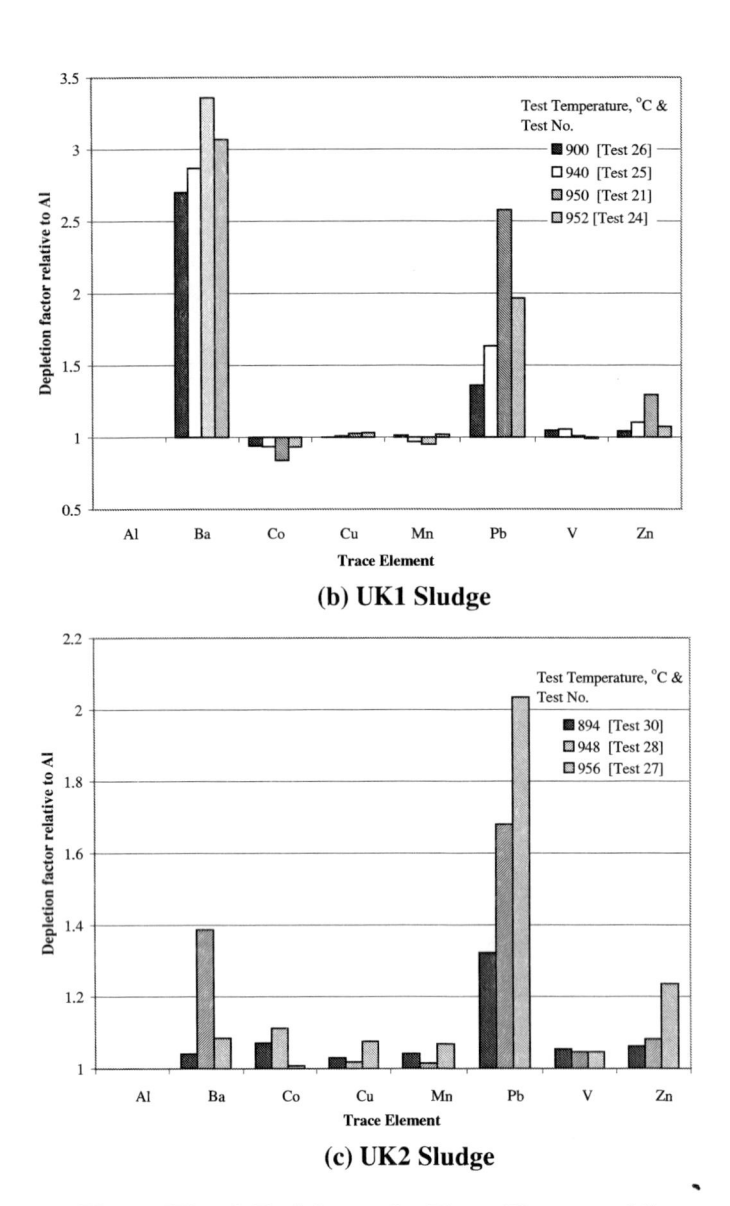

(b) UK1 Sludge

(c) UK2 Sludge

Figure 3 Flyash Enrichment for Trace Elements of Concern

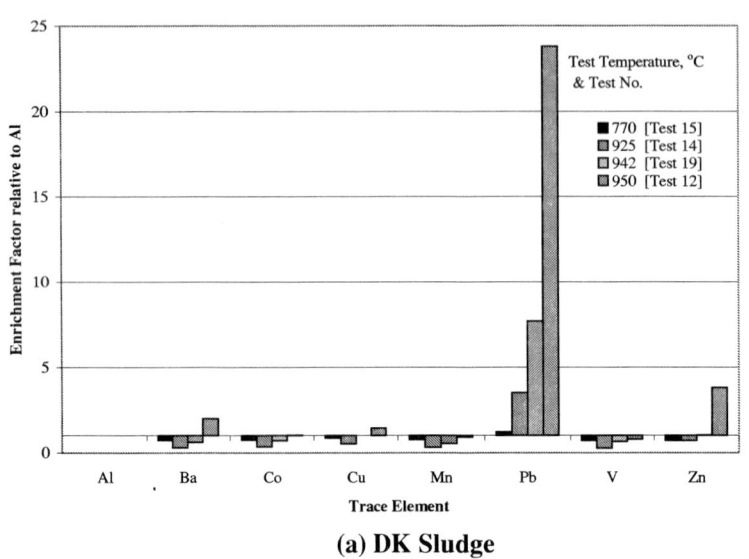

(a) DK Sludge

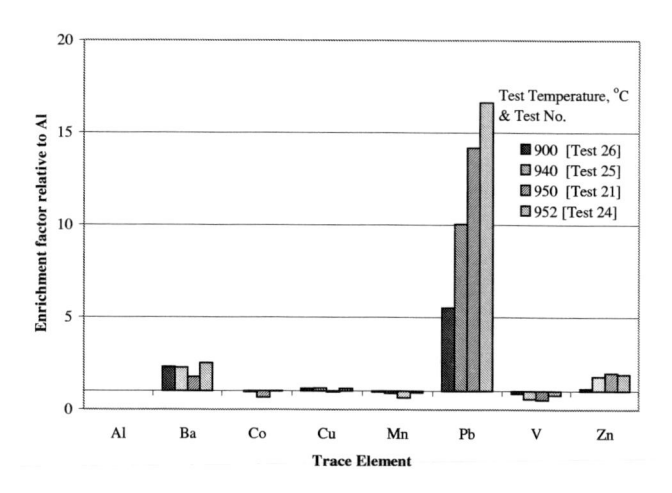

(b) UK1Sludge

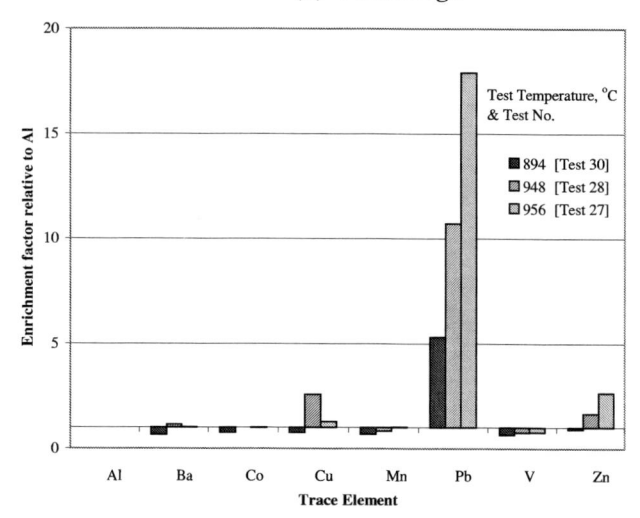

(c) UK2 Sludge

Figure 4 Predicted Gas Phase Speciation for System Composition in Table 5

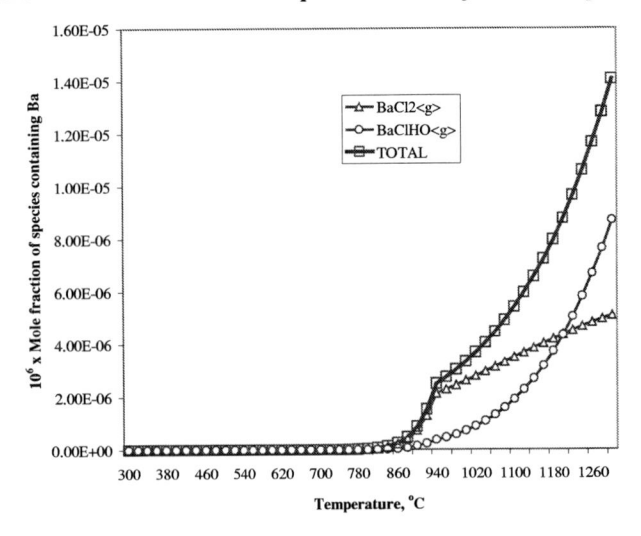

(a) Ba

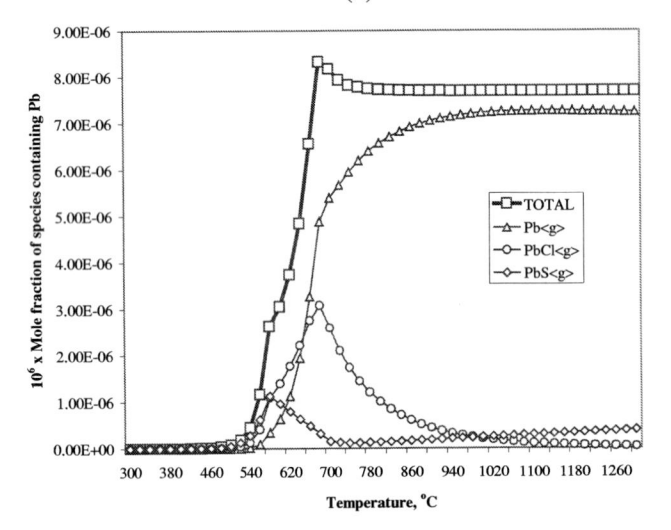

(b) Pb

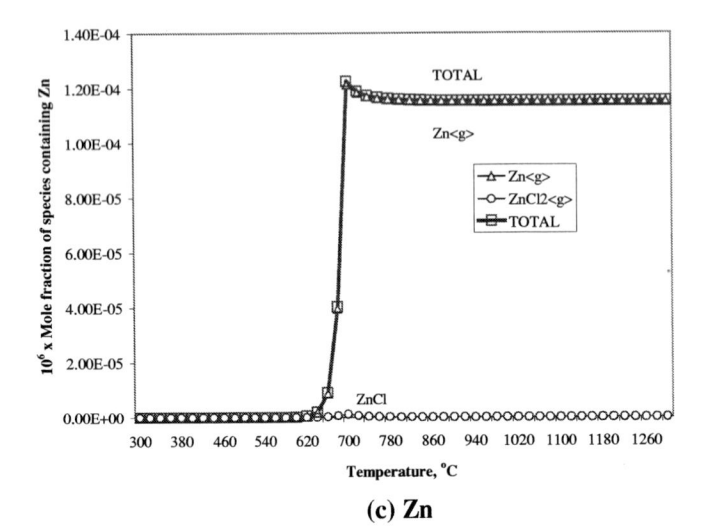

(c) Zn

PROCESSING OF SEWAGE SLUDGE: PYROLYSIS AND GASIFICATION IN A LABORATORY SCALE SPOUTED BED REACTOR

N Paterson[1], Y Zhuo, G P Reed, D R Dugwell, R Kandiyoti
Imperial College of Science, Technology and Medicine
Department of Chemical Engineering and Chemical Technology
Prince Consort Road Llondon,UK, SW7 2BY
[1] 0207 5945634
n.paterson@ic.ac.uk

ABSTRACT

The use of sewage sludge as the fuel for fluidised bed gasifiers that form part of combined heat and power systems is being studied in an EU funded project. A laboratory scale spouted bed reactor is being used to investigate the pyrolysis/gasification behaviour of a suite of sewage sludges, in the temperature range 770 – 980°C and pressures between 2 and 4 bar$_a$. High solids conversions have been measured (daf basis), which are consistent with the high volatile matter content of the sewage based fuels. It seems that most of the fuel is pyrolysed and is converted to gas and condensable tars/oils. The proportion of H_2 formed increases with the gasifier temperature and with the char bed height in the reactor, which is consistent with an increase in the extent of cracking of the tars/oils. The N content of the sewage sludge was high and this led to the high NH_3 concentrations that were measured in the fuel gas. Operating at the higher end of the temperature range tended to decrease the amount of NH_3 in the fuel gas.

KEY WORDS

Gasification, calorific value, emissions, flue gas analysis, hydrocarbons, mass balance, sewage sludge.

INTRODUCTION

Alternative routes for the disposal of sewage sludge need to be identified, tested and developed to a viable, commercial stage. Over the past decade, changes in the traditional disposal routes have been required because of the increased awareness of their impact on the environment. In 1998, disposal at sea was banned, following the implementation of European legislation [1] to protect the marine environment. This has put added pressure on the alternative disposal routes of use as landfill and in agriculture. These options are governed by other European Directives [2, 3]. Landfill sites are now facing great pressures, because of the volume of all forms of waste that need to be disposed of. Attempts to limit the volume of material going to landfill by increasing taxation, also mean that this is now a relatively expensive disposal route. Larger amounts of sewage sludge are now being sent for disposal via 'agricultural uses'. There are various ways in which this can be done and these include pressure injection into the subsoil (used for untreated material), disposal of a pelletised material on the ground surface and treatment with lime (to kill bacteria) followed by spreading on the land. One of the concerns about the medium/long term impact of using large amounts of sewage sludge on the land is the potential for the build-up of the concentration of various toxic and environmentally harmful trace elements [1].

Near term options for disposing of sewage sludge include incineration and gasification. Most other more novel options, such as supercritical water oxidation, are still at the proof of concept stage. Incineration has been used at the commercial scale, but there are concerns about the low energy efficiency and emissions to atmosphere from this type of process. Gasification has mainly been developed for coal and oil residue based applications, but this technology does provide a potentially cleaner and more energy efficient route for disposing of sewage sludge. However there are issues that need to be addressed to ensure that it provides a sustainable disposal route. These include developing a greater understanding of the fundamentals of sewage sludge gasification, including the fate of elements of environmental concern and the behaviour of process ash in the environment. In addition, the acceptability of this technology with sewage sludge depends on the development and application of viable gas cleaning systems.

An EU funded project is in progress to investigate the use of sewage sludge as the fuel for CHP applications. ICSTM is studying the use of a laboratory scale, spouted bed reactor to pyrolyse and gasify the fuel at low pressure and temperatures of up to 980°C. The reactor is a simpler version of the pilot plant developed to gasify coal and coal/waste mixtures in the Air Blown Gasification Cycle (ABGC) [4]. Coal/sewage sludge and sewage sludge alone were tested as part of the gasifier development programme of the ABGC. This showed the technical feasibility of using such fuels as part of a hybrid combined cycle power generating system, operating at elevated pressure. High fuel conversions and high CV fuel gases were obtained under the conditions used in the pilot scale ABGC gasifier. However, it was recognised that it was not a practical proposition to supply commercial scale ABGC gasifiers on sewage sludge alone. This was because of the logistical difficulties and costs of collecting sufficient material together for a continuously operated, commercial scale plant. Co-gasification with coal is the most likely process route to be followed for such large scale applications. However, the use of sludge alone as the fuel for smaller scale, locally based processes is a feasible option.

In this paper, the results obtained in the laboratory scale reactor are discussed in terms of the extents of solids conversion and quality of gas that is produced under conditions relevant to the use of the gasifier in a CHP scheme.

EXPERIMENTAL

THE LABORATORY SCALE GASIFIER

The laboratory scale reactor copies the base features of the pilot scale gasifier that was operated by British Coal in the development of the ABGC. The fuel is fed into the reactor with the fluidising gas mixture, as a high velocity spout jet, which enters the reactor at the apex of the inverted cone shaped base. A flow scheme for the equipment is shown on Figure 1. It is capable of operation at temperatures up to 1000 °C and pressures to 20 bar_a, with a continuous feed system that meters the fuel into the reactor at rates of up to 4 $gmin^{-1}$. Fluidising gases are either air/N_2 or air/steam/N_2 mixtures. With each type of mixture, the air:N_2 ratio is used to achieve the required bed temperature. Details of the fuel feed system, reactor construction, gas cleaning system and instrumentation are given elsewhere [5]. In this project, changes have been made to the mode of operation of the reactor to enable studies to be done under conditions relevant to the small scale CHP systems, using sewage sludge as the fuel. The main change has been to reduce the operating pressure to nominally 3 bar_a and to operate without added steam in the fluidising gas or sorbent addition to retain sulphur. It is noted that the moisture in the feed pellets will evaporate to form a steam concentration in the reactor of approximately 4 % of the fluidizing gas (by volume).

Each test is started with an empty reactor. The wall temperature is set at the desired bed temperature and the O_2 concentration in the inlet gas set at nominally 5%. The sewage feed (size range 200-300 micron) is started and the air/N_2 input flows adjusted slightly to achieve the desired bed temperature. There is no solids off-take through the base, so char and ash build up in the reactor during the test. A typical test duration is 30 min. After cooling, the reactor is dismantled and the bed solids and filter solids are collected and weighed. The material collected in the tar trap is separated into a chloroform soluble fraction (tar) and a chloroform insoluble fraction (dust) and the dried fractions are weighed. The weight of sewage sludge fed during the test is obtained from the initial and final weight of material in the feed hopper. A typical molar input ratio of combustibles (as C and H) to O_2 (in air) is 33, which compares with a stoichiometric ratio of approximately 2. No sintered material was found in the final beds. It is noted that there was no initial bed in the reactor at the start of these tests to enable the trace element release from the sewage sludge to also be assessed. A bed formed from the feed builds up during the test. If a different initial bed material had been used, then it could have contaminated the solid samples used to measure the trace elements.

FUEL GAS ANALYSIS

The concentrations of the major species in the fuel gas (CO_2, CO, CH_4, H_2) were measured continuously during each test. CO_2, CO and CH_4 were measured using infra-red analysers and H_2 was measured with a thermal conductivity detector. The absence of O_2 was also monitored using a paramagnetic analyzer (the presence of O_2 indicates leaks in the sampling equipment, as there is no O_2 present in the fuel gas). These analysers were calibrated at the start of each test using a certified gas mixture. C_2 and C_3 hydrocarbons were measured during selected tests by batch gas chromatography using a Carboxen ™ 1010 PLOT column. H_2S and NH_3 were detected using Draeger tubes. These tubes are primarily intended for atmosphere monitoring, but have been found suitable for monitoring fuel gas compositions. They provide an indication of the concentration of particular species that are present. NH_3 was also measured using a batch technique. This involved the absorption of the NH_3 from a measured volume of fuel gas, in deionised water and measuring the collected concentration by ion chromatography. This provided a value for the average concentration over the test period. An estimate of the steam concentration in the fuel gas was made from the weight of water collected in the gas cleaning filters.

THE FUELS
Three sewage sludges have been tested and their analyses are shown on Table 1. Sludge DK1 was from Denmark, UK 1 and 2 were obtained from water companies in the UK. UK 1 was a digested material and all had been heat treated to render them biologically inactive. All were supplied as dried materials. Both UK1 and UK 2 were supplied in pelletised form, whereas DK 1 was supplied as a ground material (in the approximate size range 50 to 1000 micron). The samples were crushed/sieved to prepare test samples in the 200-300 micron size range. The analysis data gives an insight into the likely behaviour of the fuels during gasification and pyrolysis. All of the samples had high ash and volatile matter contents. The VM contents suggest that a high proportion of the organic part of the sludges will be lost during pyrolysis. The fuel-N contents are high, which indicates high emissions of NH_3 and HCN during gasification. The S contents exhibit a range of values, which indicates that the concentrations of H_2S in the fuel gases will vary between samples.

TEST PROGRAMME
Tests were done with the dried sewage sludges over range of temperatures, pressures and sewage feed rates to gain an insight into the effect of the operating conditions on the sewage performance. It is noted that controlling a small-scale reactor, with a continuous feed and an air supply is difficult, because of the high rate of heat release in the reactor. Consequently, it is not possible to repeat tests under exactly the same conditions. Hence, the strategy adopted has been to conduct tests under a range of conditions and then draw comparisons between tests done under broadly similar conditions.

RESULTS AND DISCUSSION
MASS BALANCES
Mass balances are shown on Table 2. The ash and oxygen balances are scattered around 100%. The spread of these balances will in part be due to experimental errors. Precise oxygen balances are difficult to obtain as the oxygen content of the solids is not measured directly. It is obtained by difference and therefore the value is equal to oxygen plus errors in the other measurements. Mineralogical changes can cause poor ash balances under gasification conditions. However, the relative closeness of the values for these balances to 100%, does suggest that all of the inputs and outputs containing these species have been included in the calculations. Conversely, the C balances generally show an underbalance. This is due to the omission of condensable species in the output fuel gas in the balance calculation. These are known to be present but have not been measured yet. The balances for C given in bold font include contributions from the C_2 and C_3 hydrocarbons. Their inclusion has increased the C balance by approximately 5%.

CHANGES IN THE COMPOSITION OF THE FUEL GAS WITH TEST TIME
The concentrations of CO_2, CO, CH_4 and H_2 were measured continuously from the start of each test. The values obtained during tests at 770 (Test 15) and 950°C (Test 12) are plotted against the test time on Figures 2 and 3. The values at 770°C show that the CH_4, NH_3 and H_2S concentrations were nearly constant over the test period, the CO_2 and CO showed a minor increase, whereas the H_2 concentration showed a progressive increase, rising from 5 to 10 % during the test. The NH_3 values presented on the graph are approximate values obtained using Draeger tubes, but they are presented to show that the values were not changing over the test period. At 950°C, the extent of increase in the H_2 was greater and rose from 7 to 23 % over a similar test time. Also, at this temperature the CO showed a more marked increase in concentration with the test time and increased from 7 to approximately 11 %.

The results for H_2 are consistent with the formation of low temperature tars and oils from the initial pyrolysis of the sewage feed, which were then progressively cracked to H_2, CO and C (the final products of the cracking of the volatiles). The amount of H_2 formed during a test progressively increased because a bed built-up during the test, which provided a more effective medium for the cracking reactions. This may also explain the observed small increase in the CO concentration. The rates of the cracking reactions were greater at the higher temperature and this resulted in the higher measured concentrations of H_2 during that test. The CO concentration also showed an increase at 950°C, because at this temperature the rate of gasification by CO_2 would be higher and the extent of reaction would increase with the increase in the bed height. The same observations may be drawn from the results of each test that has been done in this project.

EFFECT OF TEMPERATURE ON THE PROCESS PERFORMANCE

There are several factors that can be used to indicate the performance of the process, including the fuel conversions (overall, C and H) and the calorific value of the fuel gas. The values are dependent on the test operating conditions, such as temperature, bed height, fuel:air ratio and pressure. In this work we have attempted to study the effect of varying the temperature, however, this has not been possible in isolation and there have also been lesser variations in the fuel:air and pressure between tests. This means that the effects that have been measured are primarily caused by changes in the temperature, but there will be lesser changes caused by the variation in the other operating conditions.

<u>CARBON AND HYDROGEN CONVERSIONS</u>

Table 3 shows the C and H conversions to gaseous species that have been calculated from the average of the gas analyses recorded during each test. The data for C and H in regular font includes C present in the fuel gas as CO_2, CO, H_2 and CH_4, whereas the values in bold include contributions from the C_2 and C_3 gaseous hydrocarbons. The data shows that, in general, the C and H conversions to gas increase with the test operating temperature. This is consistent with the increased conversion of condensibles to gaseous species at the higher temperatures. The H conversions at the highest temperatures were close to or greater than 100%, which shows that virtually all of the input H had been released into the fuel gas as gaseous species. The lower values for C show that some has remained in the solid or condensable phases. It is probable that a proportion of the non-gaseous C is present as soot (one of the final products in the pyrolysis process) and the remainder will be present as fixed C in the char.

Values of fuel C converted to C in the char (bed and fines) are also shown on the table. The values have been calculated from the solid input and output rates and their C contents. The values do not exhibit a well-defined change with the test temperature. This is in agreement with the explanation given above to explain the observed increase in the C to gas conversion. This was explained by an increase in the conversion of the condensibles to gas, and this would not cause any decreases to the C in the char itself. If any effect were to be seen, it should be an increase caused by the formation of soot (a later product of the pyrolysis of the volatiles). This is most likely to be present in the fines carried out of the gasifier and collected in the downstream filter. There are indications in the results that the quantity of fines produced did increase with the test temperature. The fines were C rich, so an increase in fines shows an increase in fine C production. Table 4 shows the amount of fines produced per g sewage fed for tests with sample DK 1. When data conducted at different temperatures are compared, for tests done under otherwise similar conditions, then the data does show an increase in the fines output at the higher temperature e.g. Tests 11 (795°C) and 12 (950°) the fines increased from 14 to 36 mg g-sewage fed^{-1}, Tests 10 (795°C) and 14 (925°C) the fines increased from 10 to 14 mg g sewage fed^{-1}.

<u>OVERALL FUEL CONVERSION</u>

The overall fuel conversion has been calculated from the solids input and output rates. The values (expressed on a dry, ash free basis), shown on Table 4, range between 85 and 94 %. The conversions were high and show that near complete conversion of the organic part of the sewage to gas is achievable. The conditions used in this study were not optimised to maximise the fuel conversion and hence there is scope to raise the conversion further. However, it is noted that it is probable that at least some of the C remaining in the solids will be soot and this could be unreactive. This may indicate that high, but not complete solids conversions are achievable.

<u>FUEL GAS COMPOSITION AND CALORIFIC VALUE</u>

The composition (average values for each test) and calorific value (calculated) of the fuel gases are shown on Table 5. For sewage sludges DK1 and UK 1, the data has been arranged into fuel:air ratio bands of 2.0-2.9, 3.0-3.9, 4.0-4.9 and greater than 5.0. Within each fuel:air ratio band, the data is in the order of increasing temperature. For sample UK 2, the data was obtained over a fairly close fuel:air ratio band (3.7 – 4.3) and this data is not separated into different fuel:air bands. The composition of the fuel gas depends on several operating parameters, including the reaction temperature, fuel:air ratio, bed height, fluidising velocity and pressure. Of these, it is the temperature and fuel:air ratio which are thought to have had the dominant influence on the data shown on the table. The pressure was mostly in the range 3 – 5 bar$_a$ and this relatively small change is unlikely to have had a significant impact on the extents of gasification and pyrolysis. The weight of char in the reactor increased from zero to between 20 and 30 g during each test. The upper part of the bed in this narrow diameter reactor will have been slugging and this would not provide an effective contacting gas/solid regime. Therefore it is assumed that the effective bed height was similar during each test. The fluidising velocity only varied over a relatively narrow band (0.07 –0.1 ms^{-1}) between tests and this will only have had a minor influence on the reactions in the gasifier. The following general observations may be drawn from the table:

- The CO and H_2 concentrations increased with temperature, which resulted in an increase in the CV of the fuel gas. The increase in H_2 is explained by an increase in the extent of cracking of the tars at the higher temperatures. The increase in the CO may also be partly explained by the increase in the cracking of the volatiles, but may also be caused by an increase in the extent of gasification by CO_2. The latter is consistent with the accompanying decrease in the CO_2 concentration in the fuel gas.

- The CH_4 concentration did not seem to be affected by the temperature, but for a given temperature, it did increase with the fuel:air ratio. The CH_4 will have been formed as part of the cracking of the volatiles and therefore the amount present is expected to increase with the fuel:air ratio. The insensitivity to temperature suggests that the CH_4 is released from the volatiles that are most easy to crack.

- For tests conducted under broadly similar temperatures and fuel:air ratios, the concentrations of H_2 and CH_4 showed some variation with the source of the sewage sludge. DK 1 produced the highest H_2 concentrations, whereas UK 2 produced the most CH_4.

- UK 2 produced fuel gases with a similar CV to DK 1, when comparisons are made under approximately similar conditions. UK 1 produced the lowest quality fuel gas, which is consistent with this sludge being digested prior to its drying and pelletisation.

POTENTIAL ENVIRONMENTAL POLLUTANTS

The NH_3 and H_2S concentrations in the fuel gas are shown on Table 6. Analyses are presented for samples DK 1 and UK 1,
This shows that the NH_3 concentration was very high (0.8 %) at the lowest temperature studied (DK 1, 770 °C). This value is equal to 19 % of the fuel-N input, the balance must remain in the solid phase, be present as other gaseous –N species (e.g. HCN or N_2) or be present in the condensible material. The concentration of NH_3 in the fuel gas decreased with increasing temperature for both samples. In other studies as ICSTM [6], this has been explained as showing the increase in the rate of movement towards the low concentration expected at equilibrium in the gas phase, for the $2NH_3 \rightleftharpoons N_2 + 3H_2$ reaction. The H_2S concentrations are regarded as approximate values, as they were determined by the Draeger tube method. However, all values were fairly low and did not show large changes with temperature. For sample DK 1, the value at the highest temperature is equal to 28 % of the S input in the sewage. Similar H_2S emissions were determined for both DK 1 and UK 1. This is surprising, as UK 1 had virtually double the S content to DK1. This may indicate that the CaO content of the ash in this material is higher and retains a higher proportion of the S (as CaS).

DISCUSSION

These studies have examined the pyrolysis and gasification of dried sewage sludge in a spouted bed reactor. A range of conditions has been used to measure the extents of reaction that were achieved, when the operating parameters were varied. The tests have shown that high solids conversions are achieved when the dried sewage sludge was heated to temperatures of between 770 and 950°C in the spouted bed reactor. The work has suggested that a bed in the reactor is needed to maximise the cracking of the primary low temperature tars and oils that are released when the initial feed is heated. The temperature should also be towards the upper end of the temperature range to maximise the extent of these reactions. The occurrence of the cracking reactions is inferred from the increases in the H_2 concentrations and improvements to the H balance, that were observed. These factors also improved the C balances and this is thought to indicate the formation of soot, which was retained within the bed char or elutriated with the fines. The reason why the increase in the H_2 concentrations is attributed to an increase in the extent of cracking, rather than an increase in the extent of gasification, is that it was not accompanied by equal changes in the CO or CO_2 concentrations. The CV of the fuel gas was greater at the higher temperatures, which was a result of its higher H_2 content. The sewage sludge contained a high N content (4.2-4.8%, as analysed) and this was reflected in the high NH_3 emissions. The NH_3 concentration in the fuel gas was lowest during the tests at the highest temperature, which is consistent with the more rapid movement towards the equilibrium concentration in the gas phase for the $N_2 + 3H_2 \rightleftharpoons 2NH_3$ reaction. H_2S emissions were fairly low (as a proportion of the total S content of the samples) and this may suggest that it was retained by the CaO present in the ash.

CONCLUSIONS

The laboratory scale trials reported in this paper have suggested that dried sewage sludge can be used to produce a fuel gas in a near ambient pressure, spouted bed, air blown gasifier. The fuel conversions were high and gases of reasonable CV were produced when a low bed was present in the reactor and the temperature was in the region of 950°C. No problems with ash sintering were noted during the tests. The use of temperatures of approximately 950°C minimises the concentration of NH_3 in the fuel gas.

REFERENCES

1. Urban Waste Treatment Directive (91/271/EEC)

2. Landfill of Waste Directive (99/31/EEC)

3. Use of Sludge in Agriculture (86/278/EEC)

4. DAWES S, MORDECAI M, BROWN D, BURNHARD K. Proceedings of the 13th International Conference on Fluidised Bed Combustion, Orlando, USA, May 1995.

5. PATERSON N, ZHUO Y, DUGWELL D R, KANDIYOTI R. *Energy and Fuels*, **2002**, *16*, 127-135.

6. PATERSON N, ZHUO Y, AVID B, DUGWELL D R, KANDIYOTI R. *Energy and Fuels*, **2002**, **16**(3), 742-751

Table 1. Analysis of the Sewage Sludge

Analysis	DK 1	UK 1 (Biogran)	UK 2
Proximate			
Total moisture, % ad	9.2	6.5	7.4
Volatile matter, % ad	50.5	50.2	52.6
Ash, % db	37.6	36.1	31.7
Ultimate, %, as analysed			
Carbon,	28.7	30.7	33.0
Hydrogen	3.9	4.1	4.5
Nitrogen	4.6	4.2	4.8
Sulphur	0.66	1.1	1.5
Oxygen (by difference)	15.4	17.3	17.1

ad: as determined in laboratory, db: dry basis,

Table 2. Mass Balances

Test No.	Sewage name	Operating Conditions			Balances, % out/in		
		Temp. °C	Feed rate, gmin^{-1}	Pressure, bar$_a$	Ash	O	C
15	DK 1	770	2.6	2.8	91	110	78
17		780	2.5	2.8	92	82	**82**
10		795	2.8	3.5	90	87	61
11		795	2.2	3.6	100	105	70
20		812	1.7	3.2	97	106	81
13		850	1.8	2.5	111	139	89
16		920	2.3	2.4	106	108	**95**
14		925	2.8	3.7	98	121	95
19		942	2.4	4.4	100	92	**79**
12		950	2.1	3.2	110	123	77
23*		951	2.3	4.2	83	111	81
18		961	2.4	4.5	116	124	81
22	UK 1	897	2.1	3.0	90	93	**71**
26		900	1.9	3.8	110	100	82
25		940	2.2	5.1	110	100	**86**
21		950	2.1	3.0	107	108	**80**
24		952	1.6	4.7	106	119	93
31	UK 2	880	2.7	2.6	100	88	72
30		894	2.4	2.9	98	99	70
28		948	2.0	3.6	112	104	93
27		956	1.8	4.5	92	106	83

Table 3. Process Performance

Test No.	SEWAGE sample	Operating Temp. °C	Fuel:Air	Overall Fuel Conversion %, daf basis	Conversions, %		
					C to gas	C to char	H to gas
15	DK 1	770	4.7	85	46	32	79
17		**780**	**3.4**	**85**	**49**	**33**	**83**
10		795	5.5	88	34	27	66
11		795	4.9	86	39	31	75
20		812	3.0	83	44	37	73
13		850	3.5	93	75	14	113
16		**920**	**3.7**	**88**	**69**	**27**	**124**
14		925	5.2	85	63	32	123
19		**942**	**4.9**	**88**	**53**	**27**	**87**
12		950	4.4	92	59	18	121
23*		951	4.9	92	66	17	119
18		961	4.0	94	67	14	118
22	UK 1	**897**	**3.4**	**92**	**55**	**16**	**78**
26		*900*	*3.1*	*87*	*55*	*27*	*75*
25		**940**	**4.4**	**86**	**66**	**29**	**97**
21		**950**	**3.6**	**89**	**57**	**23**	**100**
24		*952*	*2.5*	*91*	*74*	*18*	*95*
31	UK 2	880	4.3	89	48	24	72
30		894	4.0	89	48	22	84
28		948	3.7	84	60	33	102
27		956	3.7	88	58	26	99

Table 4. The Quantity of Fines Produced at Different Test Temperatures

Test No	15	10	11	13	14	12
Temperature, °C	770	795	795	850	925	950
Bed char, gchar/gsewage fed	0.36	0.40	0.42	0.41	0.38	0.34
Fines mgfines/gsewage fed	10	10	14	11	14	36

Table 5. The Composition of the Fuel gas and Calorific Value

Test No.	SEWAGE sample	Operating Temp. °C	Fuel:Air	Calorific Value of the Fuel Gas MJm^{-3}, dry, gross	Composition of the Fuel Gas, %, vol, dry			
					CO_2	CO	CH_4	H_2
17	DK 1	780	3.4	2.1	6.0	4.7	2.0	6.4
20		812	3.0	1.6	4.5	4.5	1.1	5.1
13		850	3.5	2.9	5.5	7.0	2.7	9.0
16		920	3.7	4.8	5.7	11.2	3.1	18.7
15		770	4.7	2.6	6.8	5.6	2.3	8.9
11		795	4.9	2.5	6.2	5.8	2.3	7.5
19		942	4.9	3.3	4.7	8.6	2.3	11.2
12		950	4.4	4.3	4.9	10.4	2.9	16.2
23		951	4.9	4.6	4.7	10.5	3.2	17.7
18		961	4.0	4.0	4.8	9.9	2.8	14.3
10		795	5.5	2.9	6.7	6.1	2.4	10.4
14		925	5.2	5.1	5.7	11.4	4.0	18.6
24	UK 1	952	2.5	2.4	3.6	6.7	2.0	6.9
22		897	3.4	2.5	6.5	5.0	3.0	5.0
26		900	3.1	2.6	4.1	7.2	2.6	6.5
21		950	3.6	2.8	3.2	7.9	2.1	8.9
25		940	4.4	3.6	4.8	8.7	3.6	9.9
31	UK 2	880	4.3	3.9	4.8	8.7	5.1	8.0
30		894	4.0	4.1	4.7	9.3	4.9	9.6
28		948	3.7	5.1	4.7	11.1	6.0	12.8
27		956	3.7	4.7	4.4	10.8	5.0	12.6

Table 6. Emissions of Sulphur and Nitrogen Compounds

Test No		15	17	13	16	14	12	18	22	21
Sewage Number		DK 1							UK 1	
Temperature, °C		770	780	850	920	925	950	961	897	950
NH_3	vpm	8000	6500	4450	4680	4100	3900	2840	5060	3570
H_2S		500	500	600	600	700	nd	500	500	400

vpm: parts per million, by volume
nd: not determined

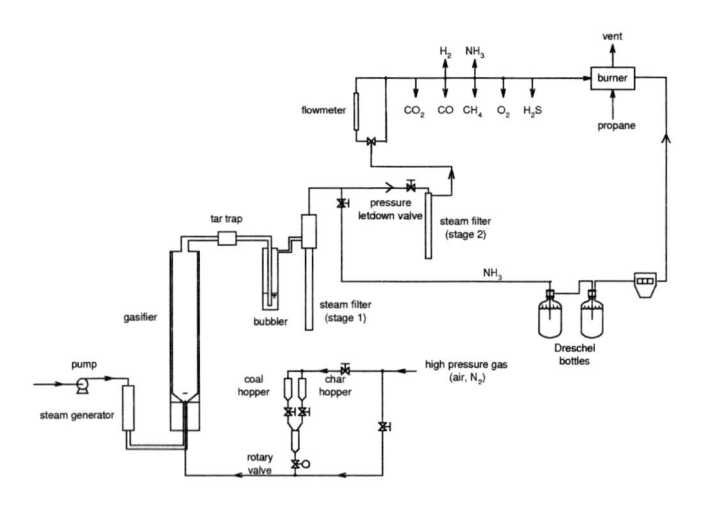

Figure 1. Schematic Diagram of the Bed Reactor

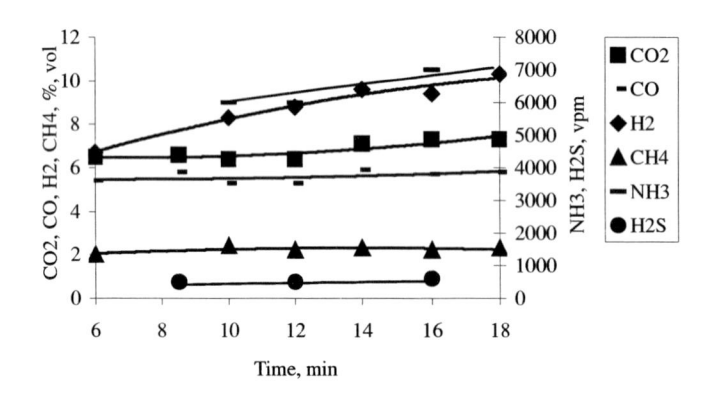

Figure 2. Fuel Gas Composition at 770°C

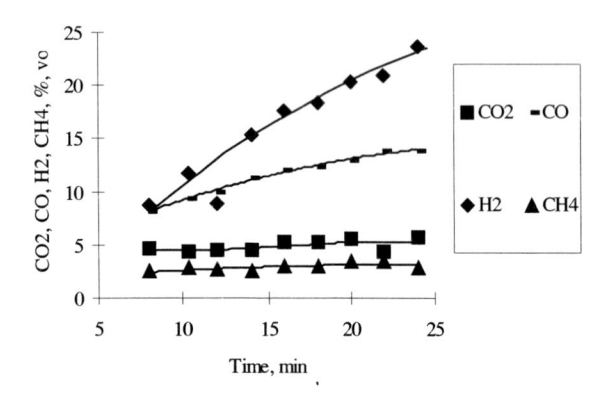

Figure 3. Fuel Gas Composition at 950°C

ENVIRONMENTAL ASSESSMENT OF ENERGY RECOVERY FROM SEWAGE SLUDGE THROUGH SUPERCRITICAL WATER OXIDATION

Magdalena Svanström and Morgan Fröling
Chemical Environmental Science, Chalmers University of Technology, SE-412 96 Göteborg, Sweden
Phone: +46 31 772 3001, Fax: +46 31 772 2999, E-mail: maggans@kmv.chalmers.se

ABSTRACT

Sewage sludge is a potential biofuel. Energy recovery from sewage sludge is today often performed through incineration, but some other new technologies exist in which the organic material is more or less oxidised. Supercritical water oxidation (SCWO) allows for quick and complete oxidation without harmful emissions. The first commercial plant for SCWO of sewage sludge has earlier been shown to be an environmentally beneficial process. The results were highly dependent on the energy system surrounding the actual SCWO plant, in that case in Harlingen, Texas, USA. This paper describes a study in which a sewage sludge SCWO plant was considered to be located in Sweden, in a different energy system than in the Harlingen case. Two different scenarios were investigated using LCA methodology. In the first, undigested sludge was fed to the SCWO process. In the second scenario, a digestion plant was considered to treat the sewage sludge before it was fed into the SCWO process and the produced biogas was used to replace natural gas in a city gas net. In the Swedish system, recovered heat was considered to replace average Swedish district heat production instead of natural gas burning as in the Harlingen case. The differences in electricity production between the Texas and the Sweden systems had a large influence on the results. All studied systems had a positive effect on the environment. SCWO treatment of raw sludge is better from a life cycle perspective than anaerobic digestion with subsequent SCWO treatment. Energy recovery from sewage sludge through the use of SCWO gives net savings for the environment, larger the larger the contribution of fossil fuels in the energy production mix that is replaced by the recovered heat.

KEYWORDS

Energy recovery, environmental aspects, LCA, life cycle assessment, SCWO, sewage sludge treatment, supercritical water oxidation

INTRODUCTION

SEWAGE SLUDGE AND SUSTAINABLE DEVELOPMENT

At the United Nation's World Summit on Sustainable Development in Johannesburg, world leaders adopted a global action plan (The Johannesburg Plan of Implementation) that commits countries to halve the number of people without access to water and sanitation by 2015 [1]. In the developing world, less than half the population have access to proper sanitation; in the rural areas, only a quarter of the population [2]. According to the World Health Organization (WHO), 1.3 million children under five years of age die each year from diarrhoea due to lack of safe water and sanitation [3]. A water transportation system for faeces and urine and different wastewaters is the most common way to solve the sanitation problem. Collected wastewaters have to be treated in order to avoid environmental effects in receiving water bodies and to ensure access to safe water for the population. Treatment produces a sludge that has to be managed properly to avoid adverse effects on health and environment and nuisances to the local population. Furthermore, sustainable development in society requires the best use of every available resource, and the sewage sludge contains many nutrients and scarce elements and a considerable amount of organic material. Recycling may be necessary to avoid deficiencies at the extraction site and contamination in the recipient.

The use of sewage sludge as a soil fertiliser and soil conditioner in agriculture makes use of the nutrients and the organic content in the sludge. Some elements in the sludge originate mainly from agricultural soils, and agricultural use is thus often regarded as recycling of these elements. However, the presence of pathogens and contaminants in the sludge from a chemically intense society pose threats to health and environment. Incineration of the sludge makes possible recovery of the inherent energy in the organic material and destruction of pathogens and organic pollutants. However, incineration costs are driven up by expensive air and water emission abatement technologies, and even more stringent regulations are expected in future. Recovery of elements from sludge incinerator ashes or from the sludge itself is possible. Several methods for phosphorus recovery exist [4-6].

For sludges containing organic material, stabilisation is necessary to avoid odours and risks associated with pathogenic organisms. In Sweden, anaerobic digestion is a common stabilisation method. Digestion of sewage sludge also facilitates dewatering and reduces the volume by up to 50%. A by-product of this treatment is biogas consisting of methane and carbon dioxide. The biogas can be used to replace natural gas. In this way, a part of the energy in the sludge is recovered as a high-value fuel.

Today, only a small part of produced sewage sludge is used in agriculture in Sweden due to a general concern among consumers, industry and organisations for unknown risks with exposure to sludge contaminants. An upcoming ban on landfilling of organic material (from 2005) makes it necessary to investigate new methods of energy and materials recovery from sewage sludge, methods without risks of spreading diseases and contaminating ecocycles. Furthermore, the Swedish Environmental Protection Agency has recommended that 20-30% of phosphorus in sewage sludge should be recycled to agricultural land each year by 2015 and 35-50% by 2025 [7]. New methods for sewage sludge treatment should aim at the following targets:

- No risk of spreading diseases
- No risk of ecocycle contamination/accumulation of persistent compounds
- No harmful emissions to air or to water
- Possibility for phosphorus recovery
- No organic material in residuals going to landfill
- Reasonable treatment cost per kg of dry sludge.

A method that allows for stack-less destruction of the organic material, recovery of the energy in the sludge and phosphorus recovery from the inorganic residuals is supercritical water oxidation. The method has been shown to have treatment costs at the same level as incineration or cheaper [8].

SUPERCRITICAL WATER OXIDATION
Supercritical water oxidation (SCWO) is a potential treatment option for sewage sludge. The principle of a SCWO reactor, Figure 1, is that oxidisable material in water above supercritical conditions, in the presence of an oxidant is quickly and completely oxidised due to the favourable properties of supercritical water. Detailed information on SCWO technology and sewage sludge SCWO can be found elsewhere [8-11]. The organic material in sewage sludge will be completely oxidised, mainly into carbon dioxide, in less than a minute in a tubular reactor. The released heat of oxidation can be recovered, excess oxidant (normally oxygen) can be recycled, carbon dioxide can be collected and sold as a by-product, the salt-containing water effluent is safe to release into normal recipients and the stable inorganic residual consists of elements in their highest state of oxidation.

A first commercial plant for sewage sludge SCWO was built in Harlingen in 2001 [9]. This plant treats undigested sludge and has been studied regarding environmental performance in a paper presented at last years Biosolids Conference [10]. The plant was shown to have a positive life cycle environmental impact. Energy related emissions were shown to be of very high importance. Since the energy system in Texas is very different from the Swedish energy system, a new study was made in which the Harlingen process was incorporated into the Swedish energy system. In Sweden, it is common to treat the produced sewage sludge with anaerobic digestion. Since many wastewater treatment plants already have a digestion facility and utilisation of the produced biogas, the impact of introducing a digestion plant before the SCWO process was also studied. The life cycle environmental impacts of moving the plant to the Swedish energy system and of introducing anaerobic digestion are reported in this paper as two different scenarios.

LIFE CYCLE ASSESSMENT OF SUPERCRITICAL WATER OXIDATION

The environmental impacts were evaluated using life cycle assessment (LCA). In an LCA study, a system is first defined, comprising a product or process under study and as much of surrounding activities as possible and appropriate, e.g. raw material extraction, electricity production and waste management. Data is then collected on all flows of natural resources, emissions, waste, energy, products etc to or from all activities. In this way, focus moves out from the product or process under study allowing a more complete view of the whole system brought about by a specific activity.

The LCA systems investigated in the environmental assessments in this study are shown in Figures 2 and 3. Figure 3 contains the activities that were added to the system in Figure 2 when digestion of the sludge was considered within the system. The LCAiT software was used to manage and analyse all data involved [12]. The Harlingen process data can be found in an earlier paper [10].

Electricity is consumed in the SCWO process by pumps used to increase the pressure of the sludge feed. In Texas, much of the electricity is produced from hard coal, oil and natural gas. Data was taken from the US Department of Energy [13]. In Sweden, electricity is produced mainly using hydropower and nuclear power. The data used are from 1995 [14]. In Figure 4, the different electricity production mixes used in the LCA are shown with pie charts. Electricity is also used to produce oxygen in a cryogenic plant. In Harlingen, oxygen is transported by truck from the production facility located 500 km away. The same distance and means of transportation was assumed for Sweden.

A gas-fired heater is used in Harlingen to raise the temperature of the sludge feed. Designs without need for external heating have been suggested [8], but the same design as Harlingen was chosen for the Sweden study.

In Texas, the recovered heat of oxidation is used to avoid natural gas heating in a nearby industry. In Sweden, an existing district heating network can be assumed in many urban areas and the recovered heat was thus considered to replace average Swedish district heat production. Data used was taken from the Swedish district heating association for 1996 [15] and is shown with a pie chart in Figure 5. "Other" includes waste heat from industries, waste incineration and a few other sources and was for the sake of simplicity not considered to have any environmental impact. Peat was considered to be a fossil resource. Electricity is from average production for Sweden as described in Figure 4.

In Harlingen, the residual inorganic solids are sprayed directly onto a dedicated land area. In Sweden, existing regulations make it necessary for the residuals to be sent to a landfill. Potential impacts from transport of the solids to a landfill and landfill management were thus included in the study with Swedish conditions. It was assumed that the transportation distance to the landfill is 20km and that transportation was made by means of a heavy truck. Full load (50% dry content) and empty returns was assumed. The landfill model was using the following assumptions [16]:
- diesel consumption for compactors and other machines = 0.04 MJ/kg waste,
- leachate production=2 litres of water /kg of waste,
- energy requirement for leachate treatment = 0.002 MJ/kg of water.

The model does not take into account material for building the landfill but the impact from this part could be considered small [17].

A study was also performed in which digestion of the sludge was added within the LCA system so that undigested sludge still enters the system, but digested sludge is treated with SCWO. The total amount of sludge that goes to the SCWO unit then becomes smaller as well as the organic content of the sludge. It is assumed that the produced biogas is used in the city gas net to replace production and consumption of natural gas. The activities added to the system presented in Figure 2 are shown in Figure 3. The anaerobic digestion unit is modelled using general performance data given in literature and information given by the Göteborg wastewater treatment plant. Assumptions used:
- In the sludge fed to the digestion unit, the organic fraction of the total solids is 60%
- During digestion, half of the organic material is metabolised into carbon dioxide and methane
- Digestion of 1 kg of organic material gives 0.6 m^3 of biogas
- 1 m^3 of biogas has a heat content of 6 kWh
- No methane leaks from the digestion unit

It was considered that no external heating is required for the digestion unit and the electricity demand for stirring was neglected. The heat output from the SCWO unit is smaller when treating digested sludge with a lower organic content. This has been accounted for in the study with a decrease proportional to the decrease in organic content. The change in consumption of electricity for pumps, gas for heating and oxygen was corrected for based on the flow of total solids through the SCWO reactor. The same flow of inorganic solids passes through the system in both cases. Since the produced methane will eventually be combusted in our scenario and replace natural gas, it is assumed that the only emission saved from combustion of methane is carbon dioxide. In methane from biogas, the carbon dioxide is of biological origin and should not be counted as a greenhouse gas according to general practice. The savings in production/extraction of natural gas, though, are fully accounted for.

All other life cycle activities have been described in detail earlier [10].

RESULTS AND DISCUSSION

The inventory results are presented here as characterisations into global warming potential (GWP) and photo oxidant creation potential (POCP) and weightings according to EcoIndicator99 and EPS2000. A description of the procedure can be found in last year's conference paper [10]. In Figures 6-9, the results are shown with columns for each major activity and one for the sum of all activities. A negative column indicates savings for the environment brought about by the suggested system and a positive column indicates an environmental impact. Both systems described in this paper (the study based on Swedish conditions and the study including digestion of the sludge) are reported in all figures and the Harlingen system is shown as a comparison. The GWP characterisation and the two weightings were chosen to be shown because the relation between the activities is representative for most characterisations and weightings that were used to interpret the results. The POCP characterisation is shown as an example of an interpretation that gives a different outcome for one scenario.

In almost all cases, the "Total" column is negative. This indicates that introduction of any of the systems is beneficial for the environment; by using the systems more emissions are avoided than are created. The major reason for this is the recovered heat replacing other means of heat production; district heat production in Sweden and natural gas heating in Texas. The possibility to recover heat is clearly very advantageous and necessary for such a positive turnout. Replacement of natural gas heating gives larger savings in greenhouse gas emissions than replacement of Swedish district heat production due to the relatively high percentage of biomass fuels in the district heat production mix, see Figure 5. It was assumed that average district heat production was replaced, but in fact it is more realistic that peak load fuels such as oil would be replaced. Even larger savings than the results from this study show could thus be expected. "Other" fuels for district heat production has been considered to have a negligible environmental impact, which also keeps the environmental impact represented by the saved district heat production lower than expected in reality. In the POCP characterisation, the heat recovered in the Texas system does not give rise to any savings at all due to the very low formation of photo oxidant precursors from natural gas combustion. This gives a completely different result for the Texas total when only photo oxidant creation is considered.

In the Swedish scenarios, the largest environmental impact is caused by gas heating of the sludge feed. The need for external heating could be avoided if another design was chosen for the SCWO process and the resulting environmental impact completely eliminated [8]. Slightly less recovered heat could then be expected. For Texas, electricity consumption in the SCWO process and for oxygen production constitutes larger environmental impacts than gas heating. This is due to the difference in electricity production between Texas and Sweden, (see Figure 4.) In Texas, electricity is produced mainly by combustion of fossil fuels while in Sweden other sources such as hydropower and nuclear power are prevalent. Transportation of oxygen is included in the "Oxygen" column, but constitutes a very small part. There are ways to change the design so that excess oxygen is recovered [8]. This would decrease the need for oxygen production and transportation and increase the electricity consumption in the SCWO process.

Solids disposal is of very low importance for the overall results. In the Sweden systems, transportation to a landfill and landfill management was added to the LCA system but the effect on the general turnout is very small.

When comparing sewage sludge SCWO in Sweden, with and without digestion, it is clear that based on the specifics reported in this paper, the sludge should be sent directly to the SCWO process, without preceding digestion of the sludge. In all interpretations of the results, the scenario with digestion yields lower environmental savings than the scenario without digestion. The savings in natural gas production and consumption due to produced biogas do not make up for the smaller amount of saved emissions from avoided heat production. The advantage should have been even larger for the scenario without digestion if a number of "best case" assumptions had not been made for the digestion process (no external heating required, electricity demand for stirring neglected, the same oxygen consumption per kg of dry sludge, the biogas does not have to be cleaned from carbon dioxide before use, no leakage of methane). In order to investigate the importance of potential methane leakages, a rough estimate was made, which indicated that even if all the methane produced leaked out this would not off-set the overall beneficial environmental performance of the system.

It must be noticed that there may be other advantages to sludge digestion before SCWO than the aspects considered in this study. Dewatering to a higher dry content is easier with digested sludge than undigested and thereby can a smaller and cheaper SCWO unit be used. Furthermore, SCWO treatment of a more homogeneous digested sludge has practical advantages. Digestion also extracts a high quality fuel from the sludge giving more flexibility in where, when and for what the energy is used. If a digestion unit already exists there may thus be reasons to continue digesting the sludge even if a SCWO unit is added to the sludge treatment system.

The two weighting methods used are based on two different value bases, giving different weights to different environmental impacts [18, 19]. As an example, human toxicity is considered relatively more important in the EcoIndicator99 method while resource depletion is considered to be of higher importance in the EPS2000 method. The methane produced in the digestion thus gives a more positive effect on the environmental impact when the results are weighted according to EPS2000 than according to EcoIndicator99. This makes the total for the two scenarios almost equal using this weighting method.

In this study, SCWO treatment of sewage sludge was not compared to other methods of sludge treatment. This should be done once the different options for the future have been identified. In this study, the functional unit on which all calculations were based was '1000 kg of sludge with a dry content of 7%'. This indicates that SCWO has been seen as a treatment method for sludge rather than a way to recover energy. If sludge SCWO was to be compared to other methods of sludge treatment or energy recovery, this must be reflected in the functional unit chosen. Another important issue, at least in Sweden, is phosphorus recovery from sludge. Phosphorus recovery from the sludge SCWO residuals is possible [5, 20] and a study should be performed in which different ways to recover phosphorus are compared from a life cycle perspective.

CONCLUSION

Sewage sludge SCWO has earlier been shown to give environmental gains in Texas and has now been studied for Swedish conditions with a positive turnout, regardless if digestion of the sludge was considered before the SCWO process or not. The main reason for the positive effect is the recovery and utilisation of the heat of oxidation. Depending on how the recovered heat is used and how electricity is produced in the area, the magnitude may differ but the overall performance is positive. In this study, digestion of the sludge before SCWO processing did not give as large environmental gains as direct SCWO processing of undigested sludge, but technical and economical considerations may still make SCWO processing of digested sludge an interesting process solution.

REFERENCES

1. UNITED NATIONS. *Plan of implementation of the World Summit on Sustainable Development.* September 23, 2002. Available at the homepage of The Johannesburg Summit: http://www.johannesburgsummit.org/html/documents/summit_docs/2309_planfinal.pdf

2. UNITED NATIONS DEVELOPMENT PROGRAMME, UNITED NATIONS ENVIRONMENT PROGRAMME, WORLD BANK AND WORLD RESOURCES INSTITUTE. *World Resources 2000-2001 – People and Ecosystems – The Fraying Web of Life.* World Resources Institute, Washington, D. C. 2000. 300-301. Also available at the homepage of World Resources Institute: http://earthtrends.wri.org/datatables/index.cfm?theme=4&CFID=65211&CFTOKEN=93267554

3. WORLD HEALTH ORGANIZATION. Brundtland starts new movement to address environmental crisis affecting children's health. Press release. Johannesburg. September 1, 2002. Available at the homepage of World Health Organization: http://www.who.int/mediacentre/releases/who66/en/

4. TAKAHASHI, M. KATO, H. SHIMA, H. SARAI, E. ICHIOKA, T. HATAYKAWA, S. MIYAJIRI, H. Technology for recovering phosphorus from incinerated wastewater treatment sludge. *Chemosphere.* 2001, 44, 23-29

5. STARK, K. *Phosphorus release from sewage sludge by use of acids and bases.* Licentiate Thesis. Department of Land and Water Resources Engineering, Royal Institute of Technology, Stockholm, Sweden. 2002

6. WOODS, N C. SOCK, S M. DAIGGER, G T. Phosphorus recovery technology modeling and feasibility evaluation for municipal wastewater treatment plants. *Environmental Technology.* 1999, 20, 663-679

7. SWEDISH ENVIRONMENTAL PROTECTION AGENCY. *Action plan for increased recycling of phosphorus from domestic/household/municipal sewage (in Swedish).* Working document, draft June 10, 2002. Available at the homepage of the Swedish Environmental Protection Agency: http://www.naturvardsverket.se/dokument/omverket/remisser/remisdok/fosfor/remfos.pdf

8. MODELL, M. SVANSTRÖM, M. Comparison of Supercritical Water Oxidation and Incineration for Treatment of Sewage Sludges, *Proceedings to the 6th Conference on Supercritical Fluids and Their Applications.* September 9-12, 2001. Maiori, Salerno, Italy

9. GRIFFITH, J W. RAYMOND, D H. The first commercial supercritical water oxidation sludge processing plant. *Waste Management.* 2002, 22, 453-459

10. SVANSTRÖM, M. FRÖLING, M. MODELL, M. PETERS, W A. TESTER, J. Life cycle assessment of supercritical water oxidation of sewage sludge, *Proceedings to the 6th European Biosolids and Organic Residuals Conference.* November 12-14, 2001. Wakefield, UK

11. KRITZER, P. DINJUS, E. An assessment of supercritical water oxidation (SCWO) – Existing problems, possible solutions and new reactor concepts. *Chemical Engineering Journal.* 2001, 83, 207-214

12. LCAiT, CIT Ekologik, a division of Chalmers Industriteknik, Chalmers Sciencepark, Sweden, http://www.lcait.com

13. UNITED STATES DEPARTMENT OF ENERGY. State Electricity Profiles 1998. Available at the homepage of the United States DOE: http://www.eia.doe.gov/cneaf/electricity/st_profiles/e_profiles_sum.html

14. BRÄNNSTRÖM-NORBERG, B–M. DETHLEFSEN, U. JOHANSSON, R. SETTERWALL, C. TUNBRANT, S. *Life-Cycle Assessment for Vattenfall's Electricity Generation.* Summary Report. Vattenfall AB, Stockholm, Sweden. 1996

15. Statistics from the Swedish District Heating Association, Stockholm, Sweden via LCAiT energy and transport database CIT-3g [12]

16. FINNVEDEN, G. JOHANSSON, J. LIND, P. MOBERG, Å. *Life Cycle Assessments of Energy from Solid Waste.* FMS 137. Department of Systems Ecology, University of Stockholm. 2000. Available at the homepage of the Environmental Strategies Research Group: http://www.fms.ecology.su.se/pdf/EurAApp6.pdf

17. CAMOBRECO, V. HAM, R. BARLAZ, M. REPA, E. FELKER, M. ROUSSEAU, C. RATHLE, J. Life cycle inventory of a modern municipal solid waste landfill. *Waste Management and Research.* 1999, 17, 394-408

18. For more information, see http://www.cpm.chalmers.se/cpm/publications/EPS2000.PDF

19. For more information, see http://www.pre.nl/eco-indicator99/ei99-reports.htm

20. Aqua Reci. Product information from Chematur Engineering AB and Feralco AB. Available at: http://www.chematur.se/download/Aqua%20Reci.pdf

FIGURES

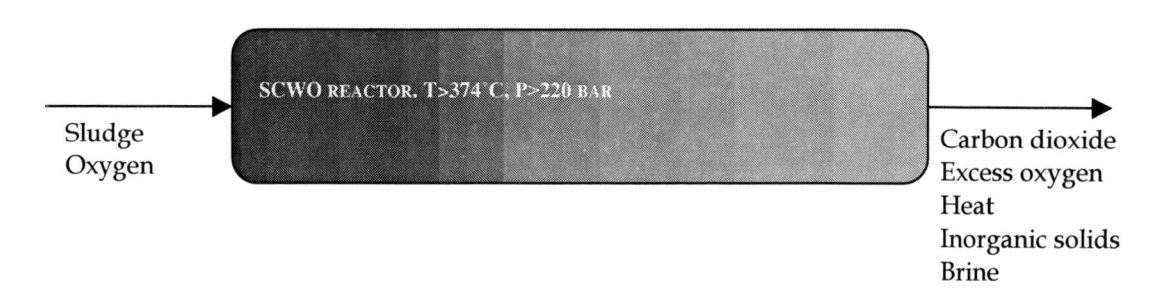

Figure 1. The general characteristics of SCWO treatment of sewage sludge.

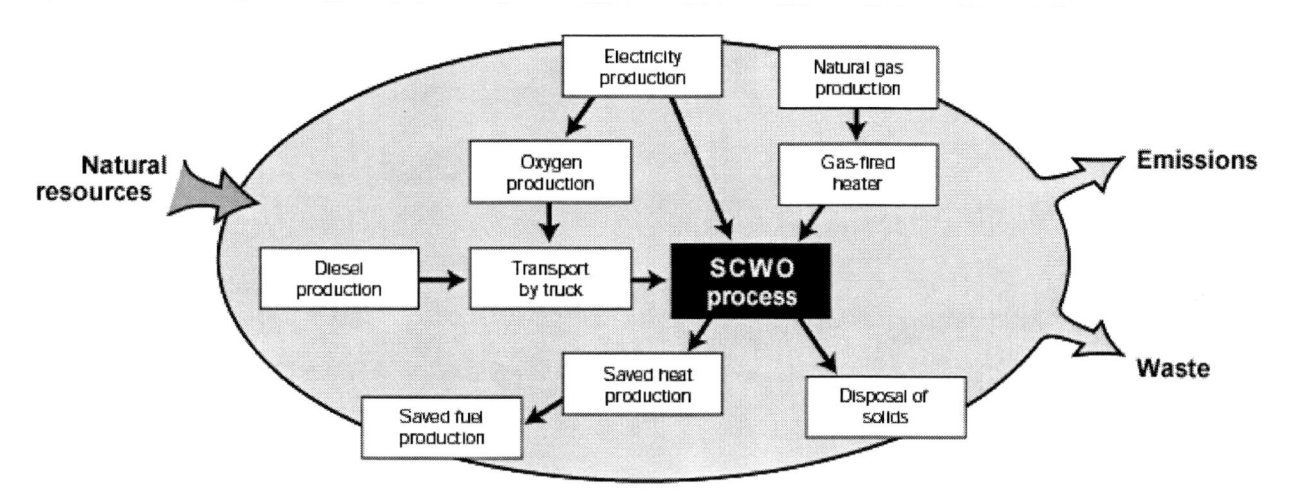

Figure 2. The LCA system used in the environmental assessment.

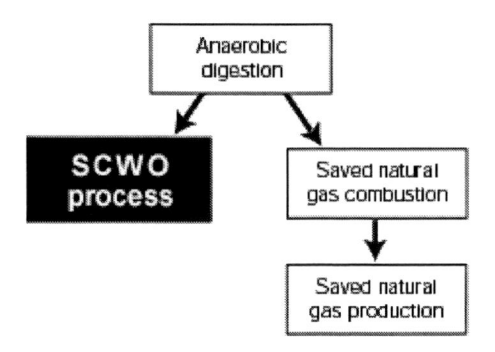

Figure 3. Activities added (white boxes) to the LCA system
in Figure 2 when considering anaerobic digestion within the system.

Texas average electricity production

- Hydropower 1%
- Nuclear power 13%
- Coal 45%
- Natural gas 41%

Sweden average electricity production

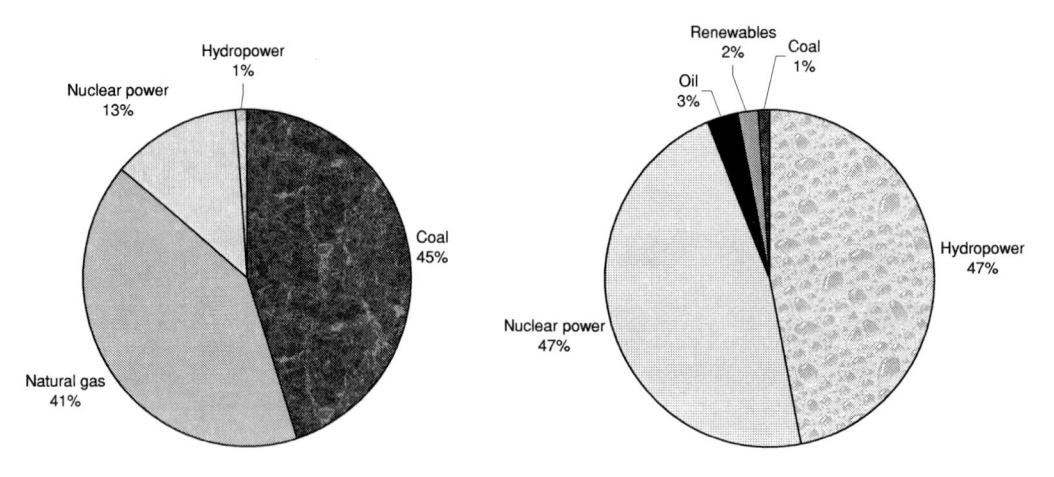

- Renewables 2%
- Coal 1%
- Oil 3%
- Nuclear power 47%
- Hydropower 47%

Figure 4. The electricity production mixes used in the study, for Texas [13] and Sweden [14], respectively.

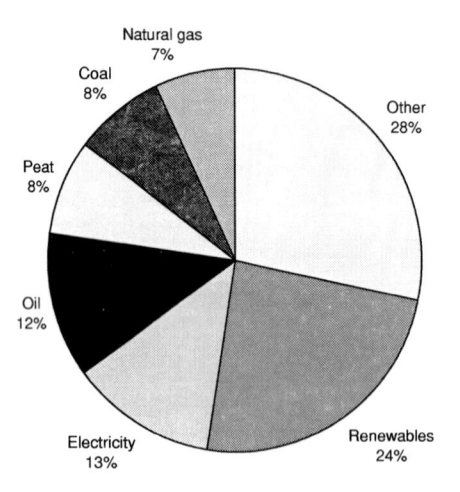

- Natural gas 7%
- Coal 8%
- Peat 8%
- Oil 12%
- Electricity 13%
- Renewables 24%
- Other 28%

Figure 5. The district heat production mix for Sweden used in the study. Statistics from the Swedish District Heating Association from 1996 were used [15].

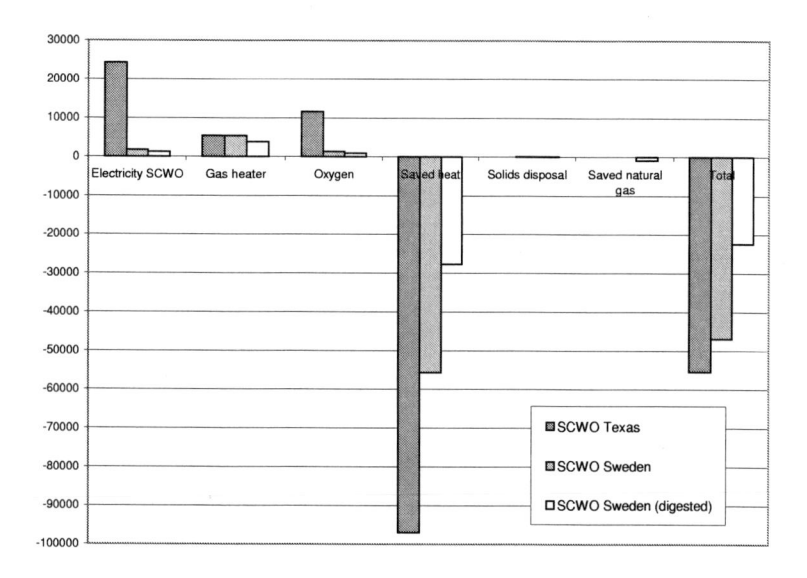

Figure 6. GWP characterisation of the LCA results. The three scenarios represent the Harlingen process with Texas energy and waste management data, the Harlingen process with Swedish data and the Harlingen process with Swedish data and digestion included in the system.
Unit: g carbon dioxide equivalents per kg dry sludge.

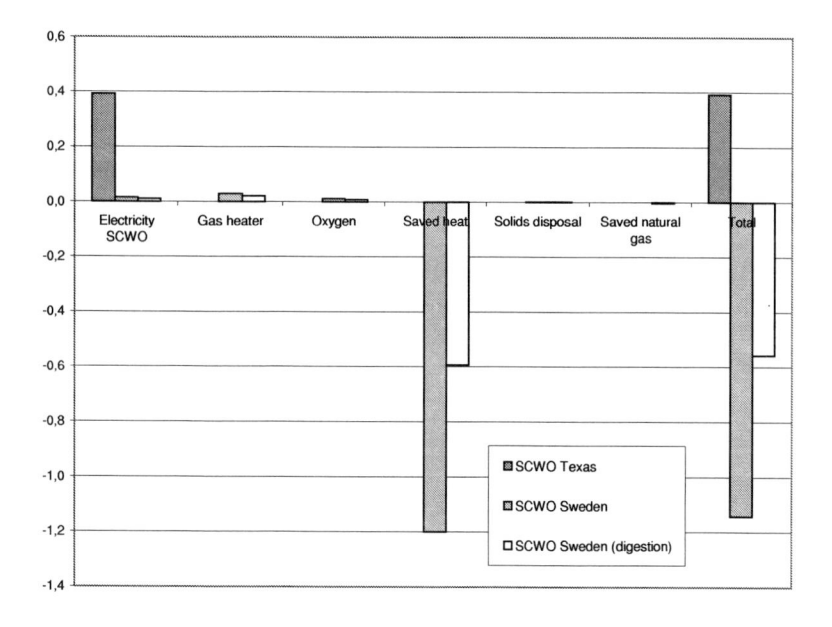

Figure 7. POCP characterisation of the LCA results. The three scenarios represent the Harlingen process with Texas energy and waste management data, the Harlingen process with Swedish data and the Harlingen process with Swedish data and digestion included in the system. Unit: g ethylene equivalents per kg dry sludge.

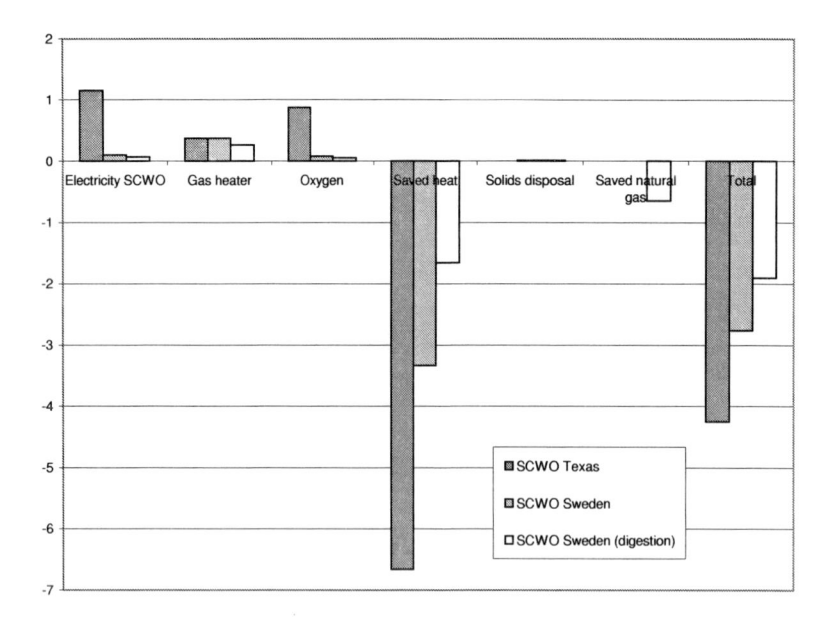

Figure 8. EcoIndicator99 weighting of the LCA results. The three scenarios represent the Harlingen process with Texas energy and waste management data, the Harlingen process with Swedish data and the Harlingen process with Swedish data and digestion included in the system. Unit: Ecopoints.

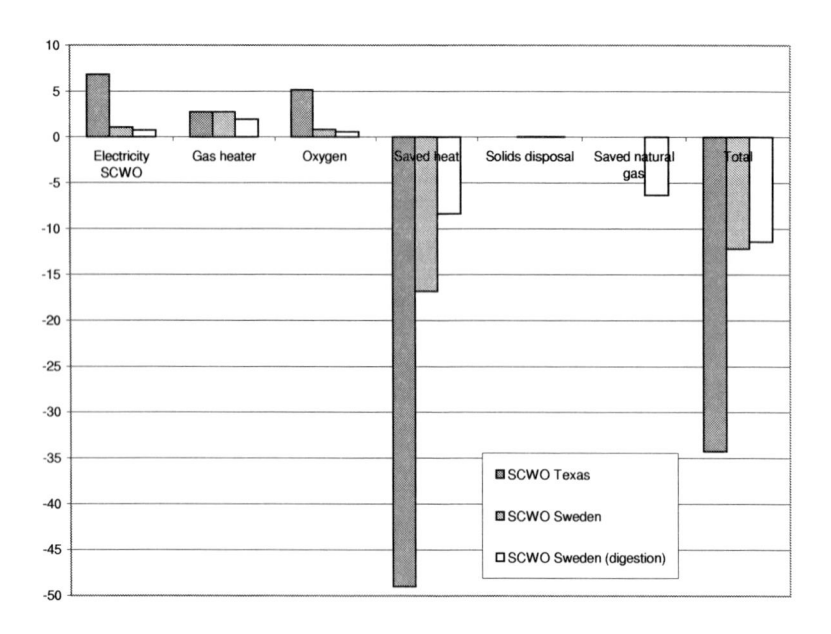

Figure 9. EPS2000 weighting of the LCA results. The three scenarios represent the Harlingen process with Texas energy and waste management data, the Harlingen process with Swedish data and the Harlingen process with Swedish data and digestion included in the system. Unit: Environmental Load Units (ELU).

WET OXIDATION USING ATHOS A WAO ALTERNATIVE; UPDATE

Gilbert, A.B.[1] Bigot, B.[1] Cretenot, D.[2]
[1]OTVB Ltd [2]Vivendi Water System (OTV SA)
OTVB Ltd, Birmingham, U.K.
Tel: +44 121 329 4000 Fax: +44 121 329 e-mail: agilbert@otvb.co.uk , bbigot@otvb.co.uk Vivendi Water Systems 4001 Saint-Maurice, Paris, France
Tel:+33 145 11 58 04 Fax: +33 145 11 58 10 didier.cretenot@vivendiwater.com

ABSTRACT

This paper details the minimisation impact of ATHOS® wet oxidation (WO) on the biosolids treated. It demonstrates stability of the solids produced, its mineral composition, which encases the heavy metal contaminants. It also details the stability of the liquid fraction its potential use as a source of carbon for carbon deficient biological processes, enabling this waste stream to replace more valuable carbon additives. ATHOS® WO, operates at temperatures and pressures below the critical point of water, is a viable minimisation disposal route for biosolids, too heavily contaminated to enter the more traditional recycle routes. The process fully contains the biosolids during their oxidisation reactions, preventing uncontrolled release of gaseous products to the environment. Also there is a significant reduction in the quantity of gaseous products with an improved quality, which greatly simplifies the exhaust gas treatment provisions needed when compared to incineration, for which such treatment is essential.

KEY WORDS

Additive, Ammonia, Biosolids, Catalysts, Dioxin/Furan, Leaching, Residual Solids, Sewage Sludge, Wastewater, Wet Air Oxidation,

INTRODUCTION

Improved wastewater collection and treatment has and will continue to substantially increase the volume of sewage sludge production and consequently quantities of biosolids requiring disposal. While some historic disposal and re-use practices are prohibited or being examined for their environmental sustainability, new disposal routes need to be identified.
Although recent research [1] has questioned the scientific values of the methodology being proposed by the new European legislation to the limitation of some of the organic residual levels in biosolids being applied to agricultural land. This legislation, whilst still permitting the reuse of biosolids as a nutrient source for crops (which is the UK Governments preferred disposal route), even before this more searching legislation takes effect, there are biosolid wastes, which are unsuitable for this use. These traditionally must be either be sent to landfill (suitably dewatered) or incinerated.
Landfill is becoming a limited and expensive resource, which should not be squandered on unreduced biosolids. The current alternative, incineration, requires the stringent operation of a sophisticated and expensive process facility with a significant number of interdependent process stages. The prospect of the construction of further incineration plants despite the Government incentives of green energy supplements (tax credits) is severely limited by neighbours (the NIMBY brigade).
ATHOS® is the solution proposed by Vivendi Water and is based on Wet Air Oxidation technology, although pure oxygen is used as the oxidant. This paper will demonstrate the viability of the ATHOS® process and provide an update on the introduction of this technology, which is an alternative disposal route to incineration.

WHAT IS WET AIR OXIDATION, OR MORE CORRECTLY WET OXIDATION?

Wet oxidation or hydrothermal oxidation is an attractive process for the destruction of biosolids from the treatment of waste streams, which are too dilute to incinerate and/or too concentrated for further biological treatment. It can be summarised as the oxidation of organic and inorganic substances in the aqueous phase, with the waste materials either in solution or suspension, using oxygen (or air) at elevated temperatures and pressures as the oxidising agent. At the operating temperatures for wet oxidation pressure is necessary to maintain water in the liquid phase (as superheated water) and to provide an over-pressurisation to maintain sufficient soluble oxygen. This enhanced solubility of oxygen in aqueous solutions at elevated temperatures provides a strong driving force for oxidation of the different species. Typical conditions for wet air oxidation range from 160°C and 20-bar to 320 °C and 140 bar. Retention times within the reactor may range from 15 to 120 minutes depending on the required degree of destruction. The chemical oxygen demand (COD) removal may typically be about 75% to 90%. By using pure oxygen, the amount of flue gases generated is 10 times less than that which would be generated by incineration of the same biosolids with air. In addition these combustion products are retained predominantly in solution and are only released to atmosphere by controlled venting through a catalytic converter.

The main components in the process comprise the reaction vessel, the feed and re-circulation pumps, heat exchangers and boiler, gas treatment, oxygen storage facility and a gravity settler.

The reaction tank is a pressure vessel of conventional design and is fitted with a re-circulation pump for mixing the contents. This serves three purposes as well as turning over the liquid contents of the reactor it also provides the motive power to an eductor which induces the pure oxygen supply into the system, whilst the treated effluent is removed as a side stream from the pumps discharge. The eductor, which, induces all the gas phases into the liquid flow passing through it, has this mixed liquid/gas phase discharged to the bottom of the reaction vessel.

The feed liquid flow passed into the eductor is a mixture of the recycled reactors contents and the thickened liquid biosolids fed. Whilst the gas stream includes components from the off-gas from the top of the reaction vessel, the oxygen supplied and the residual treated gas (from the gas stream catalytic reactor) which is not released to atmosphere via the (gas stream's) pressure regulation device.

The necessary heat required to sustain the thermal oxidation reactions is recovered from the treated effluent stream pumped out of the reactor (and removed as a side stream from the re-circulation pumps flow). The feed liquid biosolids are raised to the high pressure required by the feed pump (a positive displacement hydraulic twin ram unit) and this material is then heated by the inlet heat exchanger. The heat for this unit is recovered from the treated effluent's heat exchanger and can be supplemented by the boiler. This boiler would typically be fired, when required with biogas yielded by the upstream digestion process. To optimise the performance of the treated effluent heat recovery system, the pressure reduction for the liquid effluent from the ATHOS® process is performed down stream of the heat recovery unit.

Once its pressure has been reduced the liquid effluent containing the mineral suspension in separated into the supernatant and settled solids phases by a gravity settler (inclined plate unit). This is performed in an enclosed vessel to prevent odour release. The supernatant is returned back to the WWTP.

The ATHOS® process uses the catalytic effect of copper as will be discussed later to reduce the operating temperature. Some of the settled solids is returned to the suction side of the feed pump to minimise the quantity of fresh copper salts needed to be used as an additive to sustain the catalytic impact of the copper ion.

These settled solids are then dewatered, ideally in a combined filter press vacuum dryer unit, with low grade waste heat from the ATHOS® process being used as the heating source. The filtrate and condensate from this dewatering process are returned to the WWTP with the supernatant. The mineral solid components are removed as a dried mineralised solid which contains the heavy metals in the mineral matrix.

The result of biosolids wet oxidation is that, insoluble organic matter is converted to simpler soluble organic compounds which are in turn oxidised and eventually converted to carbon dioxide and water (if the reactions were left to completion), without emissions of NOx, SO_2, HCl, dioxins, furans, fly ash, etc. The last residual organic compounds to be oxidised are fatty acids, especially acetic acid. Organic amine nitrogen is converted to ammonia, but only a limited amount of nitrogen elimination is obtained (10-20 %) due to the solubility of ammonia.

The common feature of the existing wet oxidation processes is that by contrast with incineration, they are intrinsically energy efficient. The only energy required is the difference in enthalpy between the incoming and outgoing streams. While for incineration, the energy yield needs to supply not only the latent heat for the liquid and solids plus the heat of evaporation of the liquid, but also heat this water vapour, combustion products and excess air to the combustion temperatures of between 820 and 1100 °C, as shown in Figure 2, [2]. Therefore, wet oxidation does not require mechanical dewatering to about 35% $^w/_w$ dry solids to achieve autothermic combustion, which is common for incineration. Thickened sewage sludge's or biosolids containing about 4% $^w/_w$ dry solids can be fed to wet oxidation and treated autothermally.

Unlike Super-Critical-Water Oxidation, the temperatures and more particularly pressure of the oxidation reactions are kept in the realms of conventional pressure vessel design, so the reactor is a simple pressure vessel rather than a complicated plug flow set of very high pressure tubing.

THE HISTORY OF ATHOS®

In the early 1990s, Anjou Recherche launched a WAO sludge treatment research programme, which led to promising results in each stage of scale-up, from lab tests to pilot installations. Thus, the novel hydrothermal oxidation process (ATHOS®) was developed by Vivendi Water Systems (OTV), with a large scale demonstration plant which treats 3 m^3/hr of thickened sludge (150 kg/hr DS); i.e., the output of a 50,000 P.E. capacity plant, constructed and operated in Southern France. This demonstration unit was financed partially by the Vivendi Water group, OTV France; ADEME; LIFE (EU funding) and Agene de L'eau (the regional environmental body for) Adour Garonne. The results obtained during this process development are reported. Together with a summary of the status of the first two full scale plants being constructed.

One of the major research areas when using the ATHOS® process to dispose of biosolids was the impact this had on the potential release of heavy metals back into the environment. When the research was started residual solid organic concentration was one of most important characteristics taken into account, as demonstrated by the circular (normalised NF X 31 210 test 09/05/1999. This comprises grinding the sample and mixing these solids into demineralised water, to determine the solubility of the heavy metals.) issued by the French Ministry of Environment for slag with a low leachable fraction. Leaching tests on residual solids have been conducted with analyses of heavy metals on leachate, for which valorisation (value enhanced) properties of the solids are also being developed.

CURRENT CONSTRUCTION

Further to the ATHOS® demonstration at Toulous, a plant which treats a liquid sludge at 4 % $^w/_w$ at up to 3 m^3/h a population equivalent of 50,000 P.E., two ATHOS® plants are currently under construction.

NORTH BRUSSELS WASTEWATER TREATMENT FACILITY

The first is part of the new wastewater treatment plant to serve North Brussels. The ATHOS® plant for this works of 1.1 million P.E. has been included to ensure biosolids minimisation is available for this prestigious and sensitive scheme. The construction and operation contract's for this scheme has been won by Aquiris (Vivendi Water). The successful bid was based on a suitable biosolids disposal strategy. The site had severe constraints with the disposal of biosolids to land being very commercially sensitive due to access to the available land bank being under the control of a competitor's subsidiary organisation and the stringent controls being defined by the Belgium Departments (dioxins from chicken waste).

The site selected for the new wastewater works prohibited the use of on site incineration as a disposal route. This was considered in the project appraisal by use of an off-site facility. However, the cost effective solution identified by Vivendi Water's bidding team was the adoption of ATHOS® as the main sludge disposal strategy. The biosolids are first dewatered, to a concentration suitable for thermal hydrolysis before treatment by conventional mesophilic digestion but with improved conversion due to the hydrolysis. This is used to maximise the available energy yield and minimise the residual solids. These residual biosolids, are then directly treated by the ATHOS® process for minimisation. This scheme for Brussels is due for completion in 2006.

TRUCCAZZONA IN ITALY

This plant is being constructed at a wastewater treatment works near Milan in Northern Italy. The ATHOS® plant will treat digested bio-solids from the 350,000 P.E. plant. Although, in Italy, incineration is frequently used as a minimisation method for biosolids disposal, the driver for this scheme was the reluctance of the local municipality to introduce an incineration facility to their region. Although they will have a significant increase in the quantity of biosolids requiring disposal when the new work is completed. They have limited landfill facilities and no suitable land bank (due to the low alkalinity of the regions soils) for more conventional agricultural disposal. This works is due for construction completion in 2005 with operation following in early 2006.

RESULTS

Operating parameters for wet oxidation, are relatively high for temperature (150°- 350°C) and pressure (20 - 200 bar), with a residence time in the range (15 - 120 min). The operating pressure is fixed at a value higher than saturating pressure corresponding to the oxidising temperature to maintain condition reaction at a liquid state. In those conditions, efficiency of COD removal can attain values between 70% and 90% depending on operating parameters [3].

REACTION PRODUCTS FROM WET OXIDATION USING ATHOS®

Following wet oxidation using ATHOS® the residual organic matter is in the form of volatile fatty acid (VFA) with two to six carbon atoms. This is demonstrated in Table 1 [4], which shows the analysis of the supernatant from wet oxidation. As part of the development of ATHOS® different operating conditions with different catalytic additives have been studied. The residual VFA compositions the relationship between oxidation efficiency of the biosolids and VFA yield together with the refractory levels of COD have all been considered as indicated by the following sections.

EFFECT OF COPPER ADDITIVE AS A CATALYSTS

To optimise the temperature, pressure and residence time of wet oxidation using ATHOS® during the early test work, several heterogeneous and homogeneous catalysts were investigated [3]. Copper as soluble Cu^{2+} or supported copper oxide was found the most effective catalyst. In the presence of copper, the results for COD removal obtained at 235°C, was equivalent to that obtained at 285°C without catalyst as presented in Figure 3 [3]. This indicates removal efficiency of the incoming COD (amount of residual organics in the wastewater) against operating temperature, both with and without the catalytic action of the Copper additive.

Even at 285°C in the presence of a copper catalyst, acetic acid remained refractory to oxidation. The effect of copper was mainly to accelerate the chain reactions transforming organic compounds into acetic acid. This limited COD removal to 85 % at temperatures lower than ca. 300°C.

CHARACTERISATION OF THE RESIDUAL SOLID

Characterisation of the residual solids from wet oxidation of wastewater bisolids has been investigated using different physio-chemical methods. Compositions of the mineral constituents have been studied and are given in Table 2 [6]. Here it can be observed that the concentration of initial phosphorus contained in the sludge is converted to phosphate, which, is precipitated in the solid phase.

Analyses have also been performed with greater accuracy using Fourier infrared spectroscopy (diffuse reflection and transmission); Figure 4 [6]. The most common identified phases are the following: kaolinite, quartz, carbonates and phosphates Table 3 [6]. The Granulometric analyses show that residual solids are extremely fine and homogeneous, particle sizes are on the order of micrometers, with a median diameter of 2.7 µm and diameters T=25% of 1.5 µm, T = 70% of 5.4 µm and mean of 5.0 µm.

MEASUREMENT OF ORGANIC MATTER IN THE RESIDUAL SOLID

Analysis of organic matter in the residual solids, or evaluation of the quantity of 'unburned' material, was performed using the weight loss method at 500°C for a period of 4 hours, with samples initially dried at 103°C. The total organic carbon (TOC) is determined by the difference between total carbon, which is measured by the production of CO_2 after combustion at high temperature, and inorganic carbon, which is measured by CO_2 production after chemical 'attack' in an acid solution. Results show that the temperature of wet oxidation conditions is the critical factor, which determines the reduction of volatile organic matter in the solid phase, with residual values ranging from 37% at 195°C to 5.5% at 285°C (see Table 3). It has also been found that TOC values of these samples are always less than are indicated by the volatile suspended matter assay. This difference can be explained by the specific operating conditions of wet oxidation. In wet oxidation large amounts of CO_2 are produced after oxidation of organic matter, which leads to high CO_2 partial pressure within the reaction vessel. This high partial pressure and CO_2 content promotes carbonate and hydroxyl formation in the solid phase. These compounds can interfere in weight loss method measurements, which cannot distinguish the volatilised organic species from this mineral source of weight loss.

The concentration of the organic matter in the residual solid phase separated by settlement and filtration is dependant on the operating temperature of ATHOS®. The results from three test runs using different operating temperature are indicated in Table 4 [6] and shown in Figure 6 [6].

This differential thermal analyses performed on these three samples and indicated by the curves shows that a high exothermic reaction is observed between 290°C and 420°C for samples A and B, corresponding to wet air oxidation temperatures of 195°C and 235°C, respectively. This exothermic reaction can be attributed to the combustion of the organic matter. It can also be observed that curves B and C present endothermic characteristics around 680°C–700°C, which can be attributed to dehydroxylation or decarbonisation of minerals.

LEACHING TEST

Sludge from municipal wastewater treatment facilities always contains heavy metals, more or less concentrated, the principal constituents being lead, copper and zinc [7]. The analysis of the heavy metals contained in the sludge studied and residual solids, are presented in Table 5 [5]. The individual and average valves of the three leaching tests indicate that except for mercury, which is partially soluble in the aqueous effluent phase from wet oxidation, all the initial metals are concentrated in the solid by-product. The principal function of the process being to destroy organic material, 100 g of initial dry solids, containing 61% volatile suspended solids (VSS), leads to only approximately 30g of mineral residual after treatment, corresponding to a concentration factor around 3.3 for these heavy metals as indicated in Table 7.

Results presented in Table 5[5] from the leaching test (using the French normalised method
NF X 31 210 test), indicate that the heavy metals precipitated in to the solid phase are insoluble and thus are non-leaching in water. The operating conditions prevalent in wet oxidising using ATHOS®, with a high CO_2 partial pressure in the reactor are very favourable, both physically and chemically for the precipitation of heavy metals, which are freed during oxidation of organic substances, as metallic hydroxide and carbonate. Theses are locked in the mineral matrix of the residual solid [8].

These residual solids can be dumped in landfills without stabilisation because of the absence of soluble elements, since it is the result of treatment in an aqueous medium. It consists of a fine powder (average diameter 5 µm) mainly comprising aluminium phosphates, kaolin, quartz, calcite and an amorphous fraction, which immobilises heavy metals in the form of hydroxides, carbonates and insoluble phosphates. These tests showed that, after optimisation of sludge oxidation and dehydration of the residual solid, the latter complied with the criteria for stabilised waste products defined in the order of 18 February 1994 issued by the French Ministry of the Environment. The residual ash also meets the characteristics defined by the French circular of 9 May 1994 for slag with a low leachable fraction (category 'V').

SUPERNATANT CHARACTERISATION AND USE AS A CARBON SOURCE FOR ADVANCED NUTRIENT REMOVAL

After oxidation, most of the residual organic matter is organic volatile fatty acids (VFA) with two to six carbon atoms. The use of sludge WAO supernatant liquors as substrate for methane generation has been described [9]. Different operating conditions were studied to investigate the relationship between oxidation efficiency and VFA composition. It has been found that as the oxidation efficiency increased, the VFA composition was modified with a higher ratio of short chain molecules due to oxidation of high molecular carboxylic acids into low molecular weight acids, especially into acetic acid. In addition to VFA, oxidation products such as alcohol (methanol) and aldehydes (acetaldehyde) were formed. Biodegradability tests have been conducted on the supernatant. Only 10-15% of the COD is refractory after 23 days [10].

To identify the impact of the return-flow in the wastewater plant we studied COD biodegradability in more detail, in terms of kinetics. It is considered that, due to its high biodegradability, the return flow from wastewater biosolids wet oxidation, compared with conventional de-watering of digested sludge concentrate, is unlikely to create an additional COD load on the wastewater treatment plant.

ATHOS® WET OXIDATION SUPERNATANT AS A CARBON SOURCE

Wastewater de-nitrification utilised carbon in the influent, but total nutrient (nitrogen and/or phosphate) removal is often limited by a lack of carbon, and an additional pure carbon source is needed to achieve complete elimination. Experiments were carried out to clarify the possibility of performing de-nitrification with a BAFF (Biological Aerated Flooded Filter) process and using supernatant from hydrothermal sludge oxidation as a carbon source. Pilot plant experiments were conducted with a 1-m high, 100-mm diameter, granular expanded schist filter of particle diameter 3-6 mm. Primary sewage dosed with $NaNO_3$ adjusted to have 17mg/l of $N-NO_3$ was used as the influent. The supernatant (derived from settling the effluent from the ATHOS® process) was diluted with the influent to adjust the COD value to 70 mg/l. Water velocity was 3.8 m/h, with a temperature around 18°C.

With a $N-NO_3$ load of 1.5 kg/(m^3.d) and a COD/N ratio of 4.8, a 50% nitrate removal efficiency was obtained Figure 5 [6], which corresponds to 0.7 kg/(m^3.d). The COD reduction was also 50% of the initial concentration. This part of the COD corresponds with the VFA fraction in the influent Table 6[6].

CONCLUSION

The ATHOS® process is a more energy efficient biosolids total minimisation technology, than incineration. The optimum operating conditions for the process are at a temperature of 250°C and with a pressure of 50.0bar g. These operating conditions enable established process/mechanical design criterion to be utilised for the specific plant components of the system.

The use of pure oxygen significantly reduces the amount of off-gas, which requires treatment and helps to minimises the yield of nitrous oxides maintaining most of the organic nitrogen in a soluble state as ammonia.

Because of the fully enclosed nature of the ATHOS® process with controlled releases from the pressurised circuit the environmental impact is significantly less than that from incineration. The process stream utilises fewer interdependent treatment stages and is a much simpler treatment train when compared with the incineration route for biosolids minimisation disposal route.

If digestion (and ideally enhanced high performance digestion) is applied to the biosolids prior to wet oxidation with the ATHOS® process it is questionable whether gasification of biosolids with its associated complex treatment train will be environmentally sustainable.

REFERENCES

1. JONES, K., STEVENS, J. A Critical Evaluation of the Proposed Limit Values for Organics in EU Working Document on Sludge and Development of a Tiered Screening Process to Identify Priority Pollutants in Sewage Sludge. UKWIR 2002 in Press.

2. CHOWDURY, A.K, COPA, W.C. Wet air oxidation of toxic and hazardous organics in industrial wastewaters. *Indian Chemical Engineer* 28 (1986) 3

3. DJAFER, M., BOURBIGOT, M.M. Catalytic wet air oxidation of municipal sludge : effect of reaction parameters. *Proc. First Int. Conference on Advanced Oxidation technologies* London, Ontario, p. 207 (June 1994)

4. LUCK, F, DJAFER, M, ROSE, J. P, CRETENOT, D. ATHOS®: Hydrothermal oxidation of sludge In Proc. ICUPCT, Hong Kong, China, 1999.

5. LUCK, F., DJAFER, M., ROSE, J. P. ATHOS®: A Novel Process for Sludge Disposal, In Proc. *AWWA 18th Federal Convention*, Adelaide, Australia, (1999)

6. DJAFER, M., LUCK, F., ROSE, J. P., CRETENOT, D. Transforming sludge into recyclable solids and a valuable carbon source by wet air oxidation In Proc. *IWAQ Conference Sludge Management for the 21st century* : a value-added renewable resource, Fremantle, Australia, April (1999).

7. WIART , J., REVEILLERE, M. (1995). La teneur en éléments-traces métalliques des boues résiduaires des stations d'épuration urbaines françaises. *T.S.M.*, 12: 913-922

8. REDDY, K.J., GLOSS, S.P., WANG, L. (1994). Reaction of CO2 with alkaline solid waste to reduce contaminant mobility. *Water Research*, 28: 1377-1382.

9. KNOPS, P.C.M., BOWERS, D. L., KRIELEN, A., ROMANO, F.J. (1993). The start-up and operation of a commercial below-ground aqueous phase oxidation system for processing sewage sludge. *VerTech Treatment Systems Technical Documentation*.

10. DJAFER, M., LUCK F., WACHEUX, H., BOURBIGOT, M.M. Wet Air Oxidation of Sewage Sludge: Identification and Biodegradability of By-products. In Proceedings "Water Treatment By-Products" Conference, September 1994, Poitiers, France,Vol. 2, p. 44/1-44/7. or M. DJAFER, F. LUCK, H. WACHEUX, M.M. BOURBIGOT, Wet air oxidation : study of the by-products and biodegradation tests. *Proc. First Int. Congress on By-Products of Water Treatment*, Poitiers, France (September 1994)

11. M. DJAFER, F. LUCK, Characterization of the residual solids from wet oxidation of sludge. *Proc. Second Int. Congress on By-Products of Water Treatment*, Rennes, France (March 1997)

ACKNOWLEDGEMENTS

The author's would like to thank the following for their assistance in preparing this paper:
Victoria Ballard; Pascal Marlin; Lucie Partia and Merzak Belkhondja.

TABLES

Table 1 Mean Volatile Fatty Acid Composition in the Oxidised Supernatant

Fatty Acid	g/l of COD	Fatty Acid	g/l of COD
Acetic acid	5.6	Isovaleric acid	0.2
Propionic acid	1.9	Methyl 4-valeric acid	0.1
Butyric acid	0.25	Hexanoic acid	0.2
Isobutyric acid	0.75	Heptanoic acid	0.1
	0.087	Acetone	
Total COD from VFA			8.82
Total COD from supernatant			16.0

Table 2 Weight Analysis of Mineral Consituents in the Solid Residual form ATHOS® Treatment

Mineral Compound	SiO_2	CaO	Al_2O_3	Carbonate CO_3	Sulphate SO_4	Phosphate P_2O_5
% $^w/_w$ in mass of dry solid	38.3	16.8	12.0	3.6	2.0	13.0

Table 3 Infrared (IR-FTIR) Spectroscopy of Mean Wave Number

Wave number	Attributes
3695 and 3620 (v)	Al-OH (kaolinite <5%)
879 (v), 1426 double (v) and 1038	Ca-O, two carbonates
1263 (v) and 2926 (v)	C-H organic matter
799	Si-O quartz
604, 563 and 694	phosphate

Please also refer to Figure 3 [6]

Table 4 Concentration of Organic Matter in the Residual Soild Depending on the Oxidation Temperature

Temperature ºC & Sample Reference	VSS ($\%^w/_w$) in Solids Residual	TOC ($\%^w/_w$) in Solids Residual
195 sample A	37.0	14.8
235 sample B	8.0	3.2
285 sample C	5.5	1.2

Please also refer to Figure 5 [6]

Table 5 Results of Leaching Tests From the Solid Residuals of ATHOS® Wet Oxidation

Species	Leachate Liquid Content No 1 in (mg/l)	Leachate No 2 in (mg/l)	Leachate No 3 in (mg/l)	Leachate Total Solids fraction following drying (mg/kg DS)	Circular of 9/5/94 French Permitted Values (mg/kg DS)
Cadmium, Cd	nd*	nd*	nd*	nd*	1.0
Total Cr, as Chromium	nd*	nd*	nd*	nd*	1.5 for Cr^{6+}
Mercury, Hg	0.0013	0.0015	0.0006	0.14	0.2
Lead, Pb	nd*	nd*	nd*	nd*	10
Arsenic, As	nd*	nd*	nd*	nd*	2
Sulphates, as SO_4	40	15	5	2484	10 000
TOC	9	10	8	1117	1500

nd*: none detcetd < 20 µg/l or 0.020 mg/l

Table 6 Experimental Results for De-Nitrification, Average Values After 6 Days

COD Inlet	COD Outlet	COD Removal	% COD Removal	N as (NO_2 - NO_3) Removal	% N Removal	CON/N Ratio
70 mg/l	33 mg/l	37 mg/l	52 %	7.72 mg/l	46 %	4.8

Table 7 Analysis of Heavy Metals Contained in Municipal Sludge Before and After ATHOS® Wet Oxidation

Heavy Metal	Dry matter concentration in mg/kg		Concentration Value
	Sludge (Average values of samples)	Treated Solids (Average values of samples)	
Cadmium, Cd	10	35	3.5
Total Chromium, as Cr	40	130	3.2
Copper, Cu	376	1300	3.4
Mercury, Hg	5	2	-
Nickel, Ni	32	110	3.4
Lead, Pb	145	528	3.6
Zinc, Zn	820	2750	3.3

FIGURES

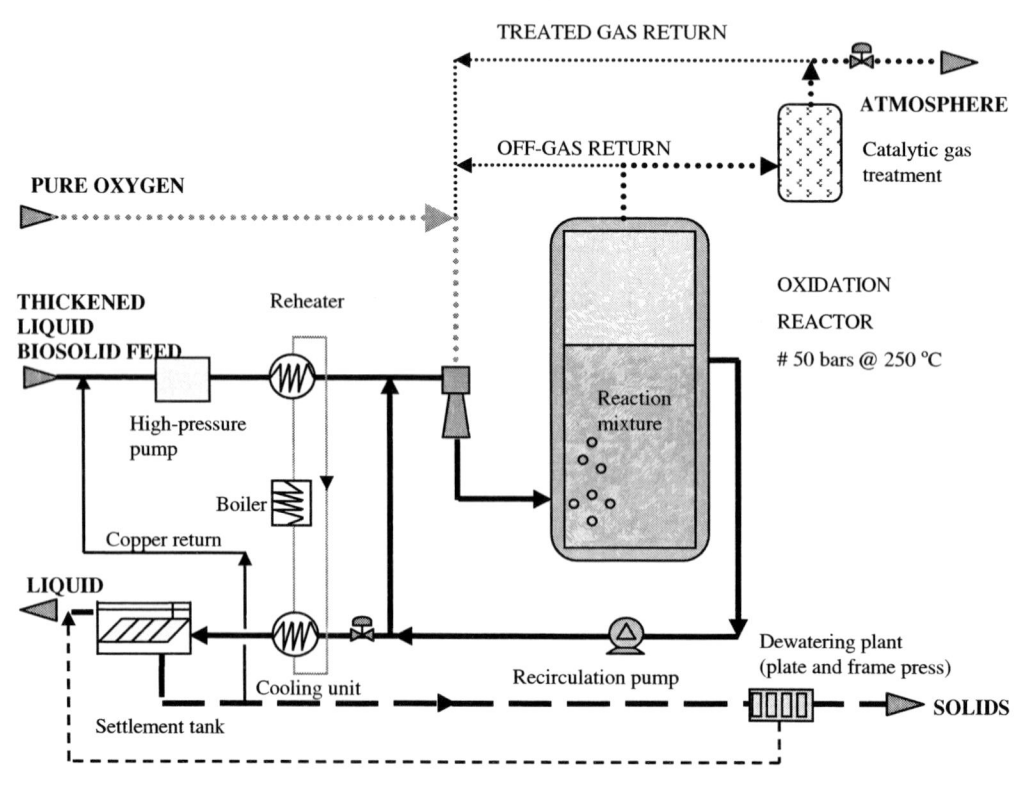

Figure 1   Process Schematic Diagram of ATHOS® Biosolids Treatment

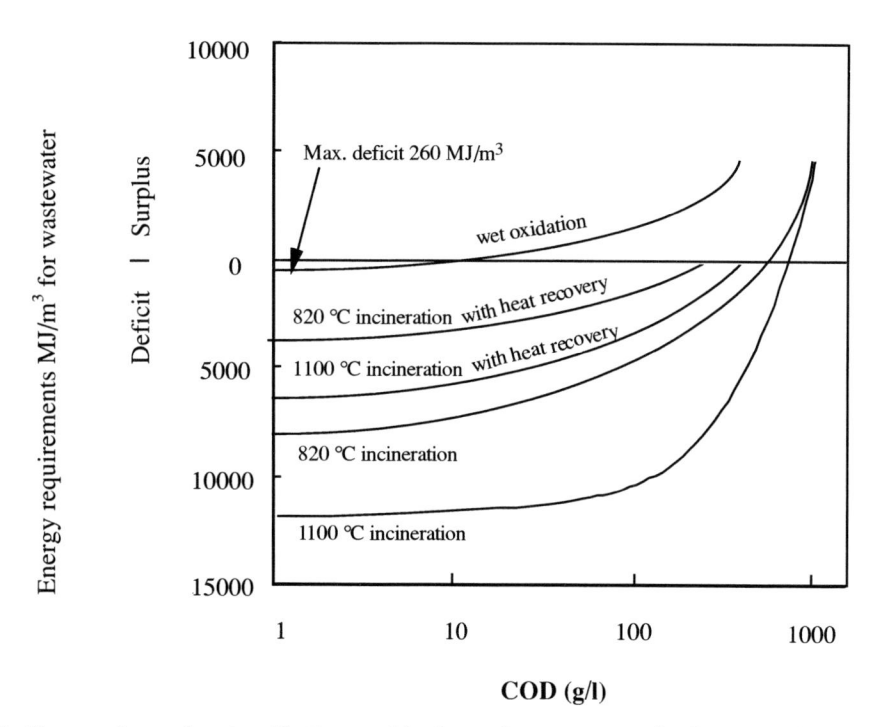

Figure 2: Comparison of wet oxidation and incineration processes for heat requirements as a function of the amount of organics in the wastewater [2].

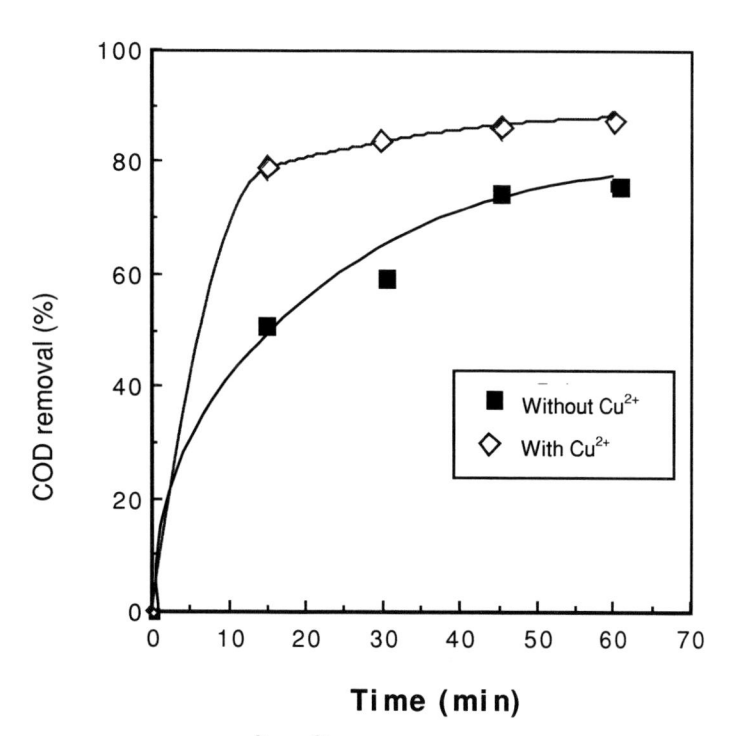

Note with Influence of Cu^{2+} (Cu^{2+}/COD = 0.01 g/g) and reaction time on the removal of COD at 235 °C.

Figure 3: Influence of Cu^{2+} and Reaction Time on Removal of COD at 235 °C [3].

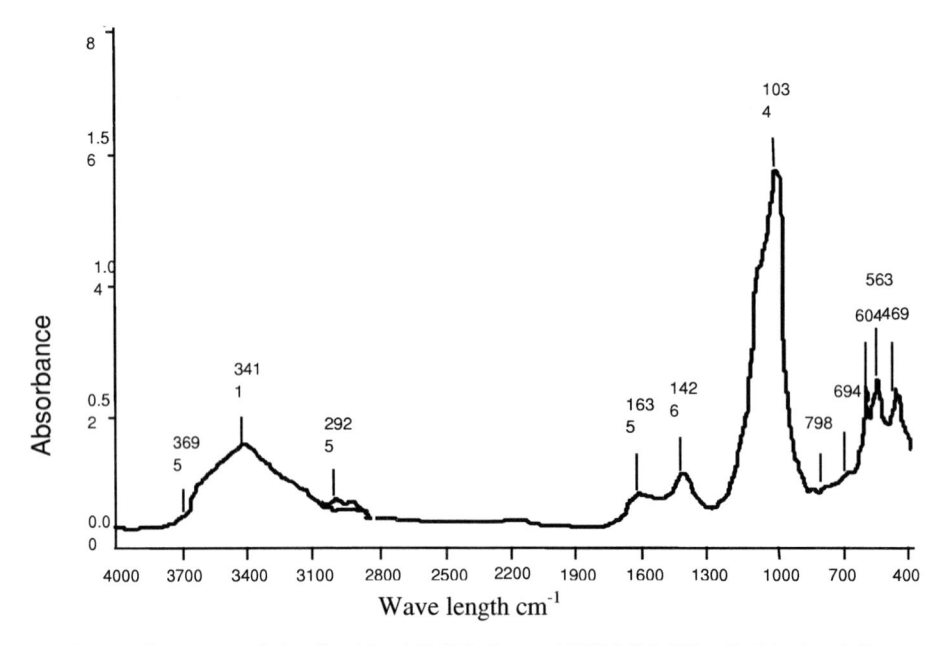

Figure 4: Infrared Septrum of the Residual Solids from ATHOS® Wet Oxidation [6].

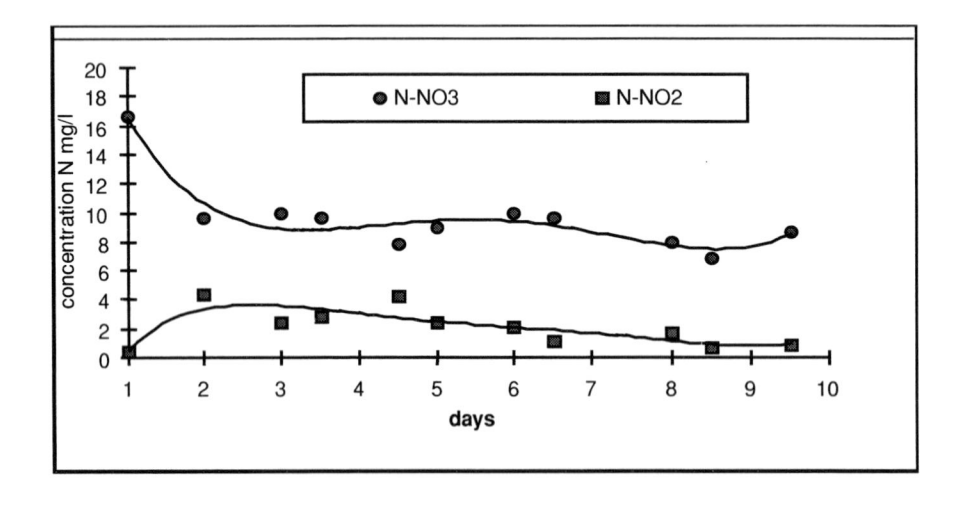

Figure 5: Time Adaption for Nitrate Removal With ATHOS® Wet Oxidation Effluent Supernatant as the Carbon Source [6]

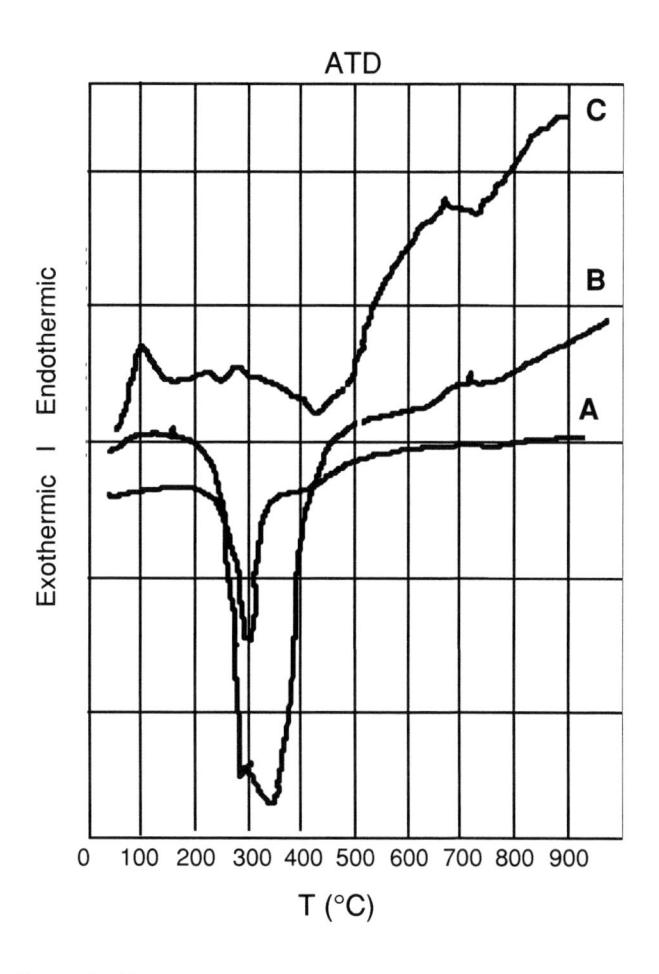

Figure 6: Differential Thermal Analysis for Three Samples A; B & C With ATHOS® Wet Oxidation [6].
Samples; A at 195 °C, B at 235 °C and C at 285 °C. Please also refer to Table 4

Figure 7: General view of plant.

FIRST OXYGEN DEBOTTLENECKING OF LARGE SCALE FLUIDISED BED FURNACE FOR SEWAGE SLUDGE INCINERATION

[1]Paul Ludwig, [2]Gerhard Gross

[1]Infraserv GmbH & Co. Höchst KG,
Industriepark Höchst, D 286, D 65926 Frankfurt am Main, Germany
Tel. +49-(0)69-305-16215 FAX +49-(0)69-305-80291 E-mail: paul.ludwig@infraserv.com

[2]Messer Griesheim GmbH
Industriegase Deutschland, D 47805 Krefeld, Germany
Tel. +49-(0)2151-37 93 87, FAX +49-(0)2151-37 97 10, E-mail: gerhard.gross@messer.de

ABSTRACT

Ever stricter carbon residual requirements for dumping municipal and industrial wastes such as sewage sludge will increase the demand for additional incineration capacity. Besides adding additional capacity by high capital expenditure debottlenecking of the existing plant can be achieved by the partial oxygen combustion process (POC) jointly developed and patented by Messer Griesheim and Infraserv Höchst. In this process, oxygen is blown directly and transversally into the fluidised bed at supersonic speed. Based on experiences of the POC process with one large scale fluidised bed reactor for iron sulphate decomposition at the Kerr McGee Pigments plant in Uerdingen Infraserv Höchst transformed the POC process the first time world wide to all two fluidised bed furnaces for sewage sludge incineration. Three years of experience with detailed operating results are achieved till today.

Each fluidised bed reactor has a grate area of 16.6m² and is followed by a heat recovery boiler, an electrostatic filter and a two-way scrubber. Up to 2,000m³/h of pure oxygen is blasted into the two furnaces transversally via the supersonic nozzles built into the walling. This results in a highly turbulent flow, which considerably improves the mixing and the combustion in the fluidised bed.

Besides industrial and residential sewage sludge with calorific values of 500 to 11,000KJ/kg (depending on the original substance (OS)) and dry-substance (DS) contents between 22 and 60%, other solid and liquid waste material suitable for fluidised bed combustion can also be processed, including bone meal, screenings, plastic granulate, spent active carbon, building timber, organic solvents and urea wastewater. Calorific values can here fluctuate between 1,500 and 42,000KJ/kg (depending on the original substance) and the dry-substance content between 5 and 100%.

The combustion capacity of the fluidised bed furnaces could be increased by the POC process from 130,000 to currently 160,000 t/a. The hourly throughput of waste material suitable for fluidised bed combustion has been increased by up to 40%. At the same time the combustion air volume has been reduced by 15% and the specific natural gas consumption by 35%. This considerably reduces the specific incineration costs calculated in Euro per ton of waste material. And even the specific emissions, e.g. carbon dioxide and nitrogen oxides per tonne of incinerated waste material are reduced, too.

KEY WORDS

Cost savings, energy savings, fluidised bed furnace, increase throughput, oxygen enrichment, oxygen injection sewage sludge incineration.

INTRODUCTION

Infraserv Höchst and Messer Griesheim have jointly developed a new process which has succeeded in increasing the throughput capacity of the two fluidised bed furnaces for sewage sludge incineration more than 40%, while at the same time reducing the specific energy demand by over 35 %.

In this process pure oxygen is injected transversally into the fluidised bed. With low investment costs, the process permits either a considerable increase of throughput capacity per unit cross sectional area or smaller physical dimensions in the case of new construction.

OBJECTIVE

The disposal of waste is controlled in Germany by waste treatment regulation laws, the so called "TA Siedlungsabfall". Especially waste containing more than five per cent annealing loss, thermal treatment will be mandatory in 2005, because the disposal is no longer admissible. This applies also to municipal and industrial sewage sludge.

Thermal treatment of sewage sludge is possible by incineration in fixed fluidised bed furnaces, operating as so called mono-incineration plants, but also by co-incineration in existing municipal waste incinerators and power plants. Because of the special design of mono sewage sludge incinerating plants not only large quantities of sludge can be incinerated but also sludge with problematic composition.

Compared with co- incinerators, which main task is electrical power generation, the highest priority of mono sludge incinerators is the safety of sludge disposal. This priority and the legal requirements by the national and European laws for flue gas cleaning are the reasons for higher disposal costs of mono incineration plants compared with co-incinerators.

From chemical processing it is known that fluidised bed reactors for decomposing iron sulphate and spent sulphuric acid from the titanium production was doubled in throughput by use of oxygen as additional oxidation agent. Instead of enriching the hot fluidising air oxygen was injected transverse by supersonic velocity directly in the fluidised bed, because of safety reasons. Such reactors are operating since 1997 at the Kerr Mc Gee plant in Krefeld, Germany [1].

To investigate the possibility of this new oxygen based process for incinerating sewage sludge, Infraserv Höchst and Messer Griesheim run tests with one of the fluidised bed furnaces. Of special interest was its effect on the feed increase, the energy consumption and emission values at the chimney.

DESCRIPTION OF THE FLUIDIZED BED FURNACE

The sewage sludge incineration plant at Infraserv Höchst consists of two independent incineration lines with a theoretical sewage water sludge throughput of 10.4 tons/hr each. Each incineration line comprises a fluidised bed furnace with secondary reaction zone, a heat recovery boiler, a three-compartment electrostatic precipitator and a two-stage wet washing process (Fig1). The flue gases produced during incineration of the sewage sludge are cooled down in the heat recovery boiler to about 200°C. Most of the 16 bar vapour produced with the waste heat is fed into the pipeline system. The furnace itself has a grate surface of 16.6m² and consists of an air distribution chamber, a fluidised bed and a secondary reaction zone (Fig. 2).

The air heated to about 800°C in the pre-combustion chamber flows into the fluidised bed furnace from the bottom after passing the air diffusion chamber and the grate. The fluidised bed material is kept in constant movement by the air pressed in. The sewage sludge falls from the top into the 850°C hot fluidised bed.

The fluidising material - silica sand - crushes the sludge and keeps it in suspense. The combustion air can now attack the sludge from all sides. The organic components of the sewage sludge are incinerated within a very short time. The inorganic components are discharged in form of ash. The water contained in the sewage sludge is vaporised. The oxygen content in the fluidised bed is regulated to an amount between one and three per cent by volume, to avoid nitrogen oxide formation. The space above the fluidised bed serves both for the boil-off of solid and gaseous flue gas components and for the separation of coarse and fine ash particles. The combustion temperatures are here between 850 and 900°C.

The post-combustion of carbon monoxide takes place in the secondary reaction zone. To that purpose the secondary air heated to about 800°C is injected via a ring channel with 8 nozzles distributed around the circumference. The combustion temperature after air supply is at least 850°C.

PROCESS MODIFICATIONS FOR THE USE OF OXYGEN

Every combustion process needs oxygen as oxidation agent, which in generally is supplied from the atmospheric air. But as everybody knows air consists of 21 % oxygen by volume but also of 79 % nitrogen as ballast which has to be heated not only at the necessary process temperature but is also a substantial part of the flue gas volume to be treated from the flue gas cleaning units. Using oxygen as additional oxidant in an incineration process the reduced nitrogen volume leads in principle in:

- a reduction of the process gas volume

and

- an increase of the combustion temperature.

Therefore using oxygen for incineration of sewage sludge in a fluidised bed furnace the reduced nitrogen volume can be replaced by flue gas volume of additional sewage sludge feed incineration without increasing the flue gas volume or fluidising speed.

Higher combustion temperatures can be kept constant for all:

- endothermic reactions

and

- exothermic reactions with heat flow

by controlling the energy input.

One important requirement is the adequate mixing of all reactants. Because of high water content the incineration of sewage sludge is typically a exothermic reaction with heat flow. When selecting the process technology for adding oxygen into the fluidised bed furnace, enrichment of fluidising air by oxygen seems to be the simplest technology on the first view. But oxygen enrichment is limited by safety reasons because of the increased reactivity of oxygen enriched air. Cold air with oxygen concentration of more than 25 % by volume is handled as pure oxygen by safety reasons.

Additionally, a fluidising bed furnace is operating normally with air preheated up to temperatures between 250 and 800°C. In these cases it is not possible to guarantee the safety and resistance of the materials used within the area of the feed line, wind box and grate.

The solution to this problem was found by separating the oxygen and fluidising air inputs by injecting the oxygen directly into the fluidising bed. To guarantee proper mixing of oxygen with the reactants in the fluidised bed, oxygen is injected transverse to the main flow with supersonic speed. Supersonic Laval-nozzles are installed in the refractory lining of the furnace wall, about 300mm above the grate (Fig.3). The number depends on the area of the grate. Additionally the high turbulent oxygen jets improve the transverse mixing or mass transfer and the radial heat transport in the fixed fluidised bed itself.

RESULTS

By means of oxygen besides municipal and industrial sludge with calorific values of 500 to 11,000 kJ/kg, in relation to original substance (OS) and with dry substance (DS) contents between 22 and 60 %, other solid and liquid wastes have been jointly incinerated, such as meat-and-bone meal, plastic granulate, filter cake, used active charcoal, organic solvents, urea wastes, detergent liquids, mother liquors, paint and lacquer sludge. The calorific values here vary between 1,500 and 42,000kJ/kg (in relation to original substance) and the dry substance content between 5 and 100 %.

The oxygen flow rate was controlled by a oxygen flow control unit depending of the sewage sludge feed rate (Fig. 4). The oxygen flow rate varies between 0 and 1600m³/hr (stp). The quantity of natural gas used for heating is automatically reduced in order to maintain the selected fluidised bed temperature. The fuel oil quantity fed into the furnace was kept constant. The air quantity was adapted accordingly by the control system.

Using a video camera installed in the furnace it was possible to permanently evaluate the quality of the fluidised bed. The sewage sludge feed was increased in accordance with the oxygen feed.

SEWAGE SLUDGE THROUGHPUT

The throughput increase of sewage sludge in a fluidised bed furnace is largely dependent on the calorific value and water content of the sludge. Fig. 5 shows the results of the incineration of a sewage sludge with a lower calorific value of 1,600 kJ/kg and 45% (DS) in relation to the quantity of oxygen. At an oxygen quantity of 1,600 m³/h, the sludge throughput (in relation to OS) rose from 8.5 to 12.8t sludge/h, i.e. by approx. 51%. From over 12,500 continous measurement data for the different sludges and suitable wastes, at a mean oxygen throughput of 1,200m³/h, a throughput increase of approx. 42% was registered. In relation to the dry substance, the mean sludge throughput increased from 3.6 to 5.1 t/h DS (Fig. 6). In principle, it can be said: The higher the calorific value and the dry substance content, the more throughput is possible with oxygen.

FUEL CONSUMPTION AND STEAM GENERATION

In order to maintain the temperature level in the fluidised bed and post reaction zone, the quantity of natural gas was altered and the oil quantity was kept constant. The adjustment of the natural gas quantity is carried out automatically by the temperature regulation system. The natural gas savings are dependent on the dry substance content and the calorific value of the sludge. All the measurement data obtained show a reduction of the specific natural gas consumption (in relation to the dry substance content) of 3,600kWh per ton of sludge to 2,340kWh/t sludge, i.e. by approx. 35% with an average throughput increase of approx. 35% (Fig.6). Thus, the use of oxygen reduces the specific demand for additional fuels. The reason for this is the reduction of the combustion air, i.e. less nitrogen ballast has to be heated up. This also has effects on the steam generation. This is not increased in proportion to the sludge throughput, since the specific natural gas consumption falls as the oxygen quantities rise (Fig 5).

TEMPERATURES

Twenty six hermocouples for temperature measurement have been installed over the whole cross section of the fluidised bed furnace, i.e., in the wind box, the fluidised bed, the furnace head and the secondary reaction zone. On the basis of the temperatures measured, it is possible for the plant operator to assess the quality of the incineration. Fig. 7, for example, shows the temperatures resulting from a sewage sludge with 45 % DS and a heating value of 1,600kJ/kg OS. The temperature profile in the furnace was homogeneous during the incineration. The temperatures measured remained constant.

When the sludge throughput is increased, the falling fluidised bed temperature is automatically compensated by increasing the quantity of natural gas. The temperatures in the post-reaction zone rise proportionally to the windbox temperature. This temperature difference is dependent on the waste constituents: In the case of a high organic waste fraction and increased sludge throughput, because of the shorter dwell time in the fluidised bed, post-burning of partially oxidised combustible gas components such as carbon monoxide takes place in the post-reaction zone. Then, the natural gas quantity in the after-reaction zone is automatically reduced according to the temperatures.

EMISSIONS

In principle, the use of oxygen concentrates all the pollutants and the CO_2 and H_2O in the flue gas, as the flue gas quantity, through the reduction of the combustion air, remains practically constant (Fig. 8).

This is of special significance for the emission of nitrogen oxides, since no special unit exists for the removal of nitric oxides from the flue gas. Even when the throughput is increased, the legal limits still have to be complied with.

In Fig. 9, the nitrogen oxide, carbon monoxide, mercury and dust emissions are shown as mean values from all 12,500 measured values. The formation of thermal NOx is not possible in the temperature ranges of approx. 850°C found in the fluidised bed and 950°C in the post-reaction zone. However, organic nitrogen compounds in sewage sludge and other wastes suitable for fluidised bed incineration are converted directly to nitrogen oxides in the fluidised bed.

Even at higher sludge quantities through the use of oxygen, nitric oxide emission is no higher than in pure air operation. It is generally found that the NOx concentrations are always below the limit value demanded by the German emission protection order (called "17. BimSchV") of 400mg/Nm³ (half-hour mean value) and 200mg/Nm³ (daily mean value). The emission of carbon monoxide remains almost constant at approx. 5mg/m³, pointing to complete combustion in the post-reaction zone. Other emissions, such as mercury and dust, for example, also show no significant increase. This means that the two-stage scrubber reliably separates the increased pollutant content. There is continued compliance with the legal limits in accordance with "17. BimSchV".

PROCESS ADVANTAGES

In terms of the environment, disposal safety, and productivity, the following advantages are found:

ENVIRONMENT
- lower specific CO2 emissions, as primary energy input is lower
- lower nitric oxide emission than before the process change
- no emission of additional pollutants
- better burnout

DISPOSAL SAFETY
- the throughput capacity of the fluidised bed furnace can be varied between 60 and 150 %
- no intermediate storage of sludge
- flexible co-ordination of the times of external sludge deliveries

PRODUCTIVITY

The experiments have shown that, with increasing utilisation, the profitability of mono-sludge fluidised bed incineration rises:
- the specific fixed costs per ton of incinerated sludge are reduced in proportion to the increase in capacity.
- the specific additional energy costs are reduced in proportion to the oxygen throughput.
- the personnel and maintenance costs are reduced in proportion to the increase in capacity.
- the investment costs for the oxygen flow control unit and the installation of the supersonic jets in the furnace are low.
- the specific quantity of steam is lower, as the specific energy consumption is reduced.
- the oxygen costs incurred per ton of sludge are low.

CONCLUSIONS

Disposal of the sewage sludge produced in Germany is presently handled as follows, about 70% is spent as fertiliser in farming, about 10 % is dumped and only 20% is incinerated. Till 2005 it will be allowed to dump sewage sludge in landfills. In 2005 the new waste regulation "TA-Siedlungsabfall" will become effective so that from thereon only ashes with an annealing loss of less than 5% are allowed to be dumped or stored in mines as a packing.

Legal regulations will be required also for sewage sludge spent as fertiliser in farming because of the increasing heavy metal accumulations in the soil.

It can be concluded that there is a trend towards incineration when speaking of future sewage sludge disposal. During the next few years an increased demand for incineration capacities can be expected.

Through employment of the partial oxygen combustion process (POC), the annual capacity of the Infraserv plant was, with the approval of local authorities and the Environmental Office, increased from 130,000 to 160,000t. Thus, with low investment costs and reduced specific energy consumption Infraserv Höchst has a virtual third furnace available, with which it can react to the requirements of the market according to demand.

REFERENCES

[1] LAILACH, G.; WÜLBECK, D.; GROSS, G., Transversale Überschallinjektion von Sauerstoff in Wirbelschichtreaktoren, gas aktuell 55, S. 10-16, 1998

[2] LUDWIG, P.; GROSS, G., Debottelnecking of fluidised bed furnace for sewage sludge incineration by oxygen, Zusammenfassung zur 16th International Conference on Fluidised Bed Combustion (Ref.Nr. 180)

FIGURES

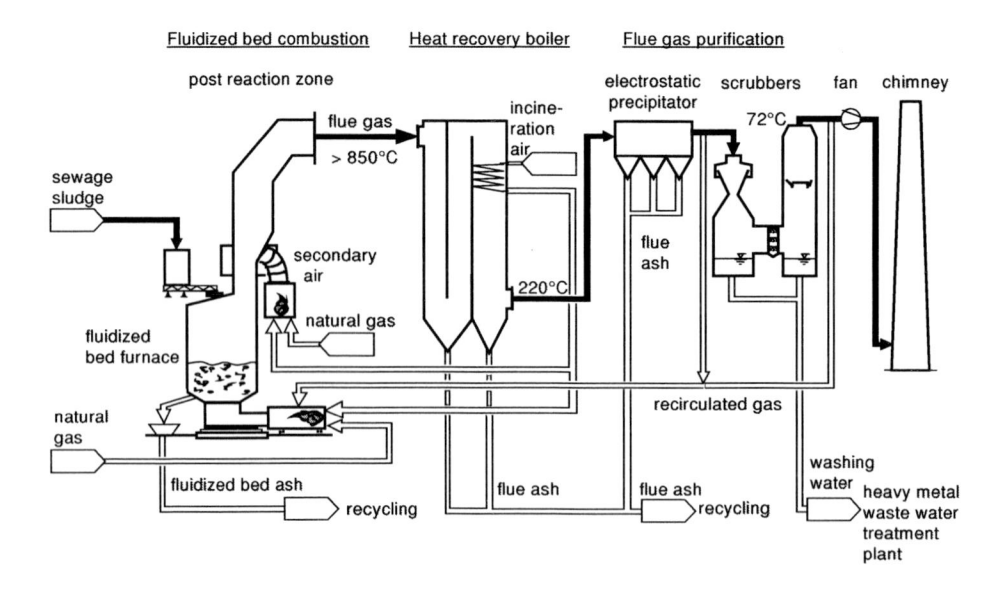

Fig.1 Flow sheet of sewage sludge incineration plant "Höchst Industrial Park" (cross sectional area of fluidised bed furnace 16,6 m²)

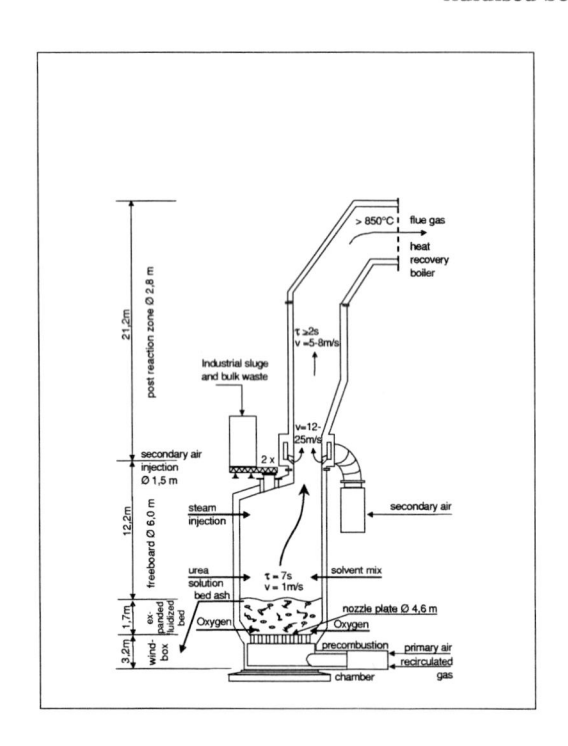

Fig.2 Schematic of fluidised bed furnace

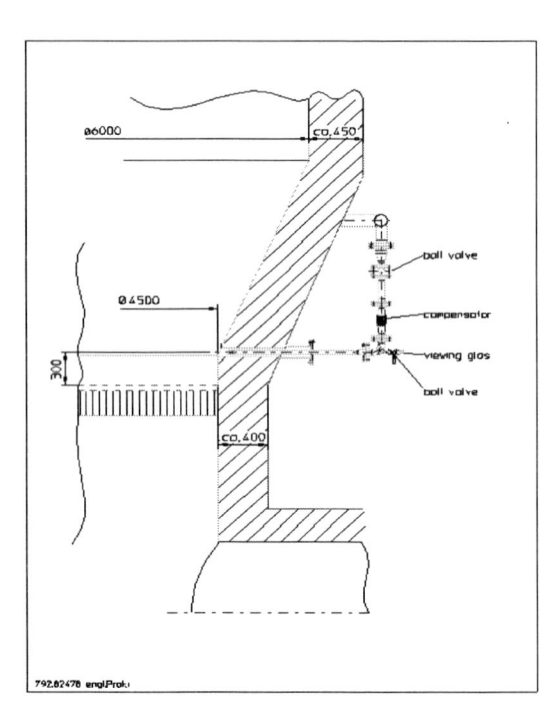

Fig. 3 Installation of supersonic nozzles in fluidised bed

Fig. 4 Automatic oxygen flow control unit

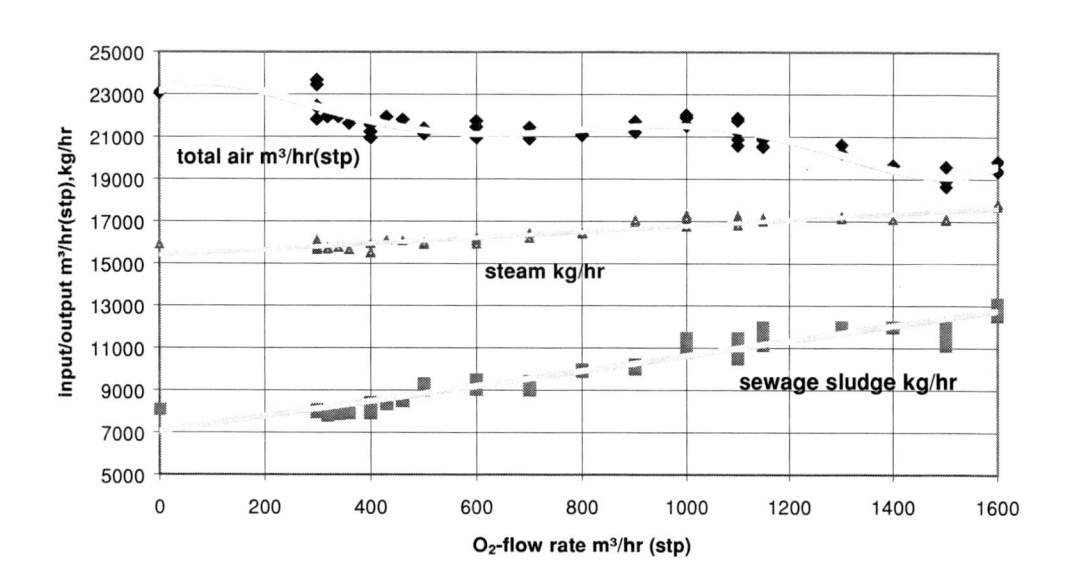

Fig.5 Sewage sludge feed rate, vapor quantity and total air flow rate as function of the oxygen flow rate

Municipal and industrial sludge mixture, total results
Dry substance DS = 44,3 %, mean lower heating value = 2.450 kJ/kgOS

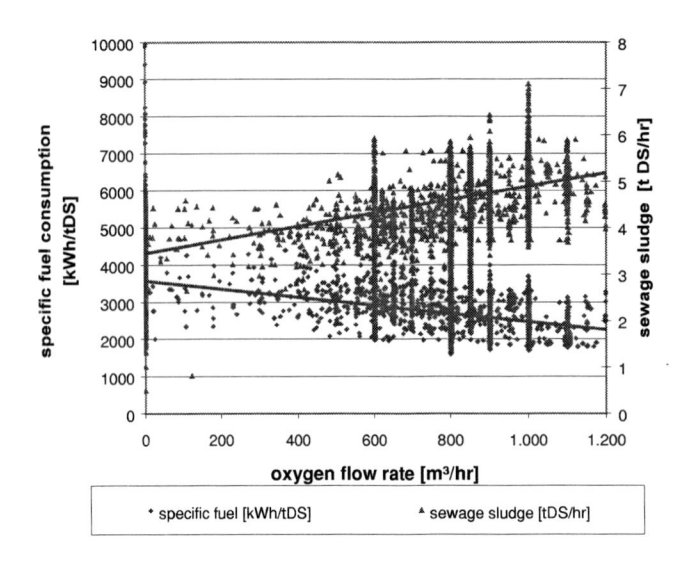

Fig. 6 Specific fuel consumption and sewage sludge feed as function of oxygen flow rate

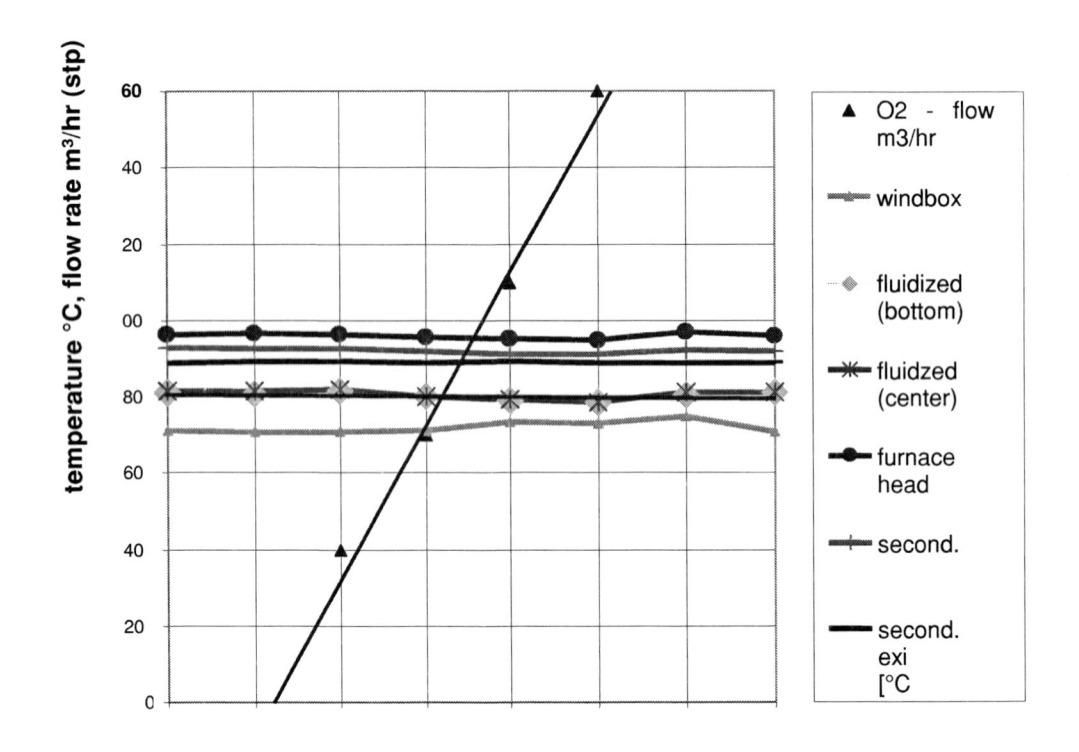

Figure 7: Furnace Temperature

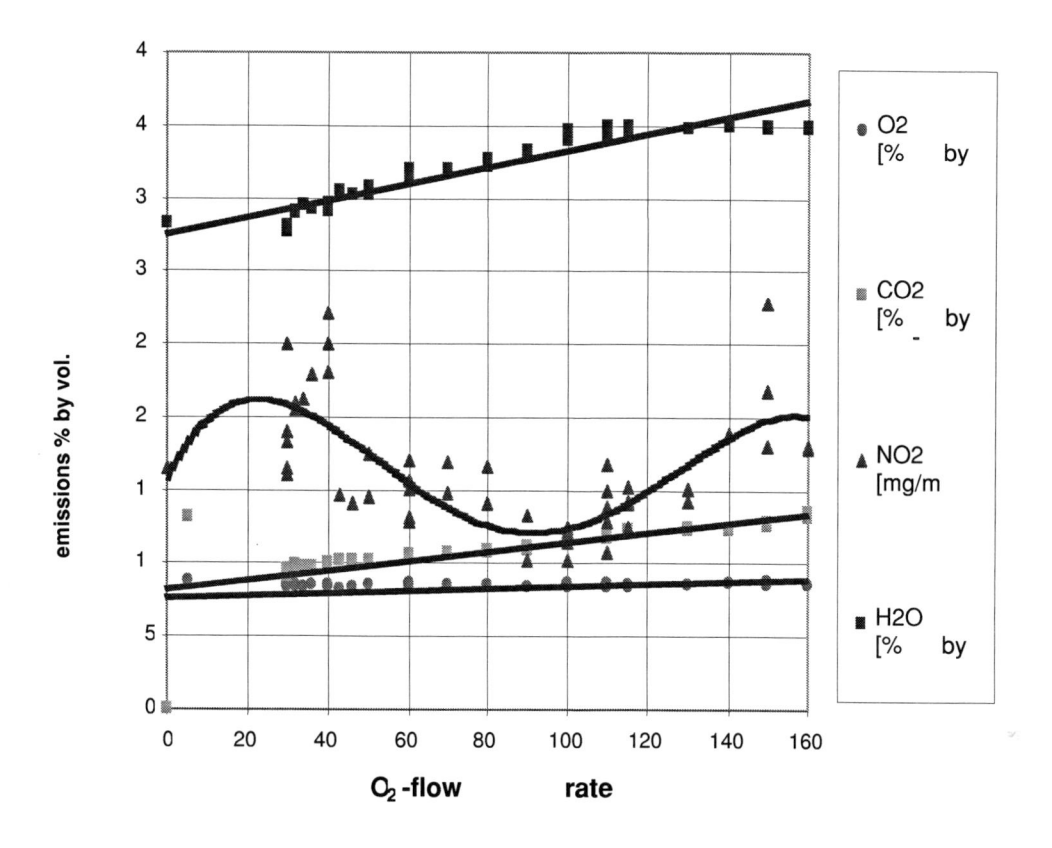

Fig. 8 Emissions at chimney as function of oxygen flow rate

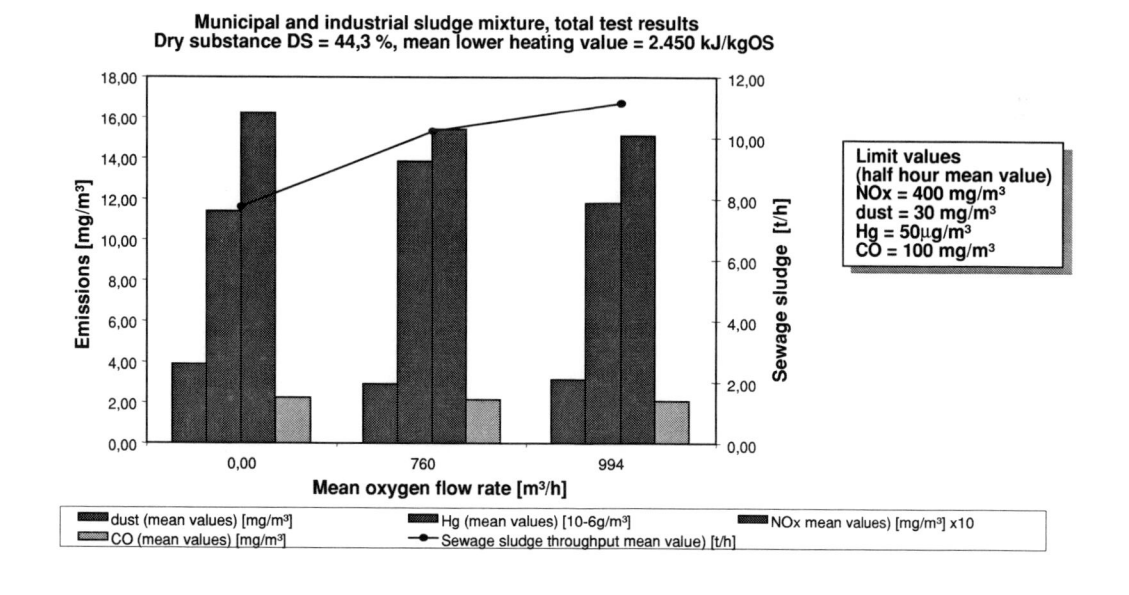

Fig.9 NOx, CO, Hg and dust emissions at the chimney

BIO THELYS®: A NEW PROCESS FOR SLUDGE MINIMIZATION AND SANITIZATION

J. Chauzy*, D. Cretenot**, P. Fernandes*, L. Patria*
*ANJOU RECHERCHE, chemin de la digue, 78 603 Maisons Laffitte, France
**VIVENDI WATER SYSTEMS, place Mongolfier, 94 417 Saint Maurice, France

ABSTRACT
The objective of sludge minimization is to optimise the activated sludge biomass growth and/or to enhance the biodegradation of the residual biomass. It will therefore always be the association of a process for organic matter hydrolysis and/or a process for biomass stress together with a biological process, either aerobic or anaerobic. The idea of all the processes that can be developed for sludge minimisation will focused on the reduction of the quantity of excess sludge leaving the wastewater treatment. One of these processes, *Bio* THELYS®, developed by Vivendi Water, associates a process of soft wet air oxidation (WAO) to the activated sludge basin of a plant. WAO is used under soft conditions (150-185°C, 10-15 bar) in order to hydrolyse the activated sludge and make it more biodegradable. *Bio* THELYS® was operated over a period of 1 year on a municipal plant (2500 PE) and reached up to 75 % reduction of sludge production. Moreover BIO THELYS® is also an opportunity to have a sanitization process for the remaining excess sludge.

KEY WORDS
Bio THELYS®, bacterial stress, biodegradation, hydrolyse,
Sludge minimisation, sludge reduction,

INTRODUCTION
Sludge from both domestic and industrial wastewater treatments is today an important point for water treatment operators, a technological challenge that must be overcome. Whereas the sludge treatment was a few years ago only a secondary element in the water treatment network, this process is now usually a decisive selection criterion when choosing an overall configuration for water treatment.

Because of demographic growth, but above all because of the increase in wastewater collection rates and the improvement in the treatment performance of plants (in France, 80% of the population are connected and 80% of pollution abatement output by 2005), annual sludge production continues to grow. Whereas in 1995, France produced 800,000 tonnes of sludge (in Dry Matter), it actually produces 1,200,000 tonnes and a figure of 1,300,000 has already been suggested for 2005.

Conventional sludge treatment or recovery processes need the support of new technologies so that local authorities and industry can be offered a range of different solutions that suit their needs and the nature of their sludge, which can often vary considerably, in particular from industry to industry.

Land disposal which, just a few months ago, took more than half quantity of sewage sludge (about 60%), is now facing some difficulties. These difficulties stem firstly from the agricultural world's mistrust and secondly from growing pressure from the food industry. Sludge storage or incineration arrive only after land disposal with 20-25% and 15-20% of the market respectively. There should be a clear reduction in the former solution as a result of a ban imposed with effect from 1 July 2002 on all waste, except final waste, being landfilled and stabilised (Law of 13 July 1992). And yet incineration will see no notable surge in activity because existing facilities are already running at full output (household waste incinerators). Only the installation of special *in situ* furnaces (Pyrofluid® type, Vivendi Water Systems process) would help to promote incineration in the sludge treatment market.

Within this context, new technologies, about which little is as yet known, are beginning to see the light of day, in particular Processes for Reducing Sludge Production (PRSP). One of these processes, *Bio* THELYS®, was developed in Vivendi Water research centre.

WHAT IS THE SCIENTIFIC BASIS FOR THESE PROCESSES?
The pollutant load (often expressed as Kg COD/day or Kg BOD/day) that reaches a treatment plant is either mineralised, i.e. essentially transformed into CO_2 and N_2 by bacterial activity (respiration, nitrification/denitrification), or used by the bacteria in the biological tank for their own growth and replication, or it is inert (refractory COD, the non-biodegradable fraction of suspended solids entering the plant). The mineralisation, biosynthesis and inert fraction proportions can vary depending on the type of biological treatment (low, average or high load). The concentration of suspended solids in the tank, also normally known as "sludge",

i.e. all living bacteria, dead bacteria and suspended solids entering the tank that have not biodegraded (fibres, cellulose, etc.), will therefore increase naturally. This sludge is not easily or not at all biodegradable by the biomass present in the biological tank, either due to the preliminary hydrolysis stage, which is kinetically limiting, or due to its refractory nature. However, a small part of this sludge, namely the dead cells, can be consumed by the active biomass: this is known as cryptic growth. Regular extraction of this sludge gives a constant sludge concentration in the biological tank and therefore helps to maintain the balance of the biological system.

The aim of PRSPs is to optimise growth of the biomass and/or its biodegradation. They combine an organic matter hydrolysis process and/or a stress process, with a biological treatment (aerobic or anaerobic). The two principles are detailed below:

Hydrolysis of particulate organic matter

This involves transforming particulate organic matter (including part of the matter that is not or is hardly biodegradable and known as "inert") into soluble organic matter that can easily be assimilated by the biomass in the biological tank. This transformation is performed firstly by cell lysis of the bacteria (destruction of bacteria by breakage of the cell wall, and release of the cell content) and secondly by reducing the size of molecules. Its aim is to increase the biodegradable fraction of sludge and therefore optimise overall mineralisation of organic pollution entering the plant (cf. *figure 1*). The efficiency of converting cell lysis products into bacteria (biosynthesis) is usually lower than that applied to municipal effluent. The mineralisation and maintenance phenomena are therefore proportionally greater. This transformation can be mechanical (pressure/decompression, grinding of sludge, ultrasound in the zone near the transmitters), chemical (acid/base treatment, oxidation by H_2O_2, ozone), biochemical (enzyme), thermal or a combination of several parameters.

Bacterial stress

This involves stressing the micro-organisms to reduce sludge production by encouraging the consumption of pollution by bacteria for their maintenance and not for their multiplication (biosynthesis). During the stress period, the micro-organisms draw on their reserves (endogenous respiration) and must then replenish their stocks, to the detriment of their replication (cf. *figure 2*). This stress is already used today both in low-load treatment plants and in membrane bioreactors (BIOSEP® type, Vivendi Water Systems process) where the sludge concentration and, implicitly, the age of the sludge are high. This is because fresh sludge produces bacteria that are highly viable and uses the energy of the catabolism for biosynthesis (anabolism), whereas old sludge encourages the appearance of bacteria that are not very viable and use the same energy for cell maintenance. This stress can be achieved in different ways, by physico-mechanical means (ultrasound in the zone away from the transmitters), by chemical means (low chlorination, low ozonation) or by biological means (increasing the age of the sludge, extended aeration/low load processes, anoxia stage).

A PRSP can also involve a combination of the two phenomena referred to above.

CONCEPT OF REDUCING SLUDGE PRODUCTION

The first and most commonly known of the processes for reducing sludge is anaerobic digestion that can eliminate up to 50% of volatile matter. The aim of PRSPs is to decrease sludge production directly at source, i.e. within the treatment plant. A conventional treatment plant which, under normal conditions, produces 100 Kg/day of DM (Dry Matter or dry waste), can, if equipped with this type of process, produce only 30 to 50 Kg/d of DM, or even less. However, it could not be said that, for a treatment plant operating under stable conditions, this reduction in sludge production could be total, i.e. equal to 100%, which would mean all the sludge had been volatilised or solubilised (this is why a conventional sludge dewatering and evacuation process, albeit smaller, is to be included in the design and cost of treatment).

The main advantages of these processes are evident, as demand on the sludge evacuation-recovery/elimination process will be lower as less sludge is produced.

PRSPs can come in different forms and involve different technologies and apparatus. However, they are invariably associated with a conventional biological treatment (activated sludge tank, digester) upstream of which they are installed, whether as supply, recirculation or as a combination of the two (cf. *figure 3*). Sludge can also be extracted at different places within the network, depending on which process is chosen.

PRSPs are different from on-site incineration processes because they transform the sludge before returning it to the top of the biological treatment. Conversely, on-site incineration (Pyrofluid® type) or land disposal processes represent a final treatment, with no interaction with the water treatment network. In this, these new PRSP processes are in a way "revolutionary" as they are totally included in the water treatment network. It is no longer a question of managing these two treatment networks independently but rather of correlating them when designing the plant.

SELECTION CRITERIA FOR CHOOSING A PRSP

As there are many various criteria to be taken into account when comparing two processes for reducing sludge production, comparison between several technologies has become extremely complex. However, one of these criteria will be of particular interest to local authorities and industry, that is the overall cost of the process.

PRE-TREATMENT

Some processes require the sludge to undergo pre-treatment, in particular thickening. This involves processes for which the hydraulic residence time is independent of the sludge concentration (thermal processes for example). The size of the reactor can therefore be reduced. The sludge can be thickened statically or dynamically, depending on the required concentration and texture of the sludge (static for concentrations under 30 g/L and dynamic for higher concentrations). This pre-treatment, when it is possible or necessary, incurs extra costs but it does have the advantage of reducing the size of structures downstream and therefore investment costs (including civil engineering).

OVERALL DIMENSIONS AND DESIGN

A number of parameters are involved in the design of a PRSP:

- **Treatment rates**: this means the quantity of sludge treated by the PRSP, compared with the quantity of sludge produced in the same time by the plant. For example, for a plant producing 1 tonne of DM per day, a PRSP with a treatment rate of 200% treats 2 tonnes of DM per day. This treatment rate is directly involved in the process design.
- **Technology**: PRSPs requiring a long residence time, in particular those involving a stress, will need a larger surface area. Processes with a biological stage generally require more space than chemical, thermal or mechanical processes.
- **Concentration of sludge treated**: all things being equal, a process treating thickened sludge is more compact than a process treating classic recirculated sludge.

RETURN OF TREATED SLUDGE TO THE TOP OF THE PLANT AND ITS IMPACT ON THE PLANT

The impact of sludge treatment on the plant is different depending on whether the PRSP uses hydrolysis or a bacterial stress. For processes involving at least one hydrolysis stage, the sludge is partly transformed into organic matter that is easily biodegradable and into nutrients for the biomass present in the biological tank. Consequently, returning the treated sludge to the biological tank represents an additional load for the plant. This extra load must be taken into account and must be compatible with the design of the biological structures and the plant's aeration capacity. This is because an increase in aeration (up to + 30%) is to be allowed for nitrification of returned sludge. In addition, depending on the treatment rate applied and therefore on the quantity of active biomass eliminated by the treatment, it may be necessary to increase the concentration of sludge in the biological tank, so as to maintain the apparent age of sludge close to or equal to that of the plant, initially (cf. *figure 4*). This is usually possible since most lysis processes, in particular thermal lysis and ozonation processes, improve the settleability of sludge (sludge index below 60-70 mL/g). The sludge index is improved insofar as the Organic Matter / Mineral Matter (OM/MM) balance of the biological tank is tipped towards MM, with OM being more biologically mineralised. Therefore the fraction of sludge volatile matter decreases and encourages better settling. Some hydrolysis processes, in particular thermal and ozonation processes, constitute an effective treatment for filament-forming bacteria and therefore remove the operating constraints linked to these micro-organisms (bulking phenomena).

On the other hand, if it is only bacterial stress, the bacteria are overactive, as it were, and returning sludge to the biological tank produces neither an overload nor a change in the age of the sludge.

IMPACT ON THE QUALITY OF TREATED WATER

First of all, it is important to specify that to correctly assess the impact of these processes on the treatment of water, the case of a treatment plant designed to treat the returns, that treated sludge can constitute, has to be considered. If this is not the case, plans for treatment of these returns upstream of the biological tank should be made so as to reduce pollution to a maximum and avoid overloading the plant.

Processes for reducing sludge production, if installed as is, generally have an impact on the quality of water treated by the plant which results in a degradation of the quality of exit water. There are a number of explanations for this degradation. The first is that the transformation of sludge, carried out within the process, can in some cases lead to the appearance of refractory soluble organic forms, such as hard (or refractory) COD and hard organic nitrogen (including in hard COD), which will hardly be or will not be decomposed by the biomass in the biological tank and will mostly be found in the treated water. The degradation of exit water is in this case almost proportional to the applied treatment rate. In addition, depending on the type of effluent to be treated and the type of process chosen, it is possible that the C/N ratio of the soluble fraction of treated sludge will not promote satisfactory denitrification, in which case an increase in nitrogen forms in the exit water might be seen. Lastly, an increase in phosphorus is usually seen in the treated water as the flow of extracted sludge, the main outlet for phosphorus, is decreased. It is therefore best to consider this aspect and, if necessary, put in place the necessary measures for compliance with current reject standards (specific treatments for nitrogen or phosphorus).

EXISTING FACILITY

Within the framework of setting up a PRSP at an existing plant, the existing infrastructures need to be taken into account when defining the most suitable process, in particular whether there is:

- a static or dynamic thickener (settling table, screws, flotation tank),
- a dewatering system,
- a heat source,
- storage of chemicals (oxygen, soda, etc.),
- a digester.

PERFORMANCE IN TERMS OF REDUCING SLUDGE PRODUCTION

Sludge production can be reduced by 70-80% with certain processes but not by as much as 100%, as this would mean that the mineral matter would disappear in the treated water or would be volatilised to the atmosphere. Performance in terms of reducing sludge production is not invariable for a PRSP, but it does very much depend on the quality of sludge to be treated. However, it is important to note that the performance of PRSPs implementing a stress is less than that of those processes which involve hydrolysis.

EVOLUTION OF THE SLUDGE QUALITY DURING TREATMENT

The sludge quality undergoes changes during the PRSP start-up stage and then stabilises after a few months.

CHEMICAL QUALITY

To understand the changes in sludge quality, you must bear in mind that the main objective of a process for reducing sludge production is to extract less sludge from the system. The fact that, regardless of which process is chosen, certain chemical forms, such as heavy metals (which have no other outlet than the extracted sludge), will be concentrated up to three times in the sludge, depending on the process performance, is therefore easily recognised. However, this concentration effect soon finds a balance once the process is in place, and the sludge quality stays in compliance with the standards for most plants. Only plants ordinarily producing sludge with heavy metal concentrations close to the Permissible Maximum Concentrations (PMC) will be penalised.

DEWATERING QUALITY

Processes involving, for example, a thermal stage or, to a lesser extent, an ultrasound stage, result in extremely good dewatering quality of the sludge yet to be extracted. They achieve excellent dryness, up to 45% in centrifugation, thus reducing storage and transport costs for excess sludge.

OPERATION

Some PRSPs are extremely simple to operate, in particular the mechanical processes. Others are trickier, in particular when they involve an additional biology stage (PRSPs that perform a stress for example), and require extremely strict monitoring. Still others require technologies that are hitherto unknown in municipal wastewater treatment, normally used in the industrial sector, such as ozonation and thermal treatment. These new technologies may require training of personnel before being put into operation.

FIELDS OF APPLICATION

"On a case-by-case basis": this best summarises the choice of solutions for a type of plant. This is even truer of industrial water treatment plants where the type of effluent, treatment and sludge quality make the cases in point very varied. However, there are key points which help in the choice of different technologies for reducing sludge production:

- size of the plant,

- quality of the effluent to be treated, sludge quality: sludge with a high concentration of volatile matter is preferable for cell lysis processes,
- dimensions of the plant in relation to the load received (without PRSP). In the case of undersized or correctly sized plants, cell lysis processes, because they return loads to the top of the plant, will need an additional biological tank to be built; the clarifier can probably stay the same as most PRSPs improve the settleability of sludge. By contrast, bacterial stress processes involve no oversizing of the aeration tank,
- possible means of disposal for excess sludge (spreading, landfill, incineration),
- existing treatment network (primary settling tank, secondary/tertiary treatment of N, P), type of biological treatment (low, average or high load, membrane bioreactor),
- type of sludge treatment in place: existence of a thickening system (settling table, static thickener or flotation tank), type of dewatering (centrifuge, band filter or press filter),
- receiving medium (reject standards),
- need for sanitation of sludge (recommendations for achieving sanitised sludge: 130°C for 18' at 3-4 bar),
- facilities in place (energy source, heat source, oxygen present on site, available space, etc.).

All these parameters specific to a treatment plant can be used to choose among the different PRSP.

COST OF TREATMENT

Among all the existing sludge treatments currently on the market, spreading is by far the easiest solution for plant operators. However, the difference in cost to other conventional treatments (incineration, landfill, wet air oxidation) has fallen in recent years insofar as the efforts made and means of persuasion used to convince farmers (communication, analytical monitoring of sludge batches, working time), which have been multiplied, can be logically credited to the cost of agricultural spreading of sludge.

In this context, processes for reducing sludge production can represent an interesting alternative from an economical point of view. In order to compare them, their treatment costs need to be calculated taking into account all the criteria. For a given plant, the overall cost of sludge treatment using a conventional process needs to be compared with the cost of treatment incurred by setting up a process or reducing sludge production.

The overall cost can be split in two: that incurred by setting up a PRSP and by its use, and that relating to the treatment and the evacuation of excess sludge.

The cost of a PRS

This cost includes the following elements:
- investment cost for a process for reducing sludge production (equipment, hydraulic structures),
- energy costs,
- cost of consumables (ozone, oxygen, enzymes, various chemicals),
- thickening costs (equipment),
- operating costs (personnel),
- cost of additional aeration in the activated sludge tank,
- cost of oversizing the biological tank or constructing a biological unit to treat returns of loaded sludge,
- maintenance costs.

THE COST OF TREATMENT AND EVACUATION OF EXCESS SLUDGE

This cost depends directly on the performance of the PRSP (cf. *table 1*). The more effective are the PRSP in terms of reducing sludge production, the smaller are the quantity of sludge to be treated and evacuated. This cost includes the following elements:
- investment cost (thickening unit, dewatering unit) in terms of hydraulic structures and equipment,
- cost of storage,
- cost of transport,
- energy costs,
- cost of consumables (coagulant, polymer, lime, etc.),
- maintenance costs,
- operating costs (personnel),
- cost of final treatment (acceptance at landfill, incineration, use in agriculture).

INTRODUCTION TO SOME PRSP TECHNOLOGIES

Some PRSPs are shown in the following table comparing their main characteristics.

THE PROCESS *BIO* THELYS®

Bio THELYS® is a PRSP developed by Vivendi Water Systems and Anjou Recherche, the Vivendi Water research centre. It has been tested during a year in the wastewater treatment plant of Witry les Reims – France, which is an extended aeration of non settled raw water. This plant is composed of two treatment lines, the first one (5000 Pop. Equ.)

is used as indicator for the reduction of sludge production, aeration and quality of water and sludge produced, whereas on the second one (2500 Pop. Equ.) is installed the process *Bio* THELYS® for reducing the sludge production.

DESCRIPTION AND PRINCIPLE

Bio THELYS® is a PRSP which associates a lysis stage to the activated sludge basin. This lysis, placed on a secondary recycling loop from the settler to the basin, is made by soft Wet Air Oxidation. The operating conditions for running are: temperature of 150-185°C, pressure of 10-15 bars and a retention time of 30-60 minutes. The sludge extracted from the settler are first screened in order to extract the fibres, then they are thickened to 60-80 g DM/L before entering the thermal reactor. Finally the treated sludge, which are more biodegradable, are returned upstream of the biological tank. The lysis has different impacts on the sludge, which are cell (dead or alive) lysis, hydrolysis of certain suspended solids and simplification of long organic linear chain. It globally enhance the biodegradability of sludge.

RESULTS

The reduction of sludge production, compared to the production of DM without *Bio* THELYS®, is possible up to 75%, depending on the treatment rate and the nature of sludge. The more organic the sludge are, the smaller the treatment rate has to be to get the same reduction. A reduction of 70% was reached on the plant of Witry les Reims. However the quality of extracted sludge is different: these sludge became more mineral. At Witry, the rate of volatile matters of extracted sludge were 75% before treatment and became 60% after. An other observation was the concentration of several components as heavy metals and phosphorous in the sludge extracted from the biological tank.

The settleability of the sludge from the biological tank rose from 2 to 3 times, and the sludge volume index decreased up to 60 mL/g. This result suggest that the settler size could be smaller. Moreover the dewatering performances were very good and the dryness could reach 45% with a centrifuge on sludge directly after thermal treatment.

Another advantage of *Bio* THELYS® is the fact that when the sludge are dewatered after thermal treatment, they are sanitized and this factor could be a key point for the treatment of wastewater from slaughterhouses against Bovine Spongiform Encephalopathy (BSE) protein.

Bio THELYS® is very efficient against the fibrous bacteria (phenomenon of bulking): microscopic observations have been done and have showed that there are less fibrous bacteria in the biological tank when *Bio* THELYS® is working. Operation problem linked to bulking on the plant are thus almost null.

The return of treated sludge by *Bio* THELYS® to the top of the biological basin is a COD and N load which generates an additional aeration from 10 to 30% equally depending on the strength of the treatment. These returns should be integrated in the design of the plant and a tertiary treatment facility should be foreseen, if necessary, if the quality of treated water, due to refractory soluble organic matters, does not conform to the standards.

CONCLUSION

Faced with the problems currently encountered by conventional sludge evacuation processes, processes for reducing sludge production (PRSP) appear to be an attractive alternative solution.

However, it must be borne in mind that these processes will not make sludge disappear, they just partly transform it. The components of sludge are transformed into different elements:

- the carbon and nitrogen load is mostly mineralised in the form of CO_2 and N_2. Part of this load is also found in the plant's treated water in an essentially soluble form (dissolved Total Organic Carbon and dissolved organic N), the rest is included in the excess sludge,
- phosphorus is not released into the atmosphere, it is concentrated in the sludge or is found in the treated water whose total phosphorus content tends to increase slightly,
- heavy metals are essentially concentrated in the excess sludge,
- mineral salts are found in the excess sludge (sludge much more mineral than conventional sludge) or in the treated water in soluble form.

This is why PRSPs must be integrated within the water treatment network, whilst still observing the quality of treated water and the chemical composition of excess sludge for possible use in agriculture or, if necessary, incineration.

Processes for reducing sludge production have many advantages and are suitable for solving a certain number of problems currently faced by plant operators. Before being launched on the sludge treatment market, they must convince local authorities, industry and operators that these technologies, usually used within the industrial sector, can find their place within a biological system, that of a wastewater treatment plant.

Bio THELYS® developed by Vivendi Water proposes on one hand to reduce significantly the sludge production of municipal and industrial plants, and on the other hand to produce excess sludge which are sanitized.

REFERENCES

CAMBI, Statistical investigation about influence of the sludge digestion process on waste water treatment, *HIAS WWTP, HAMAR, Norway*, 1997.

CANALES, A., PAREILLEUX, A., ROLS, J.L., et al. Decreased sludge production strategy for domestic wastewater treatment, *Water Sci. Tech.*, 1994, **30**, (8), p 97-106.

DOULAH, M.S. Mechanism of disintegration of biological cells in ultrasonic cavitation, *Biotechnol. Bioeng.*, 1997, **19**, p 649-660.

EDWARDS, D.G., WISEMAN, A. Novel shock-tube cell-disruptor, *Process Biochemistry*, 1971, **6**, (11), p 32-34.

HAMER, G. Lysis and 'Cryptic' growth in Wastewater and Sludge Treatment Processes, *Acta Biotech.*, 1985, **2**, p 117-127.

HARRISON, S.T.L. Bacterial cell disruption: a key unit operation in recovery of intracellular products, *Biotechnology Advances*, 1991, **9**, p 217-240.

HIGGINS, J.A., SPRINGER, A.M. Reducing excess secondary biological sludge in the activated sludge process – Result of full scale pilot testing of a lyse and recycle activated sludge system at the Portland water district's marginal way water pollution control facility, *KADY International Scarborough*, Maine.

LI, Y.-Y., NOIKE, T. Upgrading of anaerobic digestion of waste activated sludge by thermal pretreatment, *Water Sci. Tech.*, 1992, **26**, (3/4), p 857-866.

MASON, C.A., HAMER, G., BRYERS, J.D. The death and lysis of microorganisms in environmental processes, *Chem. Engng. Commun.*, 1986, **45**, p163-176.

PEIGNEN-SERALINE, P., MANEM, J. Des bactéries épuratrices, *Biofutur*, 1997, **165**, p 38-40.

ROLS, J.L., GOMA, G. Reduction of sludge production during biological wastewater treatment, *ISEB (21-23 avr. 1997), OSTENDDE*, 1997, p 71-74.

SAKAY, Y., FUKASE, T., YASUI, H., et al., An activated sludge process without excess sludge production, *IAWQ specialized conference on sludge, Poland*, 1997.

SCHUTTE, H., KRONER, K.H., HUSTEDT, H. & al, Experiences with a 20 litre industrial bead mill for the disruption of micro-organisms, *Enz. Microbial Technol.*, 1983, **5**, p 143-148.

SPRINGER, A.M., DIETRICH-VELAZQUEZ, G., HIGBY, C.M., DIGIACOMO, D., Feasibility study of sludge lysis and recycle in the activated-sludge process, *Tappi journal*, 1996, p 162-170.

WEEMAES, M., GROOTAERD, H., SIMOENS, F., HUYMANS, A., VERSTRAETE, W. Ozonation of sewage sludge prior to anaerobic digestion, *Water Science and Tech.*, 2000, **42**, (9), p 175-178.

TABLES

Process	Type	Principe	Type of sludge	[Sludge] g/L	Performance for reducing DM sludge production (%)	Advantages	Disadvantages
Anoxia	Biological	Biological stress by an anoxia stage encouraging endogenous respiration	Settled sludge	<20	<30	- cheap - improved settling - no loads returned	- hydraulic structures required, space taken up - operation difficult - redox potential control
Grinding	Mechanical	Cell lysis by impact and shearing (micro-balloons)	Thickened sludge	40-70	<30	- simple - reliable installation	- electrical consumption - loss of bearings - weak lysis
Chlorination	Chemical	Chemical stress at low dose and oxidising hydrolysis with a higher concentration	Recirculated sludge	<10	< 50	- filament abatement - compact process (short reaction time)	- formation of organo-chlorides - use of problematic Cl2 - corrosion - foam in the reactor - cost of reagents - problems with settling
Enzyme	Bio-chemical	Cell hydrolysis with thermophilic enzymes	Recirculated sludge / concentrated sludge	<20	<60	- no enzymes added	- need heat source - hydraulic structures required - space taken up
Ozonation	Chemical	Oxidising hydrolysis (cell lysis) or stress (if light treatment)	Recirculated sludge	<15	<70	- compact process - filament removal - improved settleability and dewatering of sludge	- cost of O2 consumption - biological removal of phosphates difficult - foam in the reactor
Thermal	Thermal	Thermal hydrolysis (cell lysis)	Thickened / dewatered sludge	40-150	<70	- filament abatement - sanitation - settleability and dewatering of sludge	- energy costs - maintenance
Thermal + Digestion	Thermal + Biological	Thermal hydrolysis (cell lysis) carried out on sludge with 15-16% dryness by steam, followed by meso or thermophilic digestion with very high load	Dewatered sludge	150-160	50-60	- digester takes up little space - thermal reactor takes up little space - better production of biogas	- thermal operation in batch - ammonia returns
Ultrasound	Physico-mechanical	Cavitation phenomenon causing a disintegration of floc, alteration in the cell layers and cell lysis	Recirculated sludge / thickened sludge	<20-30	<30	- compact - simple - improved settleability of sludge	- energy costs - maintenance
Ultrasound + Digestion	Physico-mechanical + Biological	Cell lysis / stress depending on how close to the transmitter and the duration of exposure	Recirculated sludge / thickened sludge	<20-30	30-50	- improved settleability of sludge - better production of biogas	- space taken up by digester

Figures

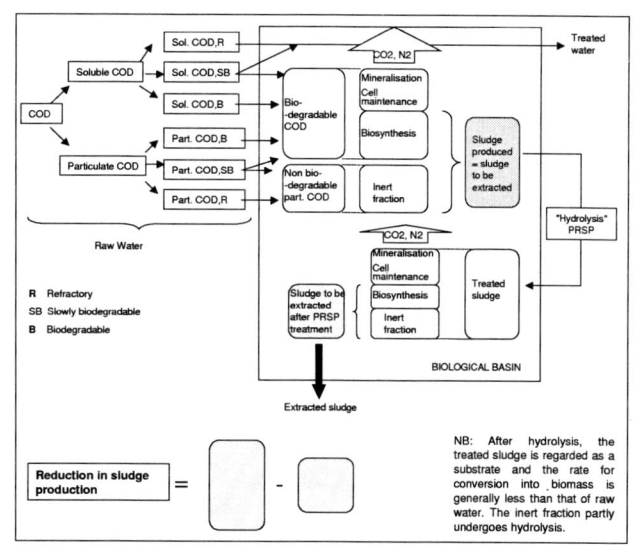

Figure 1: Principle of hydrolysis processes

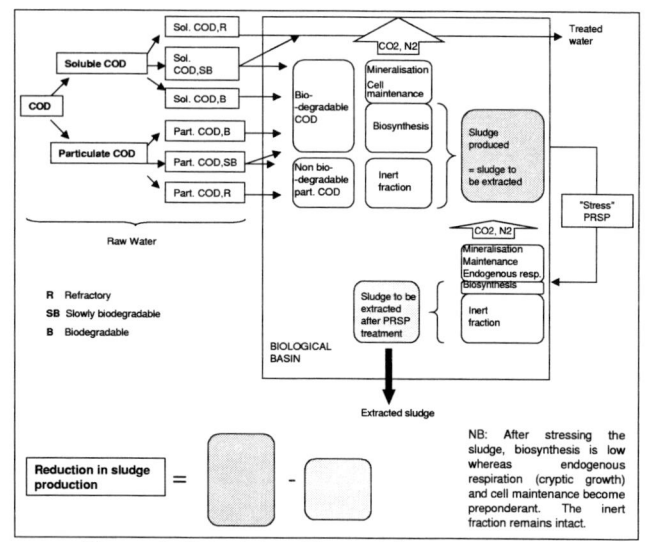

Figure 2: Principle of bacterial stress processes

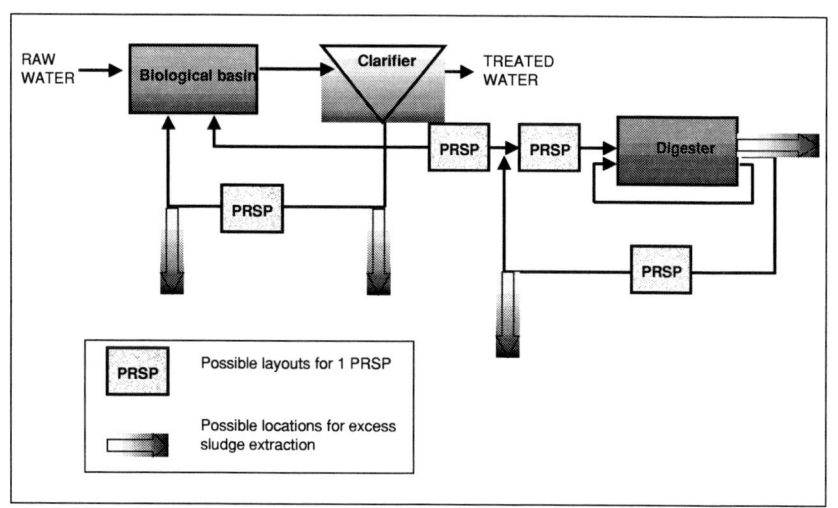

Figure 3: the various options for setting up PRSPs and residual sludge extraction

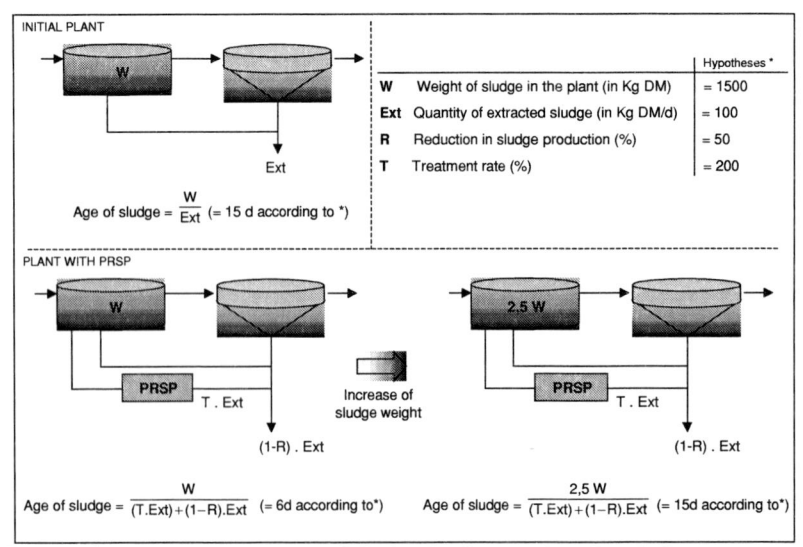

Figure 4: Preserving the apparent age of sludge by increasing the sludge concentration in the biological system

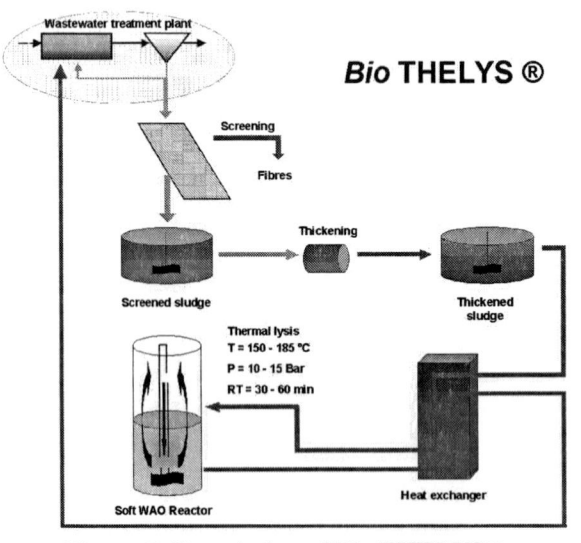

Figure 5: Description of Bio THELYS®

THE ENERSLUDGE™ PROCESS – OPERATIONAL EXPERIENCE FROM COMMERCIAL FACILITY

Tim Casey, Trevor Bridle, Peter Ashford,
Environmental Solutions International Ltd, 21 Teddington Road, Burswood, Western Australia 6100
Ph: +61 (8) 9470 4004, Fax: +61 (8) 9355 0450
Emails: timc@environ.com.au ; trevorb@environ.com.au ; petera@environ.com.au

ABSTRACT

The ENERSLUDGE™ process is a pyrolysis technology developed and marketed by Environmental Solutions International (ESI) Ltd for treatment of sewage sludges and sludges from industrial sources. The first commercial-scale plant was constructed, commissioned and operated by ESI in a joint venture with Clough Engineering at the Subiaco Wastewater Treatment Plant before being handed over to the Water Corporation of Western Australia in 2001. The objective of this paper is to;

- describe the background to development of the technology,
- provide an overview of the operational experience gained from thermal processing of sludge at the commercial-scale plant,
- describe how the knowledge and experience from commercial operations is being put to use in the design of second generation plants, and
- provide indicative capital and operating costs for a European based sewage sludge processing facility, using the ENERSLUDGE™ technology.

KEYWORDS

ash, bio-coal, bio-gas, bio-oil, ENERSLUDGE™, ESI, hot gas generator, produced water, pyrolysis, reaction water, Subiaco

WHY THE INTEREST IN THERMAL TECHNOLOGIES ?

In the domestic and industrial wastewater treatment industry, attention is increasingly being directed at development of technology for treatment of the sludge fraction. Disposal of sludge to landfill is being banned EEC wide (The EU Landfill Directive) and application of sludge to farmland (the primary disposal route in Europe) is under scrutiny. As a consequence, other technologies and disposal options are being sought by sludge producers.

In the UK these changes are particularly well demonstrated where the more traditional disposal options for sludge from wastewater treatment plants have diminished significantly over the past decade. Figures 1 and 2 indicate how these routes have changed between 1996 and 2000[1]. It is understood that the situation is likely to continue to change as a result of factors such as:

- the increasing cost of landfill due to Government impost, and
- banning of landfilling of sludge in accordance with the EU Landfill Directive.

The reduction and elimination of the traditional routes for sludge disposal have resulted in increased focus and attention on thermal technologies such as pyrolysis and gasification. One such pyrolysis process is the ENERSLUDGE™ technology, offered by Environmental Solutions International (ESI) Ltd. There is a full-scale commercial installation at the Subiaco wastewater treatment plant in Perth, Western Australia.

WHAT IS ENERSLUDGE™ AND PYROLYSIS?

The ENERSLUDGE™ technology incorporates a thermal (pyrolysis) process for the conversion of dried sludge to four fuels:

- bio-oil,
- bio-coal,
- bio-gas, and
- produced water.

The process operates at temperatures between 400 and 500°C, at atmospheric pressure in the absence of oxygen and rearranges the organic material into four products that capture the energy that was originally in the sludge. A representation of this is given in Figure 3. A distinct advantage of the process over other thermal technologies is that despite the sludge containing pollutants such as heavy metals, organochlorines, pathogens, toxins and endocrine disrupting compounds (EDCs), the products of the process have very low concentrations of contaminants because:

- the vast majority of heavy metals in the sludge partition into the char, and are retained in the ash that results from its combustion[2].

- all organochlorines are destroyed[2].
- all toxins, pathogens and EDCs are destroyed.

The partitioning of the feed sludge into clean useable products provides unique environmental, operational and commercial flexibility and the products can be used in a variety of ways. For example:
- the bio-oil can be used on-site or off-site for electricity generation or as an industrial fuel, the advantage being that it can be readily stored or transported,
- the bio-coal can be used as an industrial fuel, or possibly converted to activated carbon,
- the bio-gas and produced water can be used as fuels for on-site process heating.

Thus, the technology converts sludge to a renewable source of energy thereby reducing the quantity of greenhouse gas emissions by eliminating the need for production of power from an energetically equivalent amount of fossil fuel. Depending on its place of application, the technology will in the future attract valuable carbon credits and renewable energy credits.

In application of the technology to sewage sludge treatment, pyrolysis is conducted at 450°C and about 60 percent of the lipids, carbohydrates and proteins present in the feed are volatilized and the non-volatilized organic material remains in the bio-coal. In the process, pyrolysis is enhanced by re-contacting the vapour with the bio-coal to mediate catalysed vapour-phase reactions that refine the complex fatty acids and long chain hydrocarbons to shorter straight chain compounds. The catalysts (alumino-silicates and heavy metals) that mediate the conversion reactions are inherent in the sewage sludge and are retained in the bio-coal.

DEVELOPMENT OF THE ENERSLUDGE™ TECHNOLOGY

A brief history of the development of the technology is as follows:
- A technology for conversion of sludge to liquid fuel was conceptualised and researched at the University of Tubingen in Germany in the early 1980s.
- The first pilot-plant demonstration program was successfully conducted in Canada[3] by Environment Canada.
- ESI then bought the rights to the technology and commenced engineering development and scale-up of the process in 1987 in Australia.
- ESI built a one tonne per day pilot plant which successfully demonstrated the technical viability of the process in both Perth and Sydney[4,5].
- In Australia, the process underwent significant refinements during the early 1990s and ESI was in a position to commercially exploit the technology by the mid 1990s.
- Construction of the full-scale plant at Subiaco was started in 1997 in Australia and handover of the plant to the client was in June 2001.

The technology has also been applied at pilot-scale to a number of industrial sludges, in particular, oily sludge from refineries and tannery sludges. Extensive pilot-plant work has recently been completed on tannery sludge and high quality bio-oil has been produced.

DESCRIPTION OF THE AUSTRALIAN FACILITY

A contract to build the first commercial plant was awarded to the ESI/Clough Joint Venture by the Water Corporation of Western Australia in late 1996. The facility is a 25 dry tonne per day sludge pyrolysis plant comprising sludge dewatering, chemical stabilization, drying, pyrolysis (conversion to oil), energy recovery and flue gas cleaning.

Construction of the facility commenced in early 1997 and commissioning of sludge dewatering commenced in January 1998. Since the plant comprises numerous unit operations designed to operate in a cascade mode, commissioning of the entire plant was conducted in phases and continued to the middle of 2000.

Following successful completion of the contractual performance testing obligations, including nine months operation of the integrated facility to generate the required operational performance data, the plant was handed over to the client in June 2001.

The Austarlian facility comprises a number of unit operations, as shown in Figure 4. The unit processes of the facility are illustrated in Figure 5. A detailed description of each unit process will not be given here as that is not the intent of the paper; a full discussion can be found elsewhere[6]. In summary, the main elements of the dewatering/drying/ENERSLUDGE™ process at Subiaco are as follows:

- All raw primary and thickened excess activated sludge is strained to remove hair and fibre to increase the bulk density of the dried sludge and to minimize blockages in the dryer system.
- Strained sludge is dewatered to produce a sludge cake of between 26 and 28% TS.
- Dewatered sludge is dried in a flue gas drum dryer to produce sludge pellets of 95% TS.
- Pellets are heated in the first pyrolysis reactor at 450°C to produce bio-coal and a pyrolysis gas.
- The bio-coal and the gas remain in contact at 450°C in the second pyrolysis reactor to facilitate the catalysed vapour phase reactions that refine the raw gas, producing mainly hydrocarbons.
- The refined vapour is condensed at 55°C in a direct contact spray condenser using cooled produced water at about 35°C.
- The bio-coal is used as the main fuel in the fluidised bed hot gas generator.
- The condensed products are separated into an organic phase and an aqueous phase, with the bio-oil stored prior to use as a fuel off-site, predominately in industrial boilers for raising steam.
- The produced water is cooled using once-through secondary effluent from the sewage treatment plant and the excess is combusted in the hot gas generator.
- The vapours not condensed at 55°C in the spray condenser, are referred to as bio-gas and are combusted in the hot gas generator.
- The reactors are heated externally, using hot flue gas produced from combustion of LPG.
- The hot gas generator produces a flue gas at 850°C that transfers heat to a closed-loop of drying air, in an air-to-air heat exchanger, for sludge drying.
- Flue gas from the heat exchanger is cleaned in a venturi scrubber (for particulate removal) and SO_2 scrubber before discharge to the atmosphere, via the plant stack.

LESSONS LEARNED FROM OPERATION OF THE SUBIACO FACILITY

The Australian plant represents a number of milestones for both ESI and the technology, since (it is believed to be) it is:
- the first dryer in the southern hemisphere dedicated to sewage sludge,
- the first drying plant in the world to use the energy of the sludge or bio-coal to provide all the heat for drying (via the fluidised bed "hot gas generator")
- the first full-scale plant for conversion of sewage sludge to bio-oil.

The Australian plant presented a number of unique technical challenges during the design, commissioning and operational phases. The main ones are summarized below:

VARIATION IN SLUDGE QUALITY
Whilst the dewatering /drying/
The process is flexible, the plant operates most efficiently on a relatively consistent feed quality. This operational requirement may impose some constraints on the upstream wastewater treatment plant. Sludge quality parameters such as volatile solids content, proportion of raw and excess activated sludge in the feed, and phosphorous content of the sludge were found to be important parameters for optimal plant operability.

COMBUSTION OF BIO-COAL / DRIED SLUDGE
It is well recognized that high phosphorous content in dried sludge reduces the melting point of the ash produced the combustion of sludge or char, resulting in "sticky ash" at relatively low temperatures (900°C). However this was not an issue anticipated in the plant design since high concentrations of phosphorus in the sludge were simply not expected. However, it transpired that concentrations of phosphorus in the sludge were much worse than expected, causing excessive slagging in the hot gas generator. Stabilization of WWTP operations mitigated this possibility, as did technical modifications to the original design of the hot gas generator.

OPERATION OF REACTORS
Essentially trouble-free operation of the pyrolysis reactors has been a pleasing aspect of the conversion plant design. This had been the focus of considerable effort during the design. The success indicates that design criteria used for calculation of heat transfer surface areas and for sludge mixing were well selected. In second generation designs, the only anticipated changes will be to simplify the reactor shaft bearings and the flue gas burner control system.

VAPOUR CONDENSING
A system to adequately condense the vapours was another "first" in both design and operation. The initial indirect heat exchange system proved to be unsuitable. The redesigned reaction water spray condensing system for condensation of

vapour from the reactors has proven to be reliable and easily controllable.

OIL-WATER SEPARATION
The inclined plate oil-water separator system, commonly applied in the oil industry, has performed very well in its function of separating the reaction water from the oil but has proven prone to clogging due to the entrainment of fine solid particles. During maintenance, odour control and occupational health and safety aspects of the maintenance procedures are important.

TREATMENT OF BIO-GAS FROM REACTOR VENT EPISODES
Provision has been made for alternative treatment of the bio-gas from reactor vent episodes. For the second generation design, bio-gas will be treated by scrubbing, flaring, or by storage of the gas for later combustion.

REACTOR VAPOUR SEALS
Since the reactors have to operate at a slight positive pressure to prevent ingress of oxygen, dry sludge feed and bio-coal discharge systems need a reliable vapour lock. The dried sludge feed sealing device has proved to be suitable, but the bio-coal sealing arrangement has not been reliable. An alternative bio-coal discharge device for the second reactor has been developed for the second generation design that is based on formation of a cold bio-coal plug similar in principle to the system used for the feed sludge to the first reactor. Valuable technical data on bio-coal properties at 450°C has been accumulated. The optimum type of equipment required for this service is now also known.

POWER GENERATION
During the development stages of the technology, bio-oil was combusted successfully in a slow speed internal combustion engine to assess the feasibility of power generation from bio-oil using a high efficiency device (about 35% anticipated). A more detailed program of engine testing is currently underway. Clearly, there will be some new technical challenges with the first full-scale power generation facility.

WHAT CAN ESI NOW OFFER ?
ESI is now in a position to offer second generation facilities. In comparison with the first generation plant, the second generation technology is technically and functionally superior for the following reasons:
- During operation of the various bench-scale rigs, pilot-scale plants and the commercial Australian plant, an enormous database has been assembled that incorporates the findings associated with work conducted on a wide variety of sludge types (digested, undigested, chemically oxidized etc) from Europe and Australasia. This database is continually under review and provides a solid knowledge base for managing developments to the process.
- The database has been used to develop empirical relationships between parameters such as sludge volatile solids and bio-oil yield, bio-oil energy, bio-coal yield and bio-coal energy. These relationships are imperative for design of a thermal sludge treatment system and eliminate extensive bench-scale and pilot-plant studies.
- The first commercial plant has provided ESI with extensive operational data that has proven invaluable for the second generation design.
- The first commercial plant has provided information on process technical capability, i.e. information relating to the suitability of equipment and materials of construction. For the most part, this information was not available prior to commercial application of the technology in Australia. We now have a very good understanding of the types of materials and equipment that will and will not work.
- ESI has developed unique knowledge regarding the extent of flexibility to incorporate into design in order to manage sludge that inherently has significant variability in quality.

HOW WILL ESI MARKET THE TECHNOLOGY ?
ESI owns the intellectual property to the technology and has strong patent protection around the world. ESI is establishing a network of business/implementation partners throughout the world. Licence arrangements currently in place are:
- A non-exclusive licence agreement with Mitsubishi Electric Corporation in Japan.
- An exclusive agreement with Ondeo Degremont in France who have collaborated with ESI in submission of a number of proposals in the UK.

ESI is in discussion with several other prospective partners in Europe and North America.

The basis of a typical Licence Agreement is as follows:
- The Business Partner will drive the marketing with appropriate assistance from ESI.
- For specific opportunities identified, ESI will supply a technology package for the process tailored to specific applications. This package includes supply of some proprietary/key equipment.
- The Business Partner will design, construct and, if appropriate, operate the facility.
- ESI will remain involved throughout the process to ensure "fit for purpose" application of the technology.

WHAT DOES ENERSLUDGE™ COST ?

For a typical dewatering/drying/ENERSLUDGE™ Conversion plant, the indicative cost of disposal is as follows:

	With Power Generation	Without Power Generation
£/wet tonne (25% TS) (real term)	24	28

Key assumptions is this analysis are:
- Plant based in UK processing 20,000 dry tonnes of sludge per annum
- Scope includes all process facilities from receipt of 3% total solids sludge to handling of products of conversion
- Minimum 25% total solids content for dewatered sludge cake
- 20 year plant operating life
- Plant capital written off over operating period
- The plant takes 2 years to be brought on-stream
- Sludge disposal fee based on 0% rate of return (real term money) and is thus a break-even analysis that includes the effect of inflation.

REFERENCES

1. DEPARTMENT FOR ENVIRONMENT, FOOD & RURAL AFFAIRS. Sewage treatment in the UK. *DEFRA Publications*, March 2002.

2. BRIDLE, T.R., CAMPBELL, H.W. Oil from sludge: A technology update. In: *Proceedings 9th Annual AQTE Conference*, Montreal, 1986.

3. BRIDLE, T.R., HAMMERTON, I., HERTLE, C.K. Control of heavy metals and organochlorins using the oil from sludge process. *Wat. Sci. Tech.*, 22, 12, 249-258, 1990.

4. BRIDLE, T.R., HERTLE, C.K., and LUCEKS, T. The Oil-from-Sludge (OFS) Technology: A Cost Effective Sludge Management Option. *Water*, 1989, 16, (1), 30-33.

5. GOUGH, D.E., HAMMERTON, I.F., BRIDLE, T.R. and HERTLE C.K. Demonstration of the Oil-from-Sludge Technology at the Malabar Sewage Treatment Plant. In: *Proceedings, 14th Federal AWWA Conference*, Perth. 1991

6. BRIDLE, T.R and SKRYPSKI-MANTELE, S. Assessment of the **ENERSLUDGE**™ Process for the Management of Biosolids in the European Community. In: *Proceedings 5th European Biosolids and Organic Residuals Conference*, Wakefield, UK, 2000.

FIGURES

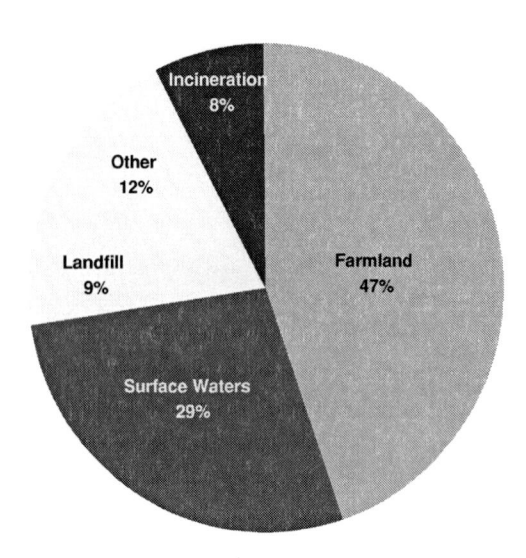

Figure 1: Disposal of sewage sludge from municipal sewage works 1996[1]

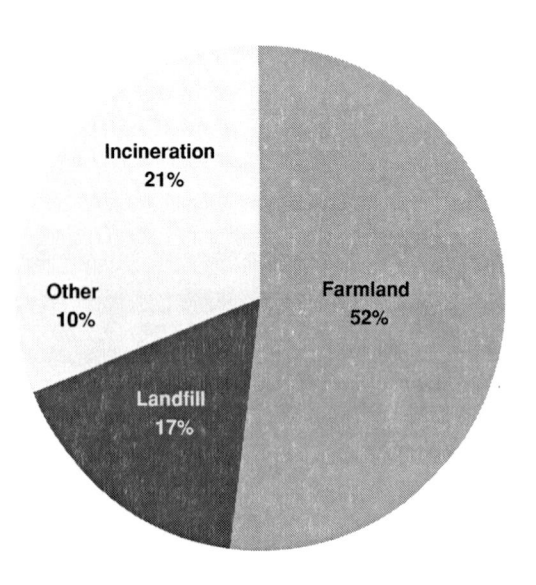

Figure 2: Disposal of sludge from municipal sewage works 2000[2]

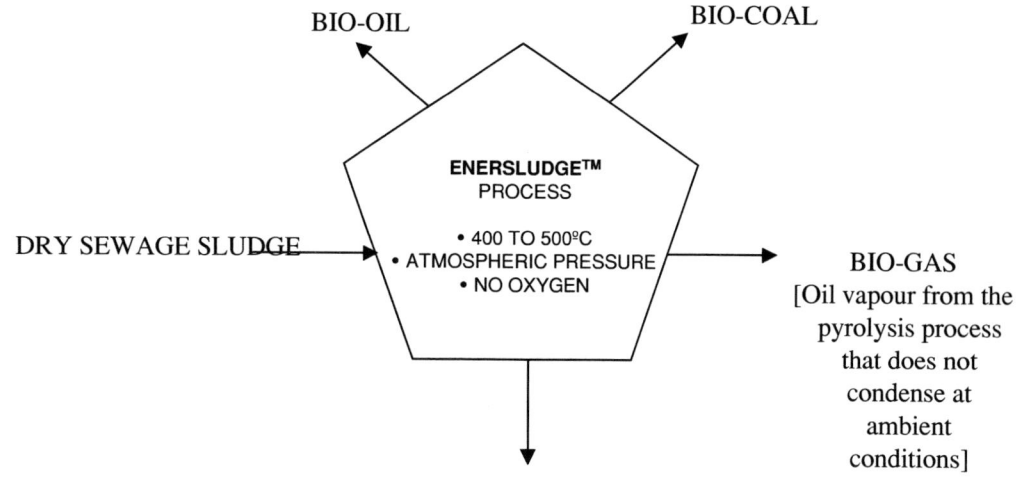

Figure 3: Products from pyrolysis of sewage sludge

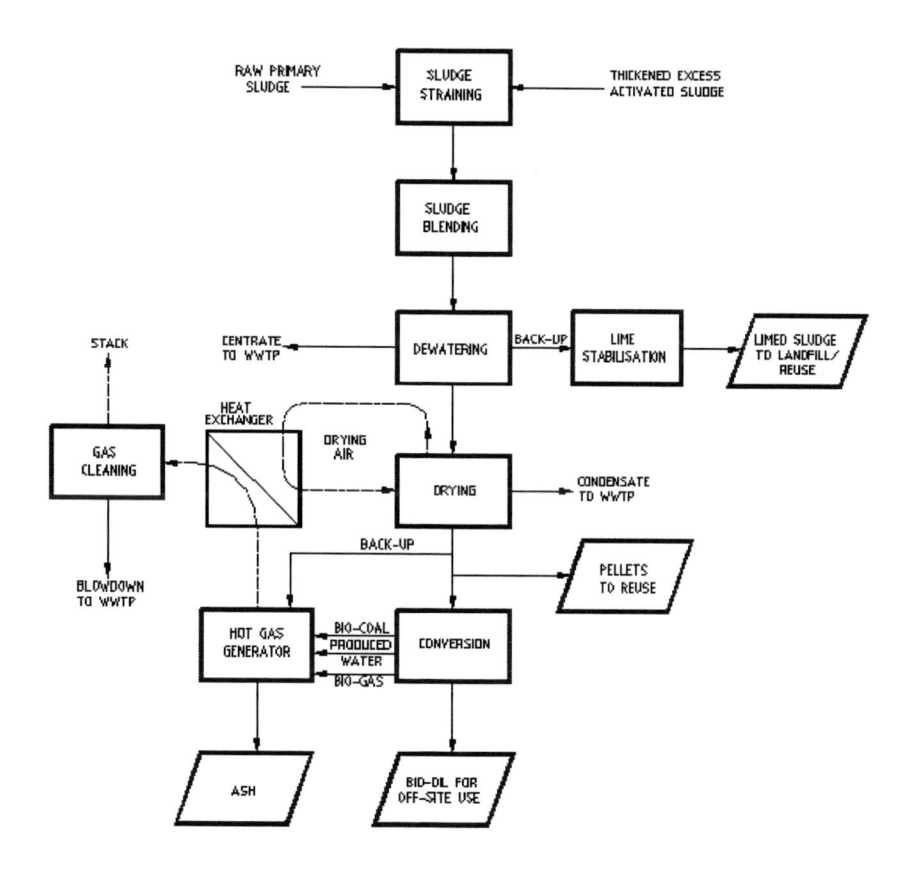

Figure 4: Integrated plant process flow diagram

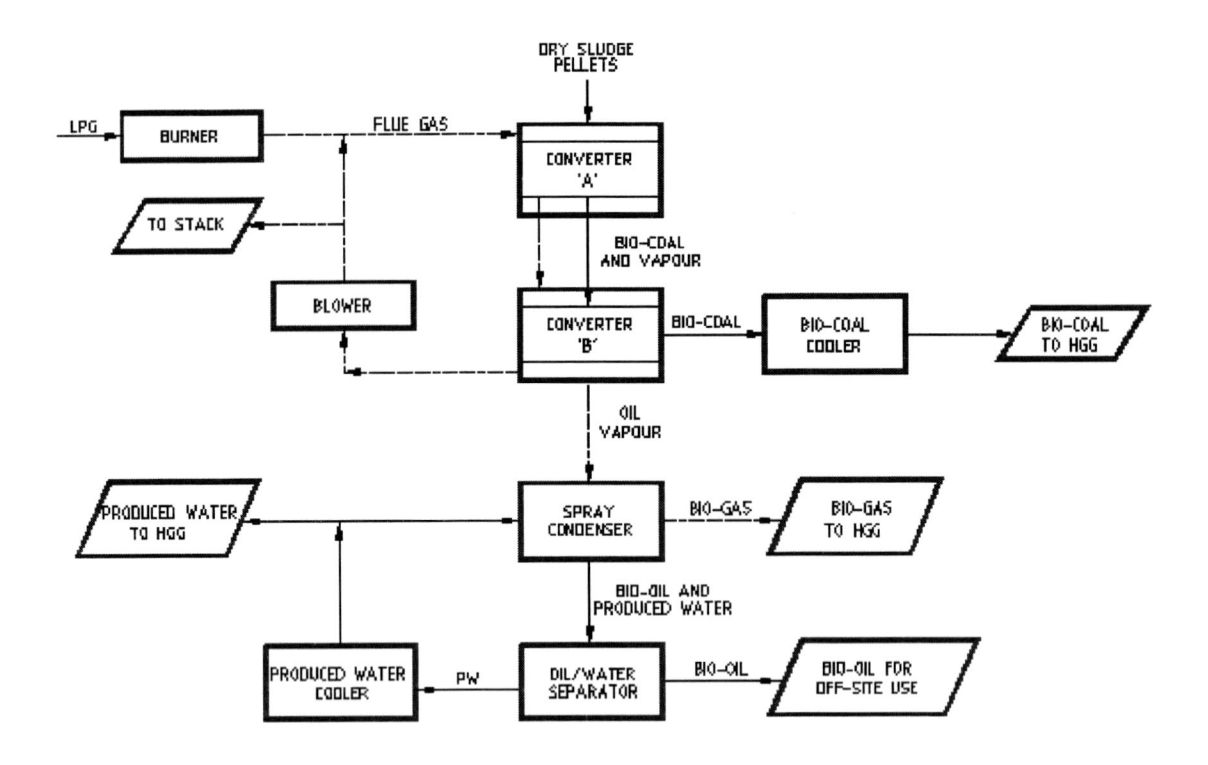

Figure 5: Conversion plant process flow diagram

OPERATING EXPERIENCE OF ABERDEEN CAMBI THERMAL HYDROLYSIS PLANT 2002

[1]Steve Wilson. [2]Keith Panter

[1]Grampian Wastewater Services Ltd Tel: +44 1 224 897816 steven.wilson@ywprojects.com
[2]Ebcor Ltd. Tel +44 1 628 524178 Keith@ebcor.freeserve.co.uk

ABSTRACT

This paper reviews the establishment of a new 50 tonne per day biosolids treatment facility. At the heart of the process is the Cambi thermal hydrolysis process. The paper describes the commissioning and operating experiences of the new installation. It goes on to review the quality of the product and the odour issues associated with the operation of the plant and how these are dealt with. The paper emphasises the importance of planned maintenance and operator requirements. Research into the use of the product and its full-scale application in an area where the local community are not accustomed to biosolids being applied to land in their area are described. The plant operation complies with a HACCP regime and as such guarantees a Class A product.

KEY WORDS

Anaerobic digestion, Cambi, commissioning, consortium, crop yields, HACCP, maintenance, operational experience, sludge cake.

INTRODUCTION

This paper reviews the establishment of a new 50 dry tonne per day biosolids processing facility at a green field site at Nigg Bay, Aberdeen Scotland. Prior to 2001 all sewage had been pumped by long sea outfall into the North Sea. Consequently there was only a minor biosolids operation in the area dealing only with small, rural works. Proposals were sought from a number of consortia for a design, build, finance, operate (DBFO) contract in 1998 by North of Scotland Water Authority (NoSWA) for the Grampian region, which includes new coastal plants at Aberdeen, Peterhead and Fraserburgh as well as many existing smaller plants inland.

NoSWA's consultant had recommended either thermal drying or incineration for the conforming bid, as there was uncertainty about the agricultural route. A consortium, Aberdeen Environmental Services (AES) including Yorkshire Water, its subsidiary - Grampian Wastewater Services Ltd (GWSL) as operator and EarthTech Engineering Ltd. as designer proposed the CAMBI digestion alternative. The consortium bid was the successful proposal due to a clear financial advantage on capital and operating costs. The inclusion of CAMBI Thermal Hydrolysis was a major contributor to the success of the project.

The advantages of CAMBI for the Aberdeen Project were:
- Absolute guarantee of Class A
- No dryer was required allowing automatic operations, using a single shift
- Digester sizing was 50% of normal.
- Ability to receive and treat a variety of imported sludge
- Net energy production of 1MW electricity
- Simple product to stack and store
- Good grassland product
- Low volume of product (<30,000 tons per year)
- By far the lowest whole life cost

Crucial to the success of the concession was NoSWA's insistence on the need for a Class A product and the inclusion in the specification of 6 known pathogens that had to be eliminated. Scotland had previously suffered a major *E Coli* 0157 outbreak with several deaths and this had led to an extreme sensitivity to pathogens. Also, the operator would have to establish a biosolids programme in a region with little history of biosolids use and with a difficult climate. Winters in North of Scotland are long, dark and can be wet or snowy.

COMMISSIONING

The plant was started up in September 2001 and handed over to GWSL in March 2001. The commissioning was described in a previous paper Jolly and Potts of EarthTech Engineering Ltd. at last years conference. GWSL carried out their own assessment of data for February and March for the final phase of the EarthTech commissioning which is reported here in Table 1.

The concession bid was based on producing about 30,000 tonnes of cake per year. However, at this stage imports are lower than predicted and there appears to be some drying occurring in the storage shed. The open porous nature of the sludge results in some product curing. Long term stored CAMBI cake develops a growth of mycelia and develops a soil like structure. The cake exits the belt press at about 32-24% DS but actual tonnes removed are lower than predicted because of further VS loss.

Overall input to the system is 18% below forecast and production of CAMBI cake product is 36% below forecast.

OPERATIONAL EXPERIENCE AND DATA

DATA

The plant has been operated directly by GWSL since April 2002. The data shown in Table 2 is based on the average of April to September data. The annual predictions are similar to those given by the February/March data. Dry solids feed has increased to over 9.5% thus raising HRT to 23.5 days. Effective digester volume has been tested using Lithium tracer and is above 95%. Again, there appears to be a loss of solids in the short time in storage. This is confirmed by the ash balance which shows that the VS% falls from 61.7% on digester exit to 57.5% on export; tonnes of ash remain constant. The plant produces more than enough gas to operate the 1MW Jenbacher engine at all times but presently has low demand at night, which falls below the minimum set point for the engine. This situation is being modified in the light of the opportunity that Renewable Obligation Credits for the export of green electricity can bring The Cambi system uses the steam from the Jenbacher waste heat boiler plus steam from a biogas boiler when the engine is not operating.

OPERATIONS

For a sophisticated form of Enhanced Treatment, CAMBI requires very little in the way of operator intervention in the steady state. The Nigg site is manned 8 hours per day Monday to Friday and 5 hours Saturday, 5 hours Sunday. During weekdays 1 operator is assigned to monitor the CAMBI process, pre and post hydrolysis de-watering belts and the boilers. Out of hours, the site is controlled by a Supervisory Control and Data Acquisition (SCADA) system, which is linked through via telemetry to the Yorkshire Water Control Room in Bradford. The controllers have the ability to call out operators or maintenance staff if required.

MAINTENANCE

In terms of skill levels it became apparent the need for greater competency than one would normally require in traditional wastewater operators. This was for fault finding and dealing with abnormal situations, as problems in such a contained system are not always clearly apparent.

A good understanding of steam and steam systems, particularly about safety is required. Because CAMBI consists of series of vessels, pumps, heat exchangers, operating under large pressures and temperatures, selection of the correct grade and specification of ancillary equipment such as pumps and macerators was paramount. The critical nature of maintenance is exacerbated by the low amount of sludge storage designed into the sludge train, with only two days liquid sludge and one day for storage of cake, up stream of the CAMBI process.

Leading edge asset management is required, with good quality and well-planned preventative maintenance, careful selection of spares to be held on site and good liaison between operations & maintenance staff. Most of all we have found the need for good procedures, prepared in advance, should any breakdown occur in order that repairs and re-instigation of operations can commence as quickly as possible, but most importantly safely.

Having said all this we have appeared to have faired better than many of our neighbours who are operating drying plants. We have had a high level of availability and all of our product has achieved class A and has been used beneficially.

ODOURS

The process of thermal hydrolysis produces a small volume of very strong odours - about 30 m^3 per hour. The "foul gas" at Nigg was measured, using olfactometry by ADAS, at 30,000,000 odour units. This off-gas is treated using a duty/standby thermal oxidiser, which combusts the foul gas in a stream of methane (biogas) or propane if biogas is not available and this process is 99.92% efficient. Making sure this system operates continually is of paramount importance. The system is designed to minimise the number of interventions, with no moving parts in the reactors and duty/standby equipment outside of the vessels. During commissioning some problems with blocked steam lances did occur which involved entry into the reactors and this led to release of the strong odour & subsequent odour complaints.

There have been no problems from the product storage either on or off site. Indeed the product has very little odour.

ESTABLISHING THE CAMBI CAKE PRODUCT

The quality of the cake and the service provided to the farmers would be paramount to the success of the whole concession. GWSL showed NoSWA a number of CAMBI installations including the longest established operation in Hamar Norway. Cake has been used here for 7 years and an agricultural programme is well established. Microbiological tests showed the product free of the pathogens of concern to NoSWA, even on extended storage.

During the development of this project the UK Biosolids market came under pressure from perception issues from the food industry in the wake of BSE, Listeria and *E-coli* 0157 outbreaks. This led to the formation of the Safe Sludge Committee comprising the British Retail Consortium, Sewage operators and ADAS - the national agricultural advisors. They coined the term - enhanced treated - which is really the same as Class A in the USA and requires pathogens to be virtually eliminated. This is particularly applicable for grassland where only enhanced treated can be used. Independently NoSWA required all sludge to be Class A. They had no significant quantities of sludge at this time due to major population centres being situated on the coast, with only preliminary treatment of the sewage, they wisely decided that if they were to launch a biosolids service it would be with the highest quality products.

NoSWA had established a relationship with SAC to test products and conduct trials with a view to establishing a sustainable biosolids programme. GWSL decided to use SAC as they were already the established extension agent. SAC have established a large database of farmers and are well trusted in the area. They are also responsible for notification of sites and storage sites to the Scottish Environment Protection Agency and local environmental health officers. SAC program the work, which is carried out by a local contractor.

In the spring of 2001 farmers were invited to witness spreading trials and at the same time a growing trial for spring barley was established. Spring barley is used by the local malsters in the production of Whiskey. It is the predominant crop along with seed potatoes and rotational grassland. The Grampian area has a very short but intense growing season due to very long daylight hours in the summer.

CAMBI cake from Norway was spread at 20 tonne/ha. It was ploughed down in February and sown in March. Grain yield was 7.5 tonne/ha at 85% dry matter. The trial showed that CAMBI cake supplied the entire phosphate requirement of the crop and 40 kg/ha of Nitrogen. The application also supplied 30 kg/ha sulphur and 1.5 kg/ha of copper. CAMBI cake from Norway gave the highest yield response of any of the products tested by SAC, which include lime products and thermally dried products.

The trial results are shown in Graph 1. Further results from the 2002 harvest are available using the cake from Nigg and will be presented at the conference, but have not been validated by SAC at time of writing.

The first biosolids were being exported in October 2001 when the Scottish Whiskey Association (SWA) decided to advise their growers against using sludge product. This edict came from a non-informed source and was "merely precautionary". Following some continuing education and outreach the WSA's situation is now neutral. Interestingly many local farmers have ignored WSA's caution despite the threat of blight to their rotational land. The other major crop user in the area Wiseman's dairies has raised no objection to the use of biosolids.

Currently around 1,600 wet tonnes per month are being recycled to agriculture. Due to its high dry solids which leads to good stacking properties and the fact that it does not absorb rainwater to any degree we have achieved 12 months of exports from site, with some on field storage prior to spreading. This has left the cake product shed on site practically empty all year round. Some pressure has been put upon the land bank by the stance taken by the malting industry however GWSL and SAC continue to work with Scottish Water (successor to NoSWA) and Water UK in educating and informing the various stakeholders about the quality product and service that is provided by GWSL and CAMBI cake.

So, despite the initial scare, the product has continued to be very popular. SAC aim to make 2,000 hectares available each year, however based on current production only 1,000 ha will be needed. The crumbly, non-smelly nature of the sludge has allowed grassland to be used. The ability to use the product on grassland has been very useful, as producers of dried granular products have not found this as market receptive to them because the product stays on the surface for so long and there is little crop response.

The product was tested for long-term stability. No pathogen re-growth occurred. In a long-term storage test the nature of the biosolids changed because of fungal curing to form a soil like product much frequented by rabbits (easy to dig into?). This opens up the possibility of getting into the topsoil market as an adjunct to agriculture.

HACCP

GWSL developed its plans for Hazard Analysis and Critical Control Points (HACCP). By measuring and recording the time and temperature (in our case pressure and hence temperature via Boyles Law due to the fact that CAMBI measures and controls primarily by pressure and the pressure measurement is more accurate), we know that every batch has met 6-bar, 165°C for 20 minutes as a minimum. From bacteriological analysis we know that this not only meets the Enhanced Treated standard but the US EPA 503 Class A product standard. Prints from the SCADA are retained for every batch demonstrating that it has met the required time & pressure, (hence temperature). Any batches not meeting this are not blown down to the flash tank but "cooked" again. We routinely measure E Coli, Salmonella, total coliforms and clostridium perfringens as well as N, P, K, pH, % D.S. and metals levels in the end product - see Table 3.

SAC operate their own QA/QC system using the Gemini database. He combined GWSL/SAC quality control system is shown diagrammatically in Figure 1.

SUMMARY

The installation of CAMBI at Nigg has more then met the requirements of GWSL. It has low operator requirement but does require a production and maintenance focus not normally applied in the water industry. However, this focus is more than justified in the payback of lower than expected volume of cake for export and profitable electricity production.

GWSL and SAC have demonstrated that with a product that is guaranteed safe, which is very stackable, stable and low in odour that a good agricultural market can be established to new customers even in a time of uncertainly and perception problems.

This is a win /win/win for the community: The farmers are getting a great product; GWSL are operating sludge management profitably; the community is getting best value.

TABLES

Table 1 . Annual Mass Balance and Average Data Based on February/March 2001

Indigenous raw dry tonnes	7552
Imported raw dry tonnes	5402
Total raw dry tonnes	12954
Digested dry tonnes	7350
HRT - days	17.5
VS loading- kg VS/m3/day	3.37
VS destruction %	57.5%
VFAs mg/l	432
Ammonia mg/l	2147
pH	7.67
DS% Feed* adjusted for VFA	9.2%
DS% digested sludge	5.12%
DS% of digested cake	32%
DS% of cake* adjusted for drying in shed	38%
Tonnes of cake	19160
Energy Value of Biogas - MWh	31765

Table 2. Annual Mass Balance and Average Data Based on April-September 2002

Indigenous raw dry tonnes	7069
Imported raw dry tonnes	5145
Total raw dry tonnes	12214
Digested dry tonnes	6891
HRT – days	23.5
VS loading- kg VS/m3/day	3.29
VS destruction %	56%
pH of digester	7.95
DS% Feed* adjusted for VFA	9.85%
DS% digested sludge	5.55%
DS% of digested cake	32%
DS% of cake* adjusted for drying in shed	36%
Tonnes of cake	18845
Energy Value of Biogas – MWh	30485

Table 3. Average CAMBI Cake Analysis April –September 2002

Dry Solids	31.7%
Volatile Solids	57.5%
Total Nitrogen	3.82%
Free and Saline Ammonia	0.43%
Total Phosphate as P	3.33%
Potassium	0.21%
Sulphur	1.45%
Copper	325 mg/kg DS
Nickel	45.1 mg/kg DS
Zinc	859 mg/kg DS
Cadmium	3.3 mg/kg DS
Lead	162 mg/kg DS
Mercury	1.5 mg/kg DS
Selenium	3.4 mg/kg DS
Fluoride	163 mg/kg DS
E.Coli	88 pfu/g wet cake

Graph 1.

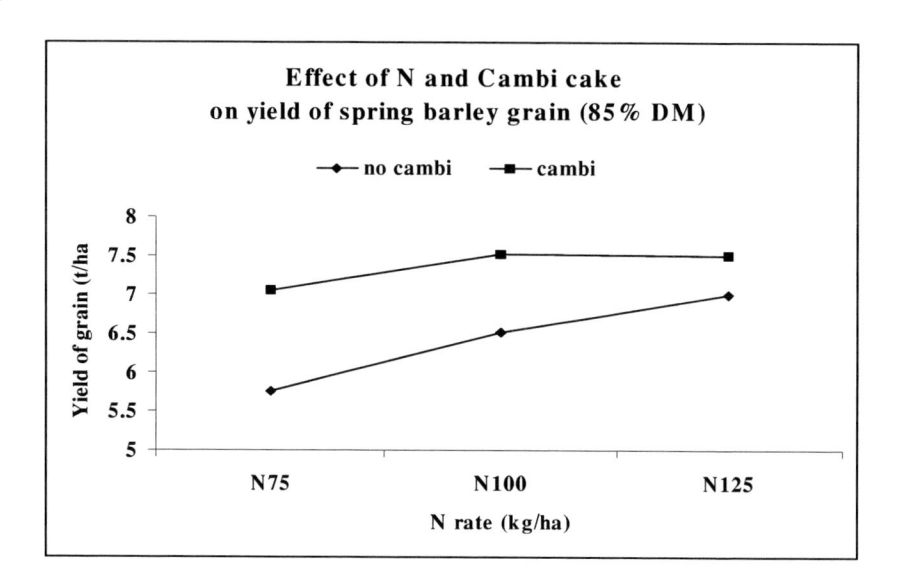

Effect of N and Cambi cake on yield of spring barley grain (85% DM)

Figure 1. SAC quality control system.

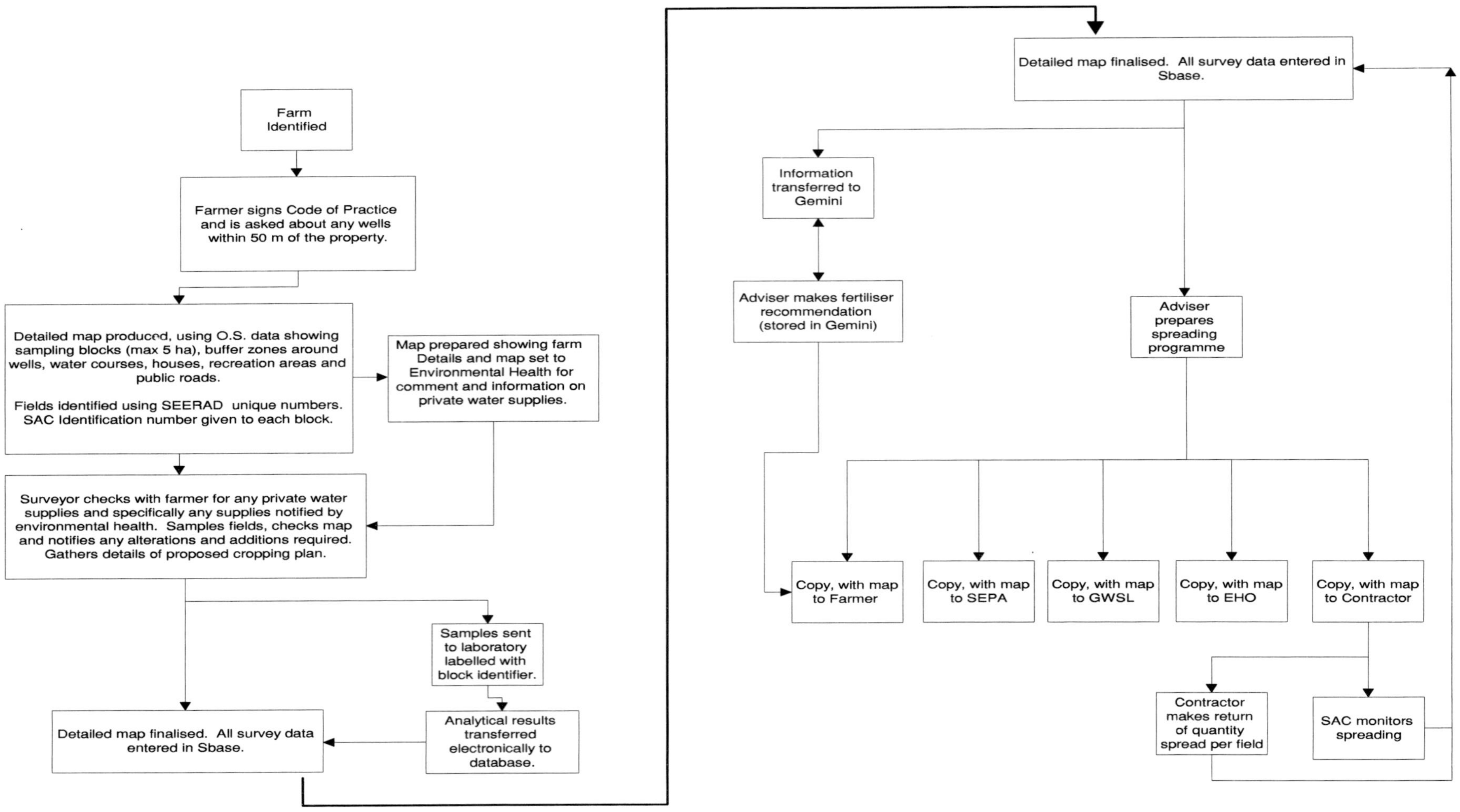

THE *MICROSLUDGE*™ PROCESS TO DESTROY BIOSOLIDS AND PATHOGENS FROM COMBINED PRIMARY AND SECONDARY SOLIDS FROM A MUNICIPAL WWTP

Rob Stephenson, Scott Laliberte, and Peter Elson
Paradigm Environmental Technologies Inc. TEL:(604) 731-9060, FAX: (604) 731-9058
rjstephenson@paradigm.prserv.net

INTRODUCTION

Wastewater treatment plants (WWTPs) worldwide are being squeezed: by treating wastewater to the higher discharge standards demanded by regulatory agencies, WWTPs generate more biosolids for disposal. To indicate the magnitude of the problem, for every million people, about 25 football fields of biosolids, each one metre deep, are generated annually.

Sludge continues to be generated at an increasing rate while in many jurisdictions environmental regulations have been tightened and the traditional biosolids disposal practices such as ocean dumping and landfilling have been eliminated (Matthews, 1997). This is occurring in the absence of effective, commercially available solutions for sludge disposal. Transforming biosolids into a safe, environmentally benign, and useful commodity is cumbersome and increasingly costly using today's practices.

The *MicroSludge*™ process radicalizes the approach to resolving this problem - a paradigm shift for the industry. *MicroSludge* addresses the squeeze placed on WWTPs in three ways:

1. By making anaerobic digesters much more effective within the limited available footprint to destroy biosolids, thus avoiding large capital expenditures for new sludge management infrastructure;
2. By destroying pathogens, reducing odours, and yielding vastly lower quantities of stabilized biosolids for disposal; and
3. By reducing the net operating costs by generating higher volumes of methane for combined heat/power generation, and by operating the digesters more efficiently, thus lowering inputs for heating and mixing.

BACKGROUND

The rate-limiting step for anaerobic digestion of waste activated sludge (WAS) is the destruction of the cell membrane of each microbe (Parkin and Owen, 1986). Anaerobic digestion of WAS is both slow and incomplete because the individual cell membranes are not significantly degraded in conventional mesophilic (35° C) anaerobic digesters that rely on enzymes to promote cell lysis.

Anaerobic digestion of WAS without pre-treatment falls short of an ideal biosolids management system for the following reasons:

1. Large quantities of undigested sludge still require disposal.
2. Partially digested sludge generates offensive odours and greenhouse gases.
3. Incomplete pathogen kill necessitates additional sludge processing before biosolids are safe to use as a fertilizer.
4. Undigested sludge is a wasted resource since the methane generation potential is not fully realized.

These shortcomings lead to high capital and operating costs for biosolids management.

DESCRIPTION OF *MICROSLUDGE* PROCESS OPERATIONS

Conventional municipal wastewater treatment typically involves mesophilic anaerobic digestion of both primary solids and secondary solids (WAS) from aerobic biological treatment to produce methane and carbon dioxide (see Figure 1). This biogas can be used as fuel in a micro-turbine to generate heat and power, both of which are used in the WWTP. In particular, waste heat can be used to dry residual biosolids after digestion to produce a dry and stable product.

The *MicroSludge* process (Stephenson and Dhaliwal, 2000) utilizes alkaline pre-treatment to weaken cell membranes, mechanical shear to reduce particle size, and an industrial scale homogenizer to provide an enormous and sudden pressure change to burst or "lyse" the cells. Figure 1 illustrates the process.

The heart of the *MicroSludge* process is an industrial scale homogenizer that provides a large and abrupt pressure drop. At 12,000 psig (82,700 kPag), sludge in the cell disruption homogenizing valve is accelerated up to 1,000 feet per second (305 meters per second) in about 2 microseconds (Pandolfe, 1993). This high velocity flow then impinges on an impact ring, disrupting the cell membranes and liquefying the cells (Middleberg, 1995).

MICROSLUDGE OPERATING CONDITIONS
A summary of the conditions that result in optimal destruction of volatile solids by mesophilic anaerobic digestion is given in Table 1.

MICROSLUDGE PERFORMANCE
Using off-the-shelf equipment, the patented *MicroSludge* process (US Patent No. 6,013,183, international patents pending) has proven to be extraordinarily effective to destroy WAS. Data presented in Table 2 are based on current bench and pilot scale testing at a municipal wastewater treatment plant in Vancouver, Canada.

COD: *MicroSludge* radically transformed the digestibility of biosolids. Soluble COD was reduced by 90% after 5 days mesophilic anaerobic digestion. Additional anaerobic digestion of total HRTs of 10 and 15 days resulted in 95% and 97% sCOD destruction respectively. In sharp contrast, conventional anaerobic digestion resulted in just 17% destruction after 15 days.

Volatile Solids: Volatile solids were reduced by 78% after 5 days of mesophilic anaerobic treatment. Methane gas was correspondingly produced. Without pre-treatment, after 15 days, conventional mesophilic anaerobic digestion resulted in 41% VS reduction.

Volatile Fatty Acids: Net volatile fatty acid levels indicated extraordinarily effective and stable digestion with *MicroSludge*, even after just 5 days HRT. Most (93%) of the VFAs were converted to biogas in the first 5 days, and almost all (97%) of the VFAs were mineralized within 10 days. Clearly, after just 5 days of mesophilic anaerobic digestion, just a small portion of biodegradable matter was remaining. Without pre-treatment, the elevated VFA concentrations indicate ineffective and perhaps unstable digestion.

BOD: BOD conversion was consistent with the other above measures of organics. The very low concentration of BOD of the filtrate from residuals de-watering was similar to that of domestic sewage. Since the volumetric flow of de-watering filtrate is relatively small, and the BOD is also small, the added load to the activated sludge plant would be minor.

Methane Generation: Higher levels of organics destruction generated a correspondingly higher volumes of digester gas. Nominally, each gram of volatile solids reduced produced one litre of gas. At an industrial scale, this would provide a substantial net energy surplus for electricity, digester heating, and drying of residuals.

Pathogen Destruction: The aggressive chemical/physical conditions applied by *MicroSludge* provided near quantitative pathogen kill to produce US EPA Class A Biosolids (see Table 3).

These pathogen results are important because Class A Biosolids are not commonly attained with mesophilic anaerobic digesters (Watanabe et al., 1997), the most common operating regime. The subsequent management of this stabilized *MicroSludge*-treated material would therefore be cheaper and far less constrained by regulations compared to biosolids with elevated pathogen levels.

QUANTITY OF RESIDUAL SOLIDS
The de-waterability of the final effluent sludge was tested through use of a centrifuge. Table 4 below summarizes these results. Table 4 confirms that the volatile solids fraction of both primary and secondary solids was largely insoluble, which is why conventional processing of biosolids is so ineffective. After 15 days of conventional processing, 44% of the sludge would require disposal compared to 12 to 14% with *MicroSludge*.

TRANSITION TO *MICROSLUDGE* FROM CONVENTIONAL TREATMENT
Figure 2 Illustrates the changes in performance of a mesophilic anaerobic digester at a 15 day HRT from conventional processing of 40% primary solids and 60% secondary solids without pre-treatment to *MicroSludge* processing.

The data show that within 8 weeks of transition to *MicroSludge*, sCOD removal was improved from 5 to 96%, BOD removal was improved from 6 to 96%, and volatile solids removal was improved from 41 to 73%. The data indicates that without pre-treatment, at a 15 day HRT the anaerobic digester was a net producer of volatile fatty acids with 51% greater than the feed concentration, whereas *MicroSludge* processing led to a net 96% reduction of the influent VFA levels. This may account for a significant amount of the odour attributed to sludge digested in conventionally operated systems.

PROCESS ECONOMICS

Shaw *et al.*, (2002) reviewed *MicroSludge* economics in some detail. The overall impact on the capital and operating costs of biosolids management are summarized as follows.

IMPACTS ON WWTP CAPITAL COSTS BY *MICROSLUDGE*.

MicroSludge processing of biosolids lowers the needed time for anaerobic digestion. This would increase the effective digester capacity, thus requiring fewer anaerobic digesters and avoiding plant expansion and associated costs. The small footprint of the process equipment could be easily integrated into existing WWTP layouts at minimal cost.

IMPACTS ON WWTP OPERATING COSTS BY *MICROSLUDGE*

The net operating costs for biosolids management using *MicroSludge* would be much lower than for all other available methods because more methane and correspondingly less biosolids would be produced that would be more de-waterable, safer, and more stable for disposal.

TECHNICAL AND ECONOMIC BENCHMARKING STUDY

Canada's National Research Council commissioned a study (Rabinowitz and Nemeth, 2001) to determine how *MicroSludge* compares with the competing means of processing biosolids: conventional mesophilic anaerobic digestion, extended thermophilic anaerobic digestion, and thermal hydrolysis. Their conclusions are summarized in Table 5 and discussed in greater detail in Shaw *et al.* (2002).

SUMMARY

The *MicroSludge* process is a simple chemical and pressure pre-treatment process that greatly increases both the rate and the extent that waste activated sludge is degraded in a conventional mesophilic anaerobic digester.

MicroSludge also kills pathogens and strips the biodegradable portion of the sludge to reduce its odour. The final sludge that results from the *MicroSludge* process is safer and more stable – and there is far less of it – than the sludge produced by conventional methods.

Conventionally operated mesophilic anaerobic digesters can be readily transitioned to *MicroSludge* without upset and with large gains in performance.

The capital and net operating costs of the *MicroSludge*™ process are far lower than for either conventional or any competing process.

REFERENCES

MATTHEWS, PETER, A Global Atlas of Wastewater Sludge and Biosolids Use and Disposal, International Association on Water Quality, Scientific and Technical Report No. 4., 1997.

MIDDLEBERG, A.P.J., Process-Scale Disruption of Microorganisms, Biotechnology Advances, 13, (3) 491 - 551, 1995.

PANDOLFE, W.D., Cell Disruption by Homogenization, APV publication, 1993.

PARKIN, G.F. AND OWEN, W.F., Fundamentals of Anaerobic Digestion of Wastewater Sludges, Journal of Environmental Engineering, 1986, 112, (5).

RABINOWITZ, B. and NEMETH, L., MicroSludge Process Technical and Economic Benchmarking Study, Earth Tech Report, November 30, 2001.

SHAW, J.N., Nemeth, L., and Henderson, G., Technical and Economic Benchmarking of the MicroSludge Process with Biological and Non-biological Disposal Options, 7th European Biosolids Organic Residuals Conference and Exhibition, Wakefield, UK, November 18 – 20, 2002.

STEPHENSON, R.J. AND DHALIWAL, H., Method of Liquefying Microorganisms Derived From Wastewater treatment Processes, US Patent No. 6,013,183, 2000.

US EPA, Protection of Environment, Chapter 40, Part 503, Standards for Use or Disposal of Sewage Sludge, 503.32 Pathogens.

WATANABE, KITAMURA, OCHI AND OZAKI, Inactivation of Pathogenic Bacteria under Mesophilic and Thermophilic Conditions, Water Science Technology, 1997, 36, (6-7), 25-32.

Table 1: Optimum *MicroSludge* Operating Conditions

PROCESSING STEP	OPTIMAL SETTING
Chemical Pre-treatment	1. <500 mg/L Na and < 200 mg/L K (added as hydroxides) in anaerobic digester
	2. WAS pH ≥ 10 to promote cell lysis
	3. Holding time of 1 hour or more to promote cell lysis
Homogenizer Pressure	12,000 to 14,000 psig (83,000 to 96,000 kPa) for maximum cell lysis
Anaerobic Digester pH	7.0 to minimize ammonia toxicity

Table 2: Effect of *MicroSludge* Processing on 40:60 Primary:Secondary Solids

Pre-Treatment	Anaerobic Digester HRT (days)	sCOD (mg/L)	VS (mg/L)	Total VFAs (mg/L)	BOD (mg/L)	NH_3 (mg/L)
40:60 Primary:Secondary Solids Feed	-	9,490	22,640	2,723	4,230	527
MicroSludge	5	920	4,940	195	415	1,160
MicroSludge	10	498	4,780	74	135	1,210
MicroSludge	15	305	6,460	159	230	1,180
Conventional Feed	-	8,330	34,269	1,601	3,420	280
Conventional Digestion	15	7,870	20,330	2,415	3,200	1,180

Table 3: Pathogen Destruction for Heat Treated Primary Solids and *MicroSludge* Processed Secondary Solids (40:60 mix)

Pre-Treatment	Anaerobic Digester HRT (days)	Faecal Coliform (MPN/g)	Salmonella sp. (Presence/absence)
Raw Feed	-	$1.0*10^8$	Present
MicroSludge Homogenized Feed	-	$2.5*10^4$	Absent
MicroSludge	5	$7.1*10^3$	Absent
MicroSludge	10	$3.5*10^2$	Absent
MicroSludge	15	$2.4*10^2$	Absent
Conventional Digestion	15	$8.6*10^7$	Present

Table 4: The Effect of *MicroSludge* on Solubilizing Solids

Pre-Treatment	Anaerobic Digester HRT (days)	Primary: Secondary Solids	Portion of Volatile Solids in Solution	% Total VS Reduced	% Particulate VS Remaining for Disposal
Primary Feed		100:0	3%	-	-
Secondary Feed		0:100	3%	-	-
MicroSludge	5	40:60	36%	78	14
MicroSludge	10	40:60	46%	78	12
MicroSludge	15	40:60	50%	71	14
Conventional Digestion	15	40:60	26%	41	44

Table 5: Summary of Technical and Economic Benchmarking Study (Rabinowitz and Nemeth, 2001)

CRITERIA	COMPARISON WITH *MICROSLUDGE* (at 15 day HRT)		
	Mesophilic Anaerobic Digestion	Thermophilic Anaerobic Digestion	Thermal Hydrolysis
Class A Biosolids*	No	Yes	Yes
Dry Solids	*MicroSludge* produces 25 to 35% less	*MicroSludge* produces 20% less	*MicroSludge* produces 15% less
Bulk Solids for Disposal	*MicroSludge* produces 45 to 50% less	*MicroSludge* produces 30 to 35% less	*MicroSludge* produces 25 to 30% less
Biogas	*MicroSludge* produces 25 to 35% more	*MicroSludge* produces 10% more	*MicroSludge* produces 10% more
Footprint	-	10 times *MicroSludge*	4 times *MicroSludge*
20 year NPV	*MicroSludge* is 40 to 55% less	*MicroSludge* is 15 to 30% less	*MicroSludge* is 20 to 30% less

* *MicroSludge* meets Class A Biosolids

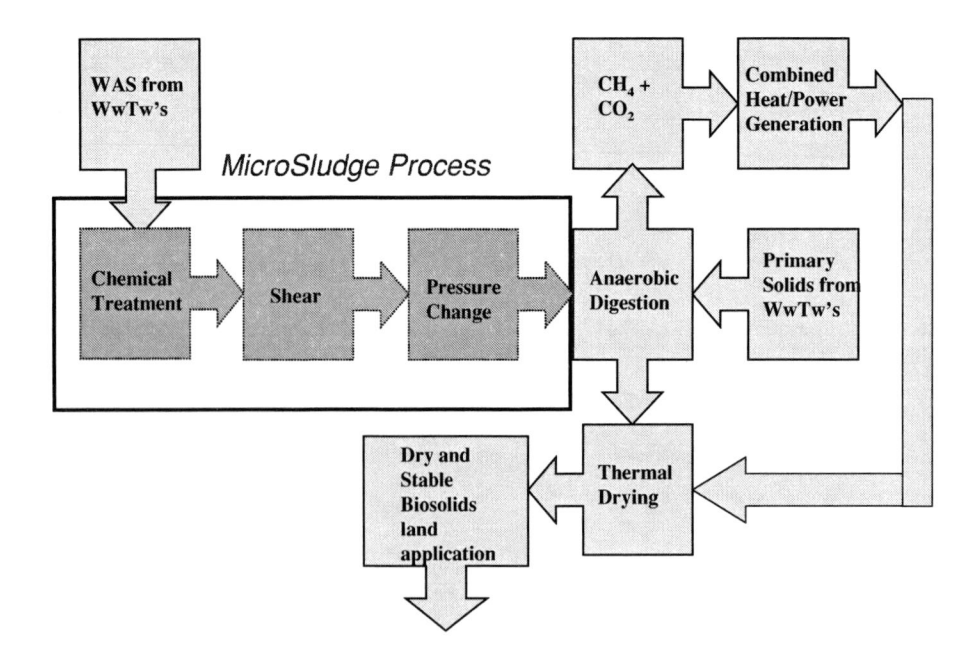

Figure 1: The *MicroSludge* Process and Conventional Anaerobic Digestion of Biosolids

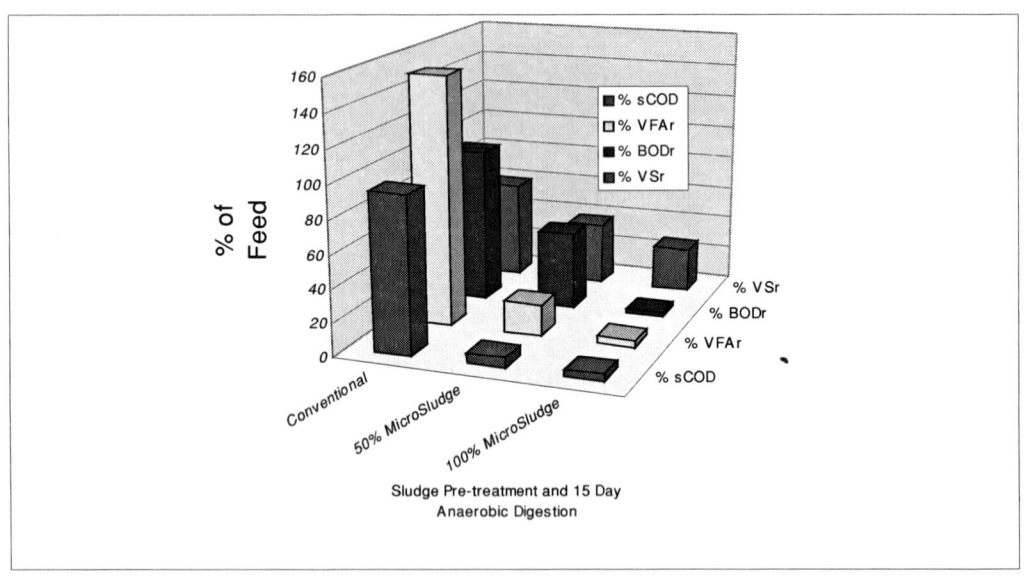

Figure 2: Transition to *MicroSludge* from Conventional Anaerobic Digestion

TECHNICAL AND ECONOMICAL BENCHMARKING OF THE MICROSLUDGE™ PROCESS WITH BIOLOGICAL AND NON-BIOLOGICAL DISPOSAL OPTIONS

Shaw, J.*, Nemeth, L**., Henderson, G***.,

*Paradigm Environmental Technologies Inc, Canada, **EarthTech Canada Inc, Canada,
***EarthTech Australia Inc, Australia

ABSTRACT

The MicroSludge process is an emerging technical option that pretreats secondary activated sludge resulting in improved solids reduction, enhanced biogas production and near qualitative pathogen reduction. The paper discussed the costs and benefits of three alternative applications of the technology. The benchmarking comparisons are including as will be a final stage in benchmarking the technology against thermal drying.

This paper failed to make the printers deadline. It should be available on the conference CD

SAPHYR® A STABILISED BIOSOLIDS TREATMENT PROCESS WITH FLEXIBLE DISPOSAL ROUTES

Gilbert, A.B[(1)]. Fernandes, P[(2)]. Baratto, G[(2)].

[(1)]OTVB Ltd Birmingham, U.K.
Tel: +44 121 329 4000 Fax:+44 121 329 4001 agilbert@otvb.co.uk
[(2)]Anjou Recherche Maisons-Laffitte, Cedex, France Tel:+33 1 34 93 31 Fax: +33 1 34 93 31 4131
paulo.fernandes@generale-des-eaux.net and gilles.baratto@generale-des-eaux.net

ABSTRACT

SAPHYR® is a liquid biosolids chemical disinfection and stabilisation process, developed primarily to permit the long term storage of biosolids destined for recycling to agricultural land (in liquid or dewatered state) without the release of offensive odours. Unlike lime stabilisation no solids are added to the biosolids and there is no appreciable increase in the actual mass of solid residual, consequently other disposal routes such as incineration remain viable for SAPHYR® treated biosolids.

The main benefit of SAPHYR® is the "odour less" nature of the stabilised biosolids with disinfection levels approaching the requirements for 'enhanced' treatment as detailed by the 'Safe Sludge Matrix' and impending revised code of practice. The SAPHYR® treatment process improves the dewaterability of the biosolids and produces drier stabilised biosoilds, with lower transport costs. The improved storage capability enables the seasonal application of biosolids to land to be more effectively managed.

KEY WORDS

Biosolids, Land Spreading, Nitrites, Odour, Recycling, Sanitisation, Stabilisation, Wastewater Sludge.

INTRODUCTION

To cater for the continued quality requirements being defined by the impending consultation on draft regulations, revision of the Code of Practice and the stack holders as demonstrated by the development of the 'Safe Sludge Matrix' [1], different methods of ensuring enhanced treatment are continually being reviewed. One such process SAPHYR®, uses low pH in combination with nitrites to disinfect and stabilise raw biosolids. The SAPHYR® biosolids treatment process has been developed as an alternative chemical stabilisation process to liming biosolids [2]. This process is applicable to liquid sludge rather than the de-watered pasty phase necessary for economic lime treatment (see Table 1). A process schematic diagram is indicated in Figure 1. Not only does the potent oxidising capacity of nitrite in an acid environment stabilise the sludge but the low pH also facilitates disinfection. Therefore the SAPHYR® treated biosolids has a significantly reduced odour potential, which is considerably less unpleasant than the raw biosolids [3].

The treated biosolids also has a significantly reduced pathogen content with up to 6-log reduction [4]. Whilst the heavy metal content of the biosolids remain substantially unaffected by the process except for the zinc content which is reduced in the solid phase, becoming more predominant in the liquid phase, separated by the down stream de-watering process. However, the de-waterability of the SAPHYR® treated biosolids is improved by the hydrolysing effect of the process.

Because there is little impact on either the calorific value of the biosolids or its available nutrient
levels [4], the SAPHYR® process is extremely suited for small remote sites, where the biosolids need to be transported to larger strategic treatment centres or recycled locally. Thus these stabilised and sterilised biosolids can be applied to agricultural land or is well suited to incineration, should there be no suitable land bank [5]. As a last resort dewatered material processed by SAPHYR® can be sent to landfill.

The relatively small size of the reaction vessel and simplicity of the process makes retrofitting or use of portable equipment an easy addition to these small regional sites where chemical stabilisation is the most expedient biosolids stabilisation technique. Unlike liming the mass of the residual solids is not increased.

The treated biosolids from SAPHYR® can be stored long term without any detrimental impact on the stabilisation provided by the process, either in the liquid phase or as a dewatered solid. This dewatered solid can be stored in vessel or in a pile, and unlike limed sludge no appreciable surface hardening, will occur [3].

The first permanent SAPHYR® process has been installed late in 2001 on a wastewater treatment plant in Italy at Cortina d'Ampezzo a 15,000 P.E. site. The next unit is currently being completed (11/02) at a 50,000 P.E. site at Cognac in France.

HOW DOES SAPHYR® WORK AND IT'S IMPACT ON THE BIOSOLIDS

PROCESS DESCRIPTION

The centre of the SAPHYR® process is based around the strong oxidising power of nitrite (NO_2), in an acidic environment, this oxidising reaction is applied to the residual biosolids fed through the process. The acidic conditions are achieved by dosing concentrated sulphuric acid into the liquid biosolids, whilst simultaneously dosing nitrites in the form of sodium nitrite solution. These are then mixed together for several minutes in a batch reactor vessel where, fermentative and pathogenic organisms are inactivated.

The process schematic diagram is presented in Figure 1. The system comprises a simple reaction vessel (stirred and baffled tank), which operates on a fill-and-draw basis. Liquid biosolids (1% to 8% $^w/_w$) composition, although ideally thickened to at least 4% $^w/_w$ are fed into the reaction vessel. These biosolids are simultaneously dosed with concentrated sulphuric acid as they enter the SAPHYR® process system. This is dosed under pH feed back control, with the pH being monitored in the stirred reaction vessel. The reaction vessel is a covered tank with forced extraction through the odour control unit. Once the correct quantity of biosolids has entered the reaction vessel determined by the liquid level, the sodium nitrite dose is commenced. This adds a fixed quantity of nitrite solution to the system to achieve the desired concentration or ratio of nitrogen as NO_2 to the mass of biosolids (which is a characteristic of the biosolids feed concentration).

The SAPHYR® process's control system then starts the contact timer. The stirrer remains operational throughout this reaction period to ensure all biosolids are exposed to the correct reaction conditions. The forced ventilation of both the reaction vessel and working area is maintained continuously.

On completion of the required contact time the reaction vessel is drained out and the stirrer stopped once low level is achieved. Because the treatment process is a physical/chemical procedure there is no requirement to retain any material to catalyses the next reaction. Consequently the reaction vessel can be emptied fully. In this way, with the fill and draw batch process the application of the SAPHYR® process can be assured for all the biosolids treated.

PROCES OPTIONS

The SAPHYR® process has high performance for both, the disinfection of the biosolids and the reduction in the olfactory power of the biosolids by stabilisation, thus preventing the uncontrolled destruction of the volatile organic component. There are two degrees of treatment available; the S process or the SH process. The difference provided is the degree of disinfection applied and final pH achieved for the treated biosolids as follows:

<u>THE S PROCESS</u>
- biosolids acidification at pH 3,
- Nitrate concentration from 4 to 8gN/gDS,
- Contact time required 30 minutes.

<u>SH PROCESS, BIOSOLIDS</u>
- acidification at pH 2,
- Nitrate concentration from 4 to 8 N/gDS, Contact time required 120 minutes.

The dose of nitrite is to achieve the required weight ratio of nitrite as N to the dry solids content of the biosolids in the reaction vessel, prior to the commencement of the reaction phase.

DISINFECTION / STERILISATION

The treatment with nitrites in the acid phase firstly disinfects the biosolids by disruption of the cell membranes, it then sterilises the biosolids by preventing fermentation for several months due to these harsh conditions presented to the biomass by the low pH in the presence of nitrite. Under these conditions the nitrite acts as a readily available source of oxygen, with its associated oxidising power. Although, the pH of the treated biosolids are low they do not exhibit high degrees of acidity, their total acidity, is comparable with the phosphoric acid content of carbonated cola soft drinks. However, the low pH of the system does require suitable selection of the materials of construction for the reaction vessel and chemical storage plant.

The degree of this disinfection and resultant sterilisation depends on the rate of treatment applied the pH reached and time of contact within the reaction vessel. Both the S and SH treatments destroy faecal pathogens, salmonella, viruses and a proportion of Helminthes eggs as can be seen in Table 4. The improved potency of the SH option also inhibits spores activity (Clostridium).

REDUCTION OF OLFACATORY IMPACT OF BIOSOLIDS

The impact of the stabilisation efficiency of SAPHYR®, has been tested using several different sludges treated with an industrial scale plant. For each site a significant quantity of dewatered SAPHYR® treated biosolids were produced and then left either in a pile or storage vessel for a period of six months. A test blank of raw untreated sludge was collected from each site. Two odour tests were conducted on each stored sample ; a fermenteccibility index and an olfactory test. The test method and facilities are briefly described below:-

SLUDGE PRODUCTION AND STORAGE.
The industrial pilot plant consists of a $1m^3$ tank equipped with a stirrer. The feed biosolids (sludge), sulphuric acid and nitrite were mixed in the tank for a contact time of 30 or 120 minutes. Then the treated biosolids was dewatered with an industrial centrifuge (capacity: 150 to 400kgDS/h) and stored outside on soil or in covered tank with a 400kg sludge capacity. Figure 5 details the design of the storage tank. The dewatered biosolids is heaped on top of a 150 mm layer of rounded gravel
(20-40 mm typical size). The base of the tank incorporates a slope that ensures that any liquid fraction which, separates can freely separate and drain. This design feature is a significant requirement for a sustainable sludge stabilisation when dewatered sludge is stored in a tank. Furthermore, holes in the tank cover permitted natural ventilation. The presented trial was based on a primary sludge treatment produced from an urban WWTP with 300,000 P.E.

*F*ermentescibility index.
The Fermentescibility Index (FI) is an analytical tool to measure sludge stability set up at ANJOU RECHERCHE (the research branch of Vivendi Water). The method consists of gas production measurement of sludge samples stored anaerobically under standard conditions (dilution, temperature and stirring). The measurement of sulphur compounds (H_2S and CH_3SH), which are the most easily detectable odorous compounds associated with unstable sludge, appeared to be a good quantitative parameter to assess overall sludge stability.
The FI is calculated taking into account; the bacteria growth kinetics (latency period, growth phase, stabilisation phase). A positive fermentescibility index indicates an unstable and a potentially malodorous sludge, the greater the number the more aggressive is this tendency. Whilst a zero FI value indicates an absence of sludge fermentation capacity and a very stable biosolid. This long term forecast method is realised at the beginning of the storage.

OLFACTORY TEST.
Olfactory tests using a panel of individuals have been used to determine olfactory nuisance of an odour sample. The method used in this trial was the "NFX43-103" experimental and specific norm to measure odour intensity. This method use an odorous reference scale realised with graduate concentration of butanol or pyridine presented in bottles (pyridine was chosen for this experiment). About 100g of sludge (biosolids) sample is presented to the odour panel. After smelling the sample and the odour reference bottles, the jury indicates which bottle approaches the sludge sample's intensity. This analysis is done at the end of the storage period with a jury comprising 13 people.
For these tests we defined four parameters to characterise the olfactory test, these are: - odour intensity; hedonistic value; discomfort value and odour characterisation. Only the odour intensity is directly related to reference values. The other characteristics are subjective and have been chosen according to our experience of odour nuisance.

- The odour intensity is based on a pyridine scale of graduating concentration
- The hedonistic value: When smelling the sample, the jury must select an hedonistic value in the scale below:

+3	very agreeable odour
+2	agreeable odour
+1	slightly agreeable odour
0	neither; agreeable or disagreeable odour
- 1	slightly disagreeable odour
- 2	disagreeable odour
- 3	very disagreeable odour

The discomfort value. The jury must select a discomfort value in the scale below:

0	not discomfort odour
- 1	light discomfort odour
- 2	discomfort odour
- 3	strong discomfort odour

- A characterisation of the odour sludge sample among current life odour (farm, flower,...). In order to help the jury, the question was "what do you associate to this odour?"

The hedonistic and discomfort values are different. An agreeable odour could be uncomfortable. On the other hand, a disagreeable odour could be smelt but not have sufficient impact to be uncomfortable.

Raw sludge olfactory analysis is managed on sludge from covered storage.

RESULTS AND DISCUSSION
FERMENTESCIBILITY INDEX.
The FI analysis on the raw biosolids at the test site was positive at FI=0.3, whilst that for the SAPHYR® treated biosolids was zero indicating fully stabilised biosolids. The raw sludge fermentescibility index was lower than the normal range (10 to 100) but this was due to the high iron content of the raw sludge, which promoted the precipitation of FeS reducing sulphurous compound emissions. For the SAPHYR® treated biosolids no sulphurous compounds (H_2S and CH_3SH) were generated during the analysis.

OLFACTORY TEST.
The test was performed after a storage period of 6 months. Samples were taken from the centre of the sludge heap, with about 100g of each biosolids being presented to the jury. The intensity value given by the jury for each biosolids sample is presented in Table 5. Inferior, medium and superior intensity were calculated using statistical analysis of the thirteen responses of the jury, using the formula of Annexe 2 of the NFX43-103 test method with 10% confidence interval.

The results from this odour study showed that the raw sludge odour was more intensive than treated sludge. In terms with comparison to pyridine concentration, raw sludge was 20 to 100 more concentrated than the SAPHYR® treated sludge. After 6 months of storage, the SAPHYR® treated sludge retained its very low odour impact or high quality. We can conclude that this result confirmed that the SAPHYR® treatment creates a long term stabilisation.

Tables 6, 7 and 8 give the odour characterisation with the hedonistic value, the discomfort value and the resemblance with current life odours.

These characterisations confirmed the result of intensity test. The raw sludge odour was considered to be disagreeable or very disagreeable and very uncomfortable. The jury characterised the odour as highly putrescent and reminiscent of nauseous smelly materials. The comparison was with piggery or stable manure, which demonstrated that nuisances resulting from wastewater biosolids and farm residues are not so dissimilar. These results also showed a high fermentation activity after 6 months of storage for the raw sludge as it was predicted by the fermentescibility index test.

By comparison, the SAPHYR® treated sludge was felt by a few of the panel as agreeable and not an uncomfortable odour. The comparison was with fungi (cellar, mushroom) or earth (dry earth, humus, potatoes). The covering of the storage material had, apparently, no impact on the stabilisation sustainability. Nevertheless, regardless of the storage method, the liquid fraction emitted from the biosolids dewatered sludge was drained either directly to the soil, or through the stone layer. It was felt that this was necessary to maintain stable conditions.

Summarising these results, zero nuisance could be guaranteed during the storage of SAPHYR® treated biosolids and or the spreading of this material on land.

Further odour treatment at a different WWTP has been undertaken using the more conventional measurement of odour as odour number (point at which only half of the panel can detect an odour, with dilution of the malodorous source with odourless test air, starting with pure test air). This trial is reported in Table 9 with the values quoted being the dilution ratio required for K_{50} only half the panel detecting an odour.

Again for this trial the panel considered the hedonistic value of the SAPHYR® treated biosolids to be neutral neither; agreeable or disagreeable and was not persistent. Whilst the odour indicator using H_2S is also indicated the SAPHYR® treated biosolids had no hydrogen sulphide present.

TREATED BIOSOLIDS DEWATERABILITY
The SAPHYR® treatment process improves the dewatering capacity of the biosolids. The average increase in dryness is between 2 to 3 percent over that of the untreated biosolids. The processed biosolids demonstrates a more fibrous texture. These properties associated with a better stability induce a better mechanical behaviour such as pilling capacity, handling feature and others (farmer) solids handling considerations.

This improved dryness can also have benefits if the biosolids are to be minimised rather than recycled, since the increased dryness improves the potential for auto-thermal combustion of SAPHYR® treated biosolids.

For the case of primary biosolids, the increase in dryness is even greater and can reach five or six percentage in dryness and requires less polymer, please see Figure 6 [4] and Table 10[2].

In addition, there is no evolution or release of organic matter content from the SAPHYR®

treated material, which represents a benefit for the biosolids auto combustion potential.

SAPHYR® TREATED BIOSOLIDS HEAVY METAL CONTENT

The heavy metal content of the biosolids treated by SAPHYR® is essentially unaffected by the process, with the exception of zinc which becomes more soluble in the liquid phase following treatment. Table 11 indicates the difference in heavy metal content between the raw biosolids and this same material once it has been treated with SAPHYR®. This analysis has been performed on the dewatered solids.

The impact of the SAPHYR® treatment process when considering its impact on the residual solid's heavy metal content does not limit the disposal route of these treated biosolids to land application.

HOW DOES SAPHYR® WORK?

REACTION KINETICS OF NO_x

The reaction of nitrite and the acid does generate a small quantity of NO_x. To prevent a build-up of this emission and uncontrolled release to the atmosphere, the reaction vessel and operating areas adjacent to the plant should be contained with forced draft ventilation. However, the yield of this NO_x is equivalent to that from a typical car's emission for each 6,000 PE (population equivalent) of Wastewater Treatment Plant's (WWTP) production of biosolids which are treated by SAPHYR®.

The nature of the reagents (H_2SO_4, NO_2), together with the iron content in biosolids and the different redox potentials, have a great impact on the NOx formation pathway. The possible reactions, which can occur, include: HNO_2 disassociation; NO_3 reduction by Fe II and HNO_2 reduction by Fe II.

These Reactions are summarised as follows:-

HNO_2 DISASSOCIATION
$3\ HNO_2 \leftrightarrow H^+ + H_2O + 2\ NO + NO_3^-$ (a)

- NO_3^-/HNO_2
$H_2O + HNO_2 \leftrightarrow NO_3^- + 3H^+ + 2e^-$ $E_1°=0,94V$

- $HNO_2/NO_{(g)}$
$HNO_2 + H^+ + e^- \leftrightarrow NO_{(g)} + H_2O$ $E_2°=1V$

NO_3 REDUCTION BY FE II
$NO_3^- + 4\ H^+ + 3\ Fe^{2+} \rightarrow NO + 2\ H_2O + 3\ Fe^{3+}$
NO_3^-/NO E=0.956 V
Fe^{3+}/Fe^{2+} E=0.771 V

HNO_2 REDUCTION BY FE II
$HNO_2 + Fe^{2+} \rightarrow NO + Fe^{3+} + H_2O$
$HNO_2/NO_{(g)}$ E=0.99 V
Fe^{3+}/Fe^{2+} E=0.771 V

THERMODYNAMIC ASPECT
The calculation of the thermodynamic constants, demonstrates that these reactions are thermodynamically favourable as can be seen in Table 2[5].

KINETIC ASPECTS
We have also studied the kinetic reactions. These experiments were undertaken in a 300 ml flask but clearly demonstrate the viability of the process. These results are reported in Table 3. Whilst figures 2, 3 and 4 show kinetics of HNO_2 disassociation, kinetics of NO_3 reduction and NO_x production mechanism respectively. In summary during the first few minutes the concentration of HNO_2 reduced rapidly. However, once the reaction reached equilibrium as indicated by the nitrite to nitrate concentration instability this decay was reduced. This instability can be attributed to the partial NO pressure. These trials show that NO_3 reduction by Fe II is a slow reaction with only 1.5 % producing NO_x. We have concluded that main source of NO_x production is from the reduction of HNO_3 by Fe II.

SAPHYR® TREATED BIOSOLIDS POSSIBLE DISPOSAL ROUTES

For biosolids, which are processed using SAPHYR® treatment there are a number of disposal routes available. These will depend on the logistics of the wastewater treatment site and quality of the raw feed biosolids, their suitability for nutrient recycling and transport requirements.

These recycle/disposal methods for SAPHYR® treated biosolids can be summarised as follows:-
- Biosolids Liquid Storage for Agricultural Liquid Recycle Route (see figure 7)
- Dewatered Biosolids for Agricultural Recycle Route (see figure 8)
- Dewatered Biosolids for Minimisation / Disposal to Incineration or Landfill (see figure 9)

From these possible disposal routes it can be seen that the SAPHYR® biosolids treatment process provides maximum flexibility to the final disposal route.

CONCLUSIONS

The SAPHYR® process is ideally suited to treating biosolids at a small remote wastewater treatment works or other similar applications (disinfection of farm wastes) were both a stabilised and odourless biosolids residuals are required. If these biosolids, from a municipal source, need to be transported then the improved dewaterability of the SAPHYR® processed biosolids will demonstrate a transport cost saving.

Because SAPHYR® operates as a batch system, utilising feed biosolids still in the liquid phase (up to 8 % $^w/_w$) and the chemical additives are dosed as true liquid solutions; the correct degree of mixing (and consequent assurance of contact and performance) can be more readily achieved than when mixing lime in to dewatered biosolids. This results in less overdosing of treatment chemicals and a lower energy requirement to mix the batches of biosolids during their reaction.

Although the biosolids are treated at low pH the total acidity of SAPHYR® processed biosolids is low, they have limited impact on the overall pH of the receiving soils when using the agricultural recycle route.

The nuisance to neighbours, both local to the wastewater treatment facility and the point of application to land recycling, from offensive biosolids odours is eliminated by the SAPHYR® process.

Biosolids treated by the SAPHYR® SH reaction are effectively disinfected to the standards required for 'Enhanced Treated Sludges' in accordance with the 'Safe Sludge Matrix' and relevant Code of Practice. This maximises the environmental worth of biosolids with recycling becoming an easily achievable route.

The impact on the heavy metal contents of the biosolids remain essentially unaffected by the SAPHYR® process except for zinc which is reduced in the final solid phase, when compared with the feed material.

Therefore the treatment of biosolids with the SAPHYR® process can readily unlock the reuse or recycle route for these materials.

REFERENCES

1 "The Safe Sludge Matrix" – Guidelines for the Application of Sewage Sludge to Agricultural Land ADAS 3rd Edition April 2001.

2 GILBERT, A. B. CLAIR, N. BIGOT, B. Advances in the Hygienic Treatment of Sludge. Proceedings of the 5th European Biosolids and Organic Residuals Conference. Wakefield, UK 20-22 November 2000.

3 CORITON, G. LEBOUCHER, G. BONNIN, C. SAPHYR: A Sludge Treatment For Odour Control in Agricultural Route. Proceedings of the Control and prevention of odours in the water industry, Conference CIWEM/IAWQ London, UK 22-24 September 1999

4 LEBOUCHER, G. BONNIN, C. CORITON, G. SAPHYR An Innovative Chemical Stabilisation of Biosolids. Proceeding of 18th Federal Convention Adelaide, Australian Water & Wastewater Association Adelaide Australia, 11-14 April 1999.

5 CORITON, G. LEBOUCHER, G. BONNIN, C. SAPHYR: A Flexible Final Route. Proceedings of the disposal and utilisation of sewage sludge: Treatment methods and application modalities, Congress IAWQ/COMMISSION EUROPEENNE DG XIII. Athens/Greece Date of congress, 13-15 October 1999.

ACKNOWLEDGEMENTS

The Authors would like to thank the following for their assistance in preparing this paper: -
Lucie Patria, Didier Cretenot; Bruno Bigot; Victoria Ballard; Claire Rousselet; Pascal Marlin and Eric Guibelin.

TABLES

Table 1: Range of Processes Available to Stabilise (and Hygienise) Biosolids from Wastewater Treatment

The Treatment Method	What the Treatment Achieves	Vivendi Water's Treatment Process	
		In the Liquid Phase (Thickened)	In the Pasty Phase (Dewatered)
Biological	Breaks down the VS according to a controlled bio-process	Digestion (MAD and TAD) Bio-pasteurisation Hydrolysis	Composting or Co-Composting
Chemical	Holds up the non controlled septic vVS destruction	SAPHYR® Nitrite in the Acid Phase	Liming
Physical	Holds up the non controlled VS degradation		Drying
	Eliminates the VS through thermal oxidation	Wet (Air) Oxidation	Incineration
	Unlocks the VS for improved Biological treatment	Thermal hydrolysis	

Table 2: Thermodynamic Constants of Reaction

	HNO3 disassociation	NO3 reduction	HNO2 reduction
K value of reaction	$10^{1.67}$	$10^{9.25}$	$10^{3.53}$

Table 3: Operational Conditions for Kinetics Study

	HNO3 disassociation	NO3 reduction	HNO2 reduction
Distilled Water (ml)	200	200	200
PH	3	3	3
NaNO$_2$ solution of 10 mgN/l (ml)	5	0	5
NaNO$_3$ solution of 9.4 mgN/l (ml)	0	10	10
FeSO$_4$.7H$_2$O (g)	0	2	2

Table 4: Reduction of Microbial Activity with SAPHYR®

	Non-treated Biosolids	SAPHYR® S	SAPHYR® SH
Feature of Process		PH=3 nitrite=75 mgN/l Removal	PH=2 nitrite=300 mgN/l Removal
E. Coli count/g wet sludge	10^5 to 10^6	5 to 6 log	5 to 6 log
Streptococcus count/g wet sludge	10^5 to 10^6	3 to 6 log	3 to 6 log
Clostridium spores count/g wet sludge	10^5 to 10^7	1 log	5 to 6 log
Helminthes eggs count/g wet sludge	100	50 to 70 %	50 to 70 %

Table 5: Intensity Values From Olfactory Test (ppm)

	Low (*) Intensity Detection (ppm)	Medium (*) Intensity Detection (ppm)	Superior (*) Intensity (ppm)
Raw sludge	26	227	1960
Treated sludge Covered storage	0.1	9.9	620
Treated sludge Outdoor storage	0.04	1.8	72

(*) The intensity value given by the jury for each sludge sample is presented in Table 5 above. Inferior, medium and superior intensity were calculated with the thirteen responses of the jury (refer to the norm for calculation NFX43-103 Annexe 2). The distribution is supposed to follow a Gauss repartition with a risk of 0.1.

Table 6: Hedonistic Values of Raw and Treated Sludge

Average value of Jury	Low Intensity Detection (ppm)	Medium Intensity Detection (ppm)
Raw sludge	-2.3	Disagreeable to Very disagreeable
Treated sludge Covered storage	+0.35	A few agreeable
Treated sludge Outdoor storage	+0.25	A few agreeable

Table 7: Discomfort Values of Raw and Treated Sludge

Average value of Jury	value	interpretation
Raw sludge	-2.5	strong discomfort
Treated sludge Covered storage	-0.2	Very light discomfort
Treated sludge Outdoor storage	0	No discomfort

Table 8: Resemblance of Odours

	characterisation
Raw sludge	Piggery, stable manure, mud, ammonia
Treated sludge Covered storage	Garret, cellar, mushroom, potatoes, earth
Treated sludge Outdoor storage	dry earth, mushroom, humus

Table 9: Odour Measurement Study After 9 Months Storage

Parameter	Non-treated sludge	SAPHYR® treated biosolids
K50	24,040	27
Odour description (Hedonistic value)	Very disagreeable	Not unpleasant
Odour indicator (ppm H_2S)	1,000	0

Table 10: Increased Dewatering Ability of Biosolids, After SAPHYR® Processing

Type of plant	Plant A ASP medium load	Plant B ASP medium load	Plant C ASP low load
Type of thickened sludge	mixed primary and secondary	secondary	mixed primary and secondary
Non-treated sludge features	SS ≈ 45 g/l VSS ≈ 65%	SS ≈ 22 g/l VSS ≈ 68%	SS ≈ 35 g/l VSS ≈ 65%
Non-treated dewatered sludge features	SS = 25%	SS = 15 to 17%	SS = 22 to 23%
SAPHYR® dewatered sludge	SS = 27 to 32%	SS = 17 to 22%	SS = 23 to 26%

ASP: Activated sludge plant

Table 11: Heavy Metal Content in Biosolids, Before and After SAPHYR® Processing

Heavy Metal	Cd	Cr	Cu	Hg	Ni	Pb	Zn
Non-treated dewatered sludge (µg/g DS)	2.0	93	217	3.0	17	140	1140
SAPHYR® treated dewatered sludge (µg/g DS)	2.0	95	210	3.1	15	160	565

FIGURES

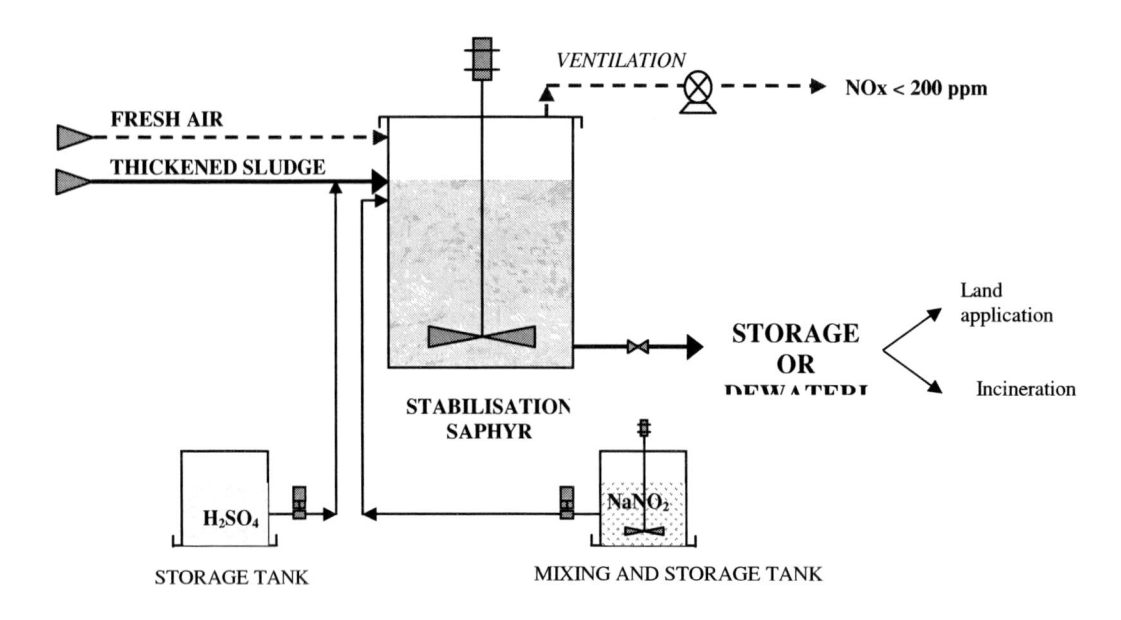

Figure 1   Process Schematic Diagram of SAPHY Biosolids Treatment

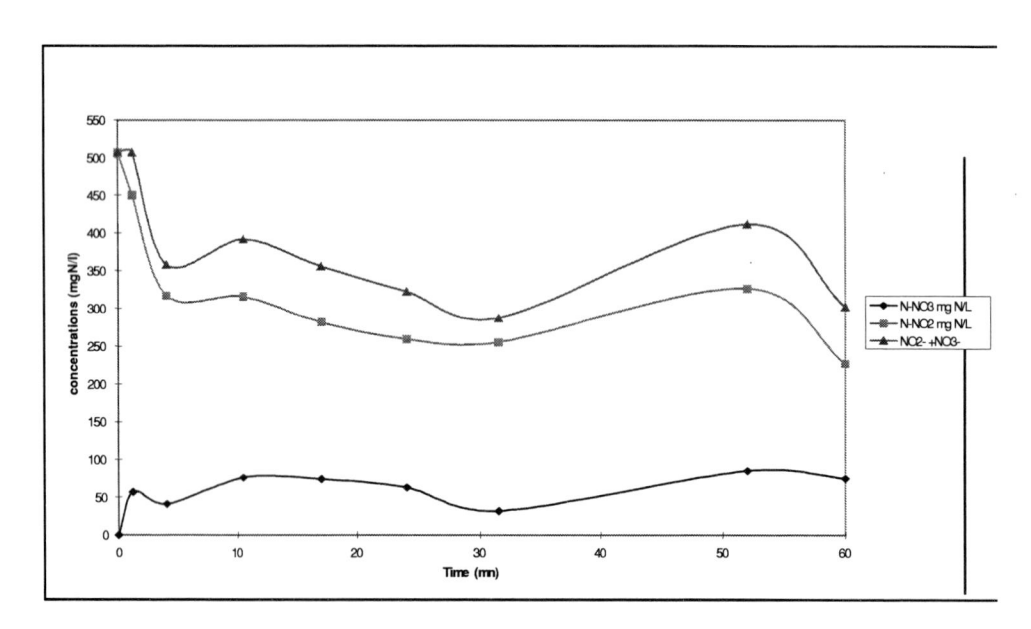

Figure 2   Kinetics of HNO_2 Disassociation

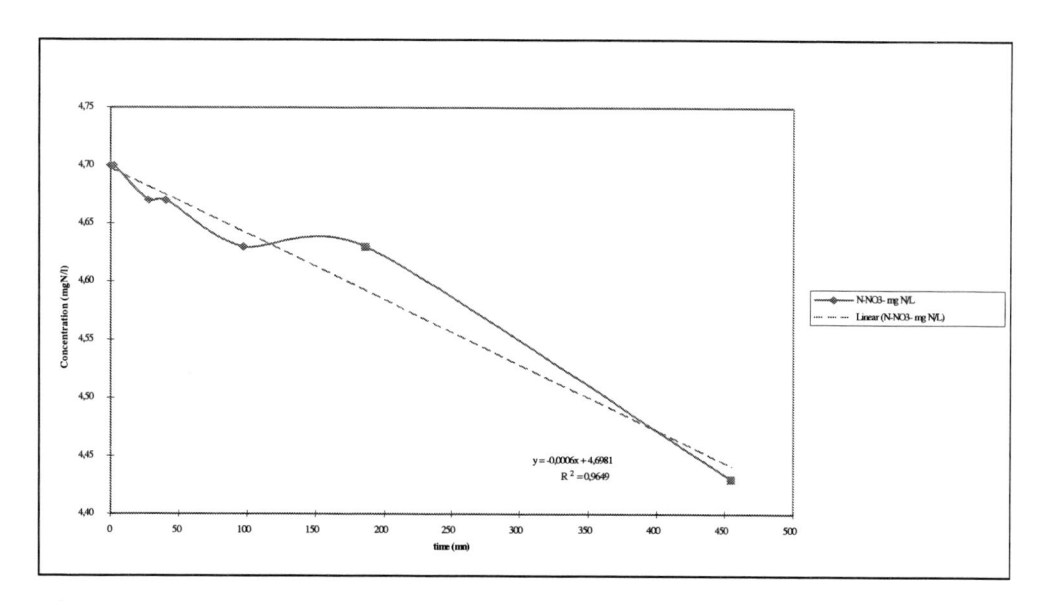

Figure 3   Kinetics of NO_3 Reduction

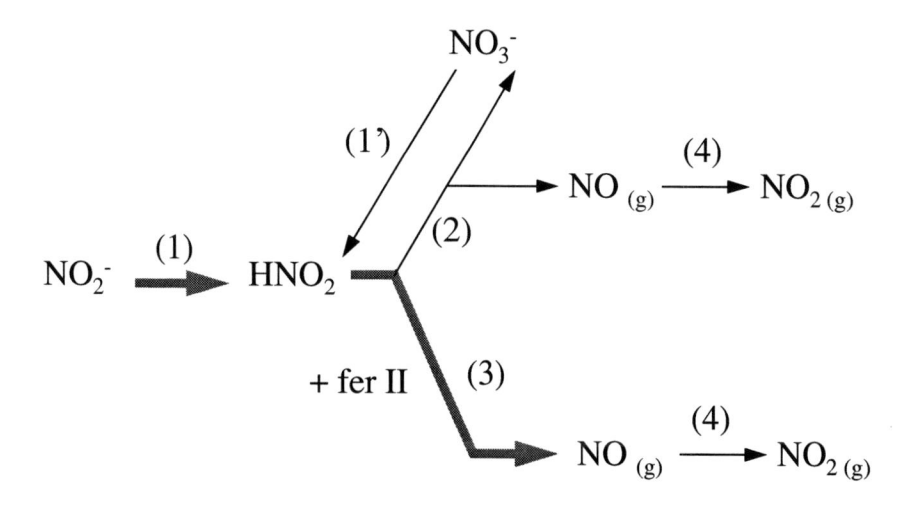

$$NO_2^- + H^+ \rightarrow HNO_2 \quad \text{(1) } HNO_2 \text{ production}$$
$$3\ HNO_2 \rightarrow H^+ + H_2O + 2NO + NO_3^- \quad \text{(2): } HNO_2 \text{ dismutation}$$
$$HNO_2 + H^+ + Fe^{2+} \rightarrow NO + H_2O + Fe^{3+} \quad \text{(3): } HNO_2 \text{ reduction by fe II}$$
$$NO_3^- + 2\ NO + H_2O + H^+ \rightarrow .3\ HNO_2 \quad \text{(1'): } HNO_2 \text{ dismutation}$$
$$NO_{(g)} + 1/2\ O_2 \leftrightarrow NO_{2\ (g)} \quad \text{(4): NO consumption in air}$$

Figure 4   NOx Production Mechanism

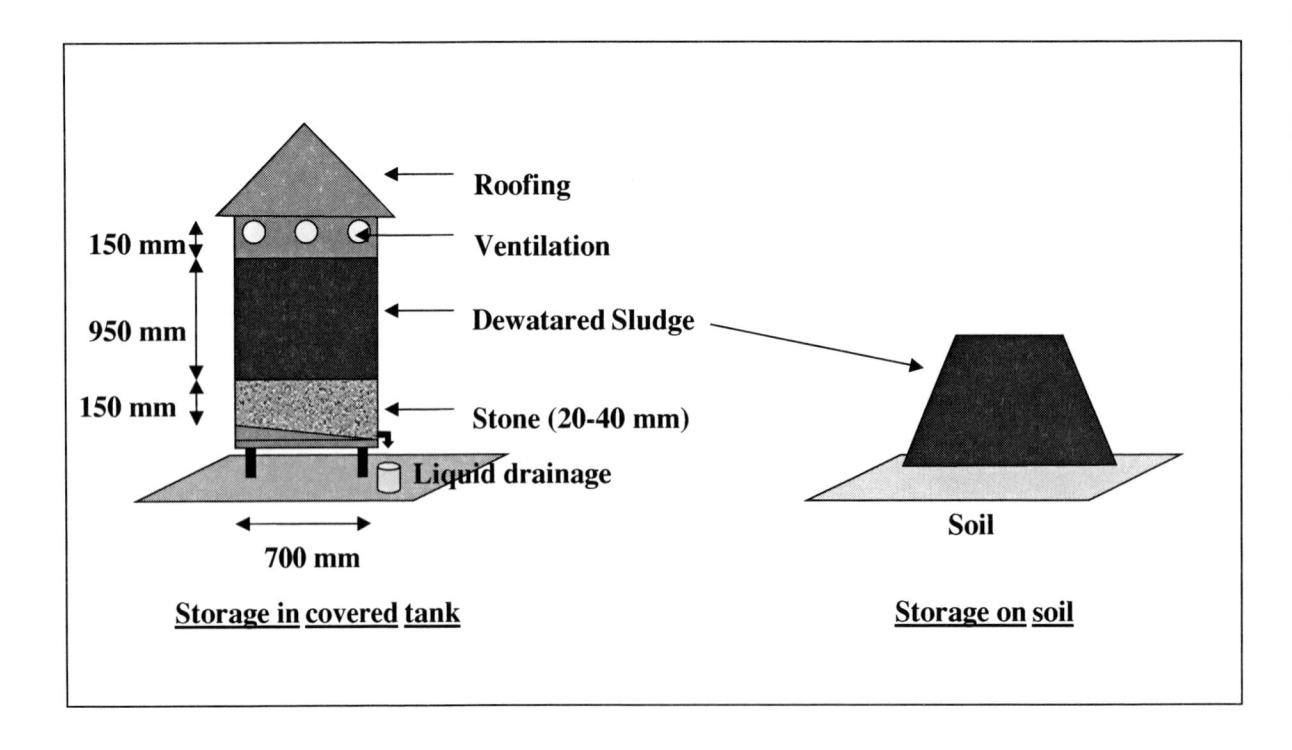

FIGURE 5 **DIFFERENT METHODS OF STORING SAPHYR TREATED BIOSOLIDS**

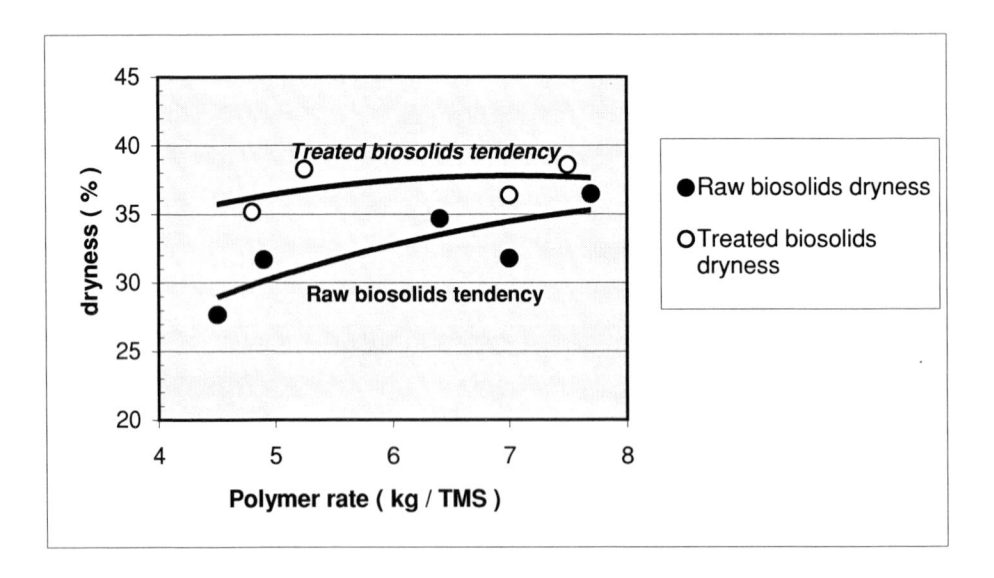

Figure 6   Impact of SAPHYR® Treatment on Primary Biosolids Dryness

Biosolids → SAPHYR → Liquid storage → Injection into soil

Figure 7: Liquid Route to Land

Biosolids → SAPHYR → Dewatering → Spread to land

Figure 8: Cake route to Land

Biosolids → SAPHYR → Dewatering → Incineration or Landfill

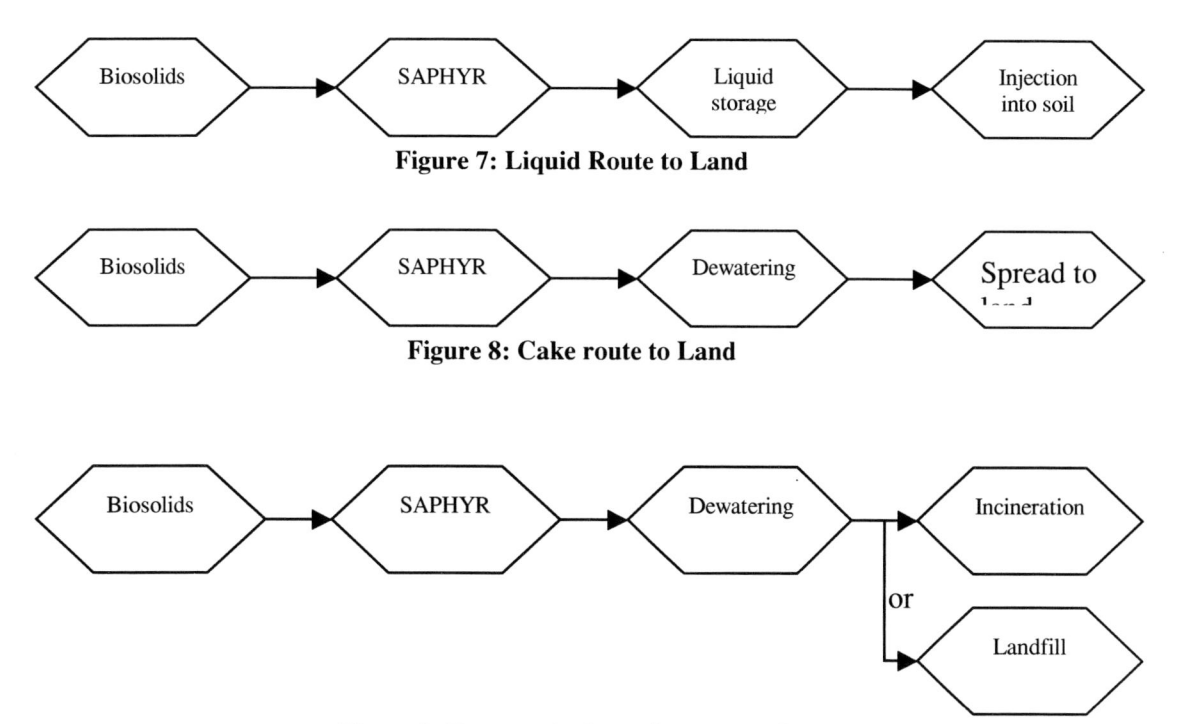

Figure 9: Route to incineration or landfill

IN-SITU DEWATERING OF LAGOONED SEWAGE SLUDGE USING ELECTROKINETIC GEOSYNTHETICS (EKG)

Jim Walker[1], Stephanie Glendinning[2]

[1] Sludge Consultant Tel: 01535 665829 Fax: 01535 665829 jimwalker@mistral.co.uk
[2] University of Newcastle Tel: 0191 222 6612 Fax: 0191 222 6613 Stephanie.Glendinning@ncl.ac.uk

ABSTRACT

Existing dewatering methods require sludge to be delivered to a machine in liquid form. EKG has the potential for dewatering sludge after it has been deposited in a tip, lagoon or windrow. The basic EKG concept combines electrokinetic phenomena (electro-osmosis and electrophoresis) with geosynthetics. The patented electrodes have reduced electrode corrosion, allowed increased current density and better electrical contact with the soil. They also provide a drainage path for the removed water. They can also function equally as cathodes or anodes, allowing polarity reversal, a crucial procedure in dewatering.

A lagooned sludge was tested in the laboratory. Within 24 days the dry solids increased from 19% to a maximum of 42% and shear strength from 1kPa to a maximum of 29kPa (sufficient to support the construction of car parking or single storey buildings with pad foundations). Further work on other sludges and to establish a full procedure for using EKG in the field is reported in the Paper.

KEY WORDS

Dewatering, electrokinetics, electro-osmosis, electrophoresis, in-situ dewatering, sewage sludge, sludge lagoon.

INTRODUCTION

It is well understood that the dewatering of sewage sludge is one of the most complex technical tasks in the field of wastewater engineering [1] and all existing methods have severe limitations. Since the effective demise of sludge drying beds in the UK all the current dewatering methods depend on feeding a liquid sludge into a machine, whether it be a centrifuge, a belt press, a plate press, etc. Once a cake has been formed, however bad the quality there are few ways of dewatering it further. A reasonable cake can be improved by stacking, given time, space and good weather, or it can be thermally dried involving high capital and operating costs. Other methods, such as lime stabilisation, essentially rely on adding more dry solids to achieve a friable product rather than actual dewatering. Once a sludge has been deposited in a lagoon, either as a thickened liquid or as a cake, it is very difficult to remove more water.

Historically the predecessors of most water companies disposed of large quantities of sludge in unengineered lagoons constructed on sewage works, this represented the cheapest possible method for 'dealing with' this sludge. Until very recently it was in fact still legal to permanently deposit untreated sludge produced at a sewage works in a lagoon on that works, although it was not legal to do so with any imported sludge. Unfortunately these lagoons did not in fact deal with the sludge, they just postponed the time when it would have to be tackled properly. For some of these lagoons this time has now come, perhaps due to pollution of a watercourse or groundwater, or where the area of the lagoon represents the only possible space to expand the works, or in a few cases where there is development potential. The EKG system has the potential to dewater existing lagoons in-situ either to facilitate removal of the sludge as a solid or to strengthen the sludge in-situ to free development potential.

ELECTROKINETICS

The ability of electrokinetic phenomena to transport water, charged particles and free ions through fine grained, low hydraulic permeability materials has been well established following their discovery by Reuss[2] in 1809.

There are five electrokinetic phenomena that occur in fine-grained material-water mixtures:
1. Electro-osmosis (movement of water under an electric field)
2. Electrophoresis (movement of particles under an electric field)
3. Ion Migration (movement of cations and anions under an electric field)
4. Streaming Potential (electric field established by the movement of water through a soil)
5. Sedimentation Potential (electric field set up by the movement of particles through water)

Phenomena 1 to 3 are illustrated in Figure 1. Of these, electro-osmosis and electrophoresis are the most relevant to sludge applications and are further discussed herein.

Electro-osmosis generally occurs in materials that contain a proportion of clay. This is because the surface of a clay crystal has a net negative charge. In a clay-water system, positive ions in the water (e.g. Calcium, Magnesium etc.) are attracted to the clay surface by electrostatics. When a direct electrical potential difference is applied across a wet soil mass, ion migration takes place. These positive ions (cations) are attracted to the cathode and repelled from the anode. As the ions migrate they drag with them their water of hydration and exert a viscous drag upon the free pore fluid around them. Thus, the net flow of water is always towards the cathode. The overall flow (q_A) generated by the application of a potential difference (D) may be expressed as

$$q_A = k_e \frac{\Delta V}{\Delta L} A \qquad (1)^{(3)}$$

where k_e is the electro-osmotic permeability of the soil; $\Delta V/\Delta L$ is the electrical potential gradient; and A is the cross-sectional area of the soil sample across which the potential difference is applied.

Electrophoresis tends to occur in slurries that contain particles with a net surface electrical charge. The fluid-like nature of the slurry allows the charged particles themselves to migrate towards the electrode of opposite electrical charge. The boundary between the two processes is somewhat gradational but can be related to percent solids content, water chemistry, clay content and surface charge.

ELECTROKINETIC GEOSYNTHETICS (EKG)

EKG technology comprises two existing technologies: electrokinetics (EK) and geosynthetics. These technologies have very different histories. Despite the ability of electrokinetics to move water in fine-grained soils at speeds up to 10,000 times faster than conventional hydraulic techniques, and its long history, so far the use of electrokinetics in soil improvement has been severely limited. This has been due primarily to difficulties with electrode corrosion, physical removal of water from the system and the inability to effect polarity reversal. On the other hand, geosynthetics have had a relatively short but successful history. Usually polymer based, they are used in conjunction with earth materials to provide drainage, separation, filtration, impermeable membranes and reinforcement. Since their introduction in the early 1970s geosynthetics have become a multi-billion dollar industry with a world wide annual growth in applications of around 14%.

EKG comprises stainless steel filaments coated and cross linked with an electrically conducting polymer. This design has overcome the problem of electrode corrosion. As can been seen from Figure 1, electrolysis of water at the electrodes always produces acidic conditions at the anode causing rapid corrosion of the historically metallic electrode. By encasing the metallic filaments in a relatively inert polymer, electrode corrosion is eliminated. By forming the electrode as a geosynthetic, EKG overcomes the problem of removing water by utilising the drainage function of geosynthetics with the additional advantages of exploiting geosynthetics' reinforcing characteristics and their ability to take on a wide variety of shapes and forms to suit different applications. By making electrodes identical, polarity reversal (critical in dewatering slurries) can be easily achieved without compromising either drainage function or electrical efficiency.

The University of Newcastle has Patents that cover the basic underlying concept, the physical configuration of the materials of EKG and the EKG structures that can be used for a variety of different applications. There are additional Patents that cover specific applications, including: introduction of conditioning materials into the ground for a variety of ground-improvement processes; improved consolidation or reinforcement of a weak substrate; and combined dewatering, decompaction, pH control, aeration and the defrosting of natural sports turf.

NON-SLUDGE APPLICATIONS OF EKG

EKG has been used in the form of an electrically conductive prefabricated vertical drain, an e-PVD, to consolidate super soft kaolin clay [4] A large test pit was filled with kaolin clay with a dry solids content of 54% to a depth of 2.4m with EKG vertical drains installed as shown in Figure 2. (A clay at 54% dry solids will be above its liquid limit and will act as a liquid). The clay was left untreated for 100 hours, resulting in a surface settlement of 20mm. An electric field was then applied for a period of 500 hours, which produced further surface settlement of 340mm and an increase in shear strength from < 1kN/m² to 15–30kN/m². To produce an equivalent result using conventional means would have required a surcharge loading of 10m of fill, which would have been impossible to place on the super soft soil.

The area around the anodes experienced the greatest reduction in water content and improvement in shear strength due to the net migration of water to the cathodes under electro-osmotic flow. In order to minimise moisture content anisotropy, the trial was completed with a phase of polarity reversal in order to draw water away from the electrodes which were acting as anodes in the first phase. Polarity reversal resulted in a more even distribution of shear strength in the soil.

The trials demonstrated that in-situ dewatering of lagoons using EKG prefabricated drains has the potential to be a rapid, effective and, in terms of energy consumption, efficient treatment technology.

The use of electrophoresis to dewater mine tailing slurries has been demonstrated at laboratory scale on a number of different materials.[5] These materials include a fine residue from the reduction of mineral sands with a dry solids content of 40% and a kimberlite diamond tailings with dry solids varying between 10 and 20%). They found that almost 50% of the volume reduction caused by dewatering occurred within 30 seconds using a direct Voltage of 10V. Subsequently they found that the efficiency of the dewatering process was facilitated by applying a pulsed rather than steady voltage across the sample.

THE SLUDGE LAGOON LEGACY

By their nature sewage treatment works are always built next to rivers, and this means that many existing lagoons are located alongside rivers, with consequent risks of pollution, either slowly by leakage, or catastrophically if a lagoon embankment collapses. A lagoon represents a large area of untreated sludge with a potential for causing smell nuisance. However in reality if a lagoon is no longer in use it will form a crust and is unlikely to cause a serious smell unless disturbed. Seeds will blow onto the crust of a lagoon and it will eventually be covered in grass and weeds and even small bushes, such that it may not be obvious to a trespasser or child that the lagoon is there at all. In this condition it is extremely dangerous, since, although the crust may appear solid, there will still be liquid underneath.

The local authority may designate lagoons as 'Contaminated Land' under Part IIA of the Environmental Protection Act 1990, which came into force in April 2000, if they 'appear ….to be in such a condition, by reason of substances in, or under the land that:
(a) Significant harm is being caused, or there is a significant possibility of such harm being caused; or
(b) Pollution of controlled waters is being or is likely to be, caused.'.

Such a designation would force the site owner, the water company, to carry out remediation works. In reality any lagoon that is bad enough to be so designated may well have already attracted the attention of the Environment Agency who would already be demanding action to improve it.

Heavy metal levels in sludges have reduced by around half in the last ten years such that almost all current sludges are fit to be recycled to agricultural land. For sludges in an existing lagoon, however, which may date back twenty or even fifty years, this will probably not be the case. In particular, levels of cadmium and lead are likely to be high in older sludges.

REASONS FOR WISHING TO DEAL WITH LAGOONS

There are basically three reasons why a water company may wish to deal with an existing lagoon: it may be leaking and so present a pollution risk, it may be occupying the only land available for the expansion of the sewage works (or in unusual cases the land may have development potential), or there may be a real or perceived smell issue from the lagoon.

If there is a risk of an embankment collapse or a leaking embankment polluting a river, rather than an issue of groundwater pollution, it will probably be cheaper to reinforce the bank locally rather than to remove the whole lagoon. Should the natural regeneration referred to above go on for too long the lagoon may become an important wildlife site with a risk that the area may effectively no longer be available for operational purposes. If a smell problem exists at a particular works it is unlikely that a sludge lagoon will be the only or even a major source, however removing the lagoon would be a highly visible attempt to tackle the problem and could improve public perception.

EXISTING MEANS FOR DEALING WITH SLUDGE LAGOONS

Dealing with an old sludge lagoon has generally meant removing the sludge altogether and clearing the site. The EKG process also opens the possibility of economically stabilising the sludge in situ.

Left to itself a sludge lagoon, deeper than say one metre, which has been deposited as a liquid, will never dry out. A crust will form on it, with plant growth that will dry out the crust to some extent, however below this the sludge will generally not thicken beyond perhaps 12 to 15% dry solids, and may well be wetter. Transpiration and evaporation will be balanced by rainfall. Attempting to remove such material is difficult since it is too thick to pump and will contain tufts of grass, etc., and is too thin to shovel since it will slump completely. There are thus two possible approaches, either to wet the sludge down until it can be pumped, or to add dry material until it can be shovelled.

Wetting would be done with a thin fresh sludge or with final effluent from the sewage works. Because of the cost of tankering the sludge offsite to another works the wetting option presupposes that the material can be fed back into the onsite sewage treatment works without overloading it or contaminating the current sludge production. The wetted sludge will almost certainly have to be screened before it enters the works to remove stones, tufts of grass or any debris that might have been thrown into the lagoon. Although the material in the lagoon has arguably been cold digested and will probably have a low pathogen count, it will not technically be 'Conventionally Treated ' according to the Safe Sludge Matrix so that it should not be taken directly to agriculture even if the metal levels are low enough to make this a possibility. If repeated sampling shows that pathogen levels are in fact low the Environment Agency may allow it to go to land on a one-off basis.

The other approach consists of creating a dry sludge which can be removed by excavator and taken to a licenced landfill site. Current legislation means that once the material in the lagoon has been disturbed it is again a waste and cannot be re-deposited on the same or another sewage works site unless that site has a Waste Licence. Obtaining such a licence is unlikely to be a realistic or economic option for a single lagoon. The best material for mixing into the sludge is clearly another waste product, ideally one for which the water company will be paid a gate fee. Construction waste could be used. Wood waste is useful in drying up the sludge but may be difficult to source without payment. It should be remembered that every extra tonne deposited in the lagoon must be dug out again and transported away, it will also have to pay a gate fee and Landfill Tax at the higher rate.

If a lot of time is available before the lagoon must be cleared it may be economic to sink a temporary pumping station at the lowest accessible point in the lagoon and pump out the topwater which must be returned to the sewage treatment works. The lagoon will also receive rainwater over time but this too will be pumped out. The topwater will take a long time to percolate through the sludge to the pumping station and it may need several months to achieve significant results. If successful however the sludge may then be stiff enough to be made into above ground piles which will dry out further. These piles should be remixed occasionally to let all the material be exposed to the air. By this method time and topwater treatment costs are substituted for the cost of obtaining, mixing in, digging out and tipping extra dry material.

STABILISING A SLUDGE LAGOON IN SITU

If the site of the lagoon is not required for an extension to the sewage works but there is a pollution or odour problem, one approach is to stabilise the lagoon until it has in effect gone and the ground can be regarded as effectively solid. The surface can then be crowned in order to shed as much rainfall as possible to minimise future leachate. Before adopting this solution it is of course important to obtain the agreement of the Environment Agency that the lagoon, once stabilised, will not pose any future problems.

Prior to the arrival of the EKG process only two methods for stabilising a lagoon in situ were available, either mixing in more and more dry material until the lagoon was effectively solid, or employing a specialist contractor to mix quicklime or cement into the sludge to thicken it. This second option would be very expensive and would be unlikely to be economic unless the site of the lagoon had development potential. Any such development would still need to be designed with precautions against the accumulation of methane.

EKG AND SLUDGE LAGOONS

Tests were performed on a lagooned sewage in order to establish the ability of the sludge to support the physical process of electro-osmosis. The aim of the tests was also to gain a preliminary assessment of the amount of dewatering that could be achieved by the electro-osmotic treatment of a lagooned sewage sludge and to establish the effect of the dewatering on its physical properties.

METHODOLOGY

A single polarity electro-osmotic consolidation test was performed using the equipment developed at Newcastle University and pictured in Figure 3. A direct voltage of 7.7V (which equates to 0.05V per mm length of sample at the start of the test) was used throughout the test. Following the termination of the test, the dewatered sludge was removed from the test cell and measured for moisture content and shear strength in accordance with BS 1377 (1990).[6]

RESULTS

The current and supernatant volumes with respect to time produced by the electro-osmotic treatment are presented in Figure 4. The water content and shear strength measurements are shown in Table 1. The condition of the sludge in the test cell during and after the test is shown in Figure 5.

TEST CONCLUSIONS

The results show that even with only single polarity treatment, electro-osmosis is effective at increasing the percentage dry solids from an initial value of about 19% to a maximum of 28% adjacent to the cathode and up to 42% adjacent to the anode. This was accompanied by a significant increase in shear strength from less than 1kPa to 12kPa adjacent to the cathode and 29kPa adjacent to the anode.

The volume of the total sample reduced from 2,795ml at the start of the test to 1,195ml at the end i.e. a 57% reduction. The supernatant accounted for about 30% of the total. The remainder is accounted for in terms of hydrolytic evolution of gas at the anode and cathode and an unknown quantity of evaporation caused by an air leak after about 22 days. In spite of the latter, the reduction in volume of the sample due to electro-osmosis is significant.

In practice, dewatering using electro-osmosis would be conducted using phases of reversed polarity to reduce the variations seen in moisture content and shear strength between the cathode and the anode. Figure 4 shows that the dewatering process was still removing water from the sample at an approximately linear rate after 14 days and further dewatering and improvements in strength could have been expected if the test had been allowed to continue longer. It is considered that in practice an average shear strength of the order of 30kPa and a dry solids content in the order of 40% could be achieved using the process if it was operated with periods of reversed polarity over a minimum period of 1 month. A shear strength of 30kPA would be sufficient to support the construction of car parking or single storey buildings with pad foundations.

PRACTICAL APPLICATION

It is envisaged that EKG could be installed into lagoons where a crust has formed by firstly creating a working platform over the surface of the lagoon using a stiff geogrid overlain by a granular capping. From this working platform EKG 'wicks' will be lanced vertically through to the base of the lagoon at 1-2m centres. Every second 'wick' would be connected as an anode and the intermediate ones connected as cathodes. Water will be drawn to the surface at the cathodes and collected at a sump from where it will be discharged. The current would then be reversed making the former anodes into cathodes. This process would be repeated several times. The design of an installation such as this will require initial laboratory characterisation of the sludge, followed by an on-site pilot-scale trial.

EKG AND COMPOSTING

In addition to its use in dewatering lagoons EKG can in principle be applied to the additional dewatering of compost windrows. The technique envisaged is to form and thoroughly mix the windrows then to insert cathode electrodes across the bottom and anode electrodes near the top. The passing of the current would cause water to move downwards to the lower electrodes which would form channels for it to leave the windrows. One water company is interested in carrying out full-scale trials of this technique.

Similar tests to those described above were performed on two materials taken from a composting site. One of the materials, which was taken from an existing windrow, contained a proportion of sawdust, the other was sludge alone. A pressure of 50kPa was applied to all samples as a means of 'squeezing out' water. This is a necessary part of using the test apparatus. The volume of water extracted was compared to that obtained when a voltage of 1V per cm length was applied to the samples. The results from both materials are illustrated in Table 2 and Figure 6. These show that the material with the sawdust had a higher hydraulic permeability than the material without sawdust.

This was not surprising as the coarse sawdust would assist the passage of water through the sludge. They also show that in both cases, the application of a voltage caused a significantly greater volume of water to be extracted than could be achieved with pressure alone. It may also be seen that the greatest advantage was gained from electrokinetic dewatering of the material without sawdust. This, however, may have been because the sawdust increased the initial dry solids content from 16% to 24% as both had similar k_e values.

FUTURE POSSIBILITIES

The EKG process could potentially be adapted to increase the removal of water from simple sludge thickening tanks which do not have stirrers using electrophoresis. In such a case there would be a fixed network of cathode electrodes on the floor of the tank while the anode electrode would be in the form of a floating mat which moved up and down with the sludge level in the tank. Topwater would be withdrawn via the normal valve system. Work proposed by one water company will allow this possibility to be assessed further. Preliminary results indicate that electrophoresis of sludge is not as efficient as electro-osmosis however the financial rewards from tankering away a thicker sludge are such that this option deserves further work.

Work at the university of Oxford[7] has found that application of an electric field causes greater degradation of organic contaminants in soils due to enhanced biological activity. Funding is being sought for a collaborative project between Oxford and Newcastle Universities to investigate the potential of using the system described above for dewatering windrows to enhance biodegradation in biopiles. The aim is to speed up the rate of clean-up and to increase the amount of organic contaminants destroyed.

CONCLUSIONS

1) Many water companies have historic sludge lagoons which may be leaking and presenting pollution risks, or may be occupying the only land available for the expansion of the sewage works, or may be a real or perceived smell issue.
2) At present removing an existing lagoon involves either wetting down and screening the sludge to remove it as a liquid or adding dry material until it can be removed as a solid. Both approaches are expensive.
3) All current sludge dewatering methods involve pumping a liquid into a machine. Once in a lagoon or windrow there are no simple methods for removing additional water from the cake.
4) Electro-osmosis has been known to be capable of in-situ dewatering for very many years but has not been much exploited because of technical problems.
5) The patented EKG process overcomes these problems by combining electrokinetic phenomena (Electrophoresis and Electro-osmosis) with geosynthetics and forming the electrodes using geosynthetic materials.
6) Such electrodes reduce anode corrosion, allow increased current density and provide better electrical contact with the soil. They provide a drainage path for the removed water and they can also function equally as cathodes or anodes, allowing polarity reversal, a crucial procedure in dewatering.
7) Tests with a lagooned sludge sample have shown that sewage sludge is a suitable material for electro-osmotic dewatering with a material with a dry solids content of 19% increasing to a maximum of 42% in 24 days.
9) The technique envisaged for dewatering lagoons is the installation of EKG 'wicks' vertically through to the base of the lagoon at 1-2m centres. Water will be drawn to the surface and collected at a sump from where it will be discharged to a sewage works for treatment.
10) Potentially EKG electrodes can also be used for dewatering compost windrows by inserting upper and lower electrodes into the windrows. Tests on a sample from an existing windrow have shown that electrokinetics can offer an efficient means of dewatering this material. One water company is interested in a full scale trial of this technique.
11) The EKG process also has potential for accelerated thickening in existing simple sludge tanks, and for improving the destruction of organics in compost systems.
12) This new process is still being developed, particularly in terms of actual operational techniques. More information will be available at the Conference in November 2002.

REFERENCES

1) METCALF, L and EDDY, H P, *Wastewater Engineering: Treatment, Disposal and Re-Use.* Mcgraw-Hill, 1991.

2) REUSS, F F. Sur un nouvel effet de l'électricité galvanique. *Mémoires de la Société Impériale des Naturalistes de Moscou*, 1809 **2**, 327-337.

3) MITCHELL, J K. *Fundamentals of Soil Behaviour*. 2nd Edn., Pub. John Wiley & Sons Inc, New York, USA. 1993.

4) JONES, C J F P. GLENDINNING, S. & SHIM, G.S.G. Soil Consolidation using electrically conductive geosynthetics. *7th International Conference on Geosynthetics*, Nice, France, 22 - 27 September 2002.

5) FOURIE, A B. PAVLAKIS, J. & JONES, C J F P. Stabilisation of mine tailings deposits using electro-kinetic geotextiles. *7th International Conference on Geosynthetics*, Nice, France, 22 - 27 September 2002.

6) BRITISH STANDARDS INSTITUTION. *Soil classification tests in Civil Engineering. BS 1377*. H.M.S.O. 1990.

7) JACKMAN, S A. MAINI, G. SHARMAN, A K. SUNDERLAND, G. & KNOWLES, C J. Electrokinetic movement and biodegradation of 2,4-dichlorophenoxyacetic acid. *Biotechnol. Bioeng*. 74. 40-48. 2001

FIGURES

Table 1 Lagoon Sludge - Shear Strength And Moisture Content Results

Location of sample		Shear strength (kPa)	Moisture content (%) *	Dry solids (%)
Cathode	*max*	11.8	279	26.4
	min		254	28.0
	mean		267	27.2
Middle	*max*	16.7	214	31.8
	min		193	34.0
	mean		204	32.9
Anode	*max*	29.4	138	42.0
	min		171	37.0
	mean		155	39.5

* Measurement of 'Moisture Content' as used in soil mechanics practice
  = % of water by weight compared with weight of dry solids taken as 100%
  eg A 'Moisture Content' of 100% equals 50% Dry Solids in sludge terms.

Table 2 Results Of Electrokinetic Treatment Of Composted Sewage Sludge

	With Sawdust	Without Sawdust
Voltage gradient	1 V/cm	1V/cm
Test duration	75 hours	75 hours
Average ke	1.17×10^{-5} cm^2/sec-V	1.47×10^{-5} cm^2/sec-V
Average k	6.45×10^{-6} cm/sec	5.75×10^{-6} cm/sec
Volume of water extracted (static load)	180ml	193ml
Volume of water extracted (static load + D.C. voltage)	305ml	407ml
Initial water content (dry solids)	310% (24%)	510% (16%)
Final water content (dry solids) (static load)	Anode: 297% (25%) Cathode: 229% (30%)	Anode: 460% (18%) Cathode: 340% (23%)
Final water content (dry solids) (static load + D.C. voltage)	Anode: 224% (31%) Cathode: 238% (30%)	Anode: 218% (31%) Cathode: 397% (20%)

NOTES
1) For explanation of water content and *dry solids* see Note for Table 1.
2) The only way water can leave the cell is via the cathode (see Figure 3). Under <u>static load only</u> the dry solids in the area of both cathodes significantly increased while the dry solids at the anodes did not, showing that water had not flowed through the samples under pressure alone.
3) Under <u>static load plus DC voltage</u> the dry solids at the anodes also increased showing that water had flowed through the samples to 'escape' via the cathodes.

FIGURES

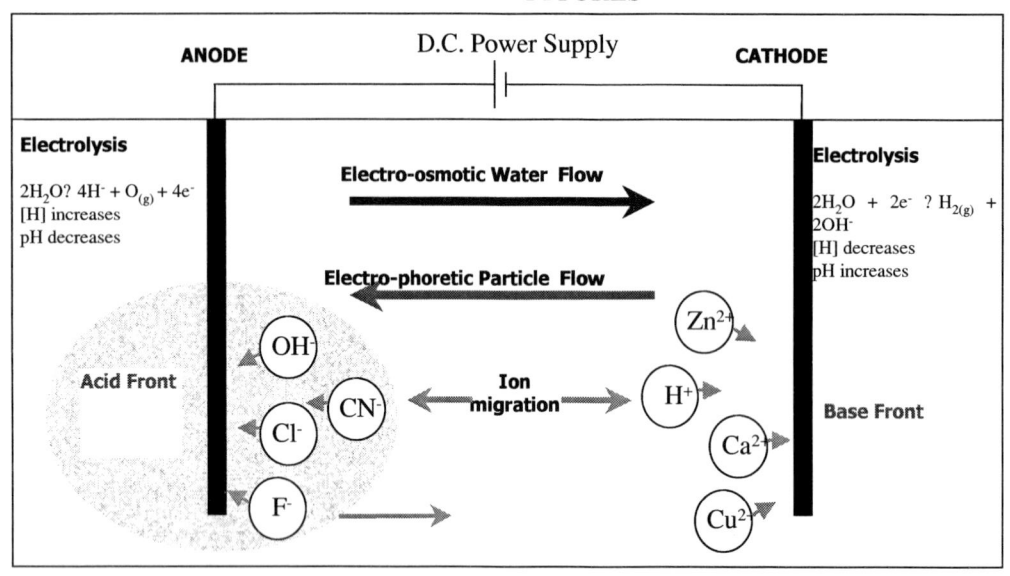

Figure 1: Electrokinetic Effects

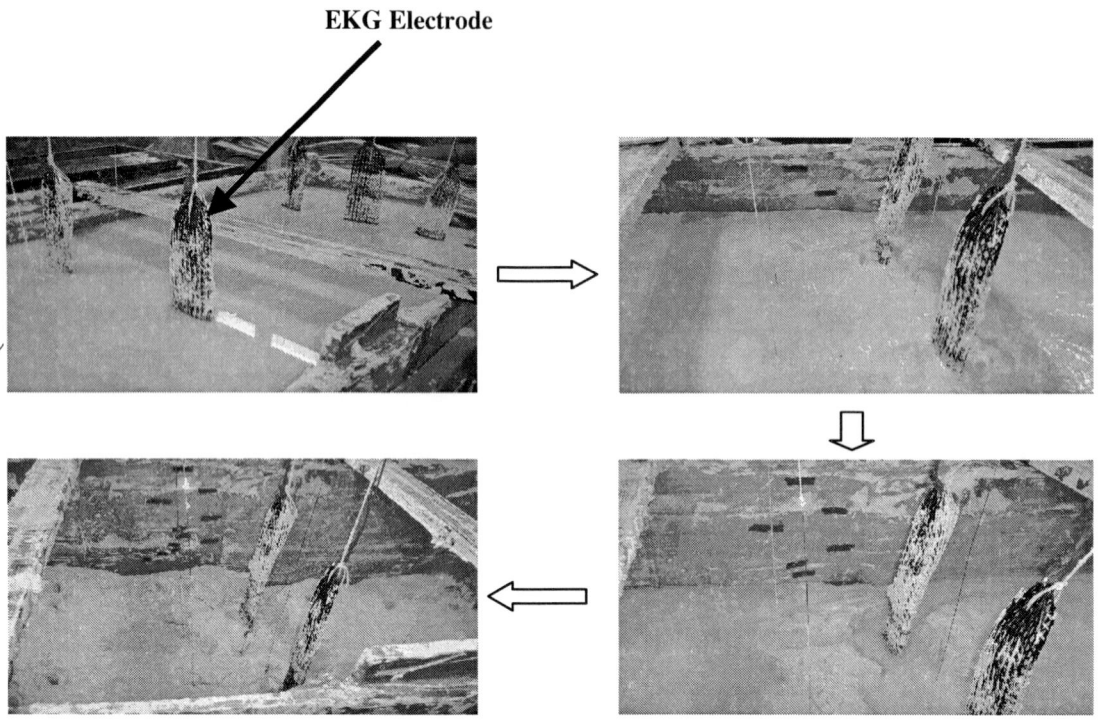

Anode electrolysis: $2H_2O \rightarrow 4H^+ + O_{2(g)} + 4e^-$, [H] increases, pH decreases

Cathode electrolysis: $2H_2O + 2e^- \rightarrow H_{2(g)} + 2OH^-$, [H] decreases, pH increases

FIGURE 2: EKG ELECTRODES INSTALLED IN TEST PIT SHOWING PROGRESSION OF DEWATERING

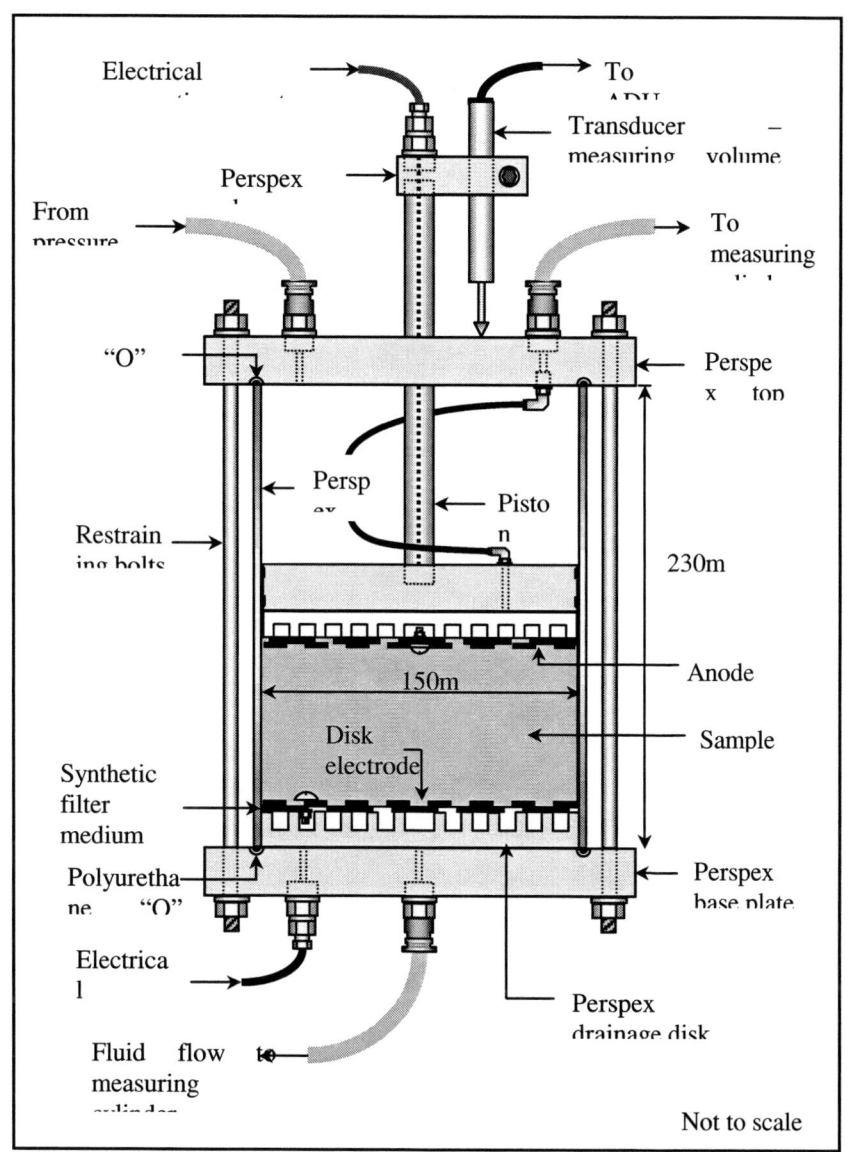

FIGURE 3 EXPERIMENTAL TEST RIG

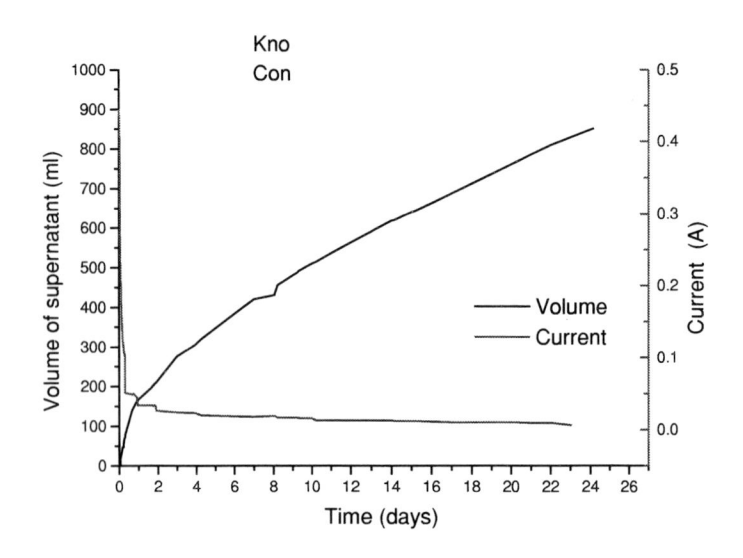

FIGURE 4

CURRENT AND SUPERNATANT VOLUME VS TIME

FIGURE 5

SECTION OF SAMPLE OF LAGOON SLUDGE AFTER REMOVAL FROM THE TEST CELL ALONGSIDE UNTREATED SLUDGE

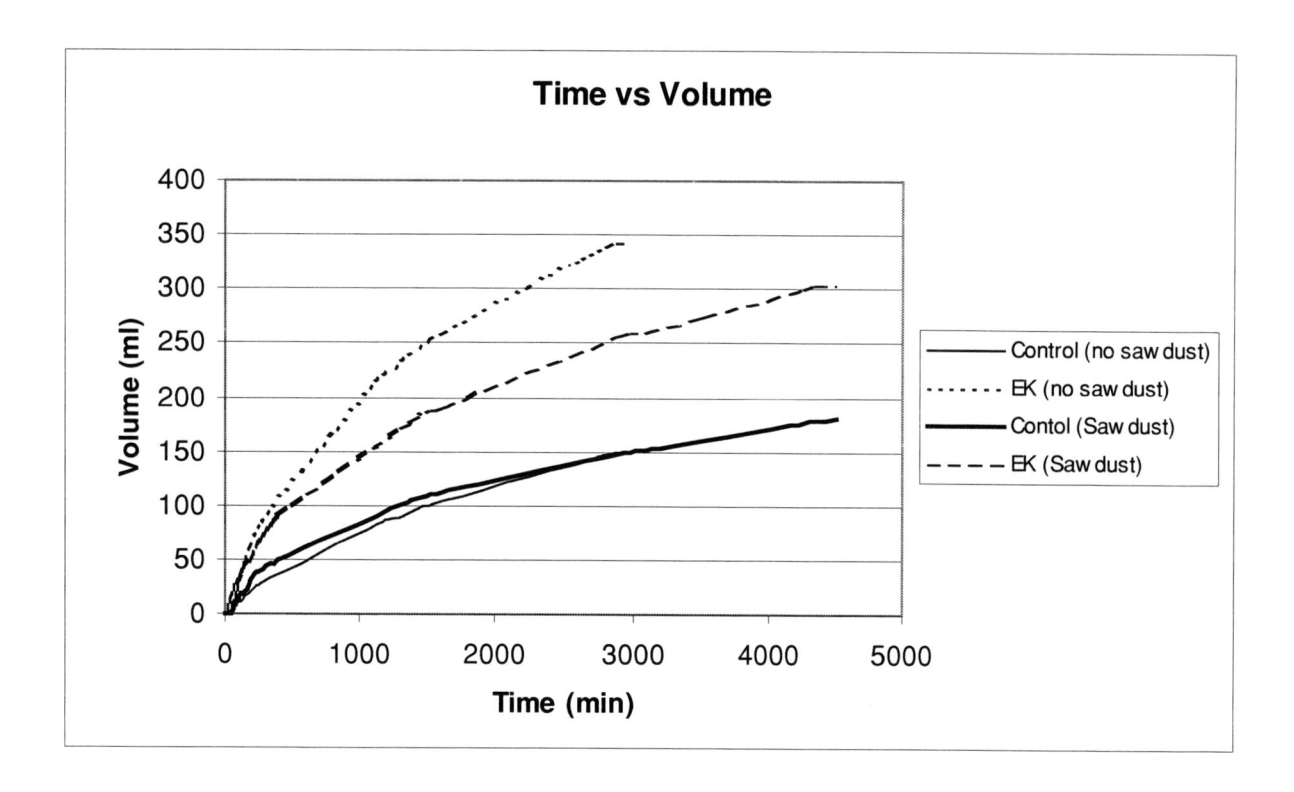

FIGURE 6

ELECTRO-KINETIC (EK) DEWATERING OF WINDROW SLUDGE

BIOLYSIS® : CUTTING EDGE TECHNOLOGY FOR THE REDUCTION OF SLUDGE QUANTITIES IN ACTIVATED SLUDGE PLANTS

Ranjit Mene* & Thierry Lebrun**

*Ondeo Degrémont Ltd, Houghton Hall Park, Houghton Regis, Dunstable, Bedfordshire LU5 5TD, UK. email: ranjit.mene@ondeo-degremont.com

**Ondeo Degrémont SA, CERDEG, 87, chemin de ronde, 78290 Croissy sur Seine, FRANCE email: thierry.lebrun@ondeo-degremont.com

ABSTRACT

A new process called Biolysis® has been developed by Ondeo Degrémont to directly reduce the surplus sludge quantities being generated by activated sludge plants. The process uses either a chemical or biological action to stress the bacteria in the activated sludge so that a portion are killed and a portion are unable to multiply. Unlike conventional sludge consolidation processes, which are a separate part of a wastewater treatment plant, Biolysis® is installed as part of the activated sludge reactor.

Biolysis® O is the name of the process which uses ozone as a chemical stressing agent directly on mixed liquor from the biological basin at ambient temperatures. Biolysis® E is the name of the process which uses a thermal reactor operating at roughly 50°C directly on thickened mixed liquor. The thermophilic conditions encourage the bacteria to secrete an enzyme which effectively renders them unable to reproduce.

It is demonstrated that consistent reductions of 60 – 80% DS are achievable without increasing the global cost of sewage treatment on a site. Additionally, no significant deterioration in the treated water or residual sludge quality is observed.

KEY WORDS

Activated Sludge, Biolysis, Enzymes, Ozone, Sludge Reduction, Thermophilic

INTRODUCTION

The production and subsequent disposal of sludge is a necessity for all wastewater treatment plants. All biological purification processes lead to an increase in the biomass quantity which must be removed and treated[1]. The treatment and disposal of the sludge is a major on-going cost for all operators of water treatment facilities. In fact, it is estimated that almost 10 million tonnes of sludge as dry solids is produced annually just from municipal water treatment plants in Europe alone[2].

Against this background, a technology designed to reduce the quantity of sludge being produced at source from wastewater treatment plants would mean lower sludge treatment and disposal costs and a consequent increase in the environmental credentials of plant operators.

As part of it's on-going research programme, Ondeo Degrémont has recently launched a process, called Biolysis®, specifically designed to reduce the amount of sludge being produced from activated sludge plants. Since activated sludge processes feature in the majority of wastewater treatment plants around the world[3], the technology should appeal to most plant owners and operators.

Two different versions of the Biolysis® process are available; one uses ozone and is based on a physico-chemical action, the other uses enzymes and is a purely biological process. Both versions use the principle of bacterial stress at the heart of the technology.

BIOLYSIS O

Biolysis O is the name given to the sludge reduction process using ozone[4]. It involves taking mixed liquor directly from an activated sludge basin and injecting ozone into it before putting it through a contact tower. The ozonated mixed liquor is then returned back into the activated sludge basin. Ozone production is carried out in-situ in the normal way using oxygen. The principle features of the process are illustrated in Figure 1.

The ozone has the effect of stressing the cellular material in the biomass such that a proportion of the bacteria is destroyed and another proportion is rendered unable to reproduce as their cell walls have been disrupted.

A demonstration Biolysis O plant was operating at the works at Aydoilles in the East of France for more than 1 year from the beginning of 2001. The existing works at Aydoilles is a 1,000 p.e. activated sludge plant and it was used to confirm the various operating parameters of Biolysis O. The operational period of the Biolysis plant was spilt up into different distinct phases. Figure 2 shows the evolution of the sludge reduction rate during the different phases by plotting the cumulative sludge production against time for the conventional plant at Aydoilles compared to the plant equipped with Biolysis O. Extensive historical data was used to determine the conventional sludge production figures. Tables 1 & 2 show the main results on the final effluent and the residual sludge quality respectively from this plant. They are based on results from phase 3 onwards of the plant operation and on an average sludge reduction rate of 60% DS. The average ozone dose over this period was 0.15 g O_3/g DS destroyed

The following points were observed during the year-long operation of the Biolysis O process at Aydoilles:

- No major decrease in the agricultural value of the residual sludge was observed.
- The excess sludge resulting from the plant could be dewatered to a higher dry solids level using less polymer per tonne DS compared to a conventional AS plant.
- Sludge settlement characteristics improved dramatically after the application of the Biolysis process. In fact, the incidence of filamentous bacteria observed on site at Aydoilles decreased.
- Application of the Biolysis process resulted in a slight degradation of the final effluent quality. This was primarily due to the solubilization of the biomass[5] and the release of intra-cellular phosphorus.

A result of the final point above was that the concentration of total phosphorus in the treated effluent increased after Biolysis O had been commissioned. Hence it was observed that, in order to comply with a strict Total Phosphorus guarantee, the use of tertiary treatment with appropriate chemical dosing, for example a lamella clarifier[6], would be required.

Also, it was found that aeration requirements for the Activated Sludge basin at Aydoilles increased by between 20% and 30% depending on the level of sludge reduction that was required from the Biolysis process. Two factors help to explain this increase:

- For carbon removal plants, there is more soluble COD pollution to treat.
- For nitrifying plants, as there is less sludge present in the system, less nitrogen assimilation takes place in the mixed liquor, which effectively means that more $N-NH_4$ needs to be treated.

In theory, when the Biolysis process was started, the mixed liquor concentration in the AS basin should have decreased. This should have lead to a requirement to increase the total AS basin volume so that the loading rates could be kept constant. However, on site, it was observed that this was not the case. This was due to the fact that as the sludge settlement characteristics improved dramatically, the mixed liquor concentration went up. These two factors effectively helped to keep the loading rates, and hence the required biological volume, fairly constant.

BIOLYSIS E

This version of the Biolysis process uses a completely biological action to break down the bacteria in the biomass so that they are rendered unable to reproduce. It consists of drawing mixed liquor from an activated sludge basin, thickening it to around 2.5% w/w DS, and then passing it through a thermophilic, enzymatic reactor operating at about 50°C - 60°C. The enzymes released by the bacteria in the reactor break down the cells in such a way that part of them are killed and part of them are unable to reproduce. The heated, degraded sludge then passes through a heat exchanger to recover some of its' energy before flowing back to the activated sludge basin. No external enzymatic source is used. The sludge thickening allows for more concentrated biomass to be present in the reactor, hence increasing the reaction kinetics. The thermophilic bacteria used in this process are naturally occurring in biomass. However, they are only activated under aerobic conditions when the temperature approaches 50 - 60ºC.

This process was originally developed by the Japanese company Shinko Pantec[7] and is under license to Ondeo Degrémont. Figure 3 illustrates the principles of Biolysis E.

A Biolysis E plant has been operating at Verberie waste water treatment plant for over 8 months. Verberie is a 3000 p.e. activated sludge plant with an extended aeration system. After the Biolysis E process was commissioned, two distinct phases were studied to determine the operational parameters of the system. The first phase (Run 1) involved running at sludge reduction rates of between 40 and 50% while the second phase (Run 2) was intended to show the performance of Biolysis E at higher sludge reduction rates (>70%). Figure 4 shows the gradual evolution of the sludge reduction rate for the Biolysis E equipped plant compared to the control for both runs. Tables 3 and 4 summarise the results for the final effluent and residual sludge qualities respectively for each run to date. The results are based on a residence time in the thermophilic reactor of 24 - 48 hours.

The tables and graph demonstrate similar results to the Biolysis O plant for similar reasons.

PROCESS SELECTION

Having two versions of the Biolysis process enables it to be implemented in a wide variety of situations and allows total flexibility with regard to the circumstances of each installation. The choice of which version to install effectively then comes down to the situation and state of the existing or new plant. One of the other advantages of Biolysis is that it can also be used as a temporary sludge reduction method. This means that plants which have undersized sludge treatment lines because of frequent influent overloads do not need additional sludge treatment facilities. Table 5 summarises the different options available to the plant operator.

CONCLUSIONS

The tightening of effluent standards and increased public pressure has meant an increase in the amount of sludge produced and thus the cost of sludge disposal for most plant managers around the world. The Biolysis process offers managers an easy and reliable way of reducing the quantities of sludge produced from activated sludge plants by up to 80% without increasing the global cost of sewage treatment[8]. Two flexible, yet fundamentally different versions of the process are available, each offering its own advantages and benefits.

REFERENCES

1. BARKER, P.J. AND DOLD, P.L. Sludge production and oxygen demand in nutrient removal activated sludge systems. *Wat. Sci. Tech.*, 1996, 34, (5-6), 43-50

2. FROST, SULLIVAN, *European Sludge Treatment Equipment Market Report*, 2001, 4 - 5

3. COOPER, P.F., DOWNING, A.L., Milestones in the development of the Activated Sludge Process over the past 80 years, *Proc. of Activated Sludge into the 21st Century Conference*, 1997, 1-29

4. YASUI, S. An innovative approach to reduce excess sludge production in the activated sludge process. *Wat. Sci. Tech.*, 1994, 30, (9), 11-20

5. DELERIS, S., PAUL, E., AUDIC, J.M., ROUSTAN, M., DEBELLEFOUNTAINE, H., Effect of ozonation on activated sludge solubilisation and mineralisation, *Ozone Sci & Eng.*, 22, 473 – 486

6. FUGGLE, R., MENE, R., Effluent re-use, the Ondeo Degrémont experience at the Langford recycling scheme, *Proc. 2nd Biennial Conference on Management of Wastewaters*, 2002, 2, 211-220

7. Shinko Pantec Brochure, Complete decomposition of biological waste sludge by thermophilic aerobic bacteria, 2000

8. MENE, R., LEBRUN, T., A novel technology for the reduction of sludge quantities in activated sludge plants, *Proc. 2nd Biennial Conference on Management of Wastewaters*, 2002, 2, 33-38

TABLES
Table 1 : Average Treated Effluent Removal Rates - Biolysis O

PARAMETERS	TREATED EFFLUENT QUALITY COMPARISON	
	WITHOUT BIOLYSIS O	WITH BIOLYSIS O
SS (%)	95	≤ 95
Total COD (%)	90	85
N-NH$_4$ (%)	96	98
P-PO$_4$ (%)	35	10 – 20
BOD (%)	90	87

Table 2 : Average Residual Sludge Quality - Biolysis O

PARAMETERS	SLUDGE QUALITY COMPARISON (AYDOILLES WwTW)	
	WITHOUT BIOLYSIS O	WITH BIOLYSIS O
DECANTIBILITY		
- SVI (ml/g)	200	< 80
- VM / DS (%)	70	60
DEWATERING (sludge from the storage tank)		
- Dryness (%)	22	25
- Polymer Consumption (kg/tDS)	7	5
SLUDGE QUALITY		
TKN (%)	4.93	4.30
P (%)	5.35	4.30
K (%)	1.01	0.63
Ca (%)	3.70	3.20
Mg (%)	0.72	0.50

Table 3 : Average Treated Effluent Removal Rates - Biolysis E

PARAMETERS	TREATED EFFLUENT QUALITY COMPARISON		
	WITHOUT BIOLYSIS E	WITH BIOLYSIS E	
		RUN 1	RUN 2
SS (%)	97	96	96
Total COD (%)	93	91	92
N-NH$_4$ (%)	95	96	92
P-PO$_4$ (%)	53	55	40
BOD (%)	98	95	99

Table 4 : Average Residual Sludge Quality - Biolysis E

PARAMETERS	SLUDGE QUALITY COMPARISON (VERBERIE WwTW)	
	WITHOUT BIOLYSIS E	WITH BIOLYSIS E (AVERAGE OF 2 RUNS)
DECANTIBILITY		
- SVI (ml/g)	250	110
- VM / DS (%)	73	60
DEWATERING (sludge from the storage tank)		
- Dryness (%)	15	18
- Polymer Consumption (kg/tDS)	12	12
SLUDGE QUALITY		
TKN (%)	5.33	6.0
P (%)	1.79	3.13
K (%)	0.94	In Progress
Ca (%)	8.20	In Progress
Mg (%)	1.10	In Progress

Table 5 : Biolysis Applications

Criteria	Version	Reason for Choice
Limited space available	Biolysis O	Biolysis O is more compact than Biolysis E in terms of surface area
Excess heat available on site	Biolysis E	Heat can be used in the thermophilic reactor
Oxygen available on site	Biolysis O	Oxygen can be used for ozone generation
Existing anaerobic digestors	Biolysis E	Biogas can be used to heat the thermophilic reactor
Sludge line of existing plant undersized due to frequent influent peak loads	Biolysis O	Biolysis O has a quick start-up time
Existing plant experiencing sludge bulking	Biolysis O or E	Both versions result in better sludge settlement characteristics

FIGURES

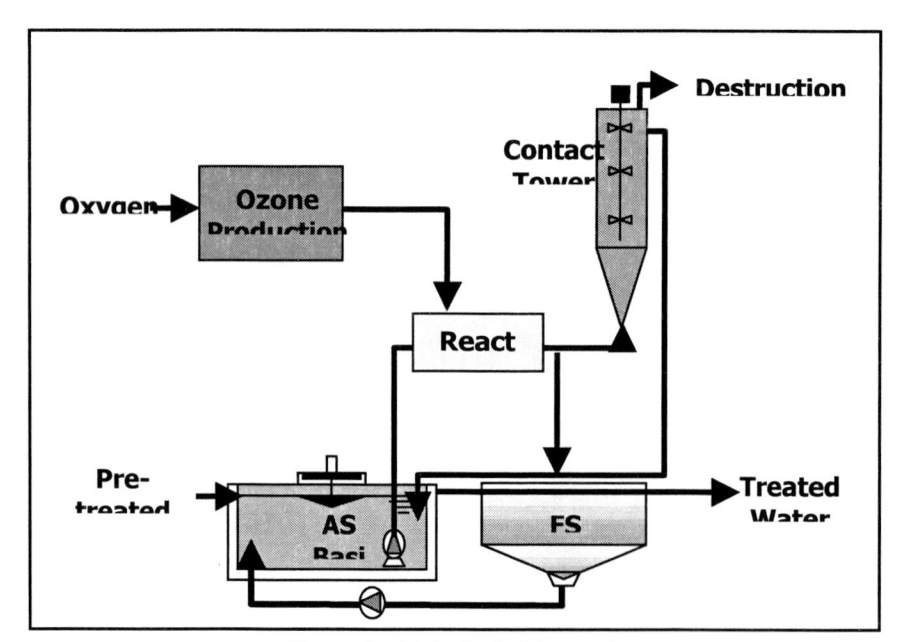

Figure 1 : Biolysis O Schematic

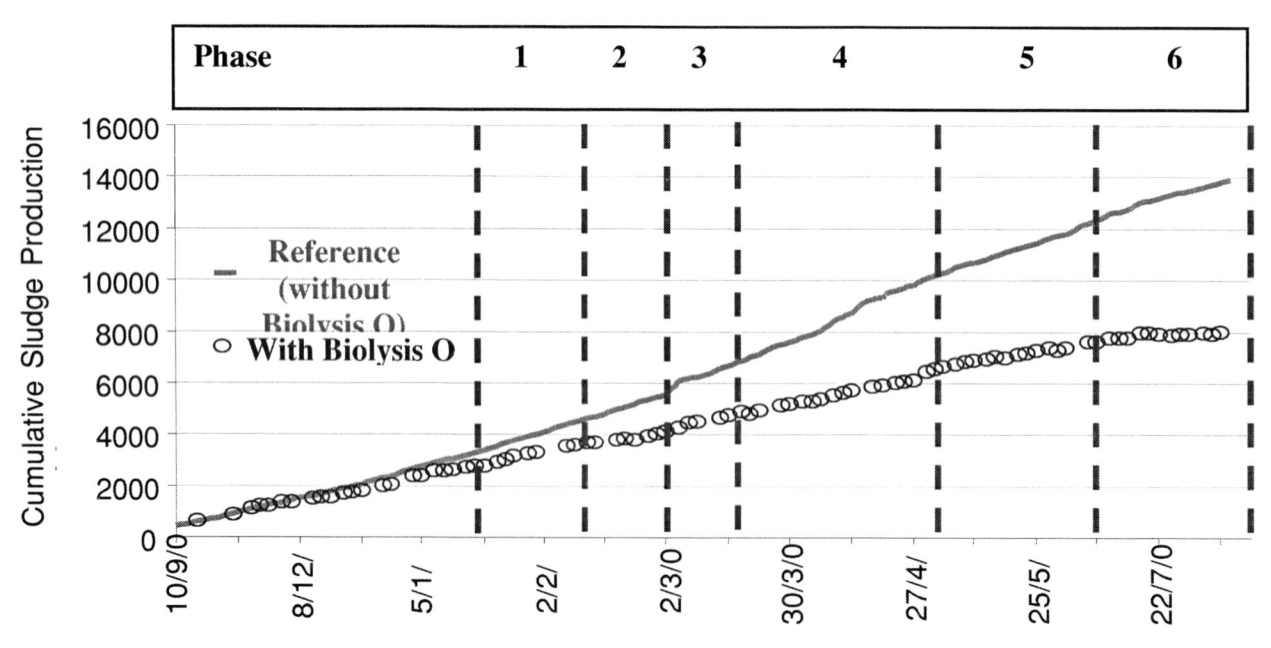

Figure 2 : Cumulative Sludge Production - Biolysis O

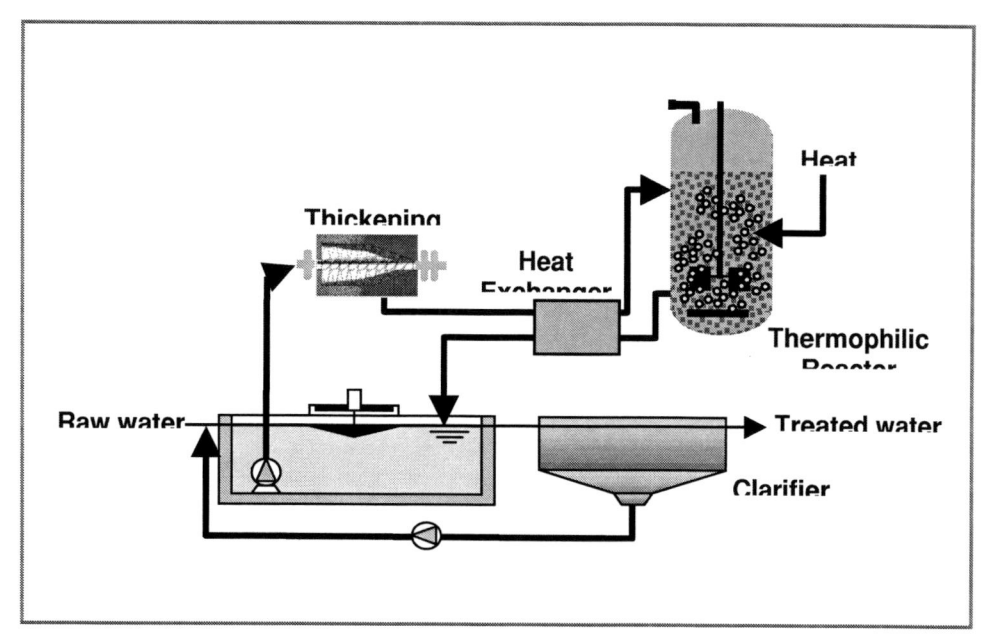

Figure 3 : Biolysis E Schematic

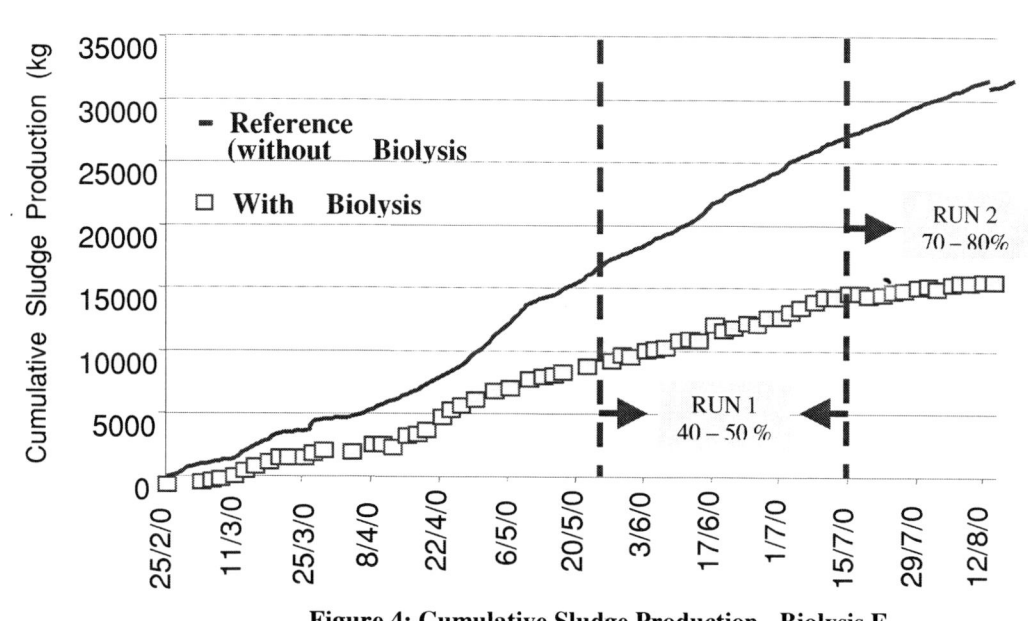

Figure 4: Cumulative Sludge Production - Biolysis E

THE REMOVAL OF NITROGEN AND PHOSPHORUS FROM THE OXIDATION OF BIOSOLIDS USING THE ADVANCED FLUIDISED COMPOSTING (AFC) PROCESS™

[1]J.E.Taylor and [2]T Donnely

[1]Environmental Performance Technologies
Telelphone +44 1865 304060; Fax +441865 304001; email jet@ptuk.com

[2]Department of Environmental Engineering, University of Newcastle
Telephone +44 191 2225899; Fax +44 191 2226669; email tom.Donnelly@ncl.ac.uk

ABSTRACT

The disposal of organic sludge, from sewage and industrial sources, to land presents a risk of contaminating the food chain. Whilst metals can be managed the risk from organic sources (pathogens, viruses etc) are less well defined and the 'risk' or 'perception' far out ways the benefit of nutrient recycle.

Incineration is an extreme but effective solution however the potential for nutrient recovery is diminished. There is clearly a need to develop a process, which can remove the risk and recover the nutrient. This investigation has indicated, on digested sludge, the potential to destroy the organic matter biologically and remove 'N & P'.

By combining the Ammonia Recovery and Removal Process (ARRP) and Advance Fluidise Composting AFC™ processes the following results were achieved:
Up to 90% removal of VSS
Up to 80% removal of TSS
Up to 50% removal of TCOD
Up to 70% removal of N
Up to 90% removal of P

KEYWORDS

Advanced Fluidised Composting, ammonia removal and recovery process, struvite.

INTRODUCTION

Sewage sludge has been disposed of as a soil conditioner and fertiliser for many years. Whilst it is an important source of Nitrogen and Phosphorus, its use in recent years has been in decline because of health concerns (or their perception) or its merely distasteful image. In the early 1990's digested sewage sludge (liquid) could be spread on land after 1 month, then it was after 3 months and now it must be stored for 6 months as a cake, reflecting these concerns and raising the cost of disposal for the industry.

Some countries have all but given up [1] or suspended the practice.

Sweden has nearly ceased sewage sludge disposal to land through a combination of regulatory pressure on contaminants in sewage and rejection of the practice by farmers for food image reasons. The UK, Germany and France have suspended the practice, during their respective BSE and Foot and Mouth crises.

In spite of the experiences referred to above there remains a diverse opinion across Europe on what should be done with sewage sludge. From Figure 3-1 [5] we can see a significant and relentless trend in Europe, as a whole, towards incineration. Whilst for example in the UK we can see from Figures 3-2 [4] and 3-3[4] that we remain committed to land spreading, a view endorsed by the UK Environment Agency.[1]

The removal of N&P from wastewater is essential to prevent eutrophication of lakes and other natural waters.

Nitrogen (nitrates, ammonia etc.) was a prime source of eutrophication 40 years ago now it is almost under control. It is usually removed biologically (N/DN) particularly in the UK, whilst it can and is recovered as a fertiliser in Germany.

Phosphorus, in the form of phosphates, is much less susceptible to leaching than nitrates; however its presence in surface waters is the prime cause of eutrophication today. Consequently the EC Urban Waste Water Directive requires the removal of 75-80% of all phosphorus in municipal waste water discharged into sensitive waters, where these are defined as all surface waters which are "eutrophic or which in the near future may become eutrophic if protective action is not taken".

Sewage contains phosphates of both biogenic and non-biogenic (usually detergents) origin, typically on a 50:50 basis. Thus modified (chemical alternatives may have a worse environmental effect) or more efficient use of detergents will only at best resolve half of the problem. So the only practicable means of reducing phosphates in our rivers is at the sewage works. There are now a number of technologies available or under development, which are based on chemical precipitation and can achieve up to 95% removal [7] and recovery. Magnesium ammonium phosphate (MAP) can be used as a fertiliser and calcium phosphate can be used in the manufacture of elemental phosphorus. Indeed 'P' recovery would address some of the phosphate industry's resource depletion issues and thus aid sustainable development. Provided the phosphate concentration can be increased to at least 60mg/l [1] it can be economically recovered [3] from sewage. The Netherlands has set a target to replace 20% of its current rock consumption with recovered phosphates of this type by 2006.

Which brings us back to 'N' and why is there not more focus on recovery? After all ammonia is produced from the synthesis of natural gas which is not a renewable resource.

In retrospect, it would be difficult or impossible to recover all of the N & P from wastewater without generating sewage sludge (biosolids) for disposal. These biosolids would encapsulate some N & P as nutrient, which could not then be recovered especially if the sludge was incinerated.

The AFC™ process (Figure 3.4) is a biological alternative to incineration, which has already been demonstrated to be able to destroy 100% of VSS with the potential to convert all nitrogenous species to ammonia.

Clearly other sanitary means of sludge disposal combining a range of technologies have to be used or developed which ideally will focus on recovery (resource renewal) and removal of risk (or perception) at reasonable cost.

METHODOLOGIES

INTRODUCTION

The investigation was carried out in the Environmental Engineering Laboratories at the Department of Civil Engineering, University of Newcastle upon Tyne (EEUN). The ARRP was designed by Rauscherts' who also fabricated the specialist components (columns and packing) in Germany. The full system was assembled and operated outside at the EEUN. The bioreactor was a standard EEUN design, which was fitted with a novel jet system designed and fabricated by Environmental Performance Technologies and assembled at the EEUN. It had been planned to use WAS from a specialist food manufacturer, to ensure the sludge would have a high VSS: TSS ratio (>9:1) and be free from inorganic debris which might block the jet system or accumulate in the reactor. Unfortunately this source was unable to supply sludge and after several attempts to find alternatives it was decided reluctantly, to use digested sludge and press liquors from a sewage works. The interest in the press liquors was that it would provide a contrast to the digested sludge, being solids free and contain high ammonia (> 800mg/l) and high soluble COD (> 20,000 mg/l) concentrations.

BACKGROUND

Biological sludge is generated from two principle sources: industry and sewage treatment. Whilst their value as a nutrient and soil conditioner has been known for many years they also carry a great risk (real and perceived) of contaminating the food chain.

When biological sludges are incinerated the risk is removed but the benefit is lost. Anaerobic digestion will yield high concentrations of ammonia (>1000mg/l), which can be recovered, but since the VSS is reduced by only 50% the risk remains.

Natural sources of phosphate are running low. There are a number of chemical processes, which can remove phosphate, now the focus is on recovery. Provided the phosphate concentration can be increased to at least 60mg/l [1] it can be economically recovered [3] from sewage.

The AFC™ process is a biological alternative to incineration, which has already been demonstrated to be able to destroy 100% of VSS with the potential to convert all nitrogenous species to ammonia.

The ideal conditions required to form MAP / struvite are the presence of magnesium, phosphorus and ammoniacal nitrogen together with a pH of at least 7.5. Like wise the reduction in any of these parameters will reduce the formation [9].

THE TASK

It was necessary to exploit the combination of several existing technologies: ARRP, AFC™ and pure oxygen on a scale small enough to produce credible results at reasonable cost and to find a stable source of sludge.

ARRP is used to remove ammonia from press liquors by raising the pH to 10.5 and the temperature to 50°C producing alkaline 'stripped' liquor. In order to evaluate this process on sludge the risk of fouling needed consideration and the subsequent construction of a unit with a 50l/h capacity.

It was not possible to provide heating, pH or flow control for the unit so it was to be supervised and operated at ambient temperature. The stripped sludge produced was to be the feed material for the laboratory scale 10litre AFC™ bioreactor. The ARRP unit was expected to operate for only 10 hours at design to produce 500l of feed, enough to last over 6 months.

The AFC™ bioreactor was to incorporate a pure oxygen based jet mixing system to enable the CO_2 produced to neutralise the alkalinity of the feed and thus prevent CO_2 toxicity and the need to use acid. The design utilised a single concentric nozzle arrangement with the liquid jet being only 2.3mm in diameter. It was important therefore to ensure that the feed was well screened to avoid blockage. The bioreactor is shown below. Dissolved oxygen measurement and control could not be provided.

Laboratory scale bioreactors tend loose heat due to the high surface area to volume ratio. To ensure thermophylic conditions the reactor has an internal hot water heat exchanger recirculated from an external heater.

A source of WAS from a food manufacturer had been identified but unfortunately supplies could not be obtained in time for the project. Reluctantly it was decided to use digested sewage sludge which, would present many challenges the most significant of which were:
- Risk of fouling the ARRP and the jet nozzle.
- Digested sludge had not been processed using either technology before.

OPERATION OF THE AMMONIA REMOVAL AND RECOVERY PROCESS

The system was clean water tested and the liquid recirculation rates were set using a bucket and stop watch before being commissioned on sewage sludge press liquors. The liquors had been diluted with water at the works and were therefore too weak (TCOD < 7000mg/l) to be used in the AFC™ bioreactor. It did however demonstrate the concept of ammonia removal and recovery.

A laboratory evaluation was carried out to determine the risk of fouling on the ARRP process using digested sludge and the packing media. An improvised column was assembled and a sample of sludge was decanted into a diffuser cup. The packing performed well provided the liquid flow was restricted. Whilst some coating / deposition was observed, this cleared as the flow diminished.

Problems with fouling were encountered after about 30 minutes of operation. The distribution pipe at the top of the stripping column was blocking. Increasing the flow overcame the problem but upset the hydraulic balance on the column, which was difficult to restore without a flow meter. Eventually the recirculation system blocked completely.

A mesh screen was fabricated to remove the fibres using a submersible transfer pump and the ARRP was restarted. The process was closed down after 8 hours operation when theoretically all 400l of sludge had had a single pass through the column.

THE OPERATION OF THE AFC™ BIOREACTOR

The bioreactor had been pre commissioned and operated using WAS from 06/06/02 at a temperature of 50°C. The jet system operated without problem using aspirated air. This was then replaced by a pressurised pure oxygen supply and the digested sludge at pH 10.5 was introduced on 09/07/02. The reactor temperature was increased to 60°C. The jet pump became obstructed (the material has not yet been identified) on 3 occasions creating severe concentration, temperature and pH gradients in the reactor (figure 5-6). Using a Challenger Respirometer the thermophylic biomass was found to be resilient and active in spite of these events.

RESULTS AND DISCUSSION

Samples were taken from the feed, reactor and final effluent and analysed for key parameters using standard procedures (as detailed in the American Public Health Association (1998) *Standard Methods for the Examination of Water and Wastewater* and as detailed in the *"University of Newcastle upon Tyne, Environmental Engineering Laboratory Methods"*) and at regular intervals for routine analysis (page 57) at each stage. The tests included:
- Soluble Chemical Oxygen Demand (SCOD)
- Total Chemical Oxygen Demand (TCOD)
- Total Suspended Solids
- Non Volatile Suspended Solids
- Volatile Suspended Solids
- Total Kjeldahl Nitrogen * (TKN)
- Ammoniacal Nitrogen *
- Total Dissolved Solids *
- Total Non Volatile Dissolved Solids * (TNVDS)
- Total Volatile Dissolved Solids) * (TVDS)

* Denotes non routine

Temperature and pH were measured in situ using probes (adapted for immersion) and which were calibrated using a portable reference probe. Dissolved oxygen could not be measured due to temperature and restricted access to the reactor. The off gas from the reactor was oxygen at saturation with ammonia.

AMMONIA REMOVAL AND RECOVERY PROCESS
PRESS LIQUORS

Table 5-1 shows the removal and recovery of nitrogen as TKN and NH_3-N from press liquors. A removal efficiency of between 10-20% would be expected by the designers at 15°C and on a single pass. However in view of the recirculation this would progressively reduce the inlet concentration and thus reduce the efficiency.

DIGESTED SLUDGE

Table 5-2 shows the removal and recovery of nitrogen as TKN and NH_3-N from digested sludge. A removal efficiency of between 16-20% is more in line with design given the

higher temperature of 17°C. The problem of recirculation was a common factor for both feed streams.
AFC

VSS, NVSS AND TCOD LOADING RATE

Figure 5-1 compares TCOD loading rate with MLSS (TSS + VSS) development. The objective was to grow the MLSS to about 25,000mg/l VSS by increasing the TCOD loading. Initially the VSS achieved 12,500 at very high TCOD loading rates of up to 300mg COD/l/h. This was also the acclimation phase and pH was closely monitored for alkaline break through which occurred at point 1. This was due to the jet system blocking (which re occurred at points 2 and 3). However as a precautionary step the loading was reduced to about 100mg COD/l/h thereafter with the VSS and TSS declining in comparison until after point 3.

TCOD DESTRUCTION

Figure 5-2 shows that TCOD destruction was sustained at 50 – 60% for most of the evaluation. The apparent reduction towards 50% at the final stage was due to the mixing problem rather than any reduction in bio performance.

SOLIDS DESTRUCTION

Figure 5-3 compares the cumulative VSS and TSS fed to the reactor with the VSS and TSS destroyed. The mixing incidents have been highlighted but what is more significant is the sustained destruction of VSS (80-90%) and the sympathetic reduction in NVSS, which follows the same profile.

DISSOLVED SOLIDS

The dissolved solids sample preparation proved the most difficult with slow filtration rates limiting the sample volume to 8ml. The results shown in Figure 5-4 are difficult to explain particularly those for digested sludge and sampling error cannot be ruled out.
The NaOH addition is perhaps confirmation of some analytical and / or sampling error since the post addition NVDS values have reduced rather than increased.

FATE OF TKN AND NH$_3$-N

The results shown in Figure 5-5 and in Table 5-3 confirm a progressive reduction in TKN and NH$_3$-N throughout the evaluation. The trend and overall percentage reduction in each is consistent with expectations.

RESPIROMETRY

The results of respirometric evaluations carried out on the 31/07/02 and the 12/08/02 are shown in Figure 5-6 and Figure 5-7. The sharp reduction in biomass temperature from 60 – 50°C has had no lasting effect, if any. Both results confirm the resilience of the biomass with their respirometric response in proportion to the TCOD loading applied. The corresponding f/m values of 1.15 (TCOD 200mg/l), 2.29 (TCOD 400mg/l) and 4.59 (TCOD 800mg/l) are also encouraging in that they provided confirmation of total tolerance to high pH whilst the procedure also indicated the buffering effect of CO_2.

CONCLUSIONS

The combination of ARRP and the AFC™ process has demonstrated using *digested sludge* under alkaline conditions (pH 8.0 +) the potential to: -

- remove and recover 'N' by up to 20% initially and up to 70% overall
- destroy up to 90% of the VSS and to remove the NVSS by up to 80%
- destroy up to 50% of the TCOD without acid hydrolysis
- remove 'P' by up to 90%

There is certainly a need for such a process which will reduce risk to the food chain and recover N & P.

There is also potential to combine proprietary technologies in a novel way to achieve more effective and efficient recovery of N & P.

The use of pure oxygen does not appear to have inhibited bio-performance.

It had been planned to base the analysis of total N and total P utilising pre mixed reagent kits and a colorimetric procedure. However this proved unreliable due to interference (e.g. Fe) and was abandoned thus preventing the collection of N & P data on a routine basis.

REFERENCES

(1) 'phosphates & potash INSIGHT' – SPECIAL CLEARWATER ISSUE - 2001

(2) Edge, D.R. (2001) *Implications of Nutrient Removal from Sewage for Sludge Disposal Strategies* – Aqua Enviro Technology Transfer October 2001

(3) Duley, Bill. (2001) *Recycling Phosphorus by Recovery from Sewage* - Aqua Enviro Technology Transfer October 2001

(4) Brodersen, J. Juul, J. Jacobsen, H. (2002) *Review of Selected Waste Streams* – European Environment Agency Technical Report No 69

(5) Munck-Kampmann, Birgit. *Waste Annual topic update 2000* – European Environment Agency Topic report 8/2001

(6) Chiu, Ying-Chih. et al. *Alkaline and Ultrasonic Pretreatment of Sludge before Anaerobic Digestion* – Wat.Sci. Tech. Vol 36, No 11, pp.155-162, 1997.

(7) Horan, N J. *Removal of Phosphorus and Recovery from Sludge* - Aqua Enviro Technology Transfer October 2001

(8) Piekema, P.G. Gaastra,S.B. (1993) *Upgrading of a Wastewater Treatment Plant in the Netherlands: Combination of several Nutrient Removal Processes* – European Water Pollution Control Volume 3, Number 3, May 1993

(9) Pearce, Peter. *Biological Phosphorus Removal Process Requirements and Options for Phosphorus Recovery* - Aqua Enviro Technology Transfer October 2001

ACKNOWLEDGEMENTS

I would like to acknowledge the Department of Environmental Engineering, University of Newcastle upon Tyne, my colleagues at Environmental Performance Technologies and associates at Rauschert Verfahrenstechnic for their support in this project.

TABLES

Table 1: TKN and NH_3-N: Press Liquors

AMMONIA REMOVAL and RECOVERY PROCESS - PRESS LIQUORS							
stripping column					washing column-50l of H_2O + H_2SO_4		
time (m)	pH	Temp(°C)	TKN (mg/l)	NH_3-N (mg/l)	pH	Temp	TKN (mg/l)	NH_3-N (mg/l)
0	7.75	15	218	224				
1	10.2	15			2.62	16.3	8.4	0
60	10.5	15			2.71	16.6		
120	10.5	15			2.87	16.1		
180	10.5	15			3.2	16.8		
240	10.5	15	216	213	3.61	16.6		
300	10.5	15			3.78	12.4		
360	10.5	15			5.74	14.9	36.4	28
372	10.5	15			2.44	16.9		
490	10.5	15			2.54	18.3		
550	10.5	15			2.64	18.4		
610	10.5	15			2.79	19		
670	10.5	15	202	207	2.94	19.3		
Ammonia removed (as TKN) = 500 (l) x (218 - 202) mg/l = 8000mg (9%)							
Ammonia removed (as NH_3-N) = 500 (l) x (224 - 207) mg/l = 8500mg (9.2%)							

Table 2: TKN and NH$_3$-N: Digested Sludge

AMMONIA REMOVAL and RECOVERY PROCESS - DIGESTED SLUDGE								
stripping column					washing column- 50l H$_2$O + H$_2$SO$_4$ + (NH$_4$)$_2$SO$_4$ from above			
Time (m)	pH	Temp (°C)	TKN (mg/l)	NH$_3$-N (mg/l)	pH	temp	TKN (mg/l)	NH$_3$-N (mg/l)
0	7.36	13.3	3038	1103				
1	10.71	15.7	2422	930	3.07	13.3		
25	10.5	17.3			3.24	14.5	56	67
55	10.5	17.3			3.39	16.3		
115	10.5	17.3			3.59	17.5		
175	10.5	17.3			3.85	18.5		
215	10.5	17.3	2520	907	4.18	18.9		
235	10.5	17.3			4.19	14.9		
372	10.5	17.3			4.26	19.6		
490	10.5	17.3			4.41	20		
550	10.5	17.3			4.48	20.1		
610	10.5	17.3	2562	884	4.48	20.2	64	67
Ammonia removed (as NH$_3$-N) = 500 (l) x (1103 - 884) mg/l = 109500mg (20%)								
Ammonia removed (as TKN) = 500 (l) x 3038 - 2562) mg/l = 238000mg (16%)								

Table 3: Fate of TKN and NH$_3$-N [2]

Sample	(1) DS pH7.0 05/07	(3) DS pH 10.5 05/07	Feed 07/07	Feed 31/07	Centrate 06/08	Centrate 07/08
TKN (mg/l)	3038	2422	2562	2268	938	938 (69% removal)
NH$_3$-N (mg/l)	1103	930	884	902	325	330 (70% removal)

Table 4: Respirometric response to pH

Sample	pH start	pH end
Blank	8.88	8.86
200 mg/l TCOD	8.98	8.81
400mg/l TCOD	9.03	8.81
800mg/l TCOD	9.12	8.66

FIGURES

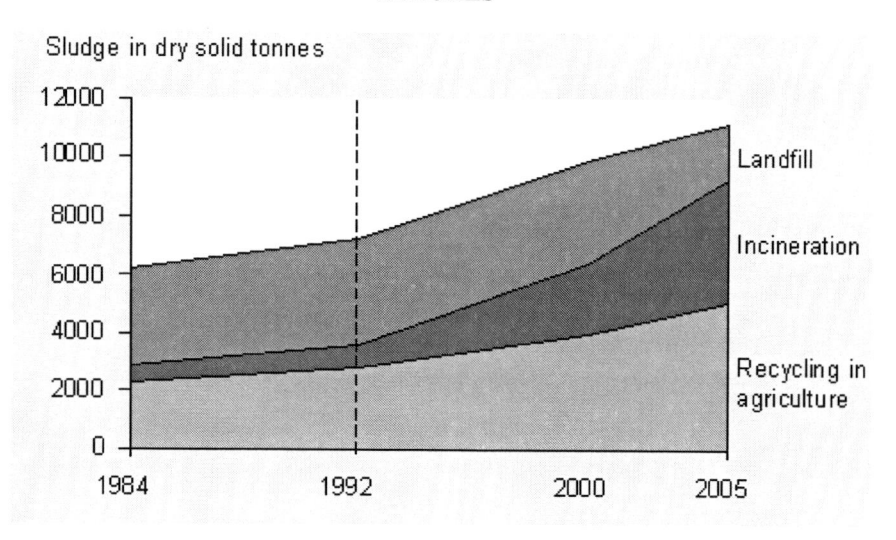

Figure 3-1 Sewage Sludge Disposal in Europe[5]

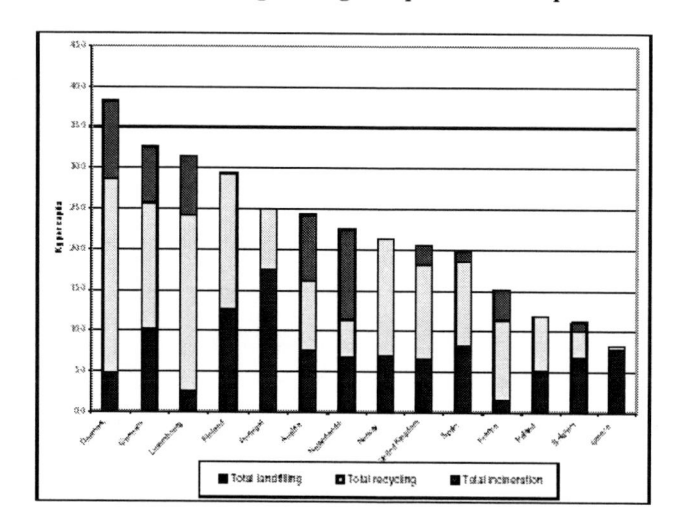

Figure 2: Treatment of sewage sludge: selected EEA member countries, 1998[4]

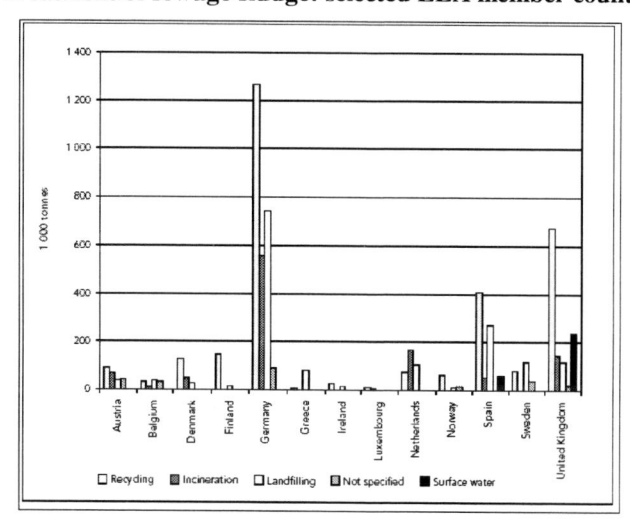

Figure 3: Treatment of sewage sludge in 1998[4]

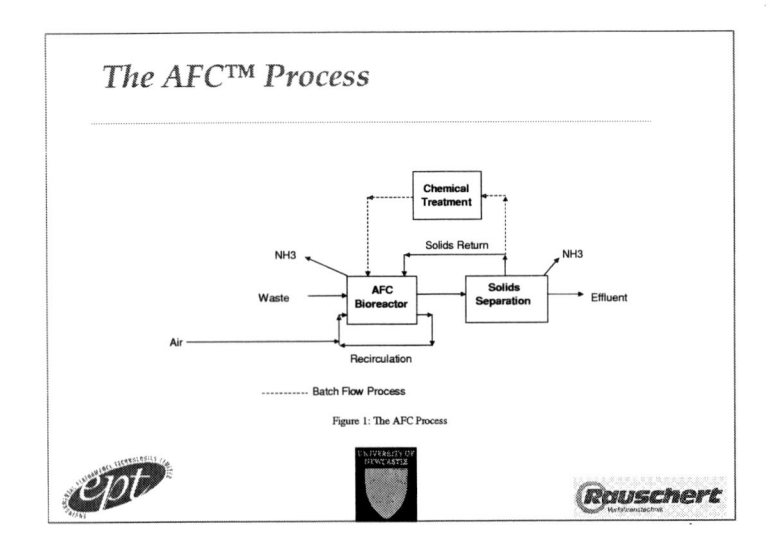

Figure 4: AFC™ schematic

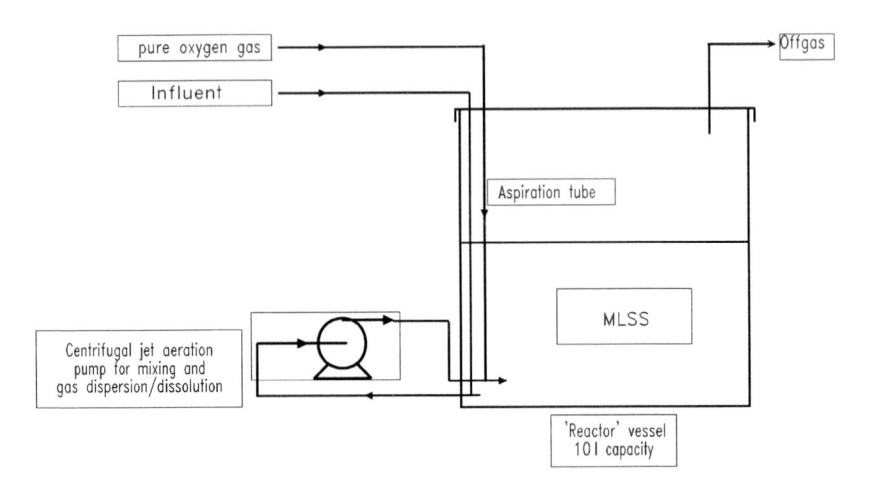

Figure 5: Schematic of AFC™ bioreactor

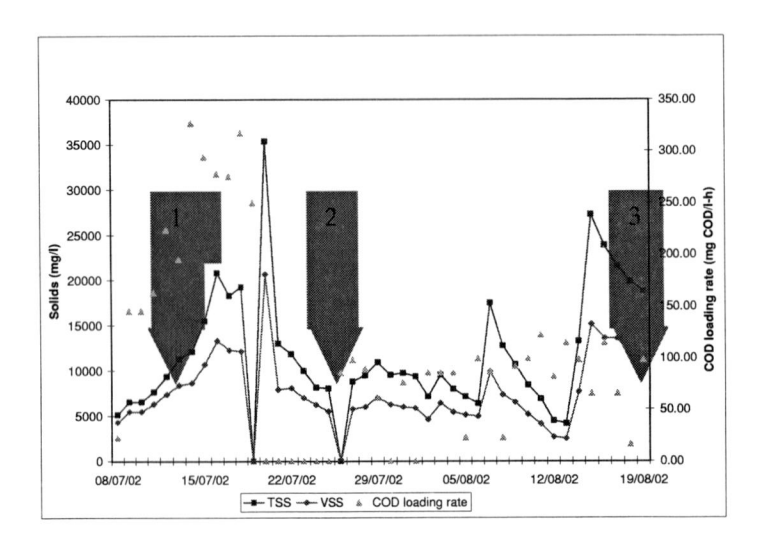

Figure 5.1 Bioreactor Performance VSS, NVSS and TCOD Loading Rate

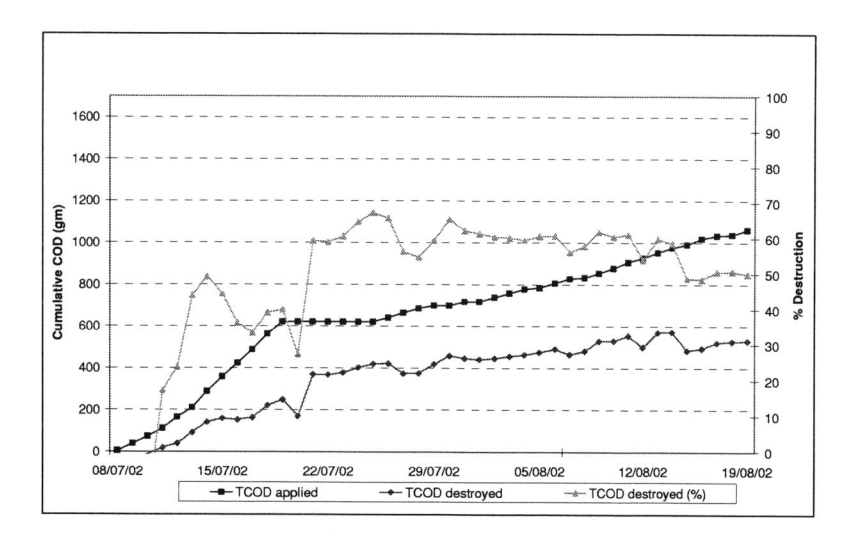

Figure 5-2 TCOD Destruction

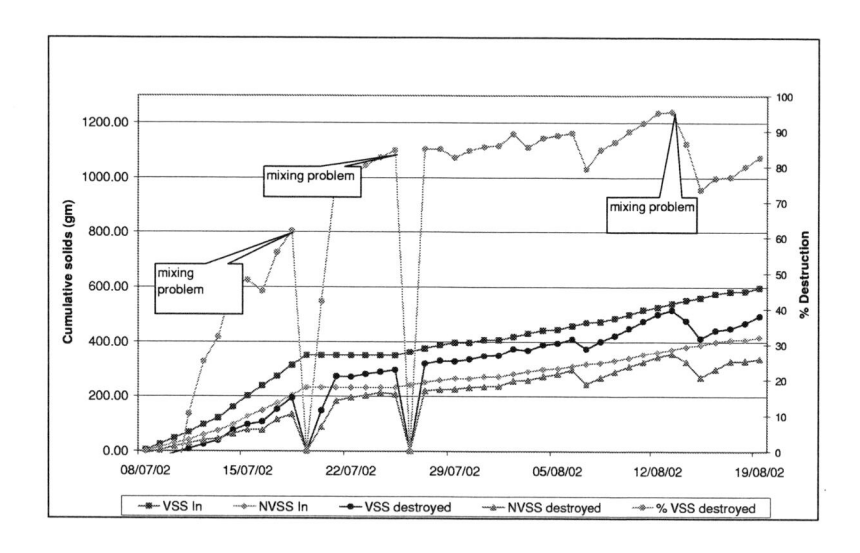

Figure 5-3 Solids Destruction

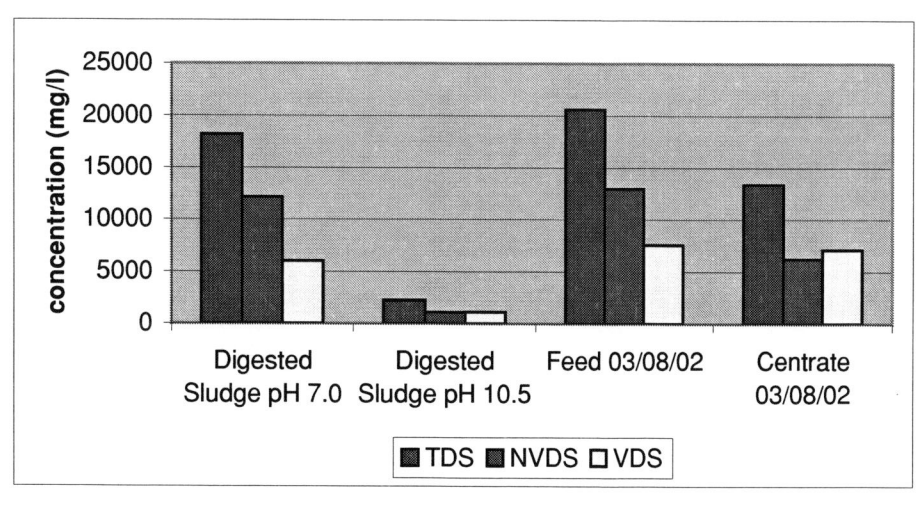

Figure 5-4 Dissolved Solids

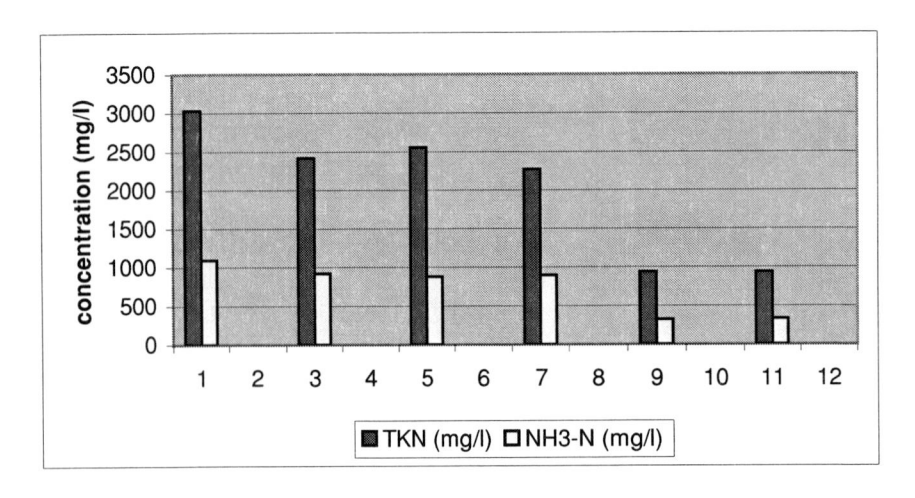

Figure 5-5 Fate of TKN and NH$_3$-N (2)

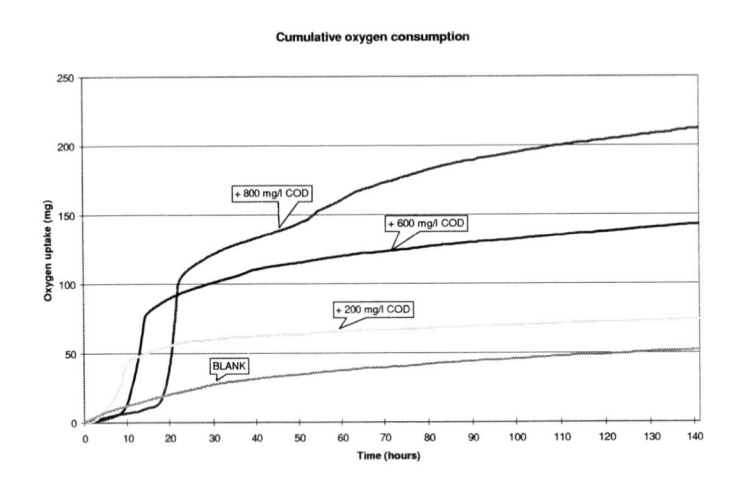

Figure 6: Respirometery 31/07/02

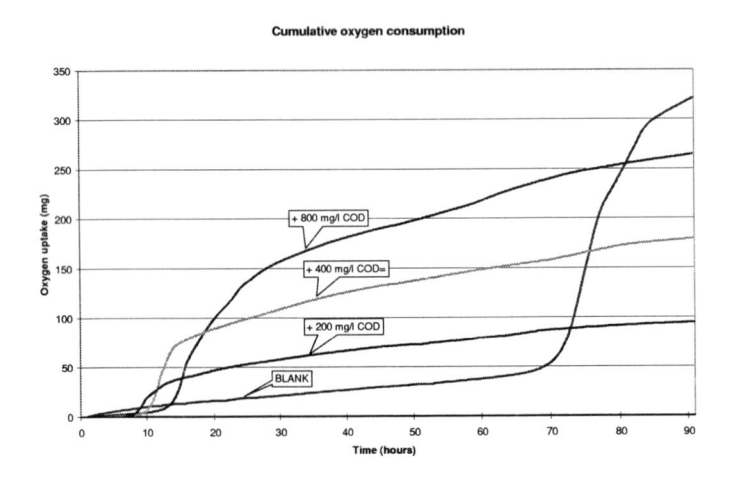

Figure 7: Respirometry 12/08/02

HAVING YOUR SLUDGE CAKE AND EATING IT?
A DISCUSSION OF THE PSYCHOLOGY OF BIOSOLIDS APPLICATION

Stephen Toogood,
Effluvium Ltd., Bassingbourn, Cambridgeshire SG8 5NX
stephen@stenches.demon.co.uk

ABSTRACT
This paper started its gestation at the Oxford Seminar in July 2002, where speaker after speaker acknowledged the problem of public attitude, but had nothing to offer in relationship to it. It is a personal view, a piece of deliberate polemic, and sets out to court controversy. As Ralph Vaughan Williams said of his 4th symphony, 'I don't like it, but it's what I meant'.

KEY WORDS
Public perception, risk, sewage sludge.

AGREEING WHAT THE PROBLEM IS
Many people in agriculture, the water industry and elsewhere have for some time been very worried about the future for land use of biosolids. If they thought the Safe Sludge Matrix was a long-term solution, and some did, then we now know it was, though an excellent step, in the end a holding operation.

We all know that public perception is the main issue, yet most of the R&D effort seems to be going into solving technical problems like stability, pathogen kill and monitoring, and few papers have been written examining the drivers of public opinion in this area. Yet this is an aspect we ignore at our peril.

NATURAL REACTIONS
It is a fact, inconvenient as it may be, that the general population is not happy with the idea of human excrement, however stable and sterile, having anything to do with the food chain, so it would be as well to establish the extent to which this is an instinct inherent in *homo sapiens*, and the extent to which it is a result of recent conditioning and influences. Indeed, unless we address this issue, then we may expect to see application to land denied us as an option for biosolids at some time in the next few years.

People do not like their own waste products; that seems clear enough. The development of a disgust for faecal matter has been discussed in several papers by Matthews and others, and I could do no better than to quote a succinct example, specifically related to biosolids[1].

We all have a ' faecal aversion barrier'. Faecal matter itself has a malodour to provide a psychosomatic warning that the origin of the odour is a microbiological hazard. We are all taught the elements of public health training in childhood and this pre-conditions us to the hazards of faecal matter. It is only when societies become stressed and impoverished that these hazards are tolerated.

This reaction is under current research, and there are two main schools of thought. One, advocated for example by Curtis and Biran[2], suggests that aversion to excreta, faecal matter, saliva, wounds and sores is genetically encoded. They undertook surveys in six different countries (in Africa and India as well as Europe) and the list of things people found disgusting had these common elements. They argue that this is evidence against the view that disgust is culturally acquired. However, some disgust was culture-specific: for example in India there was a deep-seated aversion to food cooked by menstruating women.

The other view is exemplified in the work of Mary Douglas[3] in the UK, and Paul Rozin in the USA. While both acknowledge that the mechanism of disgust (how we react and the effect it has) is in all probability genetically encoded, what we find disgusting is largely culturally determined.

Rozin's work with children suggested that the response to being offered chocolate in the shape of dog faeces was strongly age-related, with 8 as a general cut-off age. Douglas on the other hand refers to a 'dirt reaction', experienced when we see certain kinds of material in the 'wrong place'. One might cite as an example that finding a dead rat in ones kitchen provokes considerably more disgust than seeing one in the gutter. But to Douglas the seeming dichotomy between 'nature and nurture' is less important than finding out what kind of social organisation intensifies or ameliorates emotions of disgust, and what kinds of things that disgust encompasses.

IT STANDS TO REASON...
To try and unify the seemingly competitive explanations, the reaction of disgust is deeply implanted, and we tend to be disgusted by things that we think will harm us. What those things are will include things that will actually harm us, and things that we only suppose may harm us, and those lists may to a greater or lesser extent be culturally determined.

I want to suggest that greater basic scientific knowledge in the general population has increased rather than decreased the level of supposed harm. Whereas our forefathers relied on instinct and experience to base their supposition of harm, today's population is quite capable of reasoning along the lines of 'some diseases are caused by bacteria; cheese contains bacteria; therefore cheese causes disease'.

In reality we know that pathogens form only a percentage of bacteria, and of course microbiologists know many further levels of distinction, but for the general population the very word 'bacteria' rings alarm bells.

There are other influences that feed the same tendencies, as we shall see later.

A QUESTION OF TRUST?

One of the highlights of 2002 for me was the Reith Lectures, broadcast on Radio 4 in the spring. In these, Anora O'Neil examined the presumed 'crisis of trust' (for example in institutions and office-holders) in the UK, and its causes. In the first lecture she looked at the evidence for a 'crisis of trust', and found that it was in the main limited to most people saying they did not trust, when their actions indicated otherwise. People say for example that they don't trust water companies, but they're still happy to drink the water.

Perhaps we ought to quote O'Neill further.

Some sociologists have suggested that the crisis of trust is real and new because we live in a *risk society*. We do live among highly complex institutions and practices whose effects we cannot control or understand, and supposedly see ourselves as subject to hidden and incomprehensible sources of risk. It's true that *individuals* can do little or nothing to avert environmental risks, or nuclear accidents, or terrorist attacks.

All this is true, but not new. The harms and hazards modern societies impose differ from those in traditional societies. But there is nothing new about inability to reduce risk, about ignorance of its sources, or about not being able to opt out. Those who saw their children die of tuberculosis in the nineteenth century, those who could do nothing to avert swarming locusts or galloping infectious disease, and those who struggled with sporadic food shortage and fuel poverty throughout history might be astonished to discover that anyone thinks that *we* rather than *they* live in a risk society. So might those in the developing world who live with chronic food scarcity or drought, endemic corruption or lack of security. If the developed world is the paradigm of a 'risk society', risk societies must be characterised simply by their *perceptions of* and *attitudes to* risk, and not by the seriousness of the hazards to which people are exposed, or the likelihood that those hazards will actually harm them.[4]

All this suggests, if O'Neill is right, that people's 'lack of trust' is born mostly of not wanting to appear gullible, and therefore that actual lack of trust of food grown with the aid of biosolids is exaggerated. If supermarkets are wary of this food, it is not because they themselves mistrust it, but more that they see freedom from it as a potential advertising claim.

TALKING POLITICS

This is of course always dangerous ground in a technical paper, but in the end we have to acknowledge that it's an important influence. In the days of patrician politics, arguments within political parties tended to be about what was the 'right' policy in terms of the party's ideals, and electors (whatever ones views of the electoral system) chose, in theory, between competing philosophies.

In recent years the priorities have changed radically. The overwhelming priority is now to get elected at virtually any cost, and beliefs must be jettisoned if they do not accord with the reported views of vital electors. This looks on the surface all very democratic, but the statisticians and strategists have realised that it is marginal voters in marginal constituencies who make the difference between power and opposition, and public policy utterances are to a significant extent geared to pleasing this relatively small group.

How do they know what will please this target electorate? By the results of the same opinion polls whose validity O'Neill was right to question. So the assumed mistrust begins to have very real influence on public policy.

At the same time of course, government is prey to the lobby system. It is also prey to an attitude that 'if we admitted there was a problem we'd have to do something about it, and since we don't know what to do, we'd better not accept that there's a problem. Some would suggest that this attitude surfaced a few years ago over Bovine Spongiform Encephalopathy. The reassurances given as that crisis grew, and the memory of Mr. Gummer inducing, if that is the right word, his offspring to consume a nourishing burger for the television cameras, have a lasting legacy of scepticism in the public mind towards any reassurance by politicians on health or environmental effects.

HAZARDS AND DANGERS

We should now look at where these perceptions that the man on the Clapham omnibus has have come from.

IS YOURS A SAFE HOME?

A few years ago, a certain brand of disinfectant produced a series of TV adverts with the tag "Is yours a safe home? a Snibbox Home?" (only the names have been changed to protect the 'innocent'). Nothing stated explicitly of course, but the viewer is invited to conclude that a home with any bacteria in it is Not Safe. Much as we know that the primary purpose was to increase sales, we also know that advertisers prefer to do this by increasing the market size, rather than by increasing the market share.

Those adverts were of course just the beginning. Day by day the consumer is bombarded with sales pitches for disinfectant wipes, germicidal floor care, hygienic baby products, sterile this, oxidising that. And always same Enemy: Bacteria, the Unseen Menace.

How do these programmes have their effect?; by trading on insecurity. A couple of generations ago when people married younger, generally stayed in the same place, had children sooner, there was a handing down of experience and wisdom from one generation to the next - grandma was always on hand to reassure the first-time parent. These days older parents are apt to get distinctly neurotic about the health of their precious children, and tend to be unsure just how ultra-cautious to be. A soft target for the advertiser, in other words. This manifests itself in other ways too.

SECURITY RISK

I was recently involved in a Village Appraisal exercise, in which every resident in a village of 3000 population was asked among other things what their priorities and concerns were. One of the highest-scoring concerns was that people did not feel safe in the village at night. So we looked at the police crime statistics to see what grounds there were for this concern. We went back several years. A typical quarterly report listed two breaking-and-entering, one domestic violence, theft of a bicycle, one of a car, one incidence of criminal damage to a motor vehicle.

In other words in five years nobody at all had been attacked in the village, but yet so many people assumed there was a real risk of just that occurring.

Fifty years ago people's working lives were much more hazardous than they are today. Generally, hazards were accepted as a fact of life. Today, hazards are enjoyed vicariously, in books and films, but kept at arm's length, as they are at Theme Parks, where the most apparently-hazardous rides have the longest queues, yet on the rare occasions when this apparent danger becomes real, or worse still if an accident occurs, there is general outcry, loudest from those who appeared to want to put themselves into danger in the first place.

HIS HIGHNESS KING CONSUMER

The rise of 'The Consumer' is one of the defining phenomena of the late twentieth century in the developed world. In the 1950s when you wanted a loaf of bread you walked down the road to the nearest baker's. The baker knew that if he was competent the local population, and only the local population, would come to his door, and all he had to fear was a second, better baker setting up in the same street. Much of the retail trade was steady, low risk (volumes could be predicted with some confidence), but there was little to be gained outside major cities in advertising; the size of the business was more-or-less fixed.

Now, as we know, people get in their cars and travel sometimes many miles to choose a national retailer in whom they fell comfortable, and spend perhaps £200 once a week. There is so much more at stake for the retailer than there was; they want our money, their business depends on getting it, and they're going to go to almost any lengths to get it. So anything that will gain the customers' confidence is fair game. Telling the consumer what a wonderful person he is, and how we agree with all his opinions, is the flavour. "We care about your concerns, and we share them." There's real money to be had by reinforcing prejudice.

All this is aimed at getting the customer through the door. Once inside the store however, experience suggests that spending patterns no longer reflect the concerns that the marketing traded upon. Few people pick up the pack of carrots and read even what country they came from, let alone how they were grown.

SOME POSSIBLE PATHS IN THE MIRE

This may all seem dismal stuff. Ought we to give up now? No, and not only because there is too much at stake to abandon the task.

What I believe we have to do is to accept these influences, and others we might add to the list, and decide which of them have to be taken as givens, and which ones we might, if we go the right way about it, modify.

At the same time we can start to think of ways in which we might do a little opinion forming of our own.

But first, there are aspects of our own house that need to be put in order.

HAVE YOUR CAKE AND EAT IT; COST CONTROL VS QUALITY CONTROL

The history of sewage treatment and its organisation left a legacy of looking upon biosolids as a liability, and for years research was aimed at minimising the costs of 'disposal', while meeting (by no greater margin than was necessary) the demands of the regulatory framework. In the though-provoking paper I quoted earlier, Matthews lists what he terms paradoxes to consider[1]. Here is a couple of them.

We encourage farmers to use biosolids, within the context of prudent control, yet our research on treatment is to produce as little as possible for economic reasons, but this sends the message that the product is a liability.

We ask farmers to use biosolids, because they are safe, yet we often refuse to offer explicit indemnities or guarantees against changes in legislation.

We say that safety lies not only in the quality of the product, but the caution and care with which operations are managed - yet quality assurance and environmental management systems, the bed-rock of reproducibility, sustainability and safe practice, have been introduced slowly and reluctantly in some places for fear of the additional powers that this might give the Regulators. What does the public make of that ? We give the impression that we have something to hide.

In other words, the rest of the business world has been thinking for years in terms of 'added value', while the water industry seems to be stuck with 'reduced cost'.

We could years ago have introduced quoted N:P:K levels for biosolids products. We could have accepted quality standards for contaminants such as metals. In short, we could have made a product to stand in the marketplace alongside chemical fertilisers, and with quite a few technical and environmental advantages over them. But we failed to do this, because to do so would have required investment, towards which our accountants have to be dragged kicking and screaming, and because the other requirement would have been vigilance in quality control, which we never had the manpower resources to deliver.

SOIL QUALITY

One of the growth areas in food retailing has been in organically-grown produce. There has been a steady move in agriculture as well towards accreditation by the Soil Association. We know that biosolids, as long as we can control residual contaminants, is beneficial to soil. 'Feed the soil and not the plants' has been a motto of organic horticulturalists since the movement began; biosolids does this. Surely the possibility exists to cooperate with the Soil Association in promoting a properly-developed soil product that meshes with their accreditation standards.

SAFETY OF GROUNDWATER

I live in an area heavily dependent on groundwater sources for a growing population. Yet in places the topsoil has become so thin that the chalk strata are clearly visible after every ploughing. We know that nitrates are not the only problem, and we know what capital investment has become necessary in those places where nitrate levels in groundwater have risen. Is there not an opportunity to look with the EA into the scope for groundwater protection by a programme of soil enrichment?

ORGANOLEPTICS

We have in the past shown little interest in what happens to our biosolids once the farmer applies them to the land, beyond the 'hygiene factors' such as metal uptake by various crops. I failed to find any references that suggested whether crops grown with the aid of biosolids were actually better crops.

I know for example that the potatoes I grow, using animal manure, taste distinctly better than those that I buy from a shop. Are potatoes grown with biosolids bigger, or are there more small ones that cannot command the same price? Do they taste better? are there more visual defects? are they more liable to eel worm or other pests?

I suggest it is in our interests to know much more than we do about what benefits biosolids confers for the stakeholders right down the food chain.

THEY HAVE RISKS TOO

One misconception we might correct is that the use of biosolids alone carries risk to the consumer. The environmental concerns about inorganic fertilisers, just one example among many possible ones, form the basis of a competing set of risks, and these ought to be set before the public as part of the opinion-forming process. It is clear that this kind of PR cannot be seen as coming direct from the water industry, so that other bodies that can be seen as independent have an important role to play.

A HOPEFUL NOTE

One recent trend suggests that one of the tides may be beginning to turn. You may have noticed the level of advertising for bio-dairy products. These have 'beneficial bacteria'. They are 'natural'. They restore your balance (this may of course mean 'these make you feel OK about that junk food you had last night', but that's as maybe).

We could take an important cue from the way these products are presented. The makers have spent a lot of money on the research behind these campaigns. Mmm. Biosolids.

RESEARCH NEEDS

These have been just a few suggestions. In all cases we seem to come up against the lack of evidence. Some of the holes in the knowledge are alluded to above.

In whatever area we intend to work though, it seems clear that we need to do a lot more study, and to do it now. The budgets must be made available. I think we know what the consequences of inaction will be.

REFERENCES

1. Matthews, P, 'A Millennium Perspective on Biosolids and Sludge Management', presented to CIWEM symposium 'Future of Biosolids Recycling', Cambridge, March 22, 2000

2. Curtis, V A and A Biran, 'Dirt, disgust and disease: is Hygiene in our genes?', Perspectives in Biology and Medicine, 44.1 (Winter 2001)

3. Douglas, M. 'Purity and Danger: an analysis of the concepts of pollution and taboo', London: Routledge and Kegan Paul, 1966

4. O'Neill, A., BBC Reith Lectures 2002; 1. Spreading Suspicion, BBC, 2002

AWARENET:
AGRO-FOOD WASTES MINIMISATION AND REDUCTION NETWORK

Libe de las Fuentes, Brian Sanders[a], Jiri Klemeš[b]

GAIKER, Parque Tecnológico, Edificio 202, 48170 Zamudio (Bizkaia), Spain, delasfuentes@gaiker.es
[a]ADAS Consulting Ltd., Woodthorne, Wergs Road, WV6 8TQ, Wolverhampton, UK, Brian.Sanders@adas.co.uk
[b]Department of Process Integration, University of Manchester Institute of Science and Technology (UMIST), PO Box 88, Manchester M60 1QD, UK, j.klemes@umist.ac.uk

ABSTRACT

Despite of the economic importance of the food sector, during the last EU Framework Programmes no global European vision addressing the environmental problems arising from agro-food wastes has been approached. Thus AWARENET is the first Thematic Network aiming at the Prevention, Minimisation and Reduction of Wastes from the European Agro-food Industry. Funded by the GROWTH Programme of the European Commission with a global budget of 1.475 M € for a three-year duration (2001-2003), it focuses on 5 main agro-food sectors: meat, fish, dairy, wine and vegetables.

Currently the network comprises 33 members from 14 different European countries (Belgium, Denmark, Finland, France, Germany, Greece, Hungary, Italy, Poland, Portugal, Spain, Sweden, The Netherlands and the United Kingdom).

It is co-ordinated by the Spanish Technology Transfer Centre GAIKER with 32 participants from agro-food manufacturing industries (wastes producers), waste valorisation industries, equipment manufacturers, consultancies, research organisations and universities. This mix of backgrounds and expertise is a will assist the project in meeting it's main task – the formation of a network to provide a forum for information/expertise exchange and technology transfer. The three-year project is currently in its second year, having started in January 2001.

INTRODUCTION

AWARENET is a "Thematic Network" project, funded by EU DG Research, to address the problem of waste from the European Food Industry. The ever-increasing market for ready prepared and processed foods means that the potential for waste production, and therefore the environmental burden, is rising. In recognising this trend, the EU has commissioned this work to identify the opportunities for reducing wastes and recycling or upgrading waste products in five food sectors – dairy, fish, meat, vegetables and wine.

The current distribution of Network Partners in Europe is given in figure 1. The aim of project is to advise the EU Commission of the following:

- The amount and pollution potential of waste streams from food processing operations
- The level of adoption of EU Directives affecting the agro-food processing industry
- The degree to which national regulations are controlling environmental issues
- The current best practice in waste minimisation
- The main processes in the five sectors which lead to significant waste streams
- The available and potential by-product valorisation technologies
- The current and future markets for upgraded by-products
- The RTD needs derived from agro-food waste management in Europe.

The project covers only organic wastes, and inorganic wastes insofar as they are associated with organic waste streams (e.g. spent diatomeaceous earths from wine filtration). Food packaging is not part of the study. The results of this study will enable the EU to:

- Disseminate waste minimisation best practice across all EU and NAS countries.
- Design a legal framework, which will meet the future needs of the industry to maintain and improve control of environmental effects, encourage waste minimisation, recycling, and upgrading of waste streams.
- Identify potential for further savings in waste and for developing new products from waste.
- Define specific research needs to realise this potential.

The outputs and results will be published at the end of the project in a "White Book" and this will be presented at a stakeholder conference, probably in Brussels, in early 2004. In the meanwhile, there is gradual dissemination of results via an EnviroWindows Interest Group (http://ew.eea.eu.int) and through the publication of AWARENET Newsletters every six months.

In the original contract, AWARENET Network Partners were chosen to represent all EU States and, in a subsequent variation, some of the Newly Associated States. Partners were also included to represent a wide range of backgrounds, e.g. education, research, industry and consultancy. Within this group there is a very wide knowledge of the food processing industry across Europe.

The project neither expects nor has resources to address all waste streams. When selecting opportunities for the developments of new technologies and markets for waste products, the principal of *significance* will be applied. Based on the knowledge held within the Group and by applying advanced selection methodology, only those opportunities for minimisation and valorisation having the *greatest beneficial impact on the environment* and being *sustainable in economic terms* will be recommended for further research. In terms of numbers, at this stage it seems likely that a small number of opportunities in each sector will be taken forward – one initial estimate suggests a figure of 20 –30 overall.

The EU 6th Framework offers the opportunity to further develop the R&D programmes identified under AWARENET. To this end, GAIKER together with the other partners has presented an Expression of Interest for an integrated project proposal under this scheme.

AWARENET RESULTS DURING THE FIRST YEAR

The innovation of AWARENET is the development of a vision for the agro-food industries by the approaching the problem of wastes from three different and critical points of view: regulatory issues, technology and market. For this purpose, the European environmental legislation and regulations affecting the agro-food industrial wastes will be compiled in an organised and usable fashion. An inventory of agro-food waste figures and critical points in production processes will be made all over Europe with the inputs of all participating countries within the network. The current best practices used for minimisation and valorisation of wastes will be compiled all over Europe. The dissemination activities of the Network will help the European agro-food industries (with special emphasis on SMEs) by describing simple tools and rules for wastes prevention and minimisation.

Furthermore, AWARENET will identify the market demand for products from agro-food valorised by-products and define the R&D needs for minimising and valorising agro-wastes. At the end of the project a *White Book* will be published as a practical manual for European agro-food industries.

During the first year of AWARENET, activities have focused on defining the main European issues on agro-food industrial wastes. These activities have included:

- The compilation of legislation on agro-food wastes related issues at European and national level.
- The elaboration of an inventory of agro-food wastes in Europe from five industrial sectors.
- The identification of main production processes and critical points in terms of waste generation in the main production processes of these sectors and a preliminary identification of marketable products from agro-food wastes.

A software programme for the management of inventory data compiled within the Network is under development.

REGULATORY ISSUES IN EUROPE RELATED TO AGRO-FOOD WASTES

The main objectives of this task were:
- To present in an organised and usable fashion the European environmental legislation and regulations currently in place or at the proposal stage affecting the agro-food industrial wastes sector.
- To compare the legislation and regulatory issues from the different countries involved in the Network

All Members have actively collaborated in the collection of waste legislation from their corresponding countries. A template spreadsheet was designed referring the EU Directives relevant to waste, pollution, and water. The information required against each directive included:
- Corresponding national regulation
- Scope of the regulation
- Implemented (Yes/No)
- Timetable for implementation if relevant and
- Sector relevance, e.g. Meat, Dairy, Vegetables, Fish and Wine.

This information has been gathered in two main tables, one of them describing the main European regulations regarding agro-food wastes issues and another one on the individual transposition of these Directives in the different countries participating in AWARENET.

The most significant current regulations (European Directives) apart from the Waste Directive and its amendment, were: 90/667 on Disposal and Processing of Animal Waste, 91/271 on Urban Waste-water Treatment, 96/61 on Integrated Pollution Prevention Control (IPPC) and 99/31 on the Landfill of Waste. A list of EU Legislation affecting solid and liquid agro-food wastes is provided in an Annex and the end of this paper.

DEFINITION OF WASTE AND WASTE RECOVERY IN THE REGULATORY FRAMEWORK

The first question that has to be answered is "when is an industrial residue a waste?" The answer is important as in each country the processing and transportation of waste is subjected to more stringent regulations that the handling of non-waste. In *Directive 91/156/EEC* of 18 March 1991, which amended Directive *75/442/EEC (the Waste Directive),* The Commission recognised the need for a common terminology and a definition of waste. Accordingly, in Article 1 **waste** is defined, as "any substance or object in the categories set out in Annex I, which the holder discards or intends or is required to discard". Annex I lists 16 categories of waste. The list is not exclusive; point 16 states that any materials, substances or products, which are not contained in the other categories, can be waste if they comply with the definition in Article 1 of the directive. This means that everything can be waste if it is, will be or must be discarded. Waste therefore becomes defined by what is meant with 'to discard'. Article 1 states that *waste can be discarded by disposal or by recovery.*

The term **secondary raw material** is not defined at European level, but arises from the interpretation of Article 11 by the competent national and regional authorities (Fig 3). If the recovery activity is exempted from permit requirements the material may be regarded as a secondary raw material. Discrimination between waste used for recovery and secondary raw materials is then important with respect to the regulations that must be complied with.

Handling of secondary raw materials is only governed by general and sectorial production rules. Use and shipment of waste must also obey to the regulations on waste management, including permit requirement, record keeping, obligation to provide information and inherent taxes. Industrial waste treatment is subjected to both regimes, making waste more difficult and more costly to handle. Therefore it is important to know:

i) when a substance becomes waste and how?
ii) and when waste ceases to be a waste and can become a secondary raw material?

Articles 1 and 11 of 91/156/EEC answer the first question. A substance becomes waste from the moment it leaves the place of production to be disposed of or to be put through one of the operations listed in Annex II B. So, a substance cannot become waste if it is recycled at the place of production.

The second question is more difficult to answer. According to Article 11 of Directive 91/156/EEC waste can become a product if it is recycled in accordance to the rules and criteria for secondary raw materials imposed by the national or regional competent authorities and if the recovery operation is in agreement with the provisions of Article 4 with respect to public health and the protection of the environment. As it is the establishment or company that undertakes the recovery operation, which is exempted from the permit requirement, the substance is waste from the moment it leaves the place of production and becomes a secondary raw material on delivery at the recycling plant (Fig 3). If the recovery does not comply with the regulations, the substance stays waste. In that case, the recycling process can result in another waste that can be used as a secondary raw material in another recycling operation (Fig 3).

Problems arise from the criteria used by the various competent authorities to recognise non-waste and from the standpoints taken with respect to transportation of materials destined to be recycled as secondary materials. Except for Spain, all Member States took measures to comply with directive 91/156/EEC. Therefore, all Member States now use the same definition of waste. Though, due to disparity in the criteria used to define which recovery operations may be exempted from the permit requirements, there is still no uniform European waste management policy.

The Waste Management Policy Group of the **Organisation for Economic Co-operation and Development (OECD)** developed a guidance document to indicate whether or not a material can be regarded as waste. This was done in order to facilitate trans-frontier movements and to give some directions for meeting the national criteria for EU member states. Nevertheless, the document clearly states that, in cases of dispute, the interpretation of the term waste is a matter for the Courts. The full text is given in Annex I; this document gives an example of conditions under which waste is not subjected to controls normally applied to waste:

The material must be transported directly from the producer to the process in which it is to be used and the material must be directly and completely used as an ingredient in the process, and the process must not be classified as (or comparable to) a waste management process.

Furthermore the OECD Document states criteria for determining when a waste ceases to be a waste:

> *"A waste ceases to be a waste when a recovery, or another comparable, process eliminates or sufficiently diminishes the threat posed to the environment by the original material (waste) and yields a material of sufficient beneficial use. In general the recovery of a material (waste) will have taken place when:*
>
> *(i) It requires no further processing by a recovery operation.*
>
> *(ii) The recovered material can and will be used in the same way as the material which has not been defined as waste.*
>
> *(iii) The recovered material meets all the relevant health and environmental requirements".*

INVENTORY OF AGRO-FOOD WASTES IN EUROPE

The aim of this preliminary work was to get a first overview of the classes and amounts of food wastes coming from producing industries from the diary, meat, fish, wine and vegetable sectors, so as to identify the magnitude and problems of food waste management in Europe. Primary production of the raw materials (farming) and packaging and transport activities associated with product marketing and retailing are not covered by the project.

The term "by-product" was defined, according to different technical publications, as "parts of the raw material that are not included in the final product".

All AWARENET Network Partners have actively collaborated through template spreadsheets providing national data on amounts of raw materials, finished products and solid and liquid wastes, with a composition estimation, and current waste management procedures. Furthermore, official data for production volumes for different agro-food sectors and sub-sectors have been gathered from European Production and Market Statistics (EUROSTAT), and have been subsequently processed according to NACE production classification, considering CPA and ProdCom codes.

To facilitate the collation, management and dissemination of the data already gathered and subsequently incorporated, it was decided to design and develop software, which can be accessed via the Internet. This software allows the user to have information about main agro-food wastes generated in all European countries, the composition of these wastes, waste treatment cost and possible market applications for them.

Figure 4 shows the menu page of the AWARENET software.

PRELIMINARY REVIEW OF MINIMISATION TECHNOLOGIES FOR AGRO-FOOD WASTES

The objective of this Task was to get the first overview of the state of the art of the agro-food industry in terms of waste prevention and minimisation practices and technologies as the basis for future work within the Network. As a preliminary step, the definition of two basic terms for the Network, minimisation and up-grading or valorisation, seemed essential.

The OECD definition for waste minimisation was selected. Fig 5 shows the OECD definition of waste minimisation and prevention agreed on the Berlin meeting in 1996. As it appears waste minimisation is a broader term than prevention. **Waste prevention** covers 'prevention', 'reduction at source' and 're-use of products'. **Waste minimisation**, however, also includes the waste management measures 'quality improvements' (such as reducing the hazard) and 'recycling'.

Furthermore, the EPA's (US Environmental Protection Agency) definitions for this term are:
- **Minimisation:** A comprehensive program to minimise or eliminate wastes, usually applied to wastes at their point of origin.
- **Waste Minimisation:** Measures or techniques that reduce the amount of wastes generated during industrial production processes; this term is also applied to recycling and other efforts to reduce the amount of waste going into the waste stream.

Conversely, the term **up-grading** (**valorisation**) has no official definition, and, accordingly AWARENET members have agreed on the following definition: "*Increase of the technical and/or the economical value of the by-products and wastes that are generated in the different agro-food industries*".

A quite thorough description of the main production processes in the five selected industrial sectors have been described, including:
- Fish sector: tuna fish canning and frozen flat fish
- Meat sector: slaughtering
- Dairy sector: milk collection and cheese manufacturing
- Wine sector: red wine and white wine production
- Vegetable sector: potato starch and olive oil production

For each process, a flowchart identifying the sources of wastes (critical points of production) and their management procedures was prepared (Fig 6). This schematic information was supported by technical data on wastes composition and minimisation and valorisation options (Fig. 7) subsequently described.

The information so far gathered in this task was the starting point for a more thorough study of prevention and minimisation strategies for agro-food wastes, valorisation technologies for new marketable products and of BAT for food manufacturing to be approached during the second and third phases of the network.

IDENTIFICATION OF COMMERCIAL PRODUCTS FROM AGRO-FOOD WASTES

During the first year of AWARENET a preliminary study of the market associated with agro-food wastes issues was made. This took account of the commercial products currently being generated: foodstuffs, dietetic, cosmetic and pharmaceutical products, soil conditioners, fertilisers, energy, and the restrictions to be imposed to their landfill disposal.

First of all, the European market for food production was reviewed according to most recent Eurostat data (1999) on food products and beverages manufacturing from the 15 EU member states plus Norway and Iceland. In terms of product volumes, the most significant figures go for animal feed, followed by beverages and dairy products.

A preliminary categorisation of the markets providing opportunities for new products from agro-food wastes was proposed, as shown in Fig 8.

Lastly, the identification of potential commercial products coming from specific sectorial wastes was started, with the collaboration of preferably industrial companies and up-grading companies taking part in the Network. In Fig 9 an example of commercial products coming from the valorisation of winery wastes is shown.

FUTURE WORK

In the second year of AWARENET (2002) we are going deeper into production processes for the evaluation of minimisation strategies and waste valorisation technologies. In addition, a thorough study on the market for new products from agro-food wastes together with technology cost will be accomplished. A wide dissemination of results is planned during the project execution. For this purpose, every six months a Newsletter is distributed amongst interested organisations. This list is growing and requests for wider distribution are welcome. Also a web page in EnviroWindows, the service offered by the European Environment Agency, is available for public consultation at http://ew.eea.eu.int. Within this Interest Group service other documents related to AWARENET activities (technical reports, legislation, publications, etc.) are also available.

REFERENCES:

LIST OF EU LEGISLATION AFFECTING SOLID AND LIQUID WASTES

1. 75/439 Council Directive 75/439/Eec Of 16 June 1975 On The Disposal Of Waste Oils
2. 75/442 Council Directive 75/442/Eec Of 15 July 1975 On Waste. (As Amended By 91/156)
3. 76/464 Council Directive Of 4 May 1976 On Pollution Caused By Certain Dangerous Substances Discharged Into The Aquatic Environment Of The Community
4. 80/68 Measures And Restrictions For The Protection Of Groundwater Against Pollution Caused By Certain Dangerous Substances
5. 90/415 Dangerous Substances Discharged Into The Aquatic Environment
6. 90/667 Council Directive On Disposal And Processing Of Animal Waste
7. 91/156 Framework Directive On Waste 27 Of September 1994
8. 91/271 Council Directive Of 21 May 1991 Concerning Urban Waste-Water Treatment
9. 98/15/Ec (Amending 91/271 Council Directive Of 21 May 1991 Concerning Urban Waste-Water Treatment)
10. 91/689 Council Directive 91/689/Eec Of 12 December 1991 On Hazardous Waste
11. 94/3 Commission Decision Establishing List Of Wastes
12. 94/67 Hazardous Waste Incineration
13. 96/61 Council Directive 96/61/Ec Of 24 September 1996 Concerning Integrated Pollution Prevention And Control
14. 99/31 Council Directive 1999/31/Ec Of 26 April 1999 On The Landfill Of Waste
15. 2000/60 The Water Framework Directive
16. 2000/76 Directive On The Incineration Of Waste

LEGISLATION IN PREPARATION

Proposal for a regulation of the European Parliament and of the Council laying down the health rules concerning animal by-products not intended for human consumption

- Proposed directive on composting

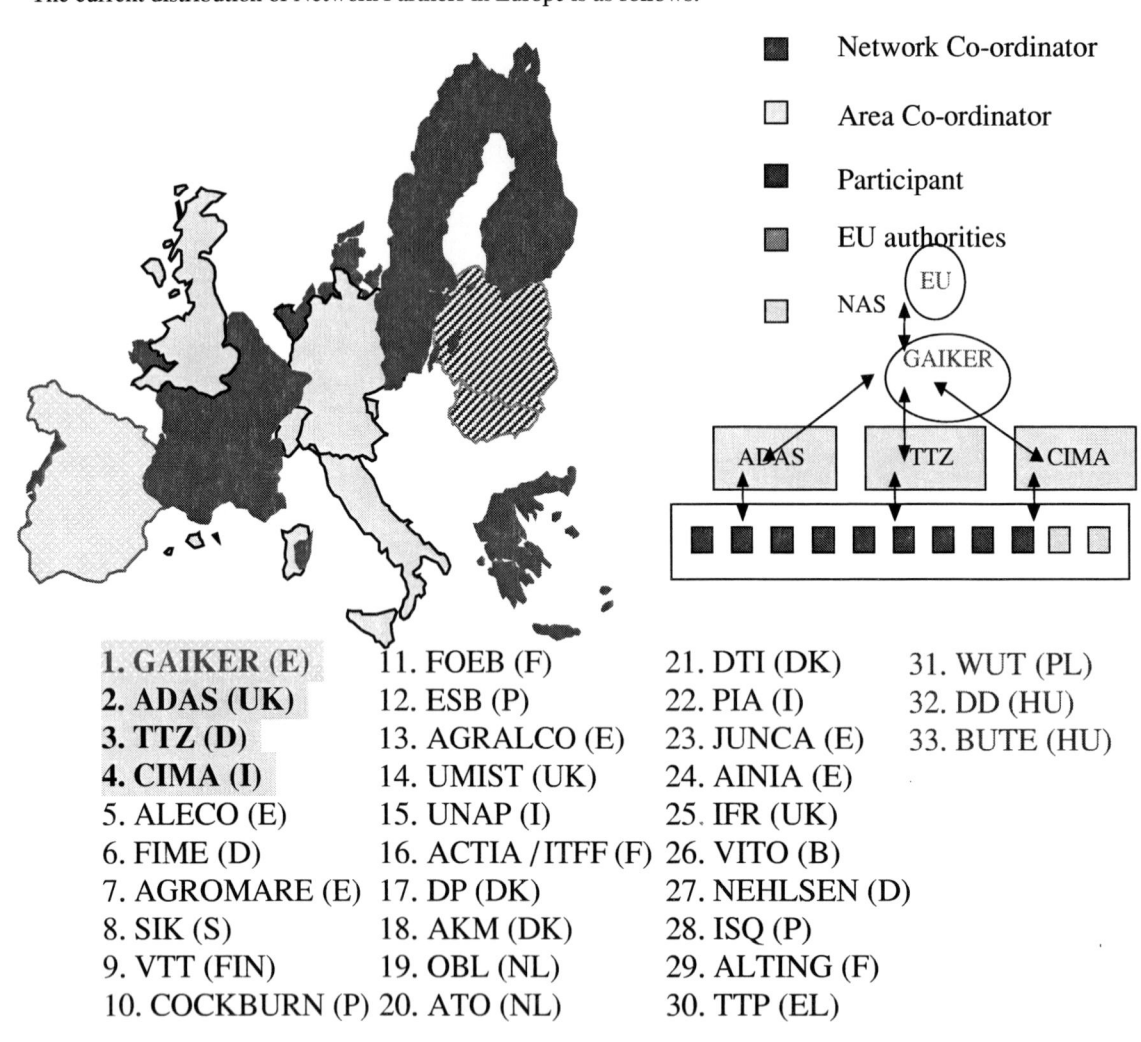

The current distribution of Network Partners in Europe is as follows:

■ Network Co-ordinator
□ Area Co-ordinator
■ Participant
■ EU authorities
□ NAS

1. **GAIKER (E)**
2. **ADAS (UK)**
3. **TTZ (D)**
4. **CIMA (I)**
5. ALECO (E)
6. FIME (D)
7. AGROMARE (E)
8. SIK (S)
9. VTT (FIN)
10. COCKBURN (P)
11. FOEB (F)
12. ESB (P)
13. AGRALCO (E)
14. UMIST (UK)
15. UNAP (I)
16. ACTIA / ITFF (F)
17. DP (DK)
18. AKM (DK)
19. OBL (NL)
20. ATO (NL)
21. DTI (DK)
22. PIA (I)
23. JUNCA (E)
24. AINIA (E)
25. IFR (UK)
26. VITO (B)
27. NEHLSEN (D)
28. ISQ (P)
29. ALTING (F)
30. TTP (EL)
31. WUT (PL)
32. DD (HU)
33. BUTE (HU)

Network coordinator
Area coordinator
NAS members

Fig 1 AWARENET Network Participants across Europe

Fig 2 AWARENET project basic scheme

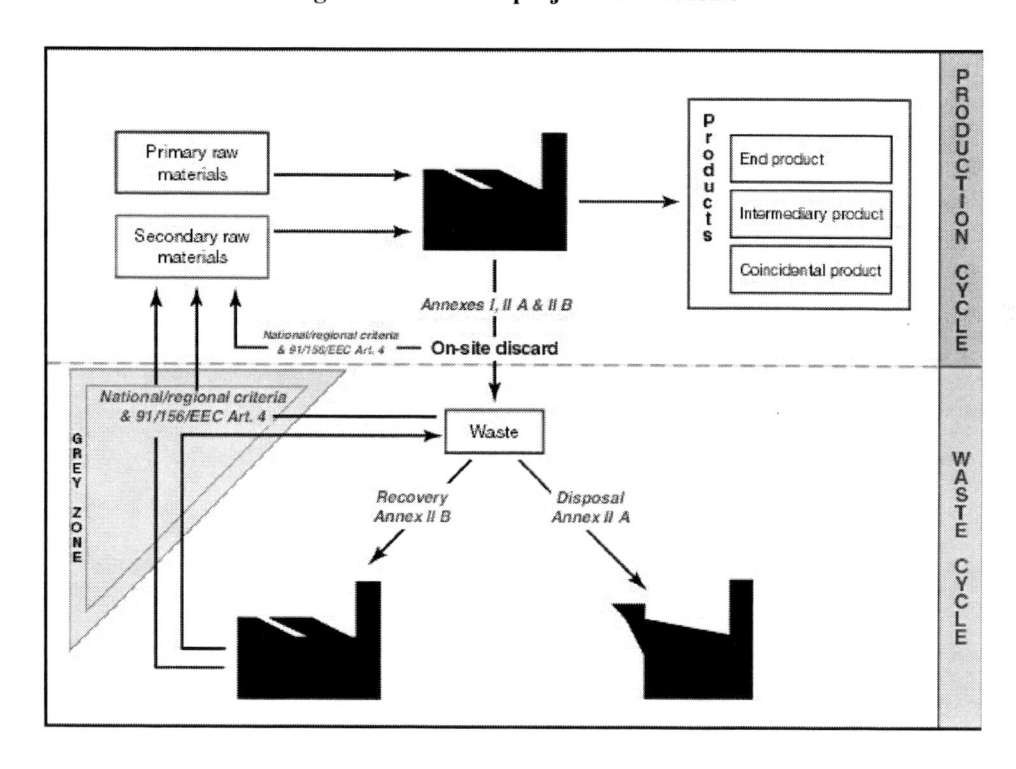

Figure 3 Flow chart showing the relationship between waste, secondary raw material and product according to EC legislation.

The 'grey zone' depicts the area of dispute among the Member States. This is partly due to the lack of a set of criteria to exempt specific recovery operations from general waste management requirements. All references are to articles or annexes of directive 75/442/EEC as amended by directive 91/156/EEC. (Source: Vito, Integrale Milieustudies 2000)

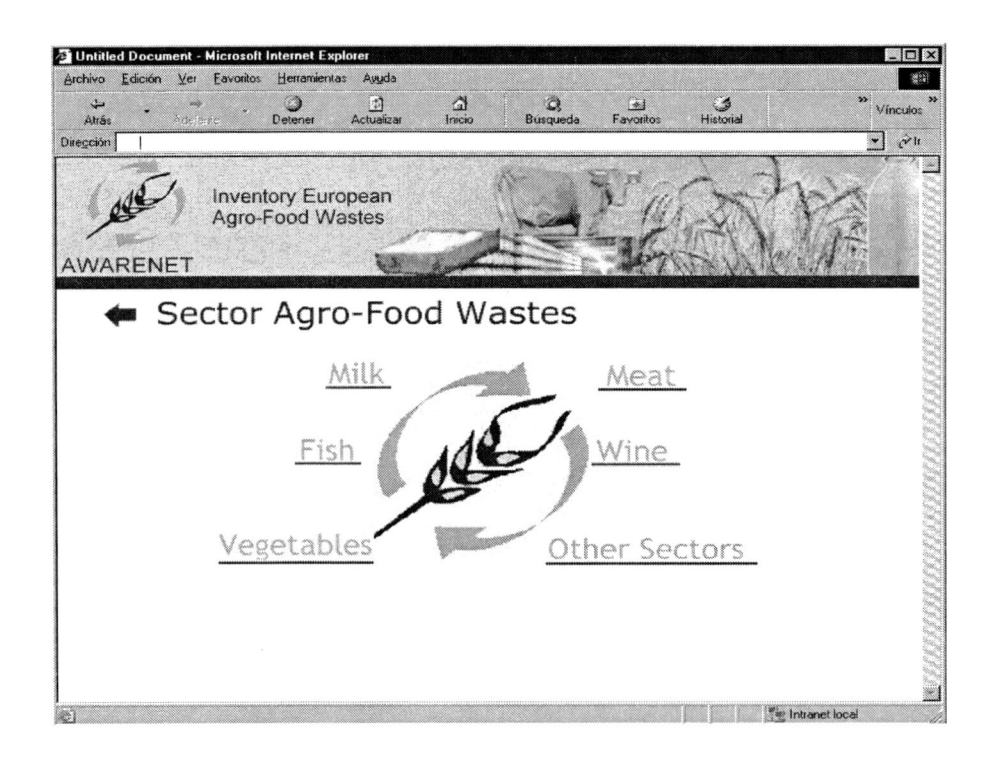

Fig 4 AWARENET software

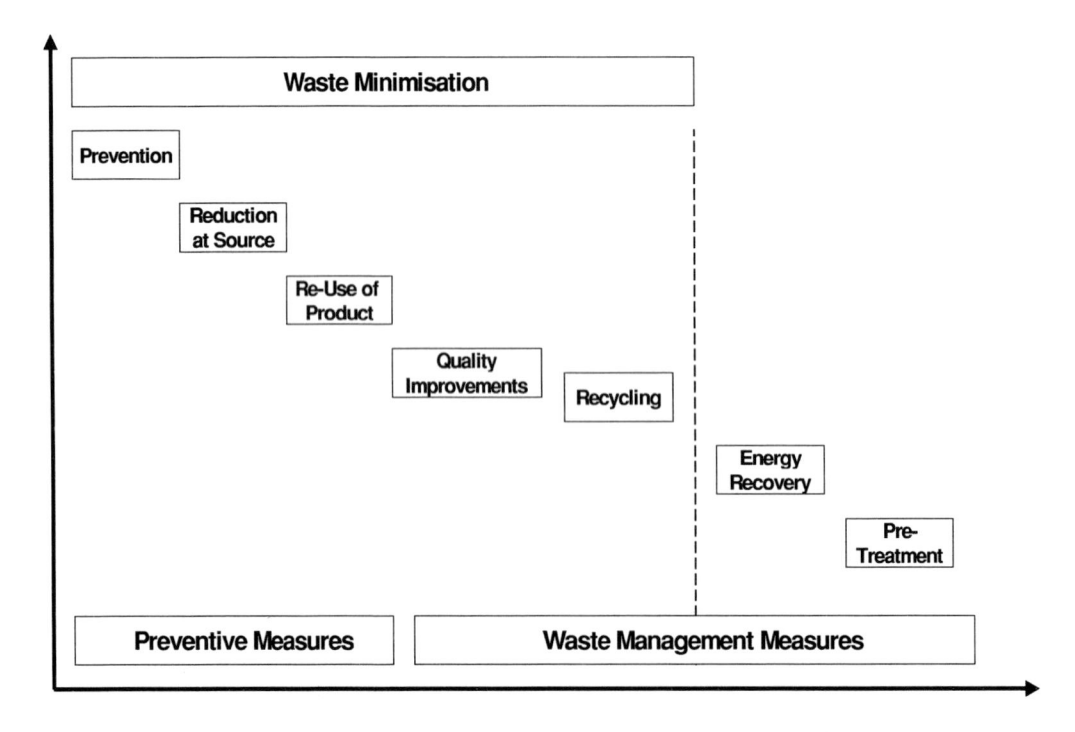

Fig 5 Waste minimisation and prevention according to OECD

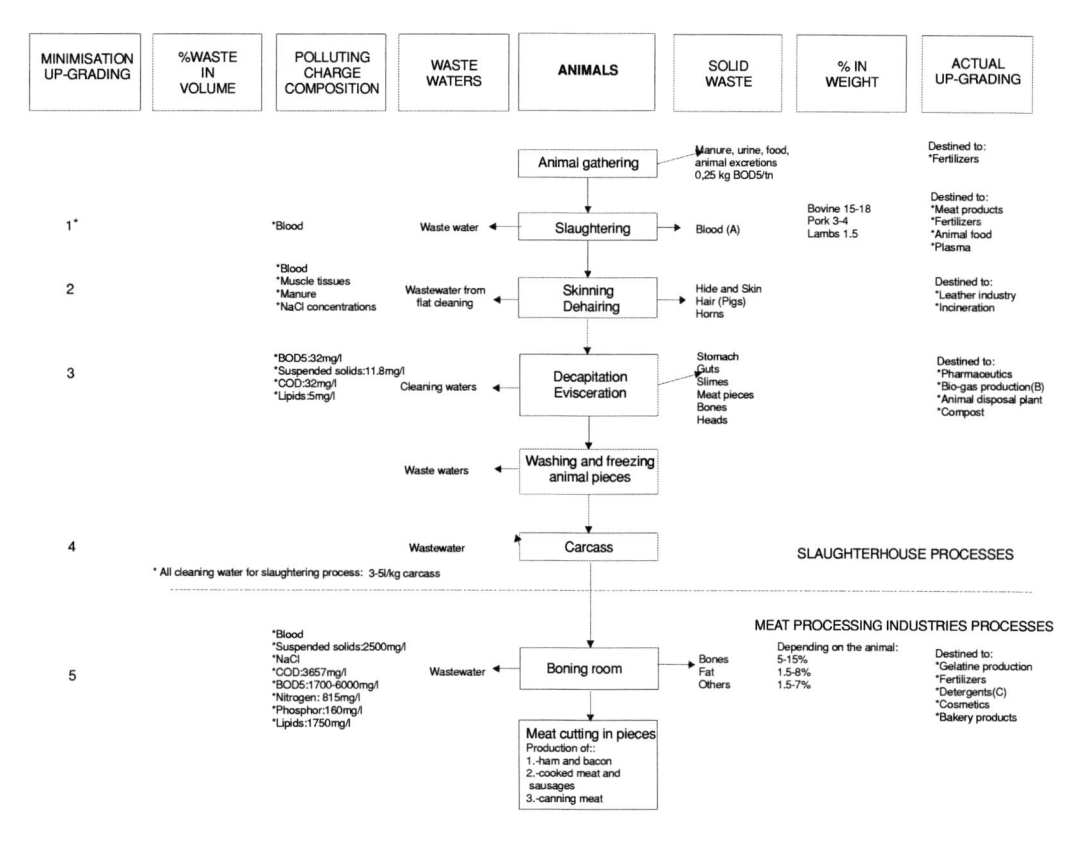

Fig 6 Flow diagram from the slaughtering to the industrial processing of meat

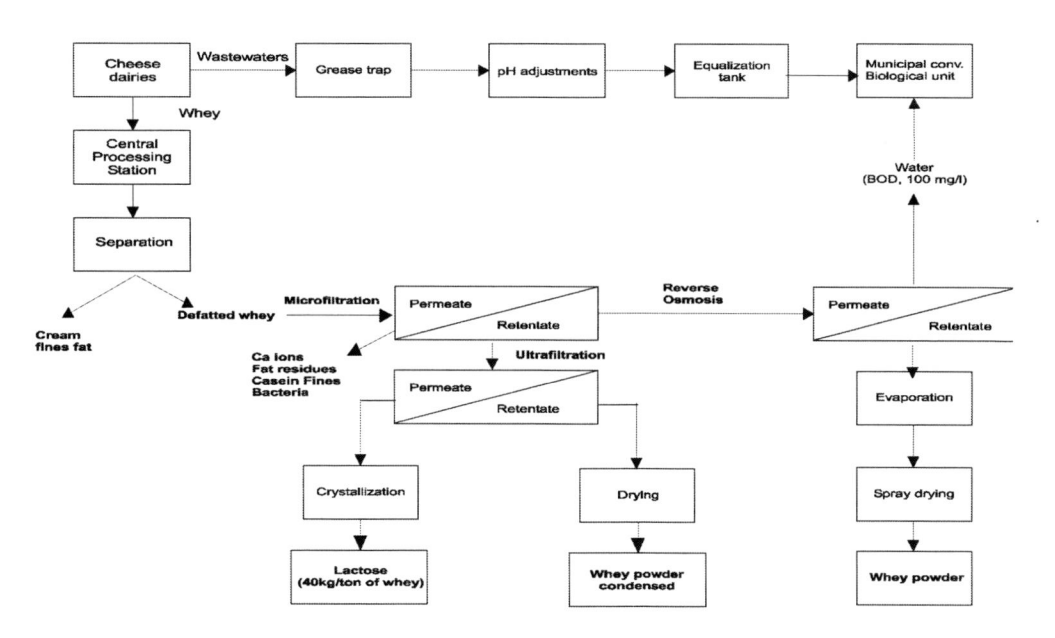

Fig 7 Lactose and whey powder production

	Examples
Established Markets (Scenarios from Inventory)	
Bioenergy: incineration, biogas,	Traditional Biogas
Biofuels: Biodiesel, Bioethanol	
Organic Materials: Compost, Byproducts	
New Markets – high tech driven (from Area 2)	
Substitution of fossil products	New Biodiesel CTER
Production of bio-plastics	
Use of bulk chemicals (Lactic acid)	
Green Markets – driven by legislation (from Area 1)	
Substitution of fossil lubricants (10-15% of consumption	Lubricants
Substitution of PVC by biopolymers	
Production of native tensides and oleochemicals	
Niche Markets –driven by demand (from Area 3)	
Enrichment of Ingredients	Grape seed oil
Production of fine chemicals	

Fig 8 Market categorisation

Agro-food sector	Raw material processed	Production Process	By-products	By-product components	Commercial products	Price and/or disposal costs (euro / Kg/l)
Wine	Grapes		Grape Pomace		Wine alcohol, protein, polysaccharides, oligosaccharides, Lactic acid	4,2-6
					Tartaric acid,	
				Seeds	Grape seed oil	4,3-19,5 Euro/Liter
					Grape seed polyphenols extracts	P100: 265, 9
					OPC (oligomericproanthocyanids)	Final product: 12000-16000
					Resveratrol	Final product: 19250-39703
					Compost, Biogas, Methane	
				Stems	Resveratrol and epsilon-viniferin	100000
				Skins	Anthocyanins	
					Polyphenols	P85 = 245
			Lees (2)		Gypsum hardening retardant	4,21-6
					Catalyst in galvanic processes	4,21-6
					Wine alcohol	
					Tartaric acid	
					Curls	
			Refused grapes		Compost	
			Stillage			
			Cleanup washwater			
			Wine wastes		Stabilized by products through anaerobic digestion	
			Fermentation and settlement tanks wastes		Alcohol	

Fig 9 Commercial products coming from wine wastes valorisation

LIFE ECOBUS: COLLECTING USED COOKING OILS TO THEIR RECYCLING AS BIOFUEL FOR DIESEL ENGINES

Navarro, R.R., Perez, J.M.R.,
Valencia City Council, Spain

ABSTRACT

Paper describes project for the elimination of used cooking oils generated in city of Valencia which cause serious pollution problems by current disposal to sewer. The project development includes citizen awareness campaign, the collecting pilot system for strategic areas, the ecodiesel process and the testing of the fuel in Valencia's city buses.

This paper failed to make the printers deadline.

PPP WASTE MANAGEMENT PROJECTS IN SCOTLAND

Christa M Reekie
Burness, Scotland
Telephone: +44 131 473 6000, Fax: +44 131 473 6006, Email: cmr@burness.co.uk

ABSTRACT

The Landfill Directive imposes ambitious waste diversion targets on Scottish Councils. Since the Councils will have difficulties in funding the implementation of the Directive, many are exploring whether they should enter into contracts with the private sector under the Government's Private Finance Initiative. The following paper deals with some of the issues which the Councils will have to address if they chose to go down the PFI route.

KEY WORDS

Area Waste Plan, Change in Law, EU targets, Income streams, Landfill Directive, PFI, Scotland, Waste Growth

INTRODUCTION

Scotland now produces 3.25 million tonnes of municipal waste per year and the amount dealt with by Scottish Councils is rising every year by between two and three percent[1]. 90% of this is landfilled (compare this with Denmark, Austria, the Netherlands and the Flemish region of Belgium where less than 35% is buried[2]) and it has been accepted for many years that for a variety of reasons this proportion has to be reduced. Sending waste to landfill pollutes the environment, is unpopular and is, ultimately, unsustainable.

On 16 July 2001 the European Landfill Directive came into effect. This requires the UK to reduce significantly the fraction of biodegradable municipal waste (BMW) that is landfilled. Currently 60% of municipal waste is estimated to be biodegradable[3].

The Landfill Directive[4] prescribes the following targets:-
a) by 2010 - reduce the amount of BMW going to landfill to 75% of the total waste produced in 1995,
b) by 2013 - reduce to 50% of the 1995 figure; and
c) by 2020 - reduce to 35% of the 1995 figure[5].

These ambitious targets will be combined with progressive increases in landfill tax, are absolute values and (intentionally) take no account of population growth or any continuing increase in the production of municipal waste.

Due to local government re-organisation in Scotland and the establishment of the Scottish Environmental Agency in 1996, no data on total levels of municipal waste in Scotland were available for the years 1995 and 1996. However, SEPA estimated in the National Waste Strategy report that the figure for 1995 was 3 million tonnes. No specific data relating to individual authority contributions to the 1995 total are available.

While the Landfill Directive seeks to divert large amounts of waste to other forms of disposal, landfilling will obviously continue to play a role, at least in the foreseeable future. Consequently, the Directive also sets out standards to be met by the remaining landfill sites. There are areas of overlap between these standards and those in the Pollution Prevention and Control (PPC) Regime that implements the Integrated Pollution Prevention and Control Directive in Scotland. The Scottish Executive is therefore proposing that the PPC controls will be extended to all landfills in order to provide a uniform approach and assist in meeting the Landfill Directive targets.

NATIONAL WASTE STRATEGY AND THE AREA WASTE PLANS FOR SCOTLAND

The Landfill Directive was one of the main drivers for SEPA's National Waste Strategy. This Strategy was adopted by Scottish Ministers, was published in December 1999, and provides a framework for a reduction of the amount of waste Scotland produces. Its key objectives are:
a) sustainability
b) economic efficiency
c) reduction of the dependence on landfill

The Strategy seeks to encourage the development of efficient and economically viable waste management systems that maximise value recovery from waste and minimise harm to the environment and to human health[6]. In achieving this goal there are four key principles to be taken into account, these being:
a) promotion of the proximity principle and self-sufficiency in the management of Scotland's waste. This principle requires the waste to be managed as close as possible to the point of production with the aim of avoiding the adverse effects of unnecessary transportation. However, there must always be a balance between the proximity principle and economies of scale.

Self-sufficiency aims for identifiable groups to deal with their own wastes.
b) application of the precautionary principle[7]. This aims to identify actions to be taken now in order avoid potential environmental damage in the future, even in cases where there is insufficient evidence to prove that damage will occur.
c) polluters should bear the full cost of their actions - "the polluter-pays" principle. This principle does not extend just to those discharging a product but also to those who manufacture it. This concept of the producer taking responsibility may require a manufacturer to take a product back for recycling or make arrangements for others to do so.
d) implementation of the waste hierarchy. This provides a framework for the most desirable waste management options:
i. waste prevention. The reduction and minimisation of waste should always be the first consideration of waste managers.
ii. re-use. This can reduce consumption of energy and raw materials, reduce disposal needs and result in cost saving.
iii. recovery. This has similar benefits to re-use. There are three main methods; recycling, composting and energy recovery (eg. incineration or gasification).[8]
iv. landfill disposal. This method of disposal will continue to play some part as there will always be some material that cannot be used for another purpose and is unsuitable for disposal in any other way.

The above hierarchy of waste options has to be considered in the context of the Best Practicable Environmental Option Guidance published by SEPA in September 2000. This sets out the decision making process to be used in order to establish how waste streams will be managed in the future.

The four principles should always be used together to meet the ultimate goal, which is to ensure that waste is disposed of through an integrated and adequate network of waste management installations without endangering human health or harming the environment.

The administrative framework for achieving the objectives is through eleven Waste Strategy Areas (WSA), with each one producing its own waste plan. Each area has a Waste Strategy Area Group (WSAG) that is responsible for detailing current systems and data and for preparing the Area Waste Plan. The WSAGs include representatives from the local authority, local enterprise companies and SEPA.

The purposes of the Waste Area Plans are:

a) to define the type of facilities required for the sustainable management of controlled waste in the Waste Strategy Area;
b) to demonstrate that these facilities are the best practical option in terms of economic, social, environmental and human health issues; and
c) to provide a framework for meeting the targets.

The plans are to be created bearing in mind not only the aims of the Landfill Directive, but also the obligations under the Environment Act[9], which also sets objectives regarding sustainability.

The Area Plans are now in their final stages and are being reviewed by SEPA with a view to integrating them in order to ensure sustainable waste management throughout Scotland. An Integrated National Waste Plan should be produced shortly. It is essential that Waste Strategy Areas do not consider local waste in geographical isolation and, while the proximity principle means that local solutions should be used where possible, there will be occasions where waste will have to be transported between WSAs. SEPA is responsible for making sure that the plans are compatible and that the systems do not conflict or overlap. The question that must always be asked is: taken as a whole, is Scotland, as far as possible, self sufficient in terms of waste disposal?

That said, Scotland has a very diverse mix of population densities and topography and careful thought will have to be given as to how waste management facilities can be developed to meet both local and national needs. For example, in central Scotland there would be sufficient waste to support a large facility, but large developments that pull in waste from far away may prove unpopular and it has been suggested that there is a need to focus on the provision of a range of smaller facilities.[10]

Any new waste facilities, regardless of size, are going to take time to plan and develop and therefore consideration must be given to future demand in order that there will always be sufficient capacity and competition in waste management services. SEPA aims to ensure that at any time the network in place exceeds current demands and has forward capacity for a period of between 5 and 10 years.

PFI (PRIVATE FINANCE INITIATIVE)

The cost of implementing the Landfill Directive will be substantial. Municipal waste management services in Scotland are currently funded by local authorities through direct funding from the Scottish Executive and council tax payments. The need to comply with the Directive, amongst other new legislation, has encouraged authorities to consider PFI (Private Finance Initiative).

The principal objective of PFI is to involve the private sector in the provision of public services, with the expectation that in this manner better services can be provided without increasing the strain on public funding or having to increase taxes. In waste management PFI projects the public sector becomes the purchaser of waste management services (usually on the basis of a 25 year contract), with the private sector assuming the responsibility for the design, construction, operation, maintenance and finance of the waste management facilities. The public sector purchases a service rather than capital assets.

The standard of the waste management services provided by the private sector is determined by law, SEPA and the purchasing authority. The private sector determines how to deliver the services and therefore any technical solution employed remains at the risk of the private sector. The public sector pays for those services, usually on the basis of a gate fee per tonne of waste. This gate fee includes payment in relation to environmental compliance, landfill tax costs[11] and disposal of the residue, as well as financing costs (senior debt and equity). If the services are not provided in accordance with the specifications the authority is entitled to make deductions from the payments made to the private sector. The deduction regime is contained in the Payment Mechanism that lies at the heart of each PFI contract.

A typical contract structure for a PFI project is set out in Appendix A.

CHANGE IN LAW

Another key driver for PFI is the allocation of risk, that is the transfer of risk to the party who is best placed to manage a particular risk. However, it is accepted that the public sector will be responsible for any changes in law that are specific to the waste management sector after the contract is signed. This means, of course, that over a 25-year period (the typical length of a PFI contract) the public sector will have to pay for such changes.

The waste management sector, more than any other sector that employs PFI (other than, perhaps, the waste water sector), will be very prone to changes in law. For example, landfill tax levels are set until 2004, but what happens thereafter is unknown. There are also a number of new waste directives such as the End of Life Vehicles Directive and the Electrical and Electronic Waste Directive that will have an impact on the disposal of waste in Scotland.

As mentioned above, the public sector authority is responsible, after contract signature, for meeting the costs of changes in waste management sector law. However, PFI rules dictate that the private sector is responsible for any changes in law that are known or foreseeable at the date of contract signature. Therefore, before the parties enter into the PFI contract, much discussion centres on the question of foreseeability. In general, the private sector will only want to be responsible for paying for foreseeable changes if such changes are quantifiable as well as foreseeable.

The role of the engineer advising the parties (including the private sector funders) is crucial because very often the legal position is not entirely clear. For example, guidance from SEPA may not be sufficiently developed to base a technical solution on it, and to price such solution. The question of foreseeability is a hotly debated issue that has a significant financial impact on the project - the battle here is really fought at the technical level!

WASTE GROWTH

As stated above, the total waste produced in Scotland is currently rising by at least 2-3% per year, but this rate of growth may change. The authorities will seek to transfer to the private sector the risk of this rate of growth being greater that that predicted. Therefore the private sector will be faced with having to meet targets despite the future volume of waste being unknown. As a consequence, any technical solution will have to be flexible enough to deal with volume fluctuations, including increases that go beyond those predicted. The challenge to the engineer is to devise a technical solution that is affordable to the public sector (after all a private sector party only wins the project after lengthy competitive bidding) but which has enough spare capacity to deal with unexpected increases.

Of course, the private sector will not be prepared to take on an open ended risk - it will not be permitted to do so by its funders and equity providers. There will therefore come a point where certain aspects of the contract will have to be re-negotiated. PFI contracts will normally include a suitable mechanism to deal with this - the engineers advising all parties should be aware of these provisions and should check them to make sure that they make sense in practice.

EU TARGETS

It is key to the PFI contract that the private sector meets EU targets. Should it fail to do so the authority will seek to penalise the private sector by making it responsible for the payment of landfill tax and the costs of tradable permits. However, the private sector will seek to share in any income from any tradable permits derived by the authority, in the event that the private sector manages to divert more waste than required by the EU targets.

SHARING INCOME STREAMS

Waste management, perhaps more than any other area, relies on a partnership between the public and private sectors, particularly in projects where the pubic authority remains responsible for the collection of household waste. Typically, the authority will seek to share in any income derived from recyclable materials particularly where it has instigated, through authority led initiatives, the collection of such materials. Similarly, it will wish to share in any income from commercial waste beyond the income forecast in the project's financial model. Changes in technologies, or in the collection of waste, that lead to savings in the cost of the project are usually also shared.

However, should circumstances alter, e.g. where a change is not sustainable throughout the contract period, there have to be coping mechanisms in place. Having said that, it is only natural for the public sector to wish to share in any upside of the project, but to leave the private sector to bear any downside in full. In general, therefore, the sharing arrangements are usually not equal.

CONCLUSION

The onward march of legislation has made it essential that Scottish waste management practices change. However government finances, or rather the lack of these, have forced local authorities to turn to the private sector to meet the new requirements, and the only currently available mechanism that permits this is PFI.

A PFI project comprises two parts, the technical solution and the contract between the public and private sectors. The contract will be long-term, typically 25-years, and must contain provision for every foreseeable circumstance and a mechanism for providing for circumstances as yet unforeseen. Negotiation of such a contract will be time-consuming and painstaking, but precedents exist and a skilled and experienced legal team will be able to minimise the time and expense involved.

REFERENCES

[1] Growth in municipal waste appears to be linked to trends in population, households and broader economic and socio economic rends. The National Waste Strategy Report (SEPA November 1999 – National Waste Strategy, Scotland) reports that the rate of growth of household waste recorded in Scotland between 1989 and 1998 has been, on average, less than 2% per annum. However, the average in household waste in higher than the rate of household growth suggesting that the average quantity of waste produced by householders is also increasing.

[2] Figure supplied by the European Environment Agency.

[3] The percentage is based on the waste composition data collected under the Environmental Agency's National Household Waste Analysis programme which was designed to better understand the composition of collected household waste. – "Local Authority Waste Management Costs Study" – Enviros Aspinwall Scottish Executive Central Research Unit 2000

[4] The Landfill Directive has been implemented by the Landfill (England and Wales) Regulations 2002 and in Scotland, it is due to be implemented by the Landfill (Scotland) Regulations 2002. Much of the associated guidance by the Environment Agency in England and the Scottish Environment Protection Agency (SEPA) remains to be completed.

[5] The Directive, Article 5, allows the target dates for achievement of the progressive BMW diversion requirements to be extended by up to four years by Member States that landfilled more than 80% of their municipal waste in the baseline year. Since the UK landfilled in excess of 80% of its Municipal Waste in 1995, it qualifies for this derogation and if the derogation is used, each of the target dates for the progressive diversion of BMW will extend by 4 years. The above figures show the derogated target dates.

[6] The National Waste Strategy, Scotland Report 1999

[7] Identified by the UN confederation on Environment and Development in the Rio Declaration as "where there are serious threats of serious or irreversible damage, lack of full scientific certainty shall not be used as a reason for postponing cost effective measures to prevent environmental degradation".

[8] Note there are already obligations to recover and recycle in place under the Producer Responsibility Obligations (Packaging Waste) Regulations 1997 as amended. The targets for 2002 are 59% recovery and 19% material specific recycling of packaging waste.

[9] Environment Act 1995, Schedule 12

[10] The National Waste Strategy, Scotland 1999

[11] This risk is usually shared with the authority taking the risk of increases in price.

FIGURES

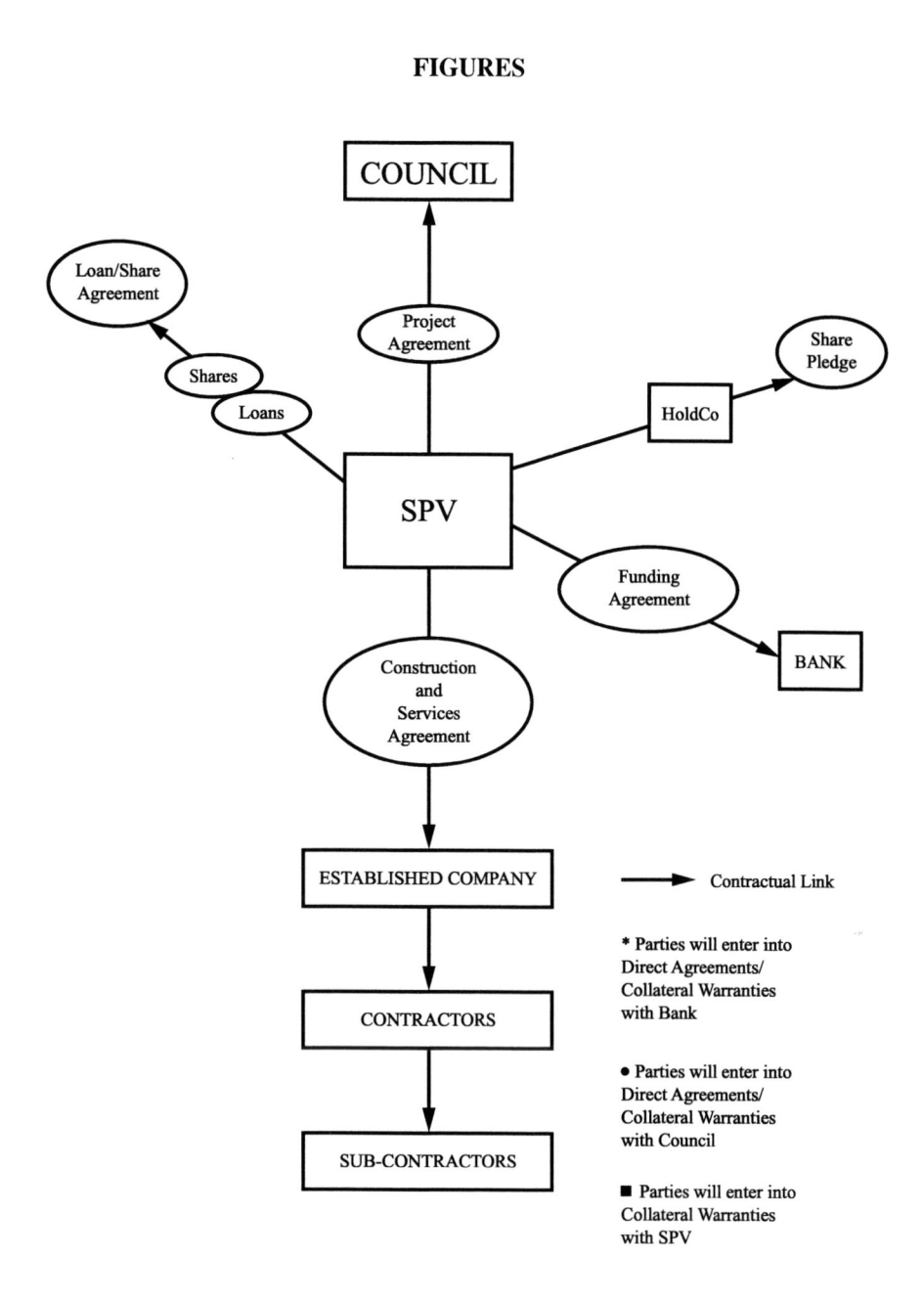

Figure 1: PFI Relationships

USE OF BIOSOLIDS TO PROTECT THE MACHAIR, AN INTERNATIONALLY IMPORTANT HABITAT.

A McQueen and M Williams
Scottish Water, Bullion House, Invergowrie, Dundee, DD2 5BB, UK
Tel. +44 (0) 1382 563231 Fax. +44 (0) 1382 563161
Email. Andy.mcqueen@scottishwater.co.uk

ABSTRACT

This paper provides an example of how a potential problem of managing sludge in a remote islands community has been transformed into a positive benefit, tailored specifically to that environment.

North West Scotland and the Western Isles host one of the most biologically diverse and rare habitats in the world. The Machair is a unique combination of dune system and associated pastureland, located at coastal margins. It is also a working environment, supporting an active crofting community and providing important common grazing lands for sheep and cattle. Machair habitats establish themselves on the highly calcareous sandy sediments, formed from shells driven onshore and over the grasslands by the prevailing water and wind action. However extreme wind erosion can create blowouts of sand, which threaten the local habitat. Scottish Water has worked closely with Scottish Natural Heritage to develop the use of biosolids to stabilise and re-instate areas of eroded Machair. This will be of long-term benefit to this internationally important habitat.

KEY WORDS

Biosolids, Erosion, Machair, Lime Treatment, Scottish Water, Sludge, Western Isles,

INTRODUCTION

The Western Isles, formerly more widely known as the Outer Hebrides, lie off the North West coast of Scotland. There is a resident population of 27,500, of which approximately 80% is connected to the public sewerage system. The population outside Stornoway, is widely dispersed in many smaller rural crofting and fishing communities. Fish and shellfish farming are also carried out in both sea and freshwater lochs. Stornoway, with a population of 7,500, is the main market town, ferry port, local government and administrative centre, with typical support services. Major trade effluents arise from two salmon processors, a woollen mill, and a shellfish processor. Wastewater treatment was provided for the first time in 2001, in accordance with the Urban Wastewater Treatment Directive. Prior to this, sewage was discharged untreated into the sea via outfalls, and therefore no sludge was produced in coastal areas. Sludge from inland public septic tanks was collected and discharged or pumped out to sea via outfalls, until 1998, when the Directive also banned the disposal of sludge to sea.

After 1998, sludge was stockpiled at two thickening centres and then treated using mobile liming equipment, to produce enhanced treated biosolids. The intention was to recycle the biosolids product to farmland. However In 1999, an innovative project was conceived to restore areas of eroded Machair land, as described later in this paper. The project benefited from grant aid from the European Regional Development Fund. Sludge production increased significantly when the Stornoway Wastewater Treatment Plant was commissioned.

Although sludge production is still relatively small within a UK context, i.e. approx. 1,000 tons p.a., sludge management still presents considerable challenges in such a remote islands environment. The same regulations have to be complied with as in more populated and accessible mainland areas. Economies of scale are non-existent, and accordingly only one full sludge treatment centre is now feasible, for an archipelago of many islands, 12 of which are inhabited, stretching in a line almost 130 miles in length. Ferry services are still required to travel the length of the islands group, and there are still significant lengths of single-track roads with passing places. It follows that transport alone incurs significant costs.

The Machair restoration route taken by Scottish Water, is more expensive than some of the alternative outlets for sludge. However, bearing in mind the hostage to fortune that biosolids has been over recent years, how does one value such wide public and stakeholder support for a sludge management practice? Many feel that this project is a leader in changing public perception and attitude towards biosolids.

THE PROJECT.

The remote coastal areas of North West Scotland and the Western Isles host one of the rarest habitats in the world. The Machair lands of the Western Isles and certain parts of the mainland are heritage sites of international renown, notable both for biological diversity and also for prehistoric and early historic archaeology. As such, the Machair lands contain many SSSI, heritage site and National Nature Reserve designations, and are proposed as one of the Scottish sites for Natura 2000 under the EC Habitats directive.

The term 'Machair' is applied to the flora and fauna occupying the dune systems and associated pastureland at the coastal margins The habitat hosts many rare species including birds such as the corncrake, rare orchids, including species endemic to the islands and other wildflowers, and insects.

In addition to its importance as a habitat of global importance, the Machair is a working environment, supporting local crofting communities and cattle grazing. Indeed, these activities have, over thousands of years, helped to promote the biological diversity and beauty of the landscape, as well as providing a living for local people. In more recent times, the Machair is of value to the community as a tourist attraction, with people choosing to visit the islands to enjoy the natural heritage of the landscape.

PROBLEMS FACING THE MACHAIR

Machair habitats establish themselves on the highly calcareous sandy sediments, formed from molluscan and crustacean shells that are driven onshore by the prevailing water and wind action. The 'shelly' sand is blown over the grasslands of the Machair plains, and the action of the wind continually erodes and rebuilds the Machair landscape.

Wind erosion is a natural and essential component of the Machair system, however extreme erosion can create blowouts of sand, which threaten the local habitat. Of particular concern are the areas surrounding sites of coastal quarrying (necessary because of the lack of suitable alternatives), which create focal points for blowouts of large areas. As the Machair is largely impacted by wind erosion rather than wave deposition, these areas seldom receive sufficient material to infill adequately.

Infilling of these blowouts with rubble and boulders is rarely successful, as they can lead to re-erosion of the site following reinstatement. Additionally, merely dumping solid material into a hole channels the wind and causes further erosion. The use of organic matter in stabilising and reinstating the land is very important as the Machair soils are typically shallow, and the erosion process removes much of the topsoil. Traditionally, Machair lands were stabilised with organic mulches such as seaweed, augmented by cattle dung, but these practices are now in decline.

USE OF 'BIOSOLIDS' TO STABILISE AND REINSTATE THE MACHAIR

Scottish Water and its predecessor organisation, the North of Scotland Water Authority, has been actively engaged over recent years in helping to protect and enhance the Machair habitat through recycling of biosolids. In seeking a local solution to the problem of what to do with the increasing quantities of sludge produced as a result of more effective sewage treatment, local operational staff recognised that the organic content of the sludge could be beneficially re-used on Machair lands. To this end, a pilot project was established to monitor the impact of recycling the sludge to areas of the Machair that had been damaged by the blowout phenomenon described above. It was established at the outset, that to ensure public acceptance, only a treated form of sludge, more commonly referred to as biosolids, could be used.

Scottish Natural Heritage (SNH) has worked closely with Scottish Water to develop this programme of using biosolids to stabilise and re-instate areas of eroded Machair in the Western Isles. This will be of long-term benefit to the environment and in particular to the special areas of Machair land which provide habitat for the internationally important wildlife of the Islands.

BIOSOLIDS

The biosolids used are produced from sludge, which has undergone an 'enhanced' level of treatment, to ensure a 6-log reduction in bacteria, and elimination of any Salmonella. A variety of mechanisms may be employed to do this, but on the Western Isles, these levels of treatment are achieved using lime stabilisation using the RDP process. The sludge is thickened on a belt press and heated to 55°C before being passed into the blending unit. Here the minimum amount of lime is added and blended with the sludge, which serves two functions. First, the exothermic reaction of hydrating the lime raises the temperature of the mixture to over 70°C, maintained for at least 30 minutes, which achieves the microbial kill. Second, the pH of the sludge is elevated to over 11.5, preventing re-infection and giving a stable, comparatively odourless, biosolids product that is easy to handle and to spread using standard agricultural pellet spreading machinery.

RECYCLING

Scottish Water and SNH jointly identify areas requiring stabilisation or reinstatement, and which might be suitable for recycling of biosolids. The decision on whether to proceed is then made after consulting with landowners, crofters, Western Isles Council archaeologists and environmental health departments, and the Scottish Environment Protection Agency. If these parties agree, then a detailed land survey is undertaken and soil samples taken, which are sent to Scottish Agricultural College for analysis.

Prior to recycling, the area is landscaped to reduce wind channelling, and to ensure the best chance of recovery of the site. Where necessary, for example because of archaeological remains, only light machinery is used on these sites. The biosolids is then recycled according to recommendations up to a maximum of 50 tonnes per hectare.

RE-SEEDING

Following recycling, the area of Machair to be re-instated is fenced off to prevent grazing animals entering. Extensive rabbit fences are also incorporated as they can wreak havoc on recovering sites. The site is then re-seeded by contractors using a seed mixture designed by SNH to mimic the natural Machair composition. Local cattle are then prevented from grazing for at least two years, and regular checks are made to ensure that rabbits do not infiltrate the site. There is optimism that the fencing will show improved breeding success of ground nesting birds such as Corncrake, as predation of eggs by hedgehogs should also be reduced.

All recycling, reseeding, fencing and monitoring costs are borne by Scottish Water.

CONCLUSION

The early trials of using biosolids to re-instate and stabilise the Machair habitat has proved to be very successful. Preventing grazing for at least 2 years enables a good turf of Machair to form, and this improves over time as Marram grass vegetatively propagates into the area. It is estimated that Marram infiltrates the re-seeded areas at a rate of 2-3 metres per annum. This will ensure that the site continues to stabilise and is less prone to being blown out again.

BENEFITS

It is apparent from the results to date that there are clear benefits to the use of the biosolids in stabilising and re-instating the Machair. This protects the integrity of the dune/coastal pasture system, and helps to reduce the impact of the blowouts. It also provides much needed organic enrichment to the soils.

In addition, recycling provides a local solution to finding outlets for the biosolids, in accordance with the proximity principle. The waste hierarchy and best practicable environmental options for sludge dictate that we should seek to re-use the waste where possible, and this is reinforced by increasing pressure to reduce volumes of waste landfilled, and reduce emissions from incineration. The ongoing monitoring, regulation by SEPA and environmental health, allied to the partnership work with Scottish Natural Heritage and archaeological interests ensures that this is a safe and sustainable outlet for Western Isles biosolids.

SLUDGE MANAGEMENT IN THE REPUBLIC OF IRELAND
Crowe, N.*, Delaney, N.**,
ENTEC Ltd, **Nicholas O'Dwyers, Ireland

ABSTRACT

The National Development Plan for the period 2000 to 2006 allocated €3.8 billion for water and wastewater projects. To date four out of approximately 30 of the main sludge treatment plants have passed tender stage. The potential market is significant and just emerging. Wastewater and water services are provided in the Republic of Ireland by 34 local authorities comprising 29 County Councils and 5 City Councils. Sludge management plans have been developed on a county by county basis. Strategies have been developed for treatment and beneficial reuse of sludges based on economics and sustainability. For the wastewater sludge these range from composting and coppicing to drying with energy recovery. These new facilities will generally take three to four years to procure and this means that many counties will require interim sludge management strategies. Procurement of the required new facilities is currently underway and is generally via design and build with 20 year operating contracts. This is a new approach in Ireland designed to pool expertise from the wider European and international market and achieve value for money. Striking the right balance of liabilities and risk transfer is important and creates new challenges.

This paper failed to make the printers deadline. It should be available on the conference CD

PARTNERSHIP FOR SUSTAINABLE USE OF ORGANIC RESOURCES ON LAND

Evans, T.D.[1] Lowe, N.R.[2]

[1] TIM EVANS ENVIRONMENT, Stonecroft, Park Lane, Ashtead, Surrey, KT21 1EU, England
[2] Enviro Consulting, 6 Tai Canol, Llangorse, Brecon Powys LD3 7UR, Wales.
Tel: 01372 272 172, Fax: 01372 272 172, email: tim.evans@messages.co.uk

ABSTRACT

Perception is reality. Compliance with regulations (whilst essential) is not enough nowadays; public and stakeholder attitudes can be of decisive importance. Sometimes policy-makers speculate what public attitudes might be without really asking them. Technically, most people agree that conserving soil organic matter and completing nutrient cycles by applying animal manures, treated organic wastes and biosolids to land is the most sustainable option in the majority of situations. There has been a huge amount of research into the hazards, and this has concluded that the risks can be managed to acceptable levels. But communicating this knowledge is often neglected, as so often in the scientific and technological arena. This paper will describe an initiative to create a partnership between the many groups with an interest in the use of organic materials on land to develop consensus on good practice and to share knowledge. It summarises the work that has been undertaken to date to establish the attitudes of more than 140 organisations. The information presented was debated at a workshop in July 2002. The paper will describe the conclusions of this study and the steps to establishment a partnership organisation, its aims and objectives, the work to date and the plans for the future. The Environment Agency considers this very important and has largely funded the work to date.

KEYWORDS

Advocacy coalition framework, biosolids, compost, manure, partnership, recycling, sludge, stakeholders, sustainable, use

INTRODUCTION

Human activity produces organic residuals that can be regarded as resources. These comprise manures and slurries from farm animals, residuals from processing crops, non-hazardous industrial wastes, the organic fraction of municipal solid waste and sewage sludge. In general, governments advocate that whenever possible these resources should be used on land (with the proviso that hazards are managed) to conserve organic matter and complete nutrient cycles.

There are many parties who directly or indirectly influence the 'plough to plate' chain. Members of Society longer accept the best technical solution being imposed upon them. Democracy and improvements in communication of information mean that the opportunities to effectively question, object and oppose are easier than they have been in the past. Some objections are well founded and very valuable, but others are be the products of lack of knowledge or misunderstanding; this does not make them any less real to the people who hold them.

Once an argument moves from rational to emotional it becomes much more difficult to change opinions. The case of the best fate for the "Brent Spa" (which was an oil storage buoy) when it reached the end of its productive life in 1991 is a good illustration. Its six tanks could collectively store 43,000t crude oil, which it received from oilrigs on the Brent oil field in the North Sea, for collection by oil tankers. Shell, the oil company that own it, analysed 6 different decommission strategies and decided that the option with least environmental impact would be to remove as much oil as possible, tow it into the Atlantic and sink it off the continental shelf. In 1995 the UK government endorsed this choice. Greenpeace campaigned that this would be very damaging and that the platform should be dismantled and processed on land. It launched a campaign for people to boycott Shell fuel stations; this was so successful that the company followed Greenpeace's wishes. It later transpired that Greenpeace's case was founded on a huge overestimate of the amount of oil and by-products that would be left on the platform, and that Shell's analysis of the relative the environmental impacts of the options was correct. On 5th September 1995 Greenpeace apologised to Shell for sampling errors and admitted to the inaccuracy of its claim. The situation would probably have been avoided if Shell had discussed its best technical solution with stakeholders in advance so that the issues could be examined objectively[1]. The rights and wrongs of the case are not relevant to this paper, but it is an example that once the emotional genie has been let out of the bottle it cannot be returned and that proactive engagement in dialogue therefore has advantages.

The Environment Agency of England and Wales launched its 'Vision for the

Environment' in January 2002[3]. The Vision focuses on environmental outcomes such as: 'Waste will be regarded by both industry and consumers as a potential resource, with the efficient reuse and recycling of materials the social norm'. Working with others and forming partnerships to create shared solutions is identified by the Vision as being essential to achieving these outcomes. To some extent inspired by the example of the National Biosolids Partnership (NBP) in America, the Environment Agency decided to investigate the potential benefits and feasibility of an Advocacy Coalition Framework (ACF) to ensure quality and to promote good practice for use of organic resources on land. There is little tradition of such ACFs in the UK but in the USA they are established as an effective mechanism of achieving concerted action. This paper will report the results of a scoping study for an AFC for organic resources in the UK and its results.

SCOPING STUDY

INTRODUCTION

It was recognised that there are some key organisations without whose endorsement an ACF could not succeed. In addition to the Environment Agency, these were the Department of Environment, Food and Rural Affairs (DEFRA) the Food Standards Agency (FSA) the National Farmers Union (NFU) the Country Land and Business Association (CLA) and the British Retail Consortium (BRC). Accordingly the idea was discussed with representatives of each of these and they gave their support in principle.

The NBP comprises the USEPA (US Environmental Protection Agency) WEF (Water Environment Federation) and AMSA (Association of Metropolitan Sewerage Agencies). It has received $1 million per year from Congress and has been exclusively devoted to biosolids and initially advocating adoption and application of the federal rule[4] in all of the 50 States and promoting good practice. More recently it has devoted much of its resource to establishing an Environmental Management System for biosolids (EMS) with independent auditing. It was not imagined that the funding of a UK ACF would be on a similar scale and it was also the intention for the ACF to encompass other organic residuals besides biosolids.. Some friends in the NBP had observed that one of its limitations is that it is a discussion "amongst the choir" it does not have representatives of the wider community of stakeholders. On the basis of this and experience described below it was felt that the proposition for ACF should be something that is more inclusive.

We already had experience of invited stakeholder engagement regarding the use of wastewater biosolids on land in the UK. This had led to a pro-forma contract between landowners and suppliers of biosolids and to the 'Safe Sludge Matrix'[5]. The use of farm animal manures is following the precedent of the sludge matrix; it is possible that composted materials and other wastes are likely to follow eventually. The negotiations that led to these agreements were conducted with 'lead stakeholder organisations' on the presumption that they represented the interests of a wider group. The participants in these discussions have honoured their agreements but some stakeholders who were not directly involved in the discussions have not bought into the outcome; this has significantly undermined confidence. We realised that an ACF needs to be both a continuing and an inclusive activity that is open to anybody with an interest in the subject.

It was interesting that the stakeholder negotiations arrived at consensus on good practice for the use of sewage sludges that were based on science and went beyond some of the prescriptions of existing legislation whilst accepting the adequacy of others. The provisions of the voluntary standards arrived at by consensus which included government and Agency representation have since been used by government as the basis for revising legislation. A vision for the ACF for organic resources is that it too might contribute to a consensus led approach to legislation, which would have the confidence of the majority because they had shared in its development. In this context the idea was discussed with the Director of Directorate-A 'Sustainable Development & Policy Support' of DG Environment at the European Commission who welcomed the initiative.

CONSULTATION

The next step in the study was to approach as many organisations as possible by phone, email and post to test their interest using a questionnaire (Table 1). The inclusion of a selection of keywords to be scored was a later revision and consequently was only sent to a minority of those questioned. 142 organisations, covering the whole of the UK, were asked to give their views on the proposed organisation. Some of the organisations were trade associations; others were individuals. Some members of trade associations said that their association could be an unwelcome filter and that they wanted to participate as full members. Indeed some went so far as to say that they feared the Partnership would be an irrelevance if membership were restricted to trade associations. All but one indicated verbally at least some interest in maintaining links with the Partnership. 43 organisations submitted completed questionnaires; these repeated the initial verbal response that they were interested in the practice of recycling organic material to land in a safe and sustainable manner. 29 proposed subjects that they believed should be included in the Aims and Objectives of any partnership these are summarised in

, together with the number of organisations that recommended them.

All but the one of the 142 indicated that they agreed that there was a need for better communication between stakeholders, and that they would be interested in involvement at some level in achieving this, including attendance at a conference or workshop. It was decided that the most effective way to review the questionnaires, formulate ideas and gauge support would be a workshop, involving a high proportion of discussion. This was held on 4-5 July 2002 at Wadham College, Oxford, and was attended by 80 people from England, Scotland and Wales representing 55 organisations covering a very wide range of interests. These interests included government departments, agencies and regulators, environmental NGOs, food retailers, water companies, waste companies, research institutes, food and drink producers, farmers, landowners, agronomists and consultants. Other organisations that expressed their support and interest but were unable to send a representative asked to be informed of the outcome. The workshop was funded by the Agency, Water UK and Waste Recycling Group, itself a promising demonstration of partnership and commitment.

RESULTS

The overall response to the questionnaire and workshop was very positive.
- Stakeholders representing all steps in the 'plough to plate' chain said that there is a very definite need for a partnership.
- They considered that consensus on good practice based on sound science will be possible by sharing objective information and building mutual trust
- Producers of organic resources considered this consensus will bring stability to and confidence in the use of organic resources on land.
- Owners and users of treated land believed that this consensus is necessary to assure the ability to sell produce from treated land.
- Members of the food and drink industries wanted this good practice to protect the quality of their produce, their brands and the health of their customers.
- Regulators wanted the promotion of good practice in order to protect health, the environment, etc. and to avoid the need for enforcement actions and proceedings against poor practitioners.

The general view was that a partnership should be formed quickly and it should make progress rapidly.

An interviewee from FSA said that it would not be able to promote an activity such as organic recycling actively. However it would be able to promote good practice and assist with the process of agreeing this by consensus[*].

The vision for the partnership was that it would work to make the use of organic resources on land welcomed and accepted as part of sustainable development. In particular the partnership should:
- be independent of vested interests and should not unduly favour any particular interest
- build mutual trust between the members,
- become a reliable source of information,
- ensure 2-way communication,
- identify gaps in knowledge,
- achieve consensus on what is good (welcomed) practice,
- build trust in that good practice and disseminate it.

Workshop delegates considered it vital that some outputs are achieved quickly so that there should be concrete evidence of the Partnership's activity. In particular it was proposed that a regular email newsletter would be very useful. It could be initiated in a short

---

[*] Until these discussions the FSA had considered, following extensive research, that since the use of organic resources on land does not pose significant risk to human health via food it should not be one of the early targets for its resources; the exception to this was animal manures. Therefore the FSA had not produced a position paper on the use of sewage sludge or municipal or industrial wastes on land. However now that the FSA does appreciate that the stakeholders regard the opinion of the FSA as the "bottom line" if a scare story should break, it is considering producing a position paper.

space of time without requiring the expenditure of a large amount of money, however it would require that a Project Officer be recruited. Delegates agreed that dissemination of information relevant to the benefits and risks of using organic resources on land is very important but that they might be too busy to find the information themselves and/or they might not have access to the information. The newsletter will provide a focussed current-awareness function. The newsletter would also be a positive benefit of membership and an incentive to join the Partnership.

It was also considered that membership criteria will have to be developed as quickly as possible in order to start to gather income and to keep the momentum that has been generated to date. However the scales of membership fee remain a dilemma because in order to achieve its goals the Partnership needs to be open to all and therefore membership fees must not be a barrier to participation.

Finally, the workshop delegates indicated that a [legal] structure for the Partnership should be devised and agreed as quickly as possible. In replies to the questionnaire those organisations that expressed a view indicated that the Partnership should be non-profit making. Several suggested that it could be set up as a company limited by guarantee and one suggested that it should seek charitable status because this would enable tax to be reclaimed as Gift Aid on donations. However a solicitor experienced in corporate law advised that revisions in rules related to charities could mean that the Partnership might not qualify. It was also possible that any tax advantages might not out-weigh the disadvantage that charitable status might be a constraint on freedom of action.

During discussions two organisations suggested that consideration be given to restricting the voting rights to trade associations and non-commercial organisations, so as to reduce the possibility of lobbying to the advantage of a single company. However others were strongly of the opinion that trade associations are often remote from the issues and that they could be a barrier or filter to the opinion of their members. This is a question of balance; the *raison d' etre* of trade associations is to represent their members, but the success of the Partnership depends on all feeling that it is open to their particular concerns and perceiving that it is independent of undue influence. If there are sufficient members, all with equal voting, it should be more difficult for any one individual to have excessive influence.

There was support for the Partnership covering the whole of the UK but it was felt that there should be a mechanism to ensure that regional issues were effectively covered. Separate Waste Policies are being developed for Scotland, Northern Ireland and Wales and these could have significant effects on the practices and policies that apply there. This ability will also be important if activities are extended to cover the rest of Europe.

The Partnership must be responsive to its members. This could entail regular stakeholder meetings to determine and/or ratify policy, and a body that includes members from Government, Regulators, industry and other stakeholders should direct activities.

The conclusion from the study was that the most suitable structure for the Partnership is a company limited by guarantee without share capital and with membership open to all. It should have a small (say 5-9) Council of Management (Board of Directors) with executive powers to direct the work. It would be advised by a Committee of Members representative of the principal organisations and funders to decide broad policy. LEAF (Linking Environment And Farming) has a similar structure and is a good model for the ACF. It was expected that much of the technical work would be undertaken by special Task Groups, which would be established to work on selected topics. Participation in Task Groups would be on a voluntary basis but they would be serviced by the secretariat.

In order to be effective it was recommended that there should be an active, and paid, secretariat. In preliminary discussions it was felt that the first appointment should be a Project Officer to produce the newsletter and service task groups and research networks. It was felt that an able postgraduate would suit this post. A breakout group at the workshop suggested that an experienced and high-powered Chief Executive should be appointed at an early stage but there was a counter view that this would be expensive and ran the risk of getting the wrong person. This should be a choice for the interim Council of Management. In any case, the view was frequently expressed that the organisation should be independent of any stakeholder or stakeholder group and to be clearly perceived as such

Funding for the Partnership is a tricky matter. The cost of subscription should not be a barrier to participation because that would inhibit some of the contributions the Partnership needs. There must be adequate resources to enable valued outputs. If there is sponsorship it must not jeopardise the independence and impartiality of the Partnership. A levy of 10p/tDS of non-farm derived organic resources would, if all producers paid, generate an annual income in excess of £200k. There is a precedent in Germany where a levy is paid by land-appliers of biosolids, but would all the producers honour a levy if it were voluntary? A combination of modest subscriptions, sponsorship and grants is the most likely. There are several funding programmes that look as if they are applicable but they take time and resources to acquire. It was considered that the current awareness electronic newsletter and the annual workshop will be incentives to enrolment. Far-sighted organisations that have regard to the objectives of the Partnership have indicated a willingness to be financial sponsors.

One of the tasks for the questionnaires and workshop was to elicit a name. There were several clever acronyms but no clear recommendation. Everybody at the workshop was comfortable with the expression "the Partnership" which was a high-scoring word, along with 'sustainable' 'resource' and 'organic' (

Table 3). Although 'network' and 'forum' scored quite highly in the questionnaire they were considered to be too passive when they were discussed at the workshop. Therefore we have recommended Sustainable Organic Resources Partnership as a working title and in a suggested logo the additional strap line 'for a better environment'.

The workshop considered that by the end of 2002 significant progress should be made on all of the following high-priority actions:
- Develop a business plan and briefing package to lobby for initial funding
- Obtain an initial sign-up & commitment in principle
- Agree the aims & objectives
- Agree a constitution & structure
- Publicise what has been achieved so far
- Recruit parties not already at the workshop
- Launch a current-awareness email newsletter
- Obtain interim funding (£15k) to employ consultant(s) to progress some of the above and develop bids for medium/long term funding.

The formation of an Interim Council of Management is required urgently so that the Partnership can be registered. It is also needed to provide a focus for action, and enable discussions on issues that have remained unresolved, including the executive staff, the benefits or otherwise of Charitable Status, and identify the areas of work of highest priority.

Active extension of the Partnership to other countries in Europe is regarded as a second phase objective that will be conducted in the medium term. However it is also hoped and expected that the Partnership will extend outside the UK by organic growth because of the perceived benefits of becoming a member. The benefits of the Partnership extending to the whole of Europe are that:-
- Environmental and food legislation come from the EU and it would be better if it were influenced by welcomed practice derived by consensus across several MS
- Many in the plough to plate chain are multinational and/or they buy in many countries
- Perception is influenced by stories from outside as well as from within the UK
- Confidence is wider than just the UK
- That consensus on welcomed practice will be achieved across the widest community and this will give the best assurance for the future of recycling

Therefore it is recommended that the Partnership should be designed so that there is no bar to it becoming an international organisation, if this is seen as in the best interests of UK stakeholders. It is recommended that stakeholder organisations (and others) outside the UK are kept informed of the development of the Partnership with an understanding that their views will be appreciated and considered.

The Environment Agency intends to continue to actively support the Partnership including funding though the establishment phase hopefully with the support of others.

CONCLUSION

There is an overwhelming belief amongst all stakeholders that an ACF is needed in order that the sustainable use of organic resources on land is secured with welcomed practices. The provisional name will be the Sustainable Organic Resources Partnership. It will be objective and independent of vested interests. All will be welcomed to the Partnership be they individuals or associations. All members will have equal voting rights including election of the executive board, whose members will retire and be eligible for re-election on a rotating basis. A representative committee selected from the membership and including the principal sponsors will advise the executive board. The main objectives of the Partnership will be to build mutual trust between members, disseminate and share information, become a trusted source of information, identify gaps in knowledge, develop welcomed practice by consensus and promote this practice. There will be an annual workshop/conference and AGM. The vision is that by this means the use of organic resources on land will be sustainable and welcomed by all in the plough to plate chain. Because the practices will be founded on science and have the confidence of the stakeholders it is hoped that they will be considered by policy makers as the basis for practicable regulation.

REFERENCES

1. Scientific Group on Decommissioning Offshore Structures, Second Report May 1998 Report by NERC for DTi.

2. http://www.nerc.ac.uk/publications/documents/decommissioning_report.pdf

3. Environment Agency Vision for the environment. http://www.environment-agency.gov.uk

4. Standards for the Use and Disposal of Sewage Sludge. 40 CFR Part 503. Federal Register 58 (32), 9248-9415. (19 February 1993) available via http://biosolids.policy.net/

5. Safe sludge matrix. www.adas.co.uk/matrix

ACKNOWLEDGEMENTS

We are grateful to the Environment Agency of England and Wales for funding the Scoping Study, to the EA, Water UK and Waste Recycling Group for sponsoring the workshop which was part of this study and to all of the organisations that have supported the work with their time and ideas.

TABLES

Table 1 The questionnaire used to assess attitudes

Does your organisation have an interest in the practice of recycling organic material to land safely and sustainably?
With regard to this subject do you think there is a need for improved communications between the various stakeholders including, but not confined to, practitioners, landowners, users of products and services from land-based industries, government at national regional and local level, regulators, environmental organisations, academia and the public?
Would you be interested in involvement at any level in a network or partnership to achieve this?
What should be the aims, objectives and priorities of such a network?
How should it be organised?
Have you any views on the content of a constitution?
Are there any activities that should be expressly excluded from its activities or interests?
Do you have any views on a suitable name?
Please rate each of the following words for their suitability for inclusion in the name for the partnership – if you think there are other key words please add them 0 = definitely should not be used 3 = very suitable work

Sustainable		Resource		Group	
Waste		Partnership		Information	
Recycling		Network		Discussion	
Recovery		Forum		Liaison	
Use		Organic		Environmental	
Residual		Beneficial			

Should there be a widespread support for establishing the network, it is proposed to hold a workshop to decide on the aims and objectives, including sources of funding and a business plan. Would you be willing to participate in such a workshop?
Do you have any other comments or suggestions?

Table 2 Subjects proposed for Aims & Objectives (summary from questionnaires)

Subject	Number of organisations
Dissemination/sharing of information	15
Good/best practice / products fit for purpose	13
Facilitating consultation with government	7
Promotion/secure long term future of recycling	7
Identifying sustainable solutions/promote sustainable development	5
Educating/informing the public	4
Legislation	4
Minimising health risk	4
Facilitating stakeholder partnerships to develop solutions	3
Identifying gaps in knowledge	3
Address pollution / environment issues	3
Oppose use of sewage sludge in agriculture	1

Table 3 Scoring of key words for acceptability in the Partnership's name

Word	Scores (highest = most favoured)											Total score
SUSTAINABLE	3	3	3	3	3	3	3	3	0	3	3	**30**
Waste	0	-	0	3	3	0	0	0	0	1	0	7
Recycling	3	1	3	3	0	3	3	3	1	3	2	25
Recovery	2	-	3	3	1	3	3	0	2	1	3	21
Use	2	-	2	3	1	1	3	0	0	1	3	16
Residual	1	-	2	-	2	1	1	0	0	1	1	9
RESOURCE*	3	3	3	-	3	3	3	*	*	*	*	**18**
PARTNERSHIP	3	2	0	-	3	2	3	0	0	3	3	**19**
Network	2	2	3	3	1	2	3	3	0	2	3	24
Forum	2	2	3	-	1	2	3	3	0	0	2	18
ORGANIC	2	3	2	0	2	2	3	3	0	3	0	**20**
Beneficial	3	-	2	-	3	3	3	0	1	3	1	19
Group	2	-	1	3	2	1	2	0	1	1	3	16
Information	3	-	2	-	2	1	2	0	1	1	0	12
Discussion	2	-	2	-	1	1	1	0	0	2	2	11
Liaison	2	-	2	-	1	2	1	0	0	2	1	11
Environmental	3	-	3	0	3	2	2	3	3	3	1	23
Reuse				3								3

Note: "-" indicates that the respondent did not score the word
* "Resource" was not included in these questionnaires because it was a late addition to the list following a respondent's suggestion, it was one of the few to score all 3s
"Reuse" only appeared on one questionnaire and was probably added by the respondent

CONTROLLING PATHOGENS IN SLUDGE USING HAZARD ANALYSIS AND CRITICAL CONTROL POINT METHODOLOGY

Petter Siljehag and Euan Low
Mott MacDonald Limited, Suite 2/3, Great Michael House, 14 Links Place, Leith, Edinburgh EH6 7EZ, UK, Tel +44(0)131 625 8700; Fax +44(0)131 625 8720

ABSTRACT
Proposed regulatory requirements for pathogen content of sludge re-used in agriculture are reviewed. Hazard Analysis and Critical Control Points (HACCP) is an established methodology in the food industry, the theory and its relevance to the control of pathogens in sludge is explained. The practical application of HACCP to satisfy the regulatory requirements is considered.

KEY WORDS
Sewage sludge, pathogen, Management Tools, HACCP

INTRODUCTION
There is increasing pressure for beneficial reuse of sludge to be applied to agricultural land, *viz*:
- the European Commission have stated that 'sludge should be used on land whenever possible and only according to relevant community or national legislation'[1],
- the Water Industry Commissioner (the Scottish equivalent of OfWAT) noted that the 'most favoured option is to reuse sludge in agricultural land'[2], and similarly
- SEPA issued a Press Statement giving their support of 'the recycling of sewage sludge to farmland where this is carried out in accordance with statutory requirements'[3].

However, there has also been concern that the re-use of sludge on agriculture may increase the hazards to health from transmission of pathogens and the pollution of soils by heavy metals and organic compounds.

This paper reviews proposed regulatory requirements to address to limit the risk of pathogen transmission and considers the appropriateness of the HACCP methodology as a managing these risks.

REVIEW OF PROPOSED REGULATORY REQUIREMENTS
In the United Kingdom, community expectations on food safety have resulted in a voluntary agreement between Water UK and the British Retail Consortium involving other parties such as the Environment Agency, the Department of the Environment, Transport and the Regions and the Ministry of Agriculture, Fisheries and Food, and the outcome has been published as the ADAS Matrix or "safe sludge matrix"[4]. This agreement is more qualitative than quantitative, in particular it does not provide detail on the microbial standards to be achieved.

Regulation to reduce the risk of pathogen transmission through the use of sludge in agriculture has been proposed by:
- the EC in a working document [1] for a proposed Sludge Directive, and
- the UK government in a proposed revision[5] of the 1989 Sludge (Use in Agriculture) Regulations.

Quantitative microbiological standards in these documents are summarised in Table 1 together with the quantitative microbiological standards from the USEPA Standards for the Use or Disposal of Sewage Sludge[6].

While the determinants and values vary between standards, there is a recognition in all three standards that: sampling and analysis of microbial determinants is only indicative rather than absolute; that there is an appreciable delay before results are available from such analysis; and that the analysis can be costly. Consequently, all three standards give consideration to the validation and continuous monitoring of the parameters that influence the possibility of pathogen survival.

For example, the proposed revision to the Sludge Regulations includes the requirements "The sludge producer shall identify sufficient critical control points in the sludge treatment process for the purpose of monitoring and control ... Sludge shall be monitored and controlled at the critical control points identified ... so as to ensure that the reduction in units of *E. coli* required to [ensure] ... enhanced treated sludge will be achieved."

INTRODUCTION TO HACCP
Hazard Analysis and Critical Control Point (HACCP) was developed by the Pillsbury Company, the American Army and National Aeronautics and Space Administration (NASA) during the 1960's[7] and in particular reduce the risk of food poisoning to astronauts. Application of HACCP methodology for the USA meat industry became a statutory obligation in 1996 and is enforced by the Food

Safety and Inspection Service of the United States Department of Agriculture[8].

HACCP methodology has very recently been brought into the UK regulatory framework in the meat processing industry[9]. As hazards from sludge reuse in agriculture may extend into food safety, there is an obvious logic of applying HACCP to the activities involved in sludge reuse where there is a perceived hazard. Furthermore, it would be prudent to consistently apply the same principles and methods being employed for the food industry to sludge treatment, handling and reuse. Indeed, it would increase the confidence of the food industry in its products. The application of standardised methods and vocabulary will be appreciated throughout the supply chain.

To draw out the key features of HACCP it is beneficial to first consider a familiar risk assessment tool, namely Hazard and Operability Studies (HAZOP) and to then compare the two tools.

In the process industry, HAZOP is a common tool for risk identification to ensure the operability of an engineering system. The HAZOP procedure is based around the generation of a series of questions regarding the operability of the design of a process that ensures all areas are comprehensively and systematically covered. The questions are aimed at identifying design faults or process deviations that could compromise the safety and operability of the designed process. This risk identification activity is ideally conducted through systematic discussion of a Process Flow Diagram by a team of individuals who are experienced in design, operation and maintenance of the designed process and draws on the benefits of a multidisciplinary group using brainstorming.

In practice, the focus of a HAZOP is commonly the process operability and safety and is typically conducted during the design of new plant or during modification to existing plant.

In the food production industry HACCP is a common risk management system for controlling the activities that may impinge on the final product safety. The HACCP methodology is based around the implementation of seven principles which are each applied in turn, with the purpose being to manage potential hazards by applying controls at points in a food production process where the hazards could be controlled, reduced or eliminated.

The seven principles of the HACCP method are summarised in the Meat Regulations[9] as:

- Identifying any hazards that must be prevented, eliminated or reduced to acceptable levels;
- Identifying the critical control points at the step or steps at which control is essential to prevent or eliminate a hazard or reduce it to acceptable levels;
- Establish critical limits at critical control points which separate acceptability from unacceptability for the prevention, elimination or reduction of identified hazards;
- Establish and implement effective monitoring procedures at critical control points;
- Establish corrective actions when monitoring indicates that a critical control point is not under control;
- Establish procedures to verify whether the measures outlines in 1 and 2 above are working effectively; verification procedures shall be carried out regularly; and
- Establish documents and records commensurate to the nature and size of the business to demonstrate the effective application of the measures outlined in 1 to 6 above and to facilitate official control.

Similar to HAZOP, the HACCP methodology should ideally be conducted by a team of individuals who are experienced in design, operation and maintenance of the designed process and draws on the benefits of a multidisciplinary group. In contrast to a HAZOP, the identification of hazards should be done by examination of a flow diagram of the activities that influence the end product quality, a 'walk through' of the process and discussion with operators. Also contrasting with a HAZOP are the requirements for the establishment of regular verification procedures and record keeping.

In practice, the focus of HACCP is the control of activities that affect product safety and should be conducted during the design of new plant or modification to existing plant and in any case should be reviewed on a regular basis. The HACCP methodology aims to identify areas of concern, including where failure has not yet been experienced and therefore is particularly useful to new operations.

Application of HACCP in sludge reuse

Application of the HACCP methodology to assist in the management of sludge re-use could address the following concerns in the following ways:

- Although there is still uncertainty of standards to be applied under the imminent regulation, examination of all activities could identify any hazards that can influence the possibility of pathogen survival and transmission and ensure that these are controlled to acceptable levels.

- In considering the acceptability of a WRc study[10] commissioned by the Directorate-General Health & Consumer Protection, the European Commission Scientific Committee on Toxicity, Ecotoxicity and the Environment (CSTEE) expressed the view that the recommendations were 'an acceptable basis for the Commission to propose appropriate treatments and user restrictions for the land spreading of sludges that will minimise risk. However, the risks are restricted to farm workers with no consideration of risks to operators in transportation and storage and little consideration of broad wildlife implications. The CSTEE believes that these risks should be given more attention.'[11]
- Fundamental to the HACCP method is that the focus is on activities and not on the process. This is illustrated in the Meat Regulations[11] that states that "the results should be used to maintain and improve the standard of slaughter hygiene. Causes of poor results may be clarified by consultation with the slaughtering staff where the following factors could be involved: poor working procedures, absence or inadequacy of cleaning and/or instructions, the use of unsuitable cleaning and/or disinfection materials and chemicals, inadequate maintenance of cleaning apparatus, and inadequate supervision".
- The focus of HACCP on activities has profound impact on how the risk to pathogen survival and transmission is managed as the scope of control will extend to handling, logistics and application to land. It should be noted that the EC Working Document on Sludge states that "there shall be a provision on producer responsibility and certification which provides that: - Producers are to be responsible for the quality of sludge supplied (even when a contractor takes care of sludge marketing and spreading) and shall guarantee the suitability of sludge for use. Producers are to analyse the sludge for ... [inter-alia] micro-organisms"
- The underlying principle of HACCP is that controllable parameters that influence the desired end-product quality are measured and controlled. Theoretically, once verified this approach makes the measurement of the end product unnecessary. This is endorsed in the UK draft Sludge Regulations "where analysis of samples over a continuous six months shows that none of them contain *Salmonella spp.* or more than 10^2 units of *E. coli*, the interval before the next sampling may be increased to three months."
- The CSTEE has strongly recommended that further consideration be given in determining the critical levels for critical control points for various sludge treatment technologies[11]. However, the concerns are predominantly associated with uncertainty to the adequacy of the various technologies for controlling pathogens and this can be addressed through implementing the HACCP methodology, in particular the verification and reporting steps.
- The HACCP methodology can contribute to the 3rd Working Document proposals for the EC Sludge Directive requirements "Producers shall implement a quality assurance system for the whole process, i.e. control of pollutants at source, sludge treatment, the way that the work is planned and the land evaluated, sludge delivery, sludge application and the communication of information to the receiver of the sludge. The quality assurance system shall be independently audited by auditors duly authorised by the competent authority."

While the proposed revision to the UK Sludge Regulations only extend to the 'sludge produced', liabilities arising from the reuse of sludge are likely to remain with the producer. The HACCP methodology provides a risk management tool that can be extended over all the associated activities and through the identification of hazards, procedures to control these hazards, verification of the efficacy of the procedures and reporting, implementing this method can be considered as delivering best practice.

The majority of the practices required under the HACCP methodology will be current requirements for example to satisfy health & safety obligations and other regulatory obligations as part of routine reporting between management and operations. Therefore, establishing a HACCP should build on existing practices wherever possible and for most operators, should not be an onerous task.

HACCP methodology as a management tool compliments environmental management and quality assurance systems. In particular,

HACCP has the potential to provide information required to manage the associated activities including the recognition of risks, identification of the critical control points, monitoring and anticipatory preparation of action plans.

CONCLUSIONS

The Hazard Analysis and Critical Control Points Methodology appears to be a robust and appropriate management tool for the control of pathogens in the reuse of sludge in agriculture.

REFERENCES

1. Working Document on Sludge [Directive], 3rd Draft, European Commission, April 2000

2. Water Quality and Standards, Scottish Executive 1999.

3. SEPA Press Statement May 2001: Sewage Sludge – Policy on spreading on land. May 2001.

4. ADAS.

5. The Disposal of Sludge Regulations, Draft. September 2001.

6. 40 Cope of Federal Regulations, Part 503, U.S. Environment Protection Agency.

7. Web: http://www.hygienemark.com/HACCP.htm

8. Guidebook for the Preparation of HACCP Plans. United States Department of Agriculture (USDA), Food Safety and Inspection Service. September 1999.

9. The Meat (Hazard Analysis and Critical Control Point) Regulations, 2002.

10. Carrington, EG Evaluation of sludge treatment for pathogen reduction – final report. European Commission Directorate General Environment September 2001.

11. European Commission April 2001. Scientific Committee on toxicity, ecotoxicity and the environment (CSTEE): Opinion on evaluation of sludge treatments for pathogen reduction.

ABLES

Table 1: Comparison of Microbiological Standards

Regulation	Microbiological Standards

The Disposal of Sludge Regulations (2001), UK Draft @ 11/9/01	"enhanced treated sludge" means sludge or septic tank sludge which has undergone a biological, chemical or heat treatment process so as significantly to reduce its fermentability and the health hazards resulting from its use, being a process of such character that it –
	(a) is designed so as to reduce the amount of Escherichia coli present in the sludge by not less than 99.9999 per cent;
	(b) is monitored and controlled in accordance with Schedule 3 to these regulations; and
	(c) satisfies the end product tests for Salmonella spp. And Escherichia coli …
	   1. "units" of *E. coli* means colony-forming units of *Escherichia coli* expressed as units per gram of dry solids; and
	   2. *Salmonella spp.* Shall be measured by reference to 2 grams of dried solids.
	Where such analysis shows that none of a set of samples contain *Salmonella spp.* Or more than 10^3 units of *E. coli*, the batch of sludge in question shall be treated as satisfying the end product test for enhanced treated sludge.

Working Document on Sludge 3rd Draft, EC	<u>Advanced treatments (hygienisation)</u> [In addition to physical, time and/or chemical constraints, the following microbial standards are given] The process shall be initially validated through a 6 Log10 reduction of a test organism such as *Salmonella Senftenberg W 775* The treated sludge shall not contain *Salmonella spp* in 50 g (wet weight) and the treatment shall achieve at least a 6 log10 reduction in *Escherichia Coli* to less than $5 \cdot 10^2$ CFU/g.
USEPA Part 503 – Standards for the Use or Disposal of Sewage Sludge	Class A – Alternative 1. (i) Either the density of fecal coliform in the sewage sludge shall be less than 1000 Most Probable Number per gram of total solids (dry weight basis), or the density of *Salmonella sp.* Bacteria in the sewage sludge shall be less than three Most Probable Number per four grams of total solids (dry weight basis) at the time the sewage sludge is used or disposed; at the time the sewage sludge is prepared for sale or give away in a bag or other container for application to the land; or at the time the sewage sludge is prepared to meet the requirements given in § 503.10 (b), (c), (e) or (f) [*i.e.* relating to the applicability to land].

HACCP IMPLEMENTATION

[1]Gareth Davies, [1]Juliet Vuong, [2]Tony Griffiths

[1]Pell Frischmann, Burrator House, Peninsula Park, Rydon Lane, Exeter EX2 7NT
Tel. +44 (0) 1392 444 345, Fax +44 (0) 1392 444 880
mail to gdavies@pellfrischmann.com, jvuong@pellfrischmann.com
[2] South West Water, Rydon Lane, Exeter. EX2 7NT
e-mail: tgriffiths@south-west-water.co.uk

ABSTRACT

The first draft revision of the Sludge (Use in Agriculture) Regulations was issued in May 2000 and formally introduced the concept of HACCP (Hazard Analysis and Critical Control Point) to the water industry. Since that time Pell Frischmann has been assisting South West Water (SWW) in developing their approach to implementing the principles and practicalities of HACCP. This paper describes the approach that SWW has adopted to implement the revised Sludge Regulations and, more specifically, the concept of HACCP contained therein across the twenty sludge treatment centres now operating. Site-specific HACCP documentation has been developed and put in place. A HACCP Team comprising Pell Frischmann working in partnership with SWW continuously monitors sludge treatment centre performance. The structured approach of risk assessment followed by quantification of sludge treatment centre performance has enabled improved treatment and the strategic refinement of the Sludge Strategy Capital Investment Programme and real cost savings.

KEY WORDS

HACCP, implementation, monitoring, pathogen, recycling, regulations, risk, sampling, sludge

INTRODUCTION

This paper describes the approach that South West Water (SWW) has adopted to implement the revised Sludge Regulations and, more specifically, the concept of Hazard and Critical Control Point (HACCP) contained therein. Throughout the implementation process Pell Frischmann (PF) has assisted in developing the SWW strategy.

HISTORY

The revised Sludge (Use in Agriculture) Regulations require water companies to demonstrate that a sludge treatment process can achieve a specified reduction in *E.coli* bacteria and produce a final sludge product to a specified hygienic standard. To demonstrate control and audibility of the sludge treatment process Hazard Analysis Critical Control Point (HACCP) principals are to be applied throughout the sludge treatment process.

HACCP was developed by NASA to supply safe food for Apollo programme and is now common in the food industry, hence the major stakeholder (British Retail Consortium) knows and trusts the concept.

HACCP is a method that was designed for routine process control with regular final product testing and uses the simple approach of identifying the points in the treatment process, Critical Control Points (CCP's) which are of key importance in achieving the desired product quality. To ensure that the product remains within specification, the CCP's must be monitored and controlled to be within a set range known as the control limits (CL's).

As part of the Water UK HACCP Implementation Group South West Water, and its HACCP Team, have played an integral role in developing the industry response to the revised Regulations.

HAZARD ANALYSIS AND RISK ASSESSMENT

Having considered the content of the draft regulations it was evidently necessary to apply the principles of HACCP to the sludge treatment processes that South West Water were planning to utilise as part of their integrated sludge strategy. An overview assessment of the application of HACCP to the generic sludge treatment processes employed by SWW was undertaken. The processes considered included anaerobic digestion, open windrow composting, alkaline stabilisation and thermal drying. The objective of the overview was to determine the potential hazards that threaten the sustainable recycling route of sewage sludge to agricultural land and to identify the generic Critical Control Points that were paramount in operating an auditable treatment process.

The hazard analysis was found to be an effective means of identifying potential problems from a variety of causes that could be expected whilst operating a sludge treatment facility. The output from the study was a long list of possible hazards and a shorter list of candidate Critical Control Points (CCP's) with generic Critical Limits to be applied for each process based on the operational parameters contained within the Code of Practice. In addition, Operational

Control Points (OCP's) and associated limits were identified to assist in the maintenance of HACCP compliance. The OCP's must be maintained within the limits to ensure that the operation of the sludge treatment process does not put the CCP's at risk of non-compliance.

PRELIMINARY HACCP AUDITS

Following the overview assessment, audit questionnaires were developed for each of the sludge treatment processes employed throughout the 17 sludge treatment centres that were being operated by SWW at the time. The audit questionnaires were designed to capture the following information:

- The current operational practices employed on the site
- Potential risks posed by the inputs to the works (domestic and trade wastes)
- The procedures/facilities that may be in place to control the import of waste/sludge
- The effectiveness of the preliminary treatment (screening and grit removal)
- Risks posed by intermediate storage of sludge
- Potential impact of use of chemicals on site
- The dependence of the treatment on thickening processes
- The main process facilities and performance (digestion, thermal drying etc)
- The facilities and operation of post treatment storage
- The collection and storage of important operational data
- The management of data and information that may be used for verification purposes

Each sludge treatment centre was subsequently visited during December 2000 and January 2001 to undertake a preliminary audit and an "interview" held with the member, or members, of the operational team appointed by the works manager to assess the risks to compliance with the generic CCP's and limits posed by the incumbent operational practices. During the interview the concept of HACCP was discussed and the questionnaire completed. Following the interview a brief tour of the treatment process was conducted to confirm details.

AUDIT FINDINGS

The completed questionnaires were then further developed into brief audit reports that drew conclusions from the information collected and made recommendations for improvements. The audits identified potential deficiencies in infrastructure and operational practices that were considered to pose risks to future compliance with the revised Regulations.

The key finding of the initial audits was that it was unlikely that any of the sites would have satisfied the requirements of a formal HACCP audit without some changes to either operational practices or plant. This was expected as there sites were not designed, or operated, to reduce sludge pathogen levels the main emphasis being adequate treatment and odour reduction to facilitate agricultural recycling. There were instances of insufficient recording of operational data to provide control, management and reporting of the process performance. Many sites had inadequate instrumentation and SCADA facilities to provide this process control, management and reporting. Cross contamination of sludge products was observed and little pathogen reduction data was available to provide verification of a "conventional treated" or "enhanced treated" final product quality. Some of the conclusions drawn at the time are presented in Table 1.

However, due to a paucity of information on microbiological concentrations in the sludge apart from some final product results, it was not possible to corroborate the audit findings with hard evidence of poor pathogen reduction.

CCP SLUDGE SAMPLING PROGRAMME

To provide verification that the deficiencies identified in the audits were affecting sludge pathogen removal performance a sampling programme was constructed that would quantify the levels of *E.coli* and Salmonella to be found throughout the many stages of the sludge treatment processes. The objectives of the sampling were as follows:

- To measure the microbiological concentrations found within the raw untreated sludge.
- To quantify the level of microbiological reduction achieved across each step of the treatment process
- To quantify the overall log reduction of *E.coli* achieved at the sites targeted.
- To make an assessment of the microbiological reduction performance of the treatment processes, as they were currently being operated (pre HACCP implementation).
- To identify benchmark processes from within the South West Water treatment work and to understand the reasons for their good performance.
- To enable site specific CCP's and critical limits to be proposed.

To achieve these objectives, sludge samples were collected from a number of locations within the treatment processes at 14 selected sites as is indicated in Figure 1. Samples were composites of a number of sub-samples where appropriate taken over a time period that enabled a representative output of the particular process stage to be sampled.

Sample results alone are important but of little strategic benefit unless vital reference information is available. Comprehensive operational monitoring was undertaken alongside the sample collection and process relevant data was recorded. This information enabled the treatment process operating conditions relating to specific samples to be recorded and hence assist in the understanding of the process performance.

SAMPLING PROGRAMME FINDINGS

The results of the programme contained a few surprises. The sampling programme results from the digestion sites are summarised in Figure 2. Overall the results were better than might have been expected with 45% of the mesophillic anaerobic digester sites achieving greater than 2-log $E.\ coli$ and 64% achieving the 10^5 $E.\ coli$ /g.ds standard consistently. There were notable exceptions with instances of the following issues being identified:
- Poor primary or secondary digester performance
- Cross contamination of digested sludge with untreated sludge due to operational practices and plant configuration
- Re-infection of digested sludge cake from final effluent being used as belt press wash water
- Low digester temperatures
- Insufficient digester boiler capacity
- Low digester (primary and secondary) retention times
- Insufficient cake storage capacity
- Lower than Sludge Strategy defined throughputs

Generally, as expected, digestion sites operating with longer digester retention times and higher digester temperatures demonstrated better pathogen reduction performance.
During storage of digested sludge cake for periods of up to six weeks a mean $E.coli$ reduction of approximately 1-log was observed as shown in Figure 3. This finding has been used to develop a short-term contingency plan in the event of digested sludge product not meeting the final product specification. The sampling confirmed that only sludge having demonstrated 2-log reduction and compliance with the final product standard was recycled to agricultural land.

Likewise, lime stabilisation using quick lime was found to be a very effective process to obtain enhanced treated sludge with lower than anticipated lime doses consistently achieving 6-log reduction and enhanced treated quality. Open windrow composting however, was found to be less effective than was hoped with very variable $E.\ coli$ reduction results proving that long retention times were required to reliably produce a conventionally treated product. However, with sufficient maturation time, enhanced treated compost appeared achievable.

CAPITAL INVESTMENT SPECIFICATION

The preliminary HACCP Audit findings and sampling programme results were utilised to refine the SWW Sludge Strategy defined capital investment programme that was at the early stages of procurement. Site specific project briefs were developed by SWW capital investment team amending the improvements that had been identified the previous year. The following areas were common to many sludge treatment centres:
- Primary digester refurbishment
- Additional secondary digester capacity
- Cake storage barns
- Lime stabilisation facilities
- Cake reception facilities
- Additional thickener capacity
- Sludge dewatering improvements

This capital investment programme is currently underway and has already yielded process improvements together with improved monitoring facilities.

HACCP DOCUMENTATION

Each sludge treatment centre has been issued with a HACCP Operational Manual that encapsulates the requirements of the Regulations and the philosophy of HACCP. Operating procedures and instructions have been developed to ensure that the treatment processes at each sludge treatment centre remain compliant. Each manual comprises five main sections:

Section 1.
Operating procedures and instructions to be used under normal running conditions, including a HACCP flowsheet identifying the CCP's, Operational Control Points and their associated limits. (See Figure 4)
Section 2.

Instructions and procedures to be followed in the event of either CCP or final product sample failure. Corrective action plans and contingency plans.

Section 3.
Forms to be completed to request authorisation for process or operational changes that may impact on CCP's.

Section 4.
The most recent HACCP Audit report

Section 5.
A record of final product sample analysis results

The HACCP Operational Manual is the single reference for HACCP information held on site. All manually recorded HACCP data and information is also kept in this file until collected for reporting purposes.

HACCP MONITORING

South West Water began routine monitoring of the sludge treatment centres against the requirements of the draft revision to the Regulations once the documentation was in place in December 2001. Following on from the production of the HACCP documentation, Pell Frischmann were awarded the contract to form a HACCP Team and develop a monitoring programme to cover each sludge treatment centre.

ROUTINE DUTIES
Working closely with the SWW Sludge Recycling Manager, his supervisors and SWW Scientific Support staff, the Pell Frischmann HACCP Team comprises two technicians driving SWW vehicles, two process engineers, a project manager and office support. Whilst it is the responsibility of the SWW operational staff to record the HACCP data and maintain compliance, the time consuming duties of data collation, interpretation and reporting have been outsourced to PF.

Each sludge treatment centre is visited once a fortnight during which time the five sub samples of final sludge product are collected together with digester feed samples, routine sludge metals samples and any follow up samples of potentially non-compliant sludge as required. Whilst on site the PF technicians check the Critical Control Points and limits from SCADA or site instrumentation, collect and check hard copies of any manually recorded data and record any significant changes in operation forecast or since the last visit. A good rapport has developed between site operatives and the PF technicians and this enables both sides to constantly improve performance and ensure full compliance.

An MS Access™ database has been compiled to record the large amount of data generated. The database automatically generates reports which are incorporated into a monthly report to the Sludge Recycling Manager. The report includes headline compliance figures for CCP and Final Product Standards together with listing of CCP record for the previous month. Additional information is provided on Final Product dry solids content, digester feed dry solids, digester capacity utilisation, applied lime doses etc. The final section of the report provides an interpretation of the long term sampling results and importantly, feedback from the site via PF HACCP Team.

NON-ROUTINE DUTIES
In addition to the routine collection of samples and HACCP data, the PF HACCP Team is able to respond rapidly to investigate problems on site and to collect additional samples when necessary. With the engineering support available in the office, the HACCP team is able to rapidly identify and specify solutions to HACCP related problems.

HACCP PERFORMANCE

Sampling has shown that not a single one of more than 2,000 final sludge product samples collected this year has failed to meet the 10^7 E. coli maximum allowable concentration. Overall, since HACCP related sampling began, the number of final sludge product samples meeting either the "conventional treated" or "enhanced treated" standards when first sampled is in excess of 90% and has improved throughout the past year. Figure 5 shows the improvements gained at the sludge treatment centres utilising anaerobic digestion.

Further improvements are expected at those sites where capital investment projects are underway and close to completion. Currently other operational sludge treatment centres are treating elevated throughputs of sludge as treatment capacity is taken off-line for refurbishment.

The study of the Critical Control Points and corresponding sludge sample results has generally shown that the Code of Practice parameters, whilst generally appropriate for achieving the required pathogen reductions, are unnecessarily strict in some areas such as secondary digester retention times in some circumstances.

RESEARCH AND DEVELOPMENT

The information gathered and reported through the implementation of HACCP and its monitoring has improved the company's understanding of the true treatment capacity of the twenty or so sludge treatment centres now being operated in the region. It has also enabled different sludge treatment processes to be combined to achieve the desired sludge standards, and provided a rapid means of assessing the pathogen removal performance of novel sludge treatment processes. Improvements in reporting are constantly being made giving new insights into sludge treatment.

CONCLUSIONS

- The implementation of HACCP philosophy has lead to the better understanding of sludge treatment plant performance.
- SWW's Capital Investment programme is already yielding improvements in sludge treatment centre performance.
- HACCP monitoring is costly, but has major operational benefits and is creating savings in sludge treatment operating costs.
- The rapid response capability of the HACCP Monitoring team is of proven benefit and is a key element of corrective action planning.
- HACCP monitoring and the interpretation of CCP data together with final product sample results has enabled the utilisation of sludge treatment centre capacities to be maximised.
- The implementation of HACCP and the Regulations is a process of continuous improvement with many benefits still to be realised.
- The structured approach of risk assessment followed by quantification of sludge treatment centre performance and continuous monitoring has enabled the strategic refinement of the Sludge Strategy Capital Investment Programme and real cost savings.

ACKNOWLEDGEMENTS

The authors would like to acknowledge South West Water and Pell Frischmann for giving permission for this paper to be submitted. Thanks are given to all members of the SWW and PF HACCP Teams and to the site operatives for providing the information presented.

TABLES

Table 1 – Preliminary Audit Conclusions

	Preliminary Audit Conclusions from January 2001
1	Many of the works will require capital investment to enable the generic operating parameters for satisfactory sludge quality to be achieved.
2	With a shortage of pathogen reduction information from sample analysis data, it was not possible to confirm that the sites were producing a product that would satisfy the "conventional treated" status (with the exception of the Thermal Drying).
3	It was a general conclusion that there was insufficient manual recording of the operational information available on each site – simple training of operatives was recommended.
4	Lack of instrumentation on some sites would hamper the collection of important operational information that would be necessary to provide the HACCP audit trail.
5	Secondary digester/digested sludge storage capacity was insufficient at most digestion sites to enable the 14 days minimum retention times at the Strategy throughputs that was deemed necessary to ensure the 2-log reduction of pathogens. Most sites were not operating, or able to operate in batch mode.
6	The use of final effluent for dewatering press belt washing was identified as a source of potential contamination. Pathogens may be reintroduced into the digested sludge and hence reduce the log reduction achieved across the process.
7	Some sites were not storing final product due to consistency/handling issues.
8	Cross contamination was occurring at sites by inadvertently introducing raw sludge into the treated sludge process.
9	On many digestion sites the importance of thickening prior to digestion was not fully appreciated. As an Operational Control Point (OCP) this process could have a direct impact on the ability to achieve a "treated" sludge product.

FIGURES

Mesophilic Anaerobic Digestion

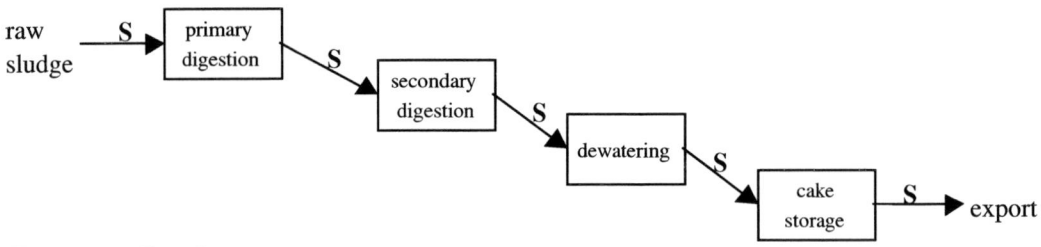

raw sludge → **S** → [primary digestion] → **S** → [secondary digestion] → **S** → [dewatering] → **S** → [cake storage] → **S** → export

Thermal Drying

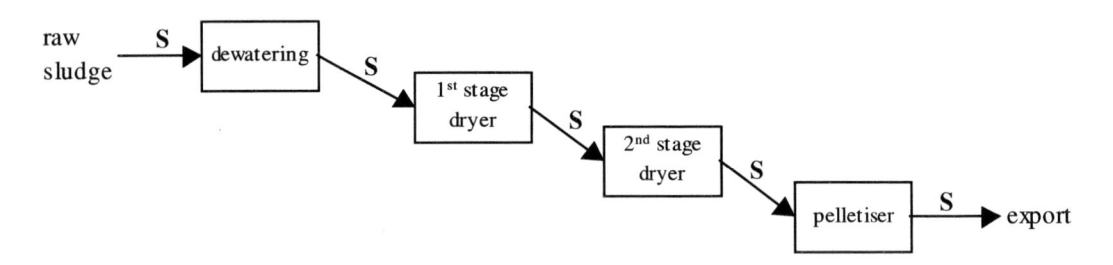

raw sludge → **S** → [dewatering] → **S** → [1st stage dryer] → **S** → [2nd stage dryer] → **S** → [pelletiser] → **S** → export

Composting

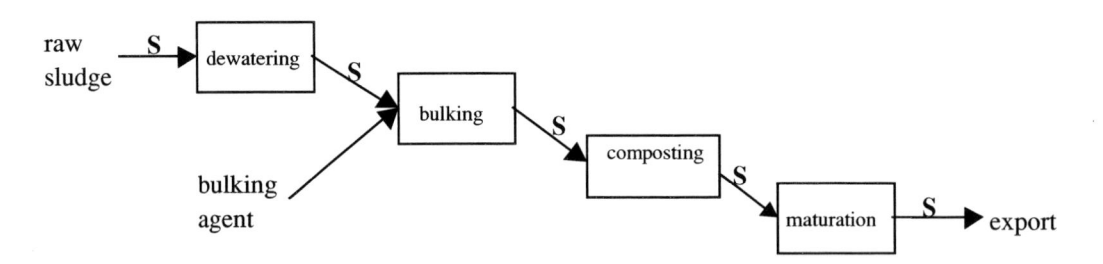

raw sludge → **S** → [dewatering] → **S** → [bulking] → **S** → [composting] → **S** → [maturation] → **S** → export

bulking agent → [bulking]

Lime Stabilisation

raw sludge → **S** → [dewatering] → **S** → [lime stabilisation] → **S** → [storage] → **S** → export

lime → [lime stabilisation]

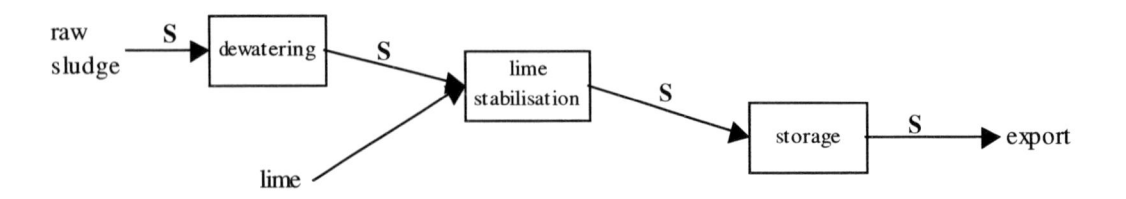

S – Sampling Point

Figure 1 - Sampling Locations

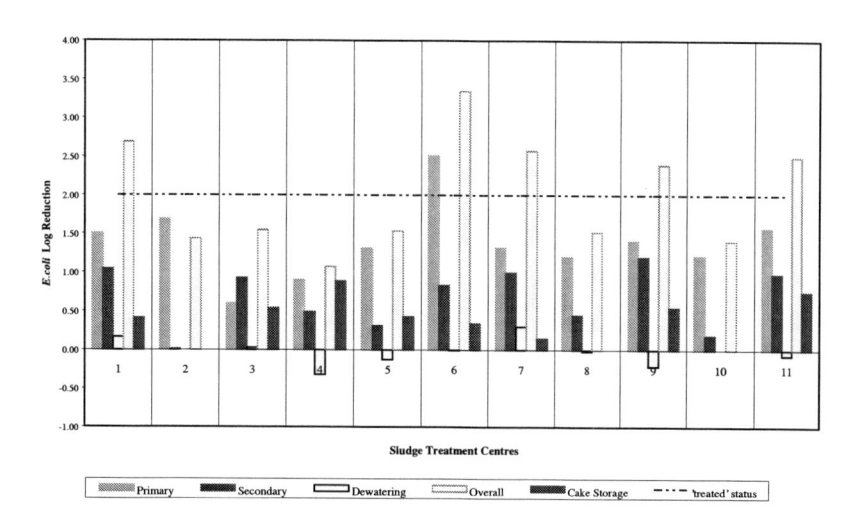

Figure 2: Pathogen Reduction Across South West Water Anaerobic Digestion Sites

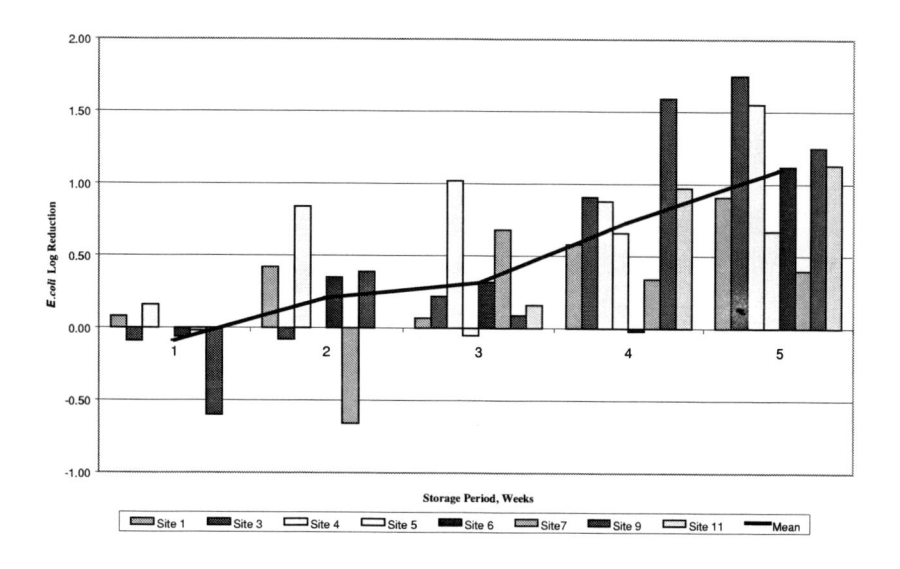

Figure 3 – Pathogen Reduction During Sludge Cake Storage

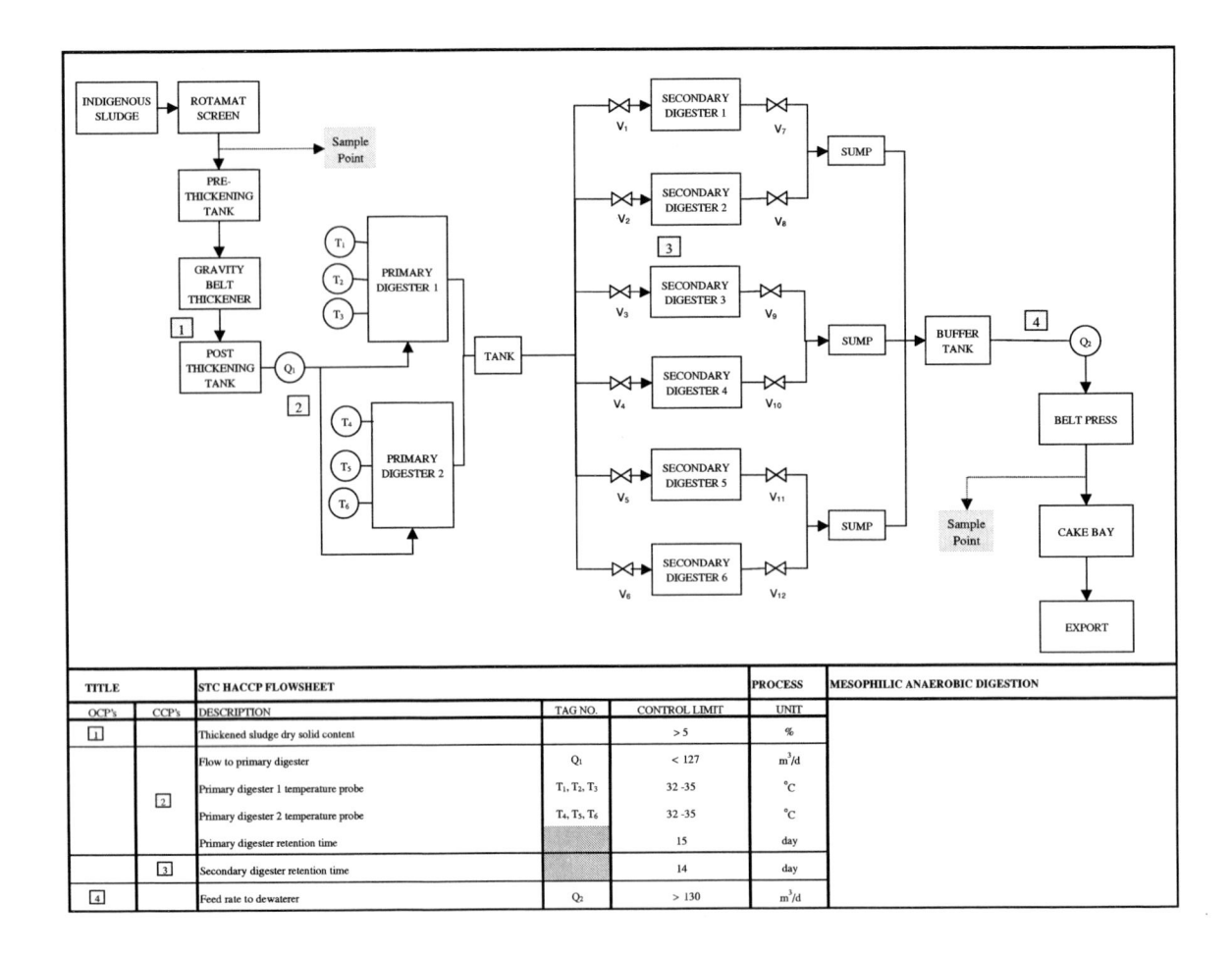

TITLE		STC HACCP FLOWSHEET			PROCESS	MESOPHILIC ANAEROBIC DIGESTION
OCP's	CCP's	DESCRIPTION	TAG NO.	CONTROL LIMIT	UNIT	
1		Thickened sludge dry solid content		> 5	%	
	2	Flow to primary digester	Q_1	< 127	m^3/d	
		Primary digester 1 temperature probe	T_1, T_2, T_3	32 -35	°C	
		Primary digester 2 temperature probe	T_4, T_5, T_6	32 -35	°C	
		Primary digester retention time		15	day	
	3	Secondary digester retention time		14	day	
4		Feed rate to dewaterer	Q_2	> 130	m^3/d	

Figure 4 – Typical Digestion Site Haccp Flowsheet

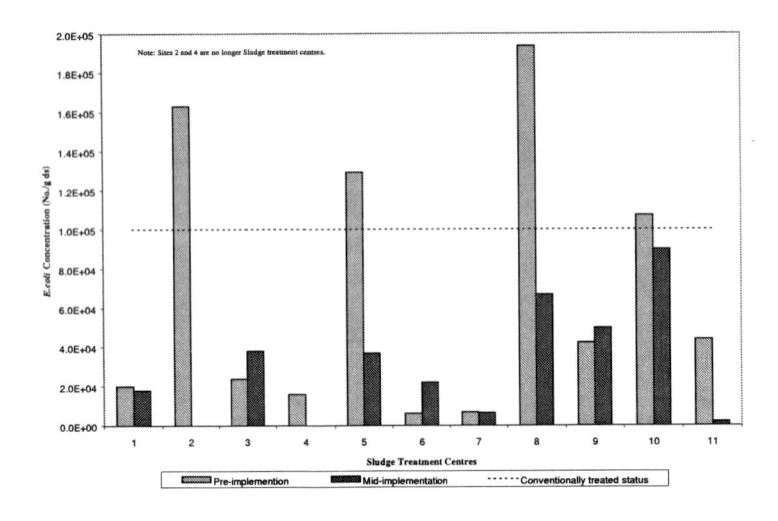

Figure 5 Pre And Mid – HACCP Implementation Final Product Sampling Results

SECONDARY STORAGE OF DIGESTED SLUDGE IMPROVES BACTERIAL KILL

Sophie Mormede, Piers Clark, Nick Pursell, Adrian Stoodley
Atkins Water, Atkins Water, Wessex Water, Wessex Water
Tel: 01372 726140; Fax: 01372 740055; e-mail: sophie.mormede@atkinsglobal.com

ABSTRACT

The present study studied the level of E-coli kill achieved when digested sludge was subjected to secondary storage. Expected benefits are improved pathogen reduction in order to meet the new Sludge Regulations (5-log E-coli on exit sludge, 2-log reduction across process stream). Fresh sludge exiting the anaerobic digesters was stored in secondary digesters for up to 12 days before being further treated. At the time of the experiment, anaerobic digestion alone was insufficient to produce compliant sludge, with 5.4 to 6-log E-coli in the exit sludge. After secondary storage, all exit sludge was compliant, with a maximum concentration of 4.3-log E-coli. Minimum guaranteed E-coli kill achieved through the secondary storage was 1 log; and 2.6 log through both anaerobic digestion and secondary storage. Therefore secondary storage might be a robust low-tech solution for those sites whose digested sludge fails the new Sludge Regulations on bacterial kill.

KEY WORDS

E-coli kill, log reduction, pathogen reduction, secondary storage, Sludge Regulations

INTRODUCTION

The objective of the present study was to determine the level of E-coli kill achieved when digested sludge was subjected to secondary storage prior to dewatering and caking. The expected benefits of this configuration are improved pathogen reduction in order to meet the new Sludge Regulations (5-log E-coli on exit sludge, 2-log reduction across process stream).

This experiment was carried out on behalf of Wessex Water at Avonmouth Waste Water Treatment Works (UK). Fresh sludge exiting the six 2,700m^3 anaerobic digesters onsite was stored in 1600m^3 secondary digesters for 6, 9 or 12 days before being further treated. Sludge before the anaerobic digesters, before the anaerobic digesters, before and after the secondary storage, centrate and cake were sampled and analysed conforming to HCCP regulations.

EXPERIMENTAL DESIGN

Three individual experiments were carried out, at respectively 6, 9 and 12 day retention time in the secondary digester. These experiments were carried out one after the other instead of simultaneously because of constraints onsite (there was not enough sludge to fill 3 secondary digesters at the same time and not enough secondary digesters to dedicate to the experiment). Therefore the quality of the feed sludge varied slightly between experiments (see results section for details). The experiments were carried out between 3rd April and 30th April 2002.

For each experiment, all the sludge exiting from the anaerobic digesters was diverted into an empty and dedicated secondary digester until full. Because of constraints onsite (namely the anaerobic digesters running on 20 days retention time instead of the typical 12 days), this process took between 3 and 4 days to be carried out. Once the secondary digester was full, it was left unmixed for its expected retention time in order to minimise heat loss and therefore bacterial activity slow-down. After the required retention time (6, 9 or 12 days), the sludge was transferred into another secondary digester and mixed before being sent to the drier. The drying of the entire sludge usually took 2 days. Therefore, the maximum retention time of some of the sludge was up to 6 days more than the minimal retention time quoted, namely 6, 9 or 12 days. This corresponds to normal HACCP procedure and guaranteed minimum retention time.

SAMPLING PROCEDURE

Sludge samples were collected from the top of the secondary digesters twice daily when they were being fed from the anaerobic digesters and emptied to the drier, representing inlet and exit sludge of the secondary digesters. Centrate and cake samples were collected twice daily during the drying process.

- In order to assess the variability of the E-coli concentrations in sludge, cake and centrate, the sampling procedure was as follows: For sludge samples (prior to and after secondary storage), a 5-litre aliquot was collected from the top of the digester twice daily (during the filling and emptying stages). This aliquot was mixed thoroughly then three individual samples taken as replicates of that sampling event.

- For centrate samples, a 5-litre composite sample was collected from the centrate lines of all three individual centrifuges whilst the sludge was being dried. As for the sludge samples, this aliquot was mixed thoroughly then three individual samples taken as replicates of that sampling event.
- For cake samples, the three individual replicates were taken at 15 minutes interval. A composite sample was not attempted due to the high solids contents of cake hence difficulty in mixing it up to representative standards.

All samples collected were taken for analysis to Wessex Water Analytical Laboratories (Saltford) within 5 hours of collection. Each individual sample was analysed for TS, VS, pH and E-coli.

Variability within pooled samples and between sampling times was assessed. The variability within pooled samples was typically between 10 and 30%, whereas that between sampling events was anything between 10 and 100%. Results summarised below represent the average over four sampling events, each with three replicates.

During the whole experiment, the maximum and minimum temperatures of the sludge in the secondary digesters were measured in order to assess heat loss and stratification of the sludge.

RESULTS AND DISCUSSION
SLUDGE TEMPERATURE
For the duration of the trial (3rd to 30th April), the weather was particularly mild. Minimum night temperatures were never below freezing, and maximum day temperatures varied between 15 and 25°C. As a consequence, the sludge temperature in the secondary digesters remained relatively high, typically cooling down from 28°C when filled to no less than 20°C. Therefore, the E-coli kill achieved in the secondary digesters during these experiments will be representative of "mild" weather, and higher than that expected to be achieved in the middle of winter.

SAMPLES ANALYSIS
The following flows were analysed for total solids, volatile solids, pH and E-coli: the raw feed to the anaerobic digesters, the sludge exiting the anaerobic digesters after 20 days retention time and fed into the secondary digesters, the sludge exiting the secondary digesters and being fed into the drier, and finally the cake and centrate exiting the drier.

Results are summarised per retention time in tables 1 to 3. These results represent the average of each flow through the filling, emptying or drying procedure. Each of these procedures took between two and three days, providing four to five individual sampling events. Each sampling event was constituted of three replicates, providing a total of 12 to 15 samples in total for each flow. The average is that of the 15 samples, and the standard deviation is that of between sampling procedures, being higher than that within replicates (see sampling procedure section for details).

The pH of the sludge before and after the secondary digesters was similar, showing the stability of the process in the secondary digester. Moreover, total solids were reduced by up to 0.4 percentage points or 10% through the batch feed phase. Volatile solids also incurred a significant reduction during the secondary storage. Therefore, the anaerobic digestion process was still being carried out to an extent during the retention of the sludge in the secondary digesters and further E-coli kill is expected during that process.

As expected, E-coli concentrations were very variable between samples. Variability was about 30% between pooled samples, and up to 100% between sampling events. However, even with such a high variability, the sludge showed a significant E-coli reduction through the batch feed at all retention times. Results of E-coli reduction are detailed in the next section.

The centrate presented high E-coli counts in all instances, but is of very low solids contents (typically less than 0.2%). Therefore total E-coli population expressed on a volume basis are negligible. Moreover, the centrate is recirculated through the works and therefore compliance is not necessary.

The cake presented higher E-coli counts than the exit sludge in all instances. However, such an effect is well known in the industry and has been reported elsewhere (personal communications with Piers Clark and Thames Water, Severn Trent, Southern Water and Anglian Water). This increase in E-coli count might be of various origins such as a contamination of the cake by the carrier of the polymer used for dewatering the sludge; or an analytical difference between the E-coli determination in liquid and solid sludge. Therefore, the E-coli reduction through the batch feed of the secondary digesters will be based on the exit sludge rather than the exit cake. It is however important to note that although the average of E-coli concentrations in cake for all three retention times was below the regulatory level of 5 log, at a maximum of 4.85 log; the variability of the data suggested possible E-coli levels up to 5.41 log, which would be non compliant.

E-COLI REDUCTION THROUGH THE SECONDARY STORAGE

E-coli concentrations in the sludge and E-coli kill through the various processes are summarised in table 4. Sludge fed into the anaerobic digesters varied between and average of 7 and 7.6-log E-coli. Anaerobic digestion at 20 day retention time reduced the average E-coli counts to between 5.4 and 6-log. Therefore, even at 20 day retention time, the sludge exiting the anaerobic digesters was never compliant.

On the other hand, sludge exiting the secondary storage, at 6, 9 or 12 day retention time presented average E-coli counts of between 3.8 and 4.1-log. Moreover, the maximum E-coli concentration never exceeded 4.3-log. Therefore, all sludge exiting the secondary storage was compliant, whatever its retention time.

The averages of E-coli kill through the secondary storage ranged between 1.5 and 2.1-log, and minimum guaranteed E-coli kill ranged between 1 and 1.5-log. No significant differences can be made between the various retention times because of the variability of the dataset, but it is expected that the longer the retention time, the better the overall E-coli kill would be.

The minimum guaranteed E-coli kill achieved through both anaerobic digestion (at 20 day retention time) and secondary storage (between 6 and 12 day retention time) was of between 2.6 and 3.1-log.

CONCLUSIONS

The objective of the trial was to determine the level E-coli kill achieved in the secondary digesters at various retention times, in order to produce compliant sludge of less than 5-log E-coli.

The results show that the primary mesophilic anaerobic digestion process alone (at 20 days retention time) at the time of the experiment was insufficient to produce compliant sludge, with 5.4 to 6-log E-coli in the exit sludge. However, after secondary storage, all exit sludge was compliant, with average E-coli concentrations of between 3.8 and 4.1-log and a maximum concentration of 4.3-log. Moreover, the log kill through both anaerobic digestion and secondary storage processes was above the required 2-log kill.

Due to the variability of the dataset, the differences between the E-coli kill at various retention times were not significant, but it is assumed that a higher retention time will provide higher E-coli kill.

Minimum guaranteed E-coli kill achieved through the secondary storage ranged between 1 and 1.5-log kill, and through both anaerobic digestion and secondary storage ranged between 2.6 and 3.1-log kill.

The present trial has shown the potential of secondary storage as a means to reduce E-coli concentrations in sludge to levels within compliance. Therefore secondary storage might be a robust low-tech solution for those sites whose digested sludge fails the new Sludge Regulations on bacterial kill.

Since the end of this trial in May 2002, a batch feed regime was set up for the entire sludge onsite Avonmouth WWTW. Weekly sampling has reproduced consistently the results obtained during the trial (Nick Pursell, personal communication).

ACKNOWLEDGEMENTS

The research presented here has been supported financially by Wessex Water, and technically by the Wessex Water staff onsite Avonmouth Waste Water Treatment Works.

TABLES

Table 1: Results for 12 Day Retention Time

	Log E-coli (log cfu/g)	E-coli *e4 (cfu/g)	VS (%)	TS (%)	pH
Feed to anaerobic digesters	6.96	913	76	6.68	5.50
Feed to secondary digesters	5.36±0.52	22.7±10	59.3±0.8	3.5±0.4	7.5±0.1
Feed to centrifuge	3.84±0.46	0.69±0.45	57.2±0.8	2.9±0.2	7.6±0.1
Cake	4.85±0.56	7.1±7.1	57.0±0.8	23.3±0.1	8.3±0.1
Centrate	5.19±0.67	16±9	31.0±12	0.15±0.01	7.8±0

Table 2: Results for 9 Day Retention Time

	Log E-coli (log cfu/g)	E-coli *e4 (cfu/g)	VS (%)	TS (%)	pH
Feed to anaerobic digesters	7.61	4080	79	4.66	5.81
Feed to secondary digesters	6.02±0.58	106±70	61.2±0.2	3.3±0.1	7.5±0.1
Feed to centrifuge	3.93±0.41	0.85±0.18	58.3±0.9	2.9±0.1	7.5±0.1
Cake	4.44±0.70	2.7±3.3	58.7±0.5	23.4±0.2	8.2±0.1
Centrate	5.02±0.45	10.5±6	39.7±4	0.18±0.01	7.8±0.1

Table 3: Results for 6 Day Retention Time

	Log E-coli (log cfu/g)	E-coli *e4 (cfu/g)	VS (%)	TS (%)	pH
Feed to anaerobic digesters	7.15	1400	83	7.24	5.67
Feed to secondary digesters	5.78±0.48	59.9±39	60.6±0.5	3.4±0.1	7.4±0.1
Feed to centrifuge	4.07±0.15	1.2±0.2	58.4±0.9	2.8±0.4	7.4±0.1
Cake	4.53±0.46	3.4±2.7	60±1	23.5±1.1	8.2±0.1
Centrate	5.40±0.58	25.1±16	31.2±11	0.15±0.02	7.8±0

Table 4: Summary of E-coli concentrations and kill

Retention time	12 Days	9 Days	6 Days
E-coli concentration in sludge fed to the anaerobic digesters (log cfu/g)	7.0±0.5	7.6±0.5	7.2±0.5
E-coli concentration in sludge fed to the secondary digesters from the anaerobic digesters (log cfu/g)	5.4±0.5	6.0±0.6	5.8±0.5
E-coli concentration of the final sludge after secondary digesters (log cfu/g)	3.8±0.5	3.9±0.4	4.1±0.2
E-coli kill through secondary storage (log cfu/g)	1.5±0.5	2.1±0.6	1.7±0.5
E-coli kill through combined anaerobic and secondary digesters (log cfu/g)	3.1±0.5	3.7±0.6	3.1±0.5

SELECTION OF A PRE-PASTEURISATION TECHNOLOGY FOR THE NEW READING SEWAGE TREATMENT WORKS

[1] Liz Bowman, [2] Patrick Coleman
[1] Target Alliance (Thames Water)
Tel: 0118 923 7406 Fax: 0118 923 7700, Liz.Bowman@Thameswater.co.uk
[2] Target Alliance (Black and Veatch), Tel: 020 8560 5199 Fax: 020 8568 5479, colemanp@bv.com

ABSTRACT

Thames Water Utilities is replacing the Manor Farm Sewage works in Reading with a new sewage works on a brownfield site. The new works consists of lamella primary treatment and biological nutrient removal (activated sludge). Because of development in the area around the new and old works, the new works is covered, odour controlled and insulated for noise. Apart from the final settling tanks and tertiary filters, the process units are housed in structures that fit in architecturally with the nearby business and technology parks. The sludge from the new works is to be thickened, pasteurised, digested, and dewatered. The final sludge product is to meet the enhanced treatment standard and will be recycled to land.

Thames Water solicited technical proposals from of a number of sludge pasteurisation suppliers. The proposals were reviewed by operations and engineering. The successful bidder demonstrated reliable operation with feed sludges greater than 7% dried solids and efficient use of available heat from CHP engines. If one part of the plant failed, the remaining components process the sludge make albeit with a reduced holding time. The plant fits within the available area in the sludge building and can be odour controlled. This paper reviews the technologies considered and explains why the Alpha Biotherm plant was chosen for this site.

KEYWORDS

Anaerobic Digestion, pre-pasteurisation.

INTRODUCTION

The new Reading Sewage Treatment Works at the Island Road site will replace the existing Manor Farm works. The new works treats the wastewater to a higher standard using sustainable technologies such as biological nutrient removal and sludge pre-pasteurisation followed by mesophilic digestion.

Pasteurisation was scoped into the project to satisfy the sludge to land regulations and to minimise the odour emissions from the site. Pasteurisation followed by mesophilic digestion produces an enhanced treated product. Therefore, there is no need for secondary digested sludge liquid or cake storage.

The purpose of this paper is to explain why the Alpha Biotherm process was selected for this site.

SYSTEM DESCRIPTION

The effluent stream consists of preliminary treatment, primary lamellas settlers and biological nutrient removal activated sludge followed by tertiary filtration. Primary sludge and scum is screened and pumped to a sludge storage tank. There are two primary sludge tanks operating in fill and draw mode. The tanks are air mixed. The primary sludge is thickened through Simon Hartley/Belmer gravity belts to approximately 9% dried solids and discharged into one of two thickened sludge storage tanks.

The surplus activated sludge is taken from the return activated sludge line. There are two pumps and sludge lines to two Belmer belts. The surplus activated sludge is thickened to approximately 6% dried solids. The thickened SAS is discharged into one of the thickened sludge storage tank.

Scum from the aeration lanes is pumped to the primary sludge storage tanks and mixed with the unthickened primary sludge.

Fats, oil and grease skimmed off the FOGG lanes is heated to above 40°C and mixed with the thickened blended sludge.

There are two thickened sludge storage tanks, operating in fill and draw mode. The tanks are air mixed. The pasteurisation plant is fed from the tank that was filled the previous day.

DESIGN INTENT

Tenders were solicited on the basis of the following design intent.

POPULATION EQUIVALENT

The sludge streams treat sludge from a population equivalence of 306 000 PE. Approximately one third of the load is trade waste. The population equivalence at start-up in 2004 is 284 000 PE.

FLOWS

The design maximum flow to the pasteurisation plant over a day is 446 m^3/d. The expected flow ranges between 284 m^3/d and 400 m^3/d.

TOTAL SOLIDS CONCENTRATION

The average solids concentration of the thickened sludge is between 7% and 8%. The pasteurisation plant shall operate with feed solids concentrations between 3% and 10%. Thin sludge may occur during start-up while thick sludge may occur if the surplus activated sludge belts do not operate on the day when the primary belts are operating.

CHARACTERISTICS
The sludge to be pasteurised will consist of:
- indigenous surplus activated sludge and scum
- indigenous screened primary sludge and scum
- fats, oil and grease from the aerated grit grease lanes (which is added to the thickened sludge upstream of the pasteurisation plant)

TEMPERATURE
The minimum feed sludge temperature is 5°C. The average ambient air temperature over a day is between -10°C and 30°C.

REDUNDANCY
The system shall consist of at least two streams. One stream will treat half of the design flow to 70°C for one hour. If one stream fails, then the plant must treat the design flow to 70°C for 30 minutes. The temperature of the sludge in the pasteurisation plant must not exceed 80°C. In batch systems, the sludge must be mixed during lock-in.

The system shall be able to operate independently of the combined heat and power engines using duty/assist/standby dual fuel boilers. The fuels are biogas and diesel. In systems that include biological process, the boilers shall be sized without taking credit for biologically generated heat.

HACCP
The plant provides the SCADA system (by others) with sufficient signals to satisfy the HACCP requirements of the proposed UK sludge to land regulations. This includes lock-in time and sludge temperature. A facility must be in place to verify the above HACCP parameters using field equipment and instruments.

If there is a risk in a system that the sludge falls below 70°C or below 30 minutes retention time, then the sludge shall not be discharged into the digesters.

The project team agreed with operations that the pasteurisation system would not be installed with a permanent raw sludge bypass. In an emergency, a pre-engineered temporary pipe could be put in place to allow thickened raw sludge to bypass around the pasteurisation plant to the digesters.

FOUL AIR
Foul air is collected and treated. Unpasteurised foul air shall not come into contact with the pasteurised sludge headspace.

DIGESTED SLUDGE
To avoid problems associated with scale (e.g. struvite, vivianite or calcium phosphate), digested sludge is not to be passed through the pasteurisation heat exchangers or reactors.

DIGESTER FEED
The system must spread the feed to the digesters evenly across a day. Each digester shall have its own feed pump and pipeline.

The pasteurisation plant shall be able to transfer the equivalent of 1 degree/day of heat into 4 digesters when the ambient air temperature is equal or greater than -10°C and the feed sludge temperature is greater or equal to 5°C.

The boilers shall be configured such that any one of the boilers in the pasteurisation plant can provide hot water to the MONSAL heat exchanger in the digester area to provide standby heating to the digesters. The pasteurisation plant supplier shall provide the isolation valves and flanges required for this purpose.

HEAT BALANCE
The pasteurisation plant shall preferentially use the heat from the CHP units.

PROCESS GUARANTEES
The purpose of specifying process guarantees at the vendor selection stage is to ensure that the systems being evaluated are comparable.

DEMANDS
- The Subcontractor shall guarantee the following:
- potable water required (excluding topping up the hot water system via the expansion vessel)
- non-potable effluent water required (excluding wash down and commissioning)
- volume and strength of odorous air to odour control plant under normal operation
- hot water requirements including instantaneous flow rate and temperature

HEAT RECOVERY AND DIGESTER TEMPERATURE CONTROL
- The plant shall recover heat from the pasteurised sludge to pre-heat the raw sludge.

- The Subcontractor shall control the temperature across the digesters in service to within one ºC.
- At no time shall the digester temperature drop below 35ºC.
- The Subcontractor shall assume that the digested sludge feed lines and digester will not lose the equivalent of more than 1 degree C of heat from 7100 m^3 of water (active digester volume).
- The pasteurisation plant control philosophy shall maximise the use of hot water from the combined heat and power engines cooling water system.

PRODUCT QUALITY

The product quality guarantees shall apply to the pasteurised feed to the digesters. The sampling procedure is appended to this paper.

Salmonella spp shall not be present in 50 g (wet weight sample) during the 28 day test period. A known wet mass of sample is obtained to determine most probable number count per 50 g of sample. The analysis is incubation of various volumes of sample in an enrichment broth (e.g. RV broth) at 37ºC for 24 hours. This is followed by selective enrichment at 41ºC for 48 hours, with sub culture after 24 and 48 hours onto selective media at 37ºC for 24 hours. The plates are examined for suspect colonies.

Escherichia Coli shall not exceed 5 x 10^2 CFU/g (dry weight) during the 28 day test period using the membrane-lactose glucuronide agar (m-LGA) method developed by D. P. Sartory and L. Howard as practised by the Water Quality Centre, Thames Water Limited.

OR (AT THE CONTRACTOR'S DISCRETION)

Escherichia Coli shall not exceed 5 x 10^2 MPN/g (dry weight) during the 28 day test period using the colilert method as practised by the Water Quality Centre, Thames Water Limited.

The above methods are being refined through a research project in Pathogens in Sludge by UKWIR. The outcome of this work will be applied to the measurement method used to determine Escherichia Coli and Salmonella spp in sludge.

ENERGY DEMAND

The Subcontractor shall guarantee the instantaneous heat energy required by the pasteurisation plant under all modes of operation listed in the specification (in terms of kW).

The Subcontractor shall guarantee that the hot water in the heating loop shall not exceed 99ºC. This is because the provision and control of the boilers is part of the pasteurisation package.

ROBUST OPERATION

The Subcontractor shall guarantee that the pasteurisation plant will not shut down on the failure of one or two CHP units.

FEED SOLIDS CONCENTRATION

The Subcontractor shall guarantee that the pasteurisation plant will not fail if the feed solids is less than or equal to 10% dry solids on a weight/weight basis.

If the pasteurisation plant fails, a sample of the sludge in the thickened sludge holding tanks will be taken. The solids concentration of this sludge shall be measured. If the Contractor deems that the failure occurred because of the thickness of the feed sludge and the feed solids concentration is less than 10% (w/w) dried solids, then the Subcontractor shall have failed this guarantee.

The method used to determine the total solids shall be Standard Method 2540 G. Total Solids Dried at 103 ºC to 105 ºC, Standard Method for the Examination of Water and Wastewater 20th Edition, or equivalent.

FIRST STAGE EVALUATION

The thickened sludge is heated by hot water from the CHP engines. To keep the system simple, the intent is that the boilers would adjust the water temperature to compensate for a short fall in heat when it occurs. Operations did not want to have system that required steam. This eliminated systems that require steam, including the Monsal Hybrid, CAMBI and the Biwater Bitherm systems.

To minimise commissioning and odour emission risks, the decision was not to consider systems that did not have a comparable sized system operating elsewhere. This eliminated the Monsal Batch, Majex Batch, and Ashbrook systems. The PURAC Puriser was not considered because it was still under development at the time.

The third decision was not to relax the requirement to treat sludge with between 7% and 8% dried solids by weight. OTV and VA TECH declined to tender.

Tenders were sought for the Alpha Biotherm and Passavant Roediger systems.

The Alpha system consists of a batch sludge-to-sludge heat exchanger and a thermophilic aerobic reactor or digester (TAD). If the reactor is aerated, soluble organics are converted to lower-energy forms through anaerobic, fermentative, and aerobic processes at thermophilic temperatures.

The Passavant-Roediger system is a continuous flow system and does not include a biological unit process.

SECOND STAGE EVALUATION CRITERIA

The design team chose to review the Alpha Biotherm and Passavant Roediger systems in more detail.

COST

The cost of the Passavant system with a 50% standby stream was less than that of the Alpha system. However, the cost of the Alpha system where any three components of two 50% streams could be operated to form one 100% stream is comparable in cost to the Passavant-Roediger system with one 50% standby stream.

REFERENCE SITES

The Alpha system is installed at Ellesmere Port WWTW (United Utilities). The system pasteurises a mixed sludge consisting of indigenous primary and activated sludge as well as some sludge imports.

The Passavant-Roediger system is installed at Lillehammer and treats primary sludge. Pasteurisation plants on mixed sludges are currently being constructed at the Galway WWTW (Ireland) and commissioned at the Swords WWTW (Ireland).

ENERGY

Integration of the Passavant Roediger system with the combined heat and power engines is straightforward because the Passavant system is not a batch system. Integration of the Alpha Biotherm system with the engines is complex because the system operates in a batch mode with uneven phase lengths.

The hot water loop is much more dynamic when supplying a batch system. This increases the level of control and maintenance required. If neither Alpha reactor requires hot water, then the heat must be dumped before sending the water back to the CHPs. This will occur at least twice per cycle. This means the valve to a reactor will open or close at least 30 times per day. The Passavant Roediger system will process the day's sludge make and then shut the hot water down. This will occur once per day.

The Passavant Roediger uses the hot water from the engines much more efficiently than the Alpha Biotherm system. Depending on the feed sludge temperature and the digester heat requirements, two Alpha Biotherm reactors may require hot water at the same time. This demand may be greater than that supplied by the CHPs and must be supplemented by a boiler. This will occur during extreme weather conditions or when the CHPs are not at full capacity.

PLAN AREA AND HEIGHT

The envelope of the sludge building was agreed as part of the planning process. Both systems could be accommodated within the sludge building envelope.

The Alpha system requires 250 m^2 of floor space while the Passavant system requires 368 m^2. However, the Alpha system requires 14 m of height while the profile of the Passavant-Roediger system is much lower. The Alpha batch sludge heat exchanger and the reactor must be lifted into the building before the roof is put on. The Passavant-Roediger system could be installed after the building is finished.

ELECTRICAL DRIVES

Both systems require a similar number of drives. The Alpha system has fewer pumps but contains process air blower and mixers in both the reactor and batch heat exchanger. The absorbed power for both systems is roughly the same. The absorbed power for the Passavant system is 91 kW while that for the Alpha system 83kW. These loads have increased by 100% during detailed design.

ODOUROUS AIR

The Passavant-Roediger system produces far less odorous air because it is a continuous system. The cost of treating odorous air from the Alpha system is eight times that of the Passavant-Roediger system. Initial costs for the Passavant-Roediger system are £1000/yr while for the Alpha system £8500/yr.

PASSAVANT-ROEDIGER SYSTEM

Under normal operation, odorous air could be discharged from the pressure relief vessels. However, this is a very small volume. Our initial estimate is that the Passavant plant produces 70 m3/h of odorous air.

ALPHA

Air displaced by the filling of the batch heat exchanger or the reactor must be treated by the odour control plant. Typically, the filling of these reactors occurs over 9 minutes meaning the instantaneous flow rate to the odour control plant is high while the Alpha plant produces 600 m3/h of odorous air (when the process air blowers are not used). The instantaneous odorous airflow rate from the Alpha plant doubles if the process air blowers are used. However, intermittent aeration will maintain the redox potential discouraging the generation of odorous compounds.

MACERATION

Both systems require maceration prior to feeding the pasteurisation plant. The primary purpose of maceration is to slice fibrous material such as hair into short lengths to prevent the formation of hairballs in the Passavant-Roediger heat exchangers or ropes on the mixers in the Alpha batch heat exchanger. Because the Altweiler macerator has proved itself at the Hias, Aberdeen and Dublin CAMBI hydrolysis plants, we chose to use the same macerator at Reading.

MAINTAINANCE

Initial estimates are that the Alpha system requires six times the operator input as the Passavant-Roediger system. This is because the Passavant system consists primarily of pumps and heat exchangers while the Alpha system includes vessels with mixers and aeration.

CRANAGE

The Alpha TAD reactor contains a mixer, which must be withdrawn from the top of the reactor. Given the profile of the sludge building roof, there was very little room above the reactor for the crane to lift the mixer sections free of the top of the reactor.

HEALTH AND SAFETY

The Alpha system requires a high-level access walkway. The Passavant-Roediger system did not require operations staff to work at such heights.

HEAT EXCHANGERS

The Passavant-Roediger system uses small bore heat exchangers to maintain the velocity in the pipework (40 mm inlet and outlet). The pipework is 65 mm. Operations expressed concern over these small diameters but were willing to accept the design because of the level of engineering expertise contained within the Passavant-Roediger organisation.

The Alpha system uses a batch sludge-to-sludge heat exchanger to transfer heat from the hot pasteurised sludge to the cold raw feed sludge. The diameter of the concentric spiral inside the Alpha heat exchanger is 100 mm. The cold sludge is mixed using a conventional mixer, which could be withdrawn from the exchanger if problems occurred.

The sludge is heated to temperature in the TAD reactor. The lower half of the reactor is heated by a series of tubes around the reactor's periphery.

DIGESTER FEED

The Passavant-Roediger system spreads the feed to the digester more evenly over the day than the Alpha biotherm system.

PASSAVANT-ROEDIGER SYSTEM

The Passavant-Roediger system operates at a fixed flow rate. The system is designed to maintain a constant velocity in the heat exchangers. When the system is not operated, the sludge is moved through the system at a higher velocity to scour the heat exchangers and reactor.

The discharge from the Passavant system would feed each digester for approximately 20 minutes, adjusting the pasteurised sludge temperature to control the temperature in the digester.

ALPHA SYSTEM

The Alpha system discharges 15 m^3 of pasteurised sludge over 9 minutes at the end of every cycle. The cycle length varies depending on the raw feed sludge temperature and the heat requirements of the digester. A typical cycle length is 90 minutes. Alpha envisioned that one batch would go to one digester.

If the feed to the two streams were staggered, then each digester would be fed every 180 minutes. The implication of this are that (a) the diameter of the digester discharge pipework would have to enlarged and (b) the gas production profile would be more variable, possibly requiring an increase the size of the gas storage bags.

DRIED SOLIDS CONCENTRATION

The Alpha system tolerates a higher dried solids concentration than the Passavant-Roediger system.

PASSAVANT-ROEDIGER SYSTEM

Passavant-Roediger guaranteed that their system would operate with sludge up to 8% dried solids.

The design team considered installing a dry solids measurement device in the thickened sludge storage tank to warn the operator that the sludge was too thick to be pasteurised. The design allows unthickened primary sludge or SAS to be discharged into the tank to dilute the thickened solids. Operations rejected this approach arguing that most solids measurement devices are unreliable.

ALPHA SYSTEM

Alpha provided data showing that the Ellesmere Port installation successfully treated sludge with up to 10% dried solids concentration.

PASTEURISATION REACTOR
PASSAVANT-ROEDIGER SYSTEM

The Passavant-Roediger system uses a patented continuous flow reactor. The flow through the reactor is in the laminar region. There was a concern that this would cause short-circuiting to occur in the reactor with flow along the walls moving at a slower rate than flow in the centre of the channels.

Passavant-Roediger-Roediger guaranteed that the sludge heated to 72°C prior to entry to the reactor vessel would enter the vessel and remain in the vessel for a period of not less than 60 minutes. The guarantee would be tested using a tracer test by injection of Lithium at the inlet point of the vessel and measurement at the outlet point of the vessel. The test would be conducted on sludge with a concentration not greater than 8% dried solids at the design raw sludge feed rate.

ALPHA BIOTHERM SYSTEM

The Alpha Biotherm system adds 15 m^3 of warmed raw sludge to 85m^3 of previously pasteurised sludge. The hot water is pumped through the reactor's water jacket to heat the reactor's contents. When the temperature reaches 70°C, the lock-in timer starts. When the sludge reaches the temperature set point (75°C), the hot water is turned off.

PREFERRED TECHNOLOGY

The primary reason the Alpha Biotherm system was chosen over the Passavant system is that the system has a proven track record in the UK pasteurising a mixed sludge with up to 10% dried solids concentration.

SYSTEM DESCRIPTION
DEVELOPMENT

Once the Alpha Biotherm system was selected as the preferred technology for Reading, the Reading design team worked with the Alpha design team to optimise the design. This process was repeated with Clarke Energy to optimise the heat loop.

SPECIAL PROGRAM

Unique to Reading, two special programmes were developed so that the Alpha Biotherm system could process 100% of the sludge make when either a batch sludge-to-sludge heat exchanger or a TAD reactor is offline. These are described below.

HEAT LOOP

When at full load, the CHPs (located in the power building) raise the temperature of the cooling water by 10°C. This hot water crosses the site to the sludge building. The pasteurisation system uses the boilers to top up the heat when required. The hot water is sent to the reactors (or the hot water to sludge heat exchanger). The hot water is sent back across site to the power building. Heat for the fats, oil and grease system, screens and building heating is taken from this hot water. If the water temperature is above 85°C, heat is dumped. The cooled hot water is sent to the CHP engines.

If heat is required to warm a cold digester or to heat a digester that is not being fed by the pasteurisation plant, one of the pasteurisation boilers can be used to send hot water to the MONSAL tube in tube heat exchanger in the dewatering building.

DIGESTER TEMPERATURE CONTROL

Alpha's original proposal was to feed one batch to one digester. This was changed so that each digester was fed a portion of each batch. This meant that the Alpha plant could only control the average temperature of the four digesters rather than the temperature of an individual digester.

RISKS
OVER OR UNDER HEATING THE DIGESTER

Unlike the CAMBI system in Dublin, the system does not have a heat exchanger that can fine tune the temperature of the sludge being fed to the digester. Therefore, there is a risk that the pasteurisation plant may either provide excess or insufficient to the digester. For this reason, the MONSAL heat exchanger can be used to either heat or cool a digester.

Similarly, the Alpha system controls the average temperature across the digesters. If the temperature in one digester diverges, then the MONSAL heat exchanger can be used to adjust the temperature. If one batch were fed to a single digester, the Alpha system would be able to control the temperature in a single digester.

FAILURE TO PASTEURISE SLUDGE

If the sludge does not reach temperature, then it will be locked in for another 60 minutes. If the system fails to maintain temperature or lock-in time, the system shall send an alarm to SCADA and the system shall stop processing sludge.

OPERATION

The pre-pasteurisation process pasteurises the feed to the digesters by heating the sludge to 70°C for at least 30 minutes before cooling the sludge and feeding the pasteurised sludge to the mesophilic anaerobic digesters. After mesophilic digestion for at least 12 days, the digested sludge will satisfy the UK standards for an enhanced treated sludge product.

The pasteurisation plant consists of
- two batch sludge to sludge heat exchangers
- two thermophilic aerobic reactors
- one standby hot water to sludge heat exchanger
- four digester feed pumps
- three dual fuel (diesel/biogas) boilers

CONTROL

The volume of sludge to be pasteurised over 24 hours will be provided to the pasteurisation package. The pasteurisation plant package processes this volume to utilise heat available from the CHPs and to spread the digester feed events evenly across the 24-hour time period to control the average digester temperature.

The changing to or from 50% to one stream (normal operation) to any of the special modes is automatic once initiated by the operator.

NORMAL OPERATION

A stream consists of a batch sludge to sludge heat exchanger and a TAD reactor. When both streams operate, their cycles are offset so that they minimise the time that they demand heat from the hot water circuit at the same time. Figure 1

BATCH HEAT EXCHANGER CYCLE

The system starts a cycle with the inner chamber of sludge-to-sludge heat exchanger full and the outer chamber full of partially cooled pasteurised sludge.
Heat is transferred from the warm pasteurised sludge into the cold raw sludge. The phase ends when either the allowed time has expired or the pasteurised sludge is cooled to the temperature required to transfer sufficient heat into the digesters.
Transfer cooled pasteurised sludge to the digesters via dedicated digester feed pumps.
Transfer hot pasteurised sludge into the outer chamber of the heat exchanger.
Heat is transferred from the hot pasteurised sludge into the warm raw sludge. This phase ends when either the allowed time has expired or the raw sludge has reached the preset recovery temperature.
The warm raw sludge is transferred to the TAD reactor.
Transfer thickened raw sludge from the sludge storage tanks to the inside chamber of the sludge-to-sludge heat exchanger via a disk-to-disk macerator.

TAD REACTOR CYCLE

The TAD reactor is 85% full at the start of the cycle, ready to receive a new batch of raw sludge.
Mixer and scum cutter start.
Receive warm raw sludge from the batch heat exchanger.
Hot water flows through the reactor's water jacket to heat the TAD reactor to above 70°C.
(OPTION) The TAD is aerated intermittently during the heating up stage.
Once the sludge temperature reached 70°C, the lock in count starts.
Once the sludge reaches the target temperature (~75°C), the hot water stops. The stirrer and scum cutter continue to operate.
Prior to transfer out, shut off mixer and scum cutter.
Transfer pasteurised sludge when the heat exchanger is ready to receive the pasteurised sludge.

SPECIAL PROGRAM 1 – LOSS OF ONE TAD REACTOR

The two batch heat exchangers feed the TAD reactor sequentially as above except the hot water to sludge heat exchanger is used to pre-heat the raw sludge further before it is pumped to the TAD.
The lock in time in the TAD is reduced to 30 minutes. Figure 2

BATCH HEAT EXCHANGER CYCLE

The system starts a cycle with the inner chamber of sludge-to-sludge heat exchanger full and the outer chamber full of partially cooled pasteurised sludge.
- Heat is transferred from the warm pasteurised sludge into the cold raw sludge. The phase ends when either the allowed time has expired or the pasteurised sludge is cooled to the temperature required to transfer sufficient heat into the digesters.
- Transfer cooled pasteurised sludge to the digesters via dedicated digester feed pumps.
- Transfer hot pasteurised sludge into the outer chamber of the heat exchanger.
- Heat is transferred from the hot pasteurised sludge into the warm raw sludge. This phase ends when either the allowed time has expired or the raw sludge has reached the preset recovery temperature.
- If temperature is below the target value, the raw warm sludge is pumped through a hot water to sludge heat exchanger and back into the batch heat exchanger until the target temperature is achieved. The

hot water will turn on at the start of this phase and stop at the end of the phase.
- The warm raw sludge is transferred to the TAD reactor.
- Transfer thickened raw sludge from the sludge storage tanks to the inside chamber of the sludge-to-sludge heat exchanger via a disk-to-disk macerator.

The only difference between Normal mode and this mode is step 5.

TAD REACTOR CYCLE

As per the normal cycle except the lock in time is reduced from 60 minutes to 30 minutes.

SPECIAL PROGRAM 2 – LOSS OF A BATCH HEAT EXCHANGER

The batch heat exchanger and the two TAD units operate in series. The warm raw sludge is pumped to the first TAD unit to be preheated. The preheated sludge from the first TAD units is pumped to the second TAD unit for pasteurisation. The pasteurised sludge is pumped from the second TAD unit to the batch heat exchanger to transfer heat to the cold raw sludge.
There is a removable section of pipe that has to be installed before the system can be operated in this mode. Figure 3.

BATCH HEAT EXCHANGER CYCLE

As per Normal operation except the warm raw sludge is transferred to the pre-heating TAD reactor and the pasteurised sludge is transferred to the heat exchanger from the pasteurisation TAD.

PRE-HEATING TAD

The TAD is at 85% volume and ready to accept a batch.
Start hot water
Start mixer and scum cutter
Transfer warm sludge from batch heat exchanger
Once transfer is finished, continue to heat vessel until target temperature is reached.
Transfer pre-heated raw sludge from pre-heat TAD to pasteurisation TAD.
Stop mixer and scum cutter.

PASTEURISATION TAD

As per normal operation, except the preheated raw sludge is transferred from the pre-heat TAD, not the batch heat exchanger.
The lock in time in the TAD is reduced to 30 minutes.

CONCLUSIONS

The decision not to use steam eliminated a number of systems that could handle thick sludges.
The decision to feed the digester a mix sludge between 7% and 8% dried solids eliminated most systems that rely on a continuous flow heat exchanger to heat the sludge to 70˚C.
The decision to thicken primary and secondary sludge separately eliminated any system that could not handle sludges above 8% dried solids.
The Alpha Biotherm system was selected for the new Reading Sewage Treatment Works because the Ellemsere installation successfully treated thick sludges (>8% dried solids).

SAMPLING PROCEDURE FOR PRODUCT QUALITY

A spot sample will be taken daily during the first and second 28-day test period from the pasteurised feed to the digester at the manifold that supplies the four digester feed pumps.
The Subcontractor shall supply this sample point and shall submit a drawing of its location and design to the Contractor. The sample shall include permanent trace heating that will operate on a timer. The use of the trace heating shall be solely at the discretion of the Contractor.
The Contractor may reduce the number and type of samples taken if they are satisfied that the Pasteurisation Plant will satisfy its performance guarantees.

- The Contractor will take the samples from a sample valve in the header to the four digester feed pumps. All persons taking samples are equipped with helmet, safety visor, protective clothing, and waterproof gloves.
- All sample bottles shall be clearly labelled.
- The sample bottle shall be at least one litre in volume.
- Sample bottles shall be sterilised and opened immediately before sampling.
- Following sampling, the lids will be replaced immediately. The sampler shall ensure that there is no contact between the lip of the bottle or the lid with any external surface.
- The samples shall be immediately cooled to below 40 degree C in a cool box.
- The samples shall be stored in a refrigerator until they are collected for transport to the laboratory.
- The samples shall be transported to the laboratory for analysis the same day the sample is taken.

- The sampling procedure shall be as follows:
- Inspect sample point to ensure that it is clean. If not, clean the sample point with potable water.
- Place a suitable bucket under the sampling port
- Open the valve slowly
- Allow at least 10 litres of sludge to flow through the port into the bucket
- Open the sample bottle ensuring that no external surfaces come into contact with either the bottle lip or the bottle lid.
- Hold sample bottle under running stream of sludge and fill bottle.
- Close valve
- Replace lid of sample bottle
- Clean the outside of the sample bottle with potable water
- Label bottle with sample location, type of sample, date sample taken and name of person taking the sample
- Put sample in cool box
- Dispose of sludge in bucket
- Wash down area and sample point

ACKNOWLEDGEMTS

The authors wish to acknowledge contributions from Target (Reading) Design Team, particularly Nick Worral, Nick Townrow, Iain Partington, Alastair Steven and Phil Weston, Thames Water Engineering and Research and Development, including Ian Cranshaw and Manocher Assadi. Warren Davies (Alpha Environmental Technology Ltd), Roy P. Smith (Passavant Roediger Anlagenbau UK) and Clarke Energy (Nick Nicolaou).

FIGURES

Figure 1

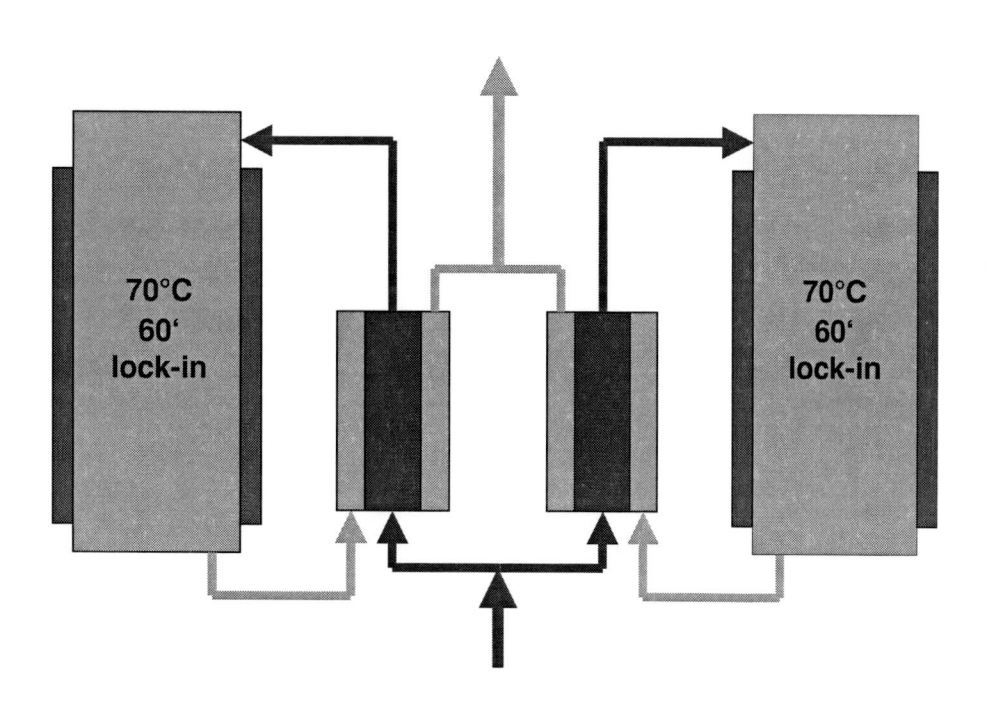

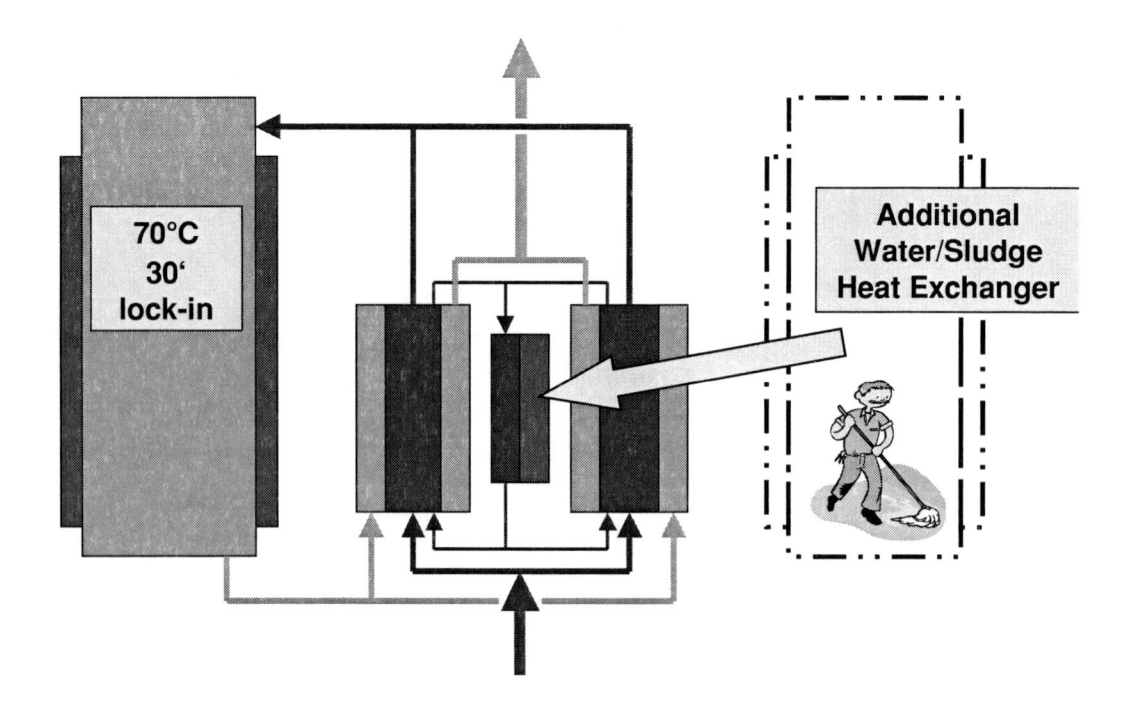

Figure 2

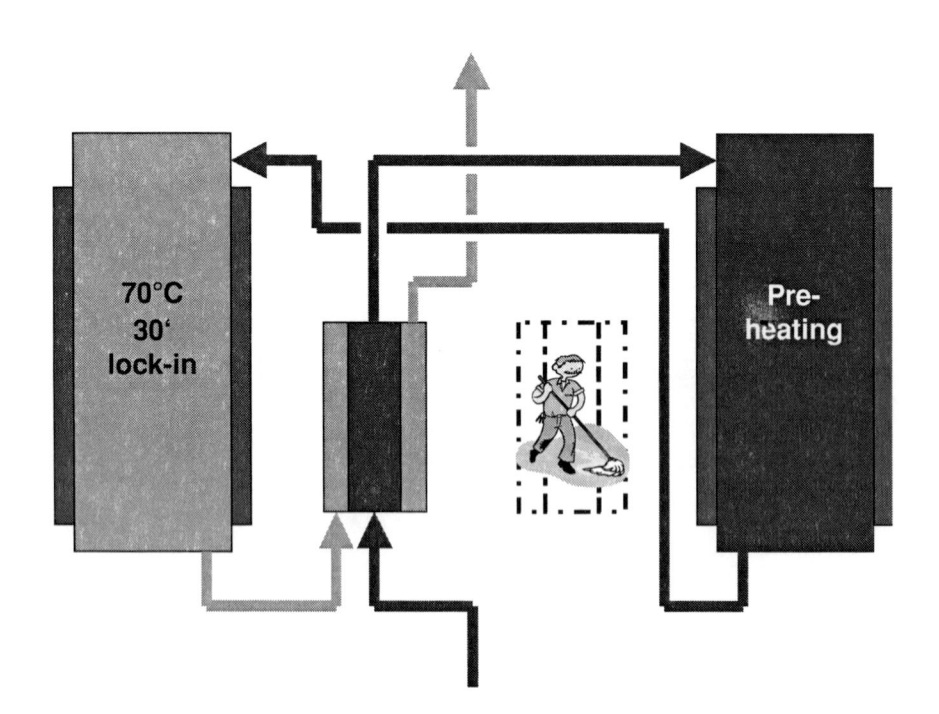

Figure 3

EXPERIENCES WITH THE ALPHA BIOTHERM BIOSOLIDS PASTEURISATION PROCESS AT UNITED UTILITIES ELLESMERE PORT SLUDGE PROCESSING CENTRE

[1]M. Mayhew, [2] W.J. Davies
[1] Biosolids Programme Team, United Utilities plc.
Maxine.mayhew@uuplc.co.uk
[2], Alpha Environmental Technology Ltd.

ABSTRACT

The proposed revisions of the 'Sludge (Use in Agriculture) Regulations' sets out two levels of sludge product with respect to *E Coli*; Treated and Enhanced Treated. This has focussed the Water Industry on a far wider range of sludge treatment technologies than previously experienced within the UK. United Utilities in anticipation of these changes commissioned the Alpha Biotherm process in December 1999, at the Ellesmere Port Sludge Processing centre, which processes 7000 tonnes DS per year. The thermophilic aerobic process, as well as producing a high quality pathogen and odour free biosolids product for agricultural usage, can produce a number of other benefits attributable not just to the higher operating temperatures (65^0C) of the process but also to the enzymic activity of the thermophilic aerobic microorganisms. The paper discusses the operation and performance to date of the Biotherm process at Ellesmere Port and the application of HACCP (Hazard Analysis of Critical Control Points) that are integral in the proposed regulations for process monitoring.

KEY WORDS

HACCP, operation and performance, pasteurisation, sludge treatment, thermophilic aerobic,

INTRODUCTION

United Utilities were the first UK Water Company to have an operational Alpha-Biotherm pasteurisation plant. The Alpha-Biotherm process was commissioned in December 1999 to achieve pasteurisation of the sludges at the Ellesmere Port Sludge Processing centre in the North West of England. Ellesmere Port had 3 existing Mesophilic Anaerobic Digesters (MAD) but insufficient secondary digestion tanks to comply with the UK Code of Practice for sludge digestion treatment. Development of the treatment solution for Ellesmere Port showed that installation of the Alpha-Biotherm plant was not only cost effective compared to the conventional secondary digestion but also returned a higher product quality and conferred processing security. Alpha Biotherm was chosen as the technology to deliver pasteurisation due to its utilisation of aerobic processing to aid the heat demand.

MAD processes have been used since the early 1900's for stabilisation and volatile solids reduction. The recent revisions to the 'Sludge (Use in Agriculture) Regulations 1989' adds another dimension to the requirements of digestion – the requirement for pathogen reduction. The routes for achievement of the new final product standard for sludge in terms of *E-coli* have been the focus of much recent research and optimisation of digestion processes and has prompted the rapid development of several new processes.

Pasteurisation processes have gained growing interest due to their ability to achieve the higher level sludge product (Enhanced Treated) and their robustness with regard to pathogen kill. Much has been reported on the use of pasteurisation technology for pathogen reduction; this paper concentrations on the operation and control of the Biotherm plant, its performance and the application of HACCP (Hazard Analysis of Critical Control Points) to the system.

ALPHA - BIOTHERM – THE PROCESS
AEROBIC-THERMOPHILIC BIOSOLIDS CONDITIONING AND PASTEURISATION

The Alpha-Biotherm process provides a retrofit capability to existing or new mesophilic digestion systems to achieve greater than 6-log pathogen kill in conjunction with beneficial biosolids conditioning. Typical values attained in UK and Swiss operational Biotherm plants are 0 Enterobacteriacea per 10 gms biosolids (CFU-Test) and absence of Salmonella per 10 gms biosolids (MPN-Test). Process efficiency is emphasised by the lack of regrowth with <100 Enterobacteriacea in 1 gm biosolids following 3 months storage in open tanks. Alpha-Biotherm plants have successfully been in operation in many wastewater treatment plants since 1985.

PROCESS DESCRIPTION
The process is of modular design and in most cases one stream serves each existing digestion plant (Figure 1). The process comprises of 2 main units the sludge/sludge heat exchanger and the Thermophilic Aerobic Reactor.

SLUDGE / SLUDGE HEAT EXCHANGER

The design has been optimised over a number of successful years of operational use within Switzerland and gives an efficient and rapid heat exchange between the incoming cold biosolids feed and the treated biosolids from the Alpha-Biotherm reactor. The efficiency is inherent in the engineering design and the process operation whereby a two-phase exchange is achieved. The inner and outer chambers are of equal capacity and represent a process batch volume (Figure 2).

THERMOPHILIC AEROBIC REACTOR

The design features achieve efficient oxygen transfer from the aspect ratio of the design and the use of a hyperbolic mixer (figure 3, 4). The hyperbolic mixer, patented and named the Hyperclassic mixer, achieves high bottom velocities (discouraging settlement of grit and heavier particles) and intimate mixing with long air bubble paths and breakage of the coarse bubbles into finer bubbles (the latter achieving high oxygen transfer rates). The reactor is supplied with heat from the biogas utilisation system (either a boiler or CHP system) delivered by circular heating coils around the body of the reactor. The design ensures that all of the heating requirements of the Alpha-Biotherm-MAD system is satisfied by this mechanically supplied heat. The exothermic thermophilic microorganisms residing in the reactor also generate heat. This heat is generated from a maximum organic breakdown of 6%. The net effect of the Biotherm-MAD system is an overall increase in biogas and organic destruction of approximately 10%.

This additional heat source ensures that:
- Heat recovery following a batch feed into the reactor is more rapid and
- During the colder winter months the Biotherm MAD system does not have to rely solely on supplementary heat.

Detention time in each reactor is approximately 12 hours.

SUMMARY OF PROCESS OPERATION AND CONTROLS
- One batch of cold raw biosolids is fed into the inner chamber of the sludge/sludge heat exchanger while hot treated sludge is fed into the outer chamber and pumped through internal concentric coils contained in the inner cold sludge chamber. A dual phase heating procedure allows the efficient exchange of heat. Cold sludge is heated to approximately $35^{0}C$. Treated sludge is cooled to approximately 36 - $50^{0}C$.
- Pre-heated raw biosolids from the inner chamber is then batched over into Alpha-Biotherm reactor. Within the reactor it undergoes mixing and aeration until the target temperature of between $65^{0}C$ - $70^{0}C$ is attained. The heating rate is enhanced by exothermic thermophilic activity. The reactor is then locked out to allow the one-hour pasteurisation period.
- Cooled treated biosolids are transferred to the downstream digester. The treated biosolids contains all of the heating requirements of the downstream digester during summer and winter conditions.
- The sequence is then repeated.
- Automatic control to provide heat to downstream MADs is by a combination of:
    - The number of batches of biosolids required to be treated and
    - The target temperature within the downstream digester
- The system is supported by telemetric alarms, PLC control, data logging/display and automatic safe shutdown is achieved in all circumstances
- The process can be adjusted simply in order to provide additional heat to the downstream MADs.

PROCESS BENEFITS
EFFECTIVE PATHOGEN KILL

The process effectively achieves greater than 6-log kill and complies with the likely requirements of future EC and UK legislation. This is in part due to the higher thermal exposure times for pathogens kill. The batch design allows for a minimum lock-in time of one hour at 65 - 70°C. However, as it is a batch process the average hydraulic retention time in the reactors is 8 - 11 hours at peak flows. This elongated average pasteurisation period will ensure thorough pathogen kills due to higher thermal exposure times. For sludges where particle sizes will vary and potentially contain high pathogen content "embedded" within dense particles, this elongated average pasteurisation period will ensure thorough pathogen kills.

BIOSOLIDS CONDITIONING TO ENHANCE THE SUBSEQUENT MAD STAGE

This is achieved via the following mechanisms:
- Reduced viscosity of the biosolids enhancing the mixing performance of the MAD stage
- Improved dewaterability of the biosolids allowing improved consolidation

characteristics of the treated liquid sludges or improved cake dry solids with reduced chemical conditioner usage of any subsequent dewatering stage.
- Provision of an aerobic hydrolysis stage with enzymatic breakdown and modification of more complex substrates to enhance the MAD performance stage. Aerobic hydrolysis takes place at a faster rate than in anaerobic processes so that the sludge substrate entering the MAD stage is more readily degraded.
- The overall effect of reduced viscosity and substrate modification leads to an overall average increase in biogas yield and organic solids destruction of 10%, reduced chemical usage in downstream dewatering and higher sludge cake yields.

PROCESS FLEXIBILITY

The process copes with emergency operational peaks in demand, fast start-up of MAD commissioning and the ability to uprate the process for future site demands.

MINIMISATION OF STRUVITE ACCUMULATIONS

The process has advantages to avoid the detrimental effects of struvite deposition: including ensuring sufficient velocities in large diameter pipework and vigorous agitation in the heat exchanger and reactor vessels.

TROUBLE-FREE HEAT EXCHANGERS

The unique sludge-sludge heat recovery exchangers, due to high turbulence and recirculation avoid sludge baking and blockage problems associated with sludge- water heat exchangers at the high pasteurisation temperatures. Blockages are minimised by the use of pipework with internal diameters in excess of 100mm. The dual phase exchange design ensures high efficiency of heat transfer/recovery due to the maximisation of temperature gradients between cold raw and hot pasteurised sludges.

TOLERANCE TO HIGH SOLIDS CONCENTRATIONS

The patented hyperbolic stirrer installed in the reactor is designed with high bottom velocities and low turbulence and power density to accommodate the mixed and heavier particle sizes associated with sludges. More dense particles will not be allowed to settle at the base of the reactor. The design of the mixer and reactor configuration has been extensively modelled.

LOW OPERATIONAL SUPERVISION AND MAINTENANCE

Maintenance effort and costs is minimal for the Alpha-Biotherm process particularly compared with other pasteurisation technologies. The sludge – sludge heat exchangers do not require frequent cleaning to combat baking and blockages as is the case with standard sludge-water heat exchangers. The plant is fully automatic and does not require a constant manning level. Operation is simple and non-demanding involving tasks such as pump and valve maintenance common within water utility businesses. Periodic cleaning once every one - two weeks for pressure transducers is required along with normal maintenance requirements for pumps and automated valves. Telemetric alarm and information signals ensure ease of central capture of information and control. HACCP monitoring is simply achieved from the logging of batches processed and the achievement of temperature targets and lock-in times.

MINIMISATION OF ODOURS

The intermittent aeration in the reactor maintains higher Redox potential levels in the contained sludge so minimising the potential for the production of hydrogen sulphide and other mal-odorous gases associated with the anaerobic detentions in alternative pasteurisation technologies.

MINIMISATION OF PHOSPHATE RESOLUBILISATION

Pasteurisation is under predominantly aerobic conditions to prevent the resolubilisation of phosphate from BNR sludges

ELLESMERE PORT BIOTHERM – DESIGN

SUMMARY OF DESIGN

The Alpha-Biotherm plant at United Utilities Ellesmere Port WwTW, was commissioned during 1999. As an illustrative example, a summary of the design of the Alpha-Biotherm plant at Ellesmere Port WwTW is shown below. The plant was designed to allow for the processing of additional biosolids at the site as well as being UU's first reference site for pasteurisation and the production of a high quality product. Details of the design parameters are given in table1.

ELLESMERE PORT OPERATION AND PERFORMANCE

GENERAL OPERATION

The plant has been in operation since December 1999. Following commissioning, all performance targets were achieved. Pathogen content from the Alpha-Biotherm unit showed an absence of Salmonellae and E.coli per gm of biosolids for the commissioning period. The energy consumption target of less than 3.0

KWhr/m^3 of biosolids treated (with 3 streams operational) was also been achieved.

The results discussed below encompass more recent data from the past year. During that time, the Biotherm plant at Ellesmere Port was not currently utilised to full capacity due to operational issues with the existing MADs which were being addressed. Due to the reduced sludge flows on site the average retention in the Alpha-Biotherm was 1d and in the MAD a high average of 25d but a minimum of 12d. The mean temperature of operation of the Alpha-Biotherm is 64.8 ± 1.3 °C and of the MAD 36.3 ± 0.6 °C (Table 2). The operating philosophy of the plant is to feed the coolest digester in order to ensure the Alpha-Biotherm unit supplies the base heat demand for MAD. The MAD temperature was maintained very consistently as indicated by the small standard deviation.

The pH of the raw sludge was typically around 5 to 6 which increases through the Alpha-Biotherm and MAD and is 7.5 in the resultant treated sludge comparable to any other digestion process (Table 3). The soluble COD of the raw sludge was seen to increase 2 fold after Alpha-Biotherm treatment suggesting that hydrolysis is occurring in the process. The soluble COD of the resulting MAD product was typically around 2000 mg/l as per a typical MAD system (CIWEM, 1996) (Table 3). A release in ammonia is seen across the Alpha-Biotherm and MAD processes, which is a normal metabolic product of anaerobic digestion reactions and levels in the MAD product

The raw sludge feed to the Alpha-Biotherm was Gravity Belt Thickened (GBT) and was typically 6 to 7% DS and 65% volatile solids (VS) (Table 4). The combined Alpha-Biotherm/MAD process achieved 44% VS destruction with 7% of the total occurring in the Alpha-Biotherm process. This equates to a 3% destruction of the raw load VS and is within the specified parameters for the operation of the plant. The thermophilic aerobic process can achieve much greater VS destruction. However, United Utilities had designed the plant such that the bulk of the VS destruction and hence gas production was within the MAD to allow harnessing of the gas for beneficial utilisation on site.

PRODUCT QUALITY

A consistent product quality with regard to pathogens is essential to ensure robust compliance with the new sludge regulations. The Ellesmere Port Biotherm plant produces an Enhanced Treated product which therefore has to achieve less than 1000 *E Coli* per g DS and an absence of Salmonella.

The raw sludge had a typical E Coli content between 10^6 and 10^7 *E Coli* per g DS and had a presence of Salmonella. The product from the Alpha-Biotherm typically showed an absence of E Coli and of Salmonella. The product of the combined Alpha-Biotherm and MAD system consistently achieved less than 1000 *E Coli* and an absence of Salmonella (Figure 1). Inefficient mixing systems in the MADs are believed to account for the slightly higher *E Coli* results in the MAD product compared to the Alpha-Biotherm. Additionally, the analytical method for *E Coli* most commonly used and employed here for the data shown (Figure 1) was the Membrane Filtration method. This method has a limit of detection of around 10 – 100 colonies per ml – resulting in a limit of detection around the 1000 *E Coli* per g DS. Thus whilst an absence of E Coli can be reported accurately low numbers can not with this method making the full evaluation of the Enhanced standard more difficult. The method utilised for analysis of Salmonella is that described in 'The Isolation and Enumeration of Salmonellae from Water, Sludge and associated materials'.

CONDITIONING EFFECTS

One of the perceived additional benefits of the Biotherm process is increased conditioning effects of the sludge downstream. Ellesmere Port WwTW dewaters the Alpha-Biotherm/MAD treated sludge using centrifuges prior to recycling to land as a caked product. Given that the centrifuges were installed after the Biotherm process, there is no site specific comparison, however, the performance of the centrifuges has been better with the Alpha-Biotherm processed sludge than compared to other UU sites with normal MAD digested sludge.

The weekly average cake %DS from Ellesemere Port was 27.5 % with a maximum of 31.2 %. The 50%ile for the cake produced was 27% and the minimum DS produced was 25% (Figure 2). A comparative process with a MAD sludge produced only 26% DS maximum with comparable polymer dosing to Ellesmere Port. It appeared that the % DS achieved was improved with the Biotherm conditioned sludge. This improvement to cake DS was not at the expense of centrate quality with centrate solids being on average 184 ± 67 mg/l (Table 5, Figure3).

MAINTENANCE

The plant is fully automated and requires minimal daily operator intervention. Daily inspections are carried out and any adjustment required to the number or size of the batches to be processed. It is essential to ensure that the pressure and temperature transducers are regularly inspected and cleaned as these are critical to the control of the process to ensure a quality product. Additionally, the odour control unit required regular cleaning to ensure optimal performance.

The heat exchangers are designed for minimal maintenance and despite the high temperature operation of the process no routine cleaning has been required. No major blockages have been experienced despite the only screening for indigenous sludges being the inlet screens. Full vessel cleaning is required every 3 years and the 1st Alpha-Biotherm line is under cleaning and maintenance as this paper is prepared.

APPLICATION OF HACCP

HACCP has been widely used in the food industry to ensure that the product quality remains consistent. HACCP is now being applied to sludge processing in support of the changes in legislation and the introduction of a final product quality on sludge to be recycled in agriculture. HACCP is a systematic approach for the evaluation, monitoring and control of microbial hazards.

The HACCP approach implemented by UU utilises 7 key steps to ensure product quality:

1. Conduct hazard analysis
2. Determine the critical control points
3. Establish critical limits
4. Establish system to monitor control of CCP
5. Establish the corrective actions to be taken when monitoring indicates that a particular CCP is not under control
6. Establish procedures for verification to confirm that the HACCP system is working effectively
7. Establish documentation concerning all procedures and records appropriate to these principles and their application

Once CCPs and their critical limits and ranges have been defined, whilst the plant operates within the given range then assurance can be gained that the product quality remains at the known monitored value. In order for HACCP to work and be monitored successfully, simple, reliable CCPs are required. The identification of CCPs for an 'idealised' digestion process typically includes primary digester temperature, primary digester retention time and secondary digester retention time. Generally, the numerical values attributed to each of these CCPs are site specific.

The sludge treatment process at Ellesmere Port consists of '3 No Alpha-Biotherm units and 3 No MAD units'. Pathogen reduction is achieved by holding untreated sludge in the ALPHA-Biotherms at 65°C for at least 1 hour. The MADs reduce the fermentability of the sludge by converting a substantial proportion of organic matter to biogas.

The Alpha-Biotherm plant enables HACCP to be applied readily and easily on agreed CCPs. Typically CCPs for the Biotherm plant would be the temperature and the retention time in the ALPHA-BIOTHERM to achieve the pathogen reduction. The measurements for the temperature CCP can be obtained from the temperature gauge in the Alpha-Biotherm reactor and can be checked by monitoring the temperature in the heat exchanger. The temperature data is displayed on a local operator interface and so can be seen, checked and recorded manually but is also fed into the SCADA system for automatic recording. The retention time CCP is monitored through the number of batches processed. The number and volume of the batches is determined by the Operator and is automatically logged through the control and SCADA system.

SUMMARY

The Biotherm process in conjunction with standard MAD treatment has delivered a robust product quality with respect to pathogen reduction. The process has simple CCPs to monitor to ensure confidence in the continuous delivery of a quality product. It has provided reliable robust processing with little daily operator input. Routine maintenance is essential but has not been too onerous. The Biotherm process appeared to have a beneficial impact on the downstream dewatering process resulting in a consistently higher DS product.

TABLES

Table 1: Selected Design Parameters of the Alpha-Biotherm System Installed at Ellesmere Port WwTW.

Number of Alpha Biotherm Modules	3
Maximum Biosolids Flow (m³/day)	1077
Max. Flow per line (m³/day)	359
Dry Solids Feed (%,w/v)	5-7
O&V (%, w/v)	65
MAD Heat Losses (KWhr/day) (summer)	1004
MAD Heat Losses (KWhr/day) (Winter)	2383
Reactor Volume (m³)	120
Reactor diameter (m)	4
Reactor Height (m)	10
Biosolids/Biosolids Heat Exchanger capacity (m³)	2 x 25
Biosolids Feed rate to heat exchanger (m3/hr)	108
Temperature recovery (mins, max.)	15
Maximum O&V destruction in ALPHA-BIOTHERM reactor (%)	3 - 6
Maximum Biological heat production (KWhr/day)	4200
Hot water supply Temperature (°C)	80
Hot water supply rate per module (l/sec)	11.6
Pasteurisation temperature (°C)	65

Table 2: Typical operating parameters

	ALPHA-BIOTHERM	MAD
Temperature °C Mean ± SD	64.8 ± 1.3	36.3 ± 0.6
HRT Average Min	1 d 16 h	25 d 12 d

Table 3: Operational performance of Biotherm and MAD system

	Raw sludge	ALPHA-BIOTHERM Product	MAD Product
PH	5.8 ± 0.1	6.4 ± 0.1	7.5 ± 0.2
Soluble COD mg/l	5561 ± 1250	12620 ± 1494	2148 ± 373

Table 4: Dry and Volatile solids of Biotherm and MAD system

	Raw Sludge	ALPHA-BIOTHERM Product	MAD Product
DS %	6.2 ± 1.2	4.9 ± 0.5	3.3 ± 0.3
VS %	65.3 ± 1.3	63.8 ± 1.7	54.2 ± 4.5
VS destruction %		7.5 ± 5.0	36.1 ± 4.5

Table 4: Centrate Quality

Parameter mg/l	Centrate Quality Mean ± SD
BOD	82 ± 34
COD	895 ± 410
Solids	184 ± 67
Ammonia	720 ± 253

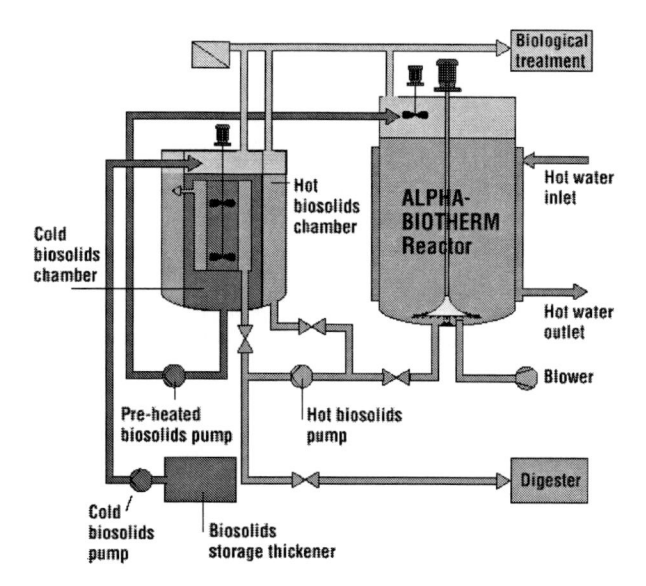

Figure 1: The Alpha-Biotherm Process

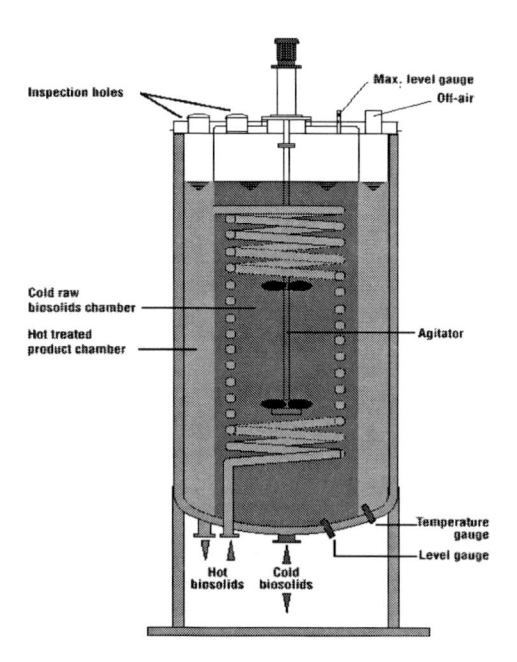

Figure 2: Sludge / Sludge Heat Exchanger

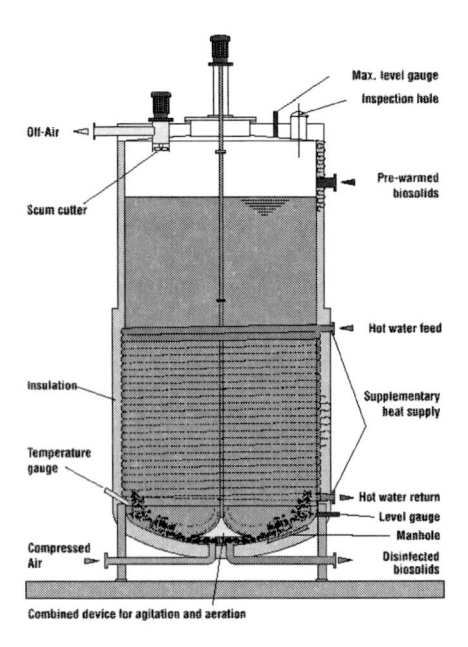

Figure 3: Thermophilic Aerobic Reactor

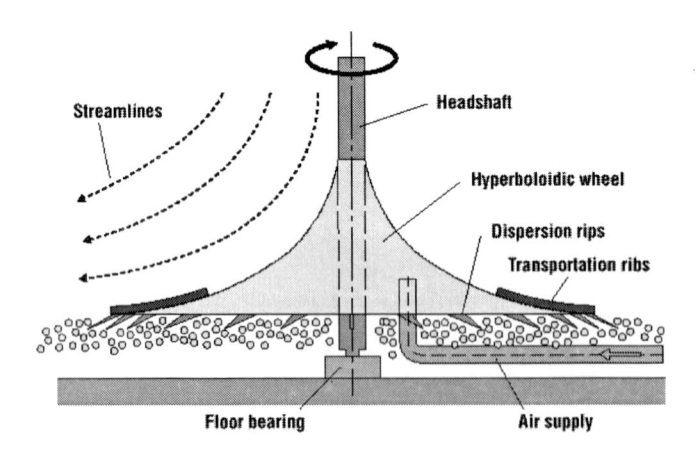

Figure 4: The Alpha Hyperolic high performance mixer / aerator

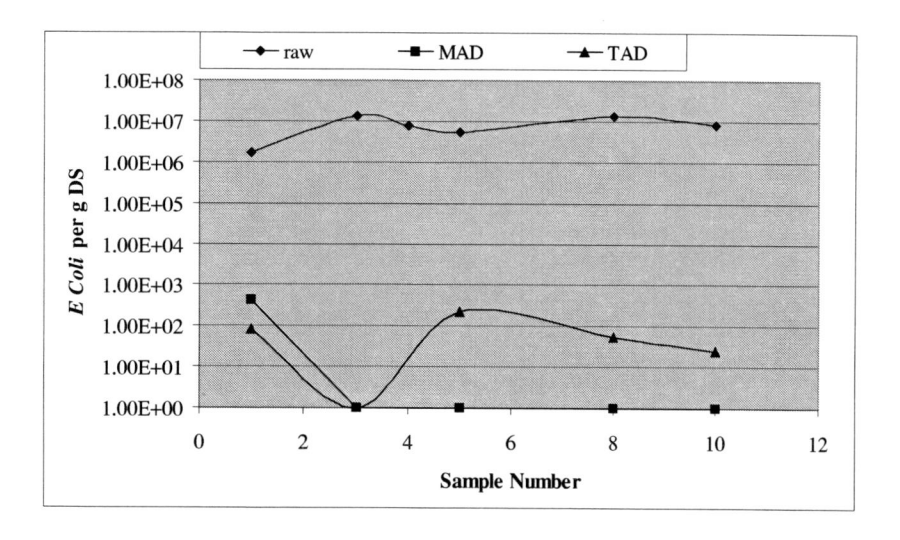

Figure 1: Typical *E Coli* per g DS in Ellesmere Port Biotherm System

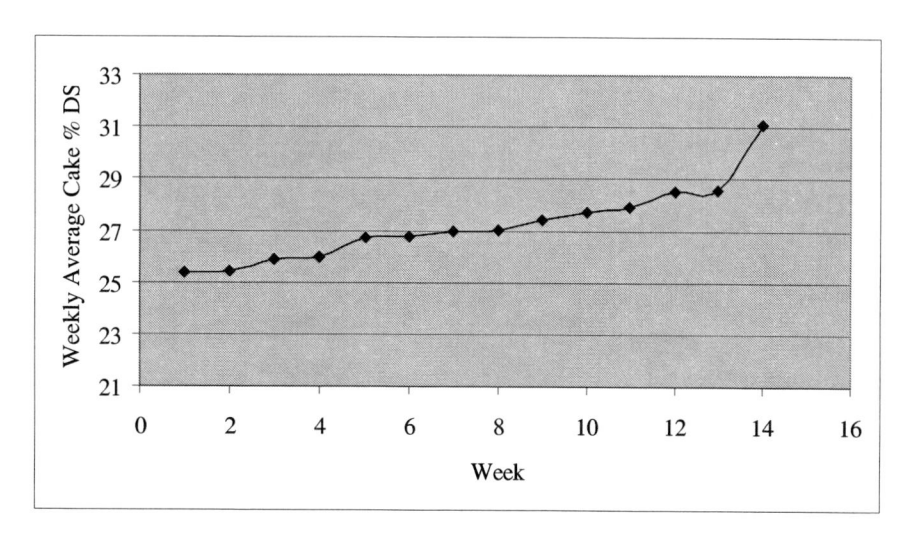

Figure 2: Typical Weekly Cake % DS as a frequency distribution plot

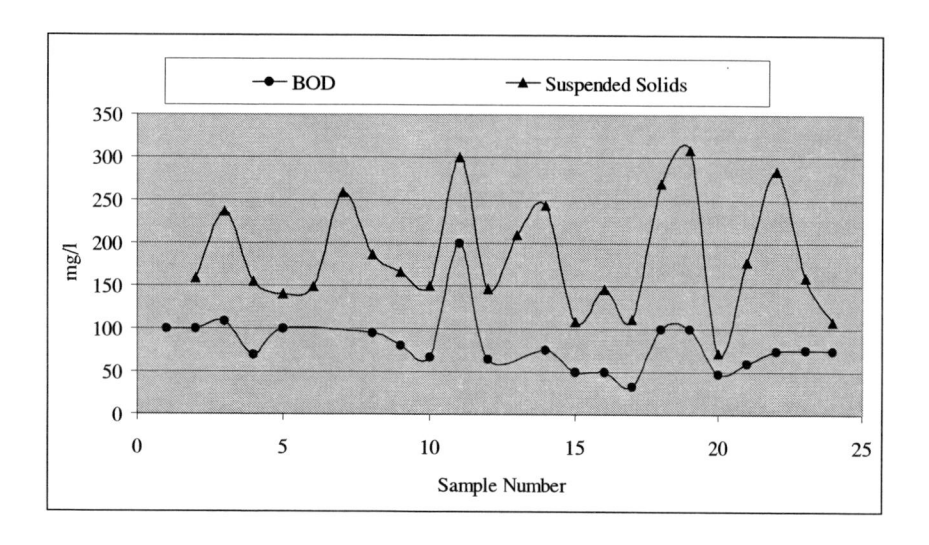

Figure 3: Typical Centrate data

COMBINED FILTER PRESS/DRYER TECHNOLOGY
-AN OPTION FOR ENHANCED TREATMENT -

[1]Mervyn Brown, [2]Kim Thompson [3]Dennis E. Bentley,
[1]Damar Engineering Ltd. [2]TVD (UK) Ltd. [3]DryVac Environmental
Phone: 01761-439111 Fax: 01761-439123
E-mail: mervyn.brown@damarnet.com, kimt@tvd.uk , Bentley@dryvac.com

ABSTRACT

At the 4th European Biosolids and Organic Residuals Conference, an initial paper was presented on the combined filter press/dryer technology [1]. In 1999, there were no full size units of this type operating in the Biosolids wastewater area as the technology was originally applied to industrial applications. This included the dewatering and drying of waste from metal hydroxide applications and of product within the pigment industry. Since 1999, the combined filter press/dryer technology has expanded outside the industrial area to various applications within the Biosolids arena. There are now more than 15 units operating or being built within the Biosolids area. While most of the units are in the United States, the tightening of environmental regulations within the industrialized world continues to increase the global potential for the application of this technology in the Biosolids area. The combined filter press/dryer technology is used in a wide range of Biosolids applications. These include industrial Biosolids, pulp/paper, municipal wastewater and water treatment. By far the largest increase in application has been in the municipal wastewater area.

In review, solids created from the processing and cleaning of municipal wastewater have typically been dewatered with centrifuges, belt presses, and filter presses. Each of the technologies has their own advantages and disadvantages. Initially, the dewatering process was chosen based on ease of operation. As environmental concerns became a higher priority, the need to produce the highest percent solids was thrown into the equation. Traditional dewatering equipment typically produces sludge in 15-25% solids range. As the desire increased to produce sludge with higher percent solids, technologies, such as thermal drying or lime stabilization became a consideration. As environmental regulations become more stringent, the wastewater plant operations have to consider all the potential negative aspects of Biosolids sludge. These include dryness of the solids, pathogen levels, odours, and final disposal/use of the sludge. As always, the economics of the overall plant process requires consideration, and in many cases will determine the treatment and product utilisation route.

The combined filter press/dryer performs the dewatering and drying in one step. The core of the unit is diaphragm filter press. After the standard filter press dewatering, the unit dries the material. This is accomplished by recirculating a heating medium behind the diaphragm while drawing a vacuum on the sludge in the filter press chambers to boil off the liquid. In 1999, the technology used hot water at 80^0C as the heating medium. The use of steam at 115^0C is now an option. This provides improved cycle times for drying and pathogen reduction while reducing the energy requirements for the process. With test units now available in the UK, wastewater facilities in the European Union have the opportunity to consider this technology as an alternative to the traditional sludge treatment methods.

KEY WORDS
Biosolids, Dewatering, Drying, Pastuerisation

INTRODUCTION

The combined filter press/dryer dewaters and dries the sludge from municipal wastewater in a single unit. The technology can dry the sludge to levels in excess of 90% solids and can obtain a log-6 reduction of pathogens. The operations are carried out with less energy and lower emissions than competing technologies. At the 4th European Biosolids and Organic Residuals Conference, an initial paper was presented on the combined filter press/dryer technology [1]. At the time, there were no full time units of this technology operating in the municipal wastewater area. There are now more than 15 units operating or being built worldwide. In addition, advances in equipment design have resulted in improved drying times and lower energy requirements.

TECHNOLOGICAL DISCUSSION
THE PROCESS

Historically, municipal biosolids have been dewatered with centrifuges, belt presses, or filter presses to produce a final sludge in the range of 15-25% solids. Pathogen reduction was typically dependent on the composting, lime stabilization, or upstream biological processes, such as aerobic and anaerobic digestion. Environmental regulations and public demand have driven the need for sludge that is drier and contains a lower pathogen content. This is especially applicable in the area of land application. This led to the consideration of a dewatering device followed by a thermal dryer and many papers, in recent years have considered this route to Enhanced Treated status [2] [3]. The combined filter press/dryer technology would be applied where the municipal wastewater facility desires the same result obtained from a dewatering device and thermal dryer combination.

The core of the combined filter press/dryer is the standard diaphragm filter press. A filter press consists of a number of plates, which are contain within a steel framework and clamped together by a hydraulic ram. The plates form a series of chambers within which the solids are accumulated. A screen or filter cloth is attached to the plates to provide the filtering mechanism. In a standard recessed filter press, the plate diaphragm along the chamber formed by the plates is stationary and an integral part of the core of the plate. In a standard diaphragm filter press, the diaphragm can be pressurized from the internal side of the plate with air or water. See Figure 1. This allows the diaphragm to expand outward from the plate into the chamber.

The initial process of the standard diaphragm filter press and the combined filter press/dryer are identical. In the case of municipal wastewater, the slurry is chemically treated to create flocculation and coagulation. This is needed to improve the "dewaterability" of the slurry. The chemical treatment is critical in sizing of the equipment since it determines the amount of solids that can be inserted into the chambers of the filter press and, therefore, the throughput capabilities of the equipment. After chemical treatment, the slurry is pumped into the chambers of the filter press. As the chambers fill, the solids are maintained in the chambers by the screens or filter cloths. The discharge of the filter press is usually to atmospheric conditions, allowing the filtrate to drain by gravity through drainage holes within the plates. The solids continue to build up within the chamber until the pressure provided by the feed pump and the resistance of the sludge to give up any more water is essentially equal. The sludge within the chambers is typically referred to as "cake". The area behind the diaphragm is pressurized to apply pressure to the cake. This can be compared to squeezing a sponge that is full of water. Air is also blown through the cake to assist in the further removal of water. See Figures 2 & 3.

In municipal applications, the cake is typically at 20-25% solids after this series of operations. The percent solids obtained in a filter press are usually higher than the percent solids obtained in a centrifuge or belt press. This has traditionally made the filter press more popular in the European Union than in the United States, where the centrifuge and belt press have been more popular in municipal wastewater applications. A similar trend toward centrifuges and belt presses versus the plate filter press has been evident in the European Union over the past 20 years.

After the "dewatering" phase, the drying technology comes into play. See Figure 4. The heat source, either hot water at 80^0C or steam at 115^0C is sent to the area behind the diaphragm. The heat source provides the energy to evaporate the water in the cake. The pressure in the heat source expands the diaphragm to:

1. Maintain direct contact between the diaphragm and the cake as the cake shrinks due to the liquid removal. This is extremely important to ensure that the heating is maintained in the conductive mode versus the radiation mode. Early attempts to implement this technology used steam heated steel plates. These plates had no diaphragm capability and, therefore, as the cake dried, it lost contact with plate surface. When this occurred, the water removal rate dropped significantly.
2. Continue to squeeze the cake to remove mechanically remove more water from the cake.

As the heat source is being applied behind the diaphragm, a vacuum is being applied to the cake. This causes the liquid within the cake to boil at a reduced temperature of approximately 40^0C. The vapour/liquid phase drawn from the cake travels through a reservoir, or knockout pot, where the liquid phase is collected. The vapour enters a condenser, where the most of the vapour is turned into a liquid. The remaining vapours and non-condensables continue through the vacuum pump where remaining vapours are condensed.

During the drying phase, the vacuum system can be secured to expose the cake to the time and temperature requirements for pasteurisation. This ensures the log-6 reduction in pathogens.

The heat and vacuum are maintained until the cake reaches the desired level of dryness. The unit is then secured, the plates separated, and the dried cakes fall from the unit into a collection bin or conveyor.

THE EQUIPMENT

Although the application of this technology is recent, the combined filter press/dryer is a collection of standard equipment. The units consist of the following:

1. A standard diaphragm filter press. These include the frame, hydraulic system, plates, and screens. The newest component is the plate used on the steam-heated units, which required the development of advanced materials to withstand the steam temperatures. Unlike a standard filter press, a piping system is attached to the frame to deliver the hot water or steam.
2. A heat source delivery system. This includes:
   - For the hot water system, the heater, pressure pump, and recirculation pump
   - For the steam system, the boiler, steam traps, and makeup tank/pump.
3. A vacuum system. This includes a liquid ring vacuum pump, condenser, knockout pot, and cooling tower, if required.

THE ADVANTAGES

The combined filter press/dryer has unique features compared to other technologies used in municipal Biosolids applications. These include:

1. Flexible Drying Levels. As requirements change, the final percent solids of the Biosolids can be changed by simply changing the length of the drying time.
2. Energy Efficient. The heating systems are a closed loop without the loss of energy to stack discharges
3. Low Emissions. The withdrawn vapour is condensed in the condenser and the vacuum pump. These two steps act as scrubbing actions to greatly reduce emissions and odours.
4. Automation. The drying of the solids within the filter press allows the solids to discharge easily from the chambers. Unlike the typical filter press where the cakes have to be manually "scraped" from the plate, this discharge can be automated.

TYPICAL PERFORMANCE

The performance parameters are based on several variables including sludge type, chemical treatment, and level of dryness desired. Typical performance specifications are as follows:

1. The energy usage to dry to 90% solids is 1,800 kWh. per dry tonne for heat energy and 120 kWh. per dry tonne for electrical energy.
2. For automated systems, the manpower requirements are 15-30 minutes per cycle, depending on the size of the unit.

RECENT IMPROVEMENTS

Since 1999, several improvements have been incorporated into the combined filter press/dryer technology. These include:

1. Development of advanced materials allows the use of 115^0C steam versus 80^0C hot water. This has resulted in the following:
   - Improved drying times. For aerobically digested sludge, the drying time for the Mountain City WWTP to reach greater than 80% solids, using ferric chloride and polymer treatment, was a 10-hour drying time [2]. As shown in Table 1, the drying time for the Rio Vista WWTP to reach the 85-90% solids, using ferric chloride and polymer treatment, has a 6-hour drying time.
   - Reduced electrical requirements. The hot water system requires pumps to provide the pressure and flow requirements. The steam system does not require these pumps.
   - Reduce pasteurization times. Due to the increased temperatures of the steam system, the time required to obtain the 70^0C in the cake is reduced.
2. Use of Polymer Treatment. Originally, it was thought that ferric chloride and polymer or polymer alone could not be used to treat the slurry due to "skinning" that occurred on

the outer surface of the cake during drying. The selection of the proper polymers and/or dosage has shown that polymer treatment can be used.

DRYVAC - POTENTIAL IN THE EU
DRYVAC, DAMAR & TVD

Towards the end of 2001 a UK based engineering company, Damar Engineering Limited became aware of the opportunity for the combined filter press/dryer technology. Damar Engineering teamed up with DryVac Environmental, Inc., a combined filter press/dryer equipment manufacturer and TVD, the sales and marketing arm of DryVac within the UK. The technology had been successfully applied to municipal sludge but the nearest reference plants were in the USA. Arrangements were made to obtain mobile test rigs for use in the UK to conduct a series of trials.

THE ISSUES

In order to assess the demand for combined filter press/dryer technology in the EU, the key issues that needed resolving at this time were as follows:

- Will the process meet the requirements of Enhanced Treatment?
- Will the dried products consistently meet Enhanced Treated standards?
- Will the treatment give rise to excessively strong filtrates?
- Are chemical conditioning and other operating costs on par with competitive sludge treatment technologies?
- What are the Health & Safety and other environmental factors?

OPERATING EXPEREINCE

TEST RIGS

Two 0.028 cubic metre capacity test rigs were delivered to Damar Engineering of Radstock during April 2002. The trailer-mounted units were positioned and operated adjacent to the Damar Engineering premises in Radstock. Most of the data associated with the combined filter press/dryer technology involved testing or processing of municipal sludge treated with inorganic conditioners, such as ferric chloride and lime, or with ferric chloride and polymer. In the EU, municipal sludge is typically treated with polymer, alone. It was, therefore, decided that the test protocol would involve chemical treatment with polymer alone.

POLYMERS

Experience using sludge conditioned with polymers was the next phase in the testing programme. Using beaker tests and CST measurements, liquid Zetag polymers provided the necessary conditioning. An initial concern, that filter cloth blinding could be a problem due to the impact of heat and vacuum on the polymer, was proved to be unsubstantiated. The manufacturer, CIBA, provided additional assurance on this point. Following some successful trial runs with sludge, conditioned with polymer from local wastewater treatment works, it was decided to conduct trials with full monitoring.

TEST RIG TRIALS

A number of runs were conducted with different types of feed sludge from three wastewater works.

Works A produced a raw co-settled sludge from conventional primary sedimentation and filter processes.
Works B produced a raw co-settled sludge from similar treatment processes.
Works C produced a raw primary and activated sludge.

TESTING DATA

The following test rig operating conditions were recorded.

- Solids and pH of the raw liquid feed sludge.
- Solids of the "cake" after feed.
- Solids of the "cake" after drying
- Cycle time including load and pressure dewatering phase and heat/vacuum phase.
- Volume of liquid polymer solution employed.

A series of test runs yielded the data shown in Tables 2, 3 and 4.

DISCUSSION ON ANALYTICAL RESULTS AND DATA

- In all the test runs a product of over 90% was produced irrespective of the nature of the feed sludge.
- For sludge types typically found in the UK, cycle times were maintained at a standard 1-hour for feeding and pressure dewatering and 4 hours for the vacuum/thermal stage.

- Feed sludge thickness ranged from 3.8 to 4.8% dry solids.
- The sludge typically had a pH between 5 and 6.
- The Zetag dosage, with a 50% activity of the product in concentrate form, ranged between . 0.23 to 0.34% active ingredient on a dry solids basis.
- The test units produced a significantly dryer cake after the dewatering phase compared with the installed plate and frame filter press at Works A (i.e. 33.1% dry solids compared to 22.4% dry solids).
- Pathogen destruction was excellent in all trials and 'enhanced treated' status was comfortably achieved.
- Although the quality of the filtrate produced in the test unit should be the same as any installed plate and frame filter press and is not affected by the addition of dryer technology to the filter press, the following results were noted:
  - Suspended solids in the filtrate were higher in the test units than the installed filter press. This can be attributable to a number of variables that need to be investigated. These include the type of polymer, polymer dosage levels, screen type, and feed cycle protocol.
  - There was high level of variability in the filtrate of the concentration of ammonia cal nitrogen. The results, however, indicate that there are no significant differences between the DryVac unit and the installed filter plate press affecting the loading returned to the Works.
  - In regards to BOD/COD, the filtrates appeared to be of the same order of strength.
  - Analysis of filtrates in the USA indicate that operating a combined plate press/dryer on ferric chloride and polymer treatment results in significantly improved BOD/COD numbers in the filtrate. However, there are no test or operating results directly comparable to the sludge types typically found in the UK. This needs to be investigated further.

CONCLUSION

Work to date demonstrates that the DryVac system of combined filter press/dryer produces a dry, quality product with no significant increase in the loading from filtrates. The pathogen destruction is, as was originally claimed by DryVac, almost total. The combined filter press/dryer is an alternative process to thermal drying, lime stabilisation, composting and pasteurisation for the production of high quality Biosolids of Enhanced Treated standard. The simplicity of the process and its operation contrasts to the traditional thermal drying process with its complex equipment and control systems, emissions, and extensive safety requirements. The advantage over lime treatment and composting is low odour and no significant additions of chemicals or other materials, which increase the tonnage for disposal or recycling.

FUTURE OF THE TECHNOLOGY IN THE EU

Combined filter press/dryer technology clearly has a lot to offer the sludge strategist but it is legislation that will ultimately drive the EU Water Industry to invest in Enhanced Processes. The full-scale experience in the USA at Rio Vista and Mountain City is encouraging. The importance of low strength filtrates cannot be over stated and conditioning with inorganic chemicals, particularly iron salts, which are already being employed for phosphate removal, should be explored. A further essential next step is to demonstrate the flexibility in terms of specifying product dryness dependant on outlet (for energy production, a dryness of 65% may be preferred) and the low operations costs of a full-scale automated plant.

REFERENCES

(1) SHAW DAVID & PRICE MIKE, New Developments in the use of Filter Press MembraneTechnology Achieving a Fully Sanitized Product in One Combined Operation: The "J-Vap" Process, Aqua Enviro – 4th European Biosolids and Organic Residuals Conference, November 1999, Paper #6

(2) SIMMS DANNY & BENTLEY DENNIS E., Relieving Wastewater Treatment Capacity Constraintsthrough Improved Processing of Aerobically Digested Solids: A Case Study, EM, November 2001, 32-36

(3) BROWN MERVYN P., Is thermal drying an effective method of recycling? IQPC London May 2001

(4) BROWN MERVYN P. & JACOBS ULRICH, Thermal drying in the UK and Germany - A shared future? Aqua Enviro - 6th European Biosolids and Organic Residuals Conference, November 2001

ACKNOWLEDGEMENTS

The authors would like to thank Wessex Water for the supply of feed sludge during the test rig trials.

FIGUERS

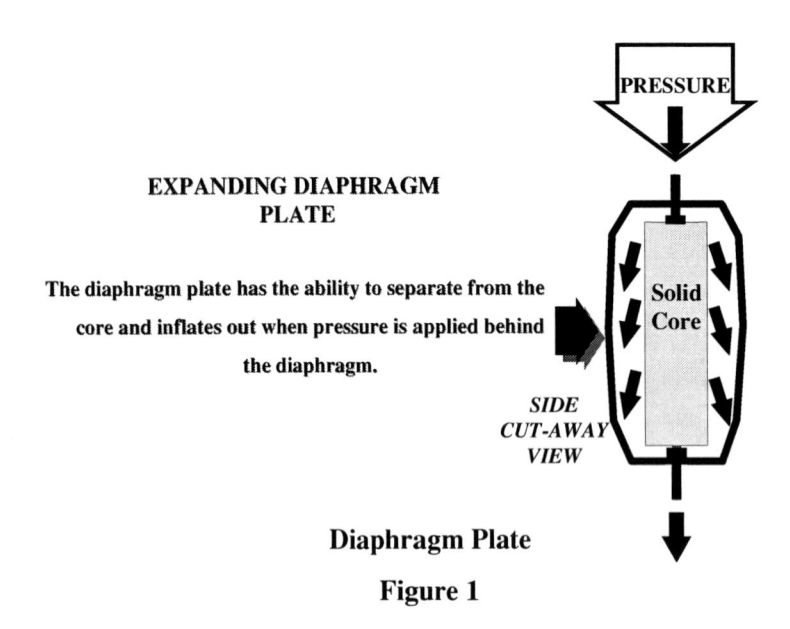

EXPANDING DIAPHRAGM PLATE

The diaphragm plate has the ability to separate from the core and inflates out when pressure is applied behind the diaphragm.

Diaphragm Plate

Figure 1

SLUDGE IS FED INTO THE CHAMBERS WITH THE FEED PUMP

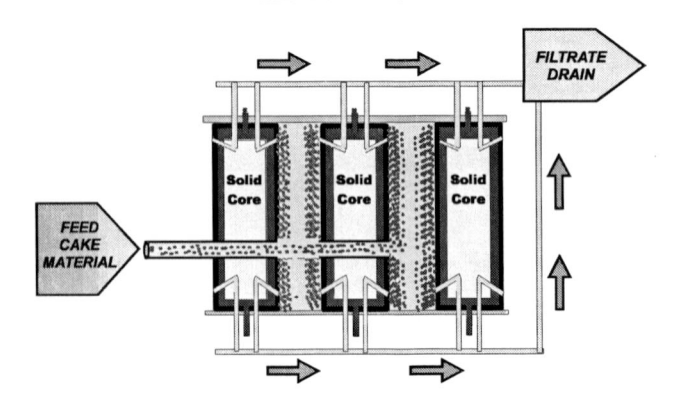

Feed Cycle

Figure 2

- **PRESSURIZED AIR OR WATER IS PUMPED BEHIND DIAPHRAGM TO SQUEEZE CAKE**
- **PRESSURIZED AIR IS BLOWN THROUGH THE CAKE TO REMOVE WATER**

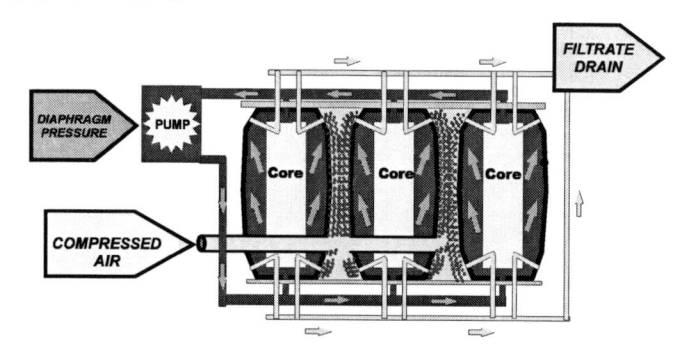

Diaphragm Cycle
Figure 3

- **STEAM IS APPLIED BEHIND THE DIAPHRAGM**
- **VACUUM IS DRAWN ON THE "CAKE"**

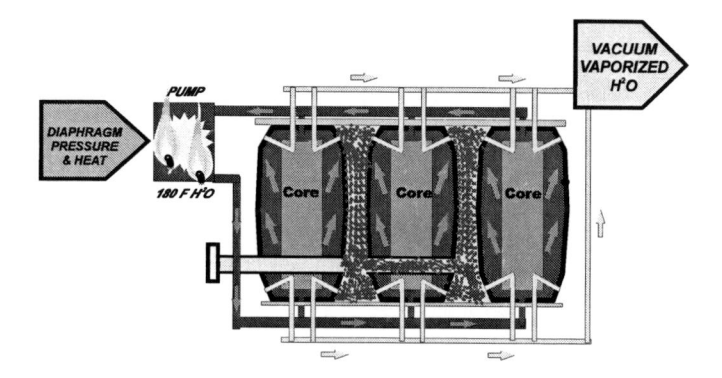

Drying Phase

Figure 4

TABLES

Table 1 - Aerobically Digested Sludge Process

Date	Run #	Feed Solids (%)	Feed Treated (ltrs)	pH	40% FeCl3 (ltrs)	pH	Polymer (mls)	Fed to Unit (ltrs)	Drying Time (hrs)	% Dry Solids
2002.8.19	2	1.00	17,400	8.4	68.0	6.3	500	17,400	6.0	*68.2
2002.8.20	2	0.99	17,010	8.3	56.7	6.5	500	17,010	7.0	96.1
2002.8.21	2	0.95	18,144	8.4	56.7	6.5	500	16,630	7.0	95.6
2002.8-22	1	1.00	16,630	8.1	54.1	6.5	500	16,630	6.0	89.0
2002.8-29	1	1.34	17,010	8.5	71.8	6.5	1150	17,010	6.0	90.0
2002.9.05	1	1.30	16,250	8.7	71.8	6.5	500	16,250	6.0	93.0
2002.9.11	1	1.01	15,500	8.8	47.3	6.3	400	15,500	5.5	79.0

*Boiler tripped off line

Table 2 - Process Conditions

Date	Source	Sludge Type	Feed-% Solids	Final - % Solids	Cycle Time (hr) *	Zetag - % DS
2002.5.27	Works A	Raw Co-settled	4.1%	94.3%	1 + 4	0.31 %
2002.5.28	Works A	Raw Co-settled	3.8%	96.6%	1 + 4	0.35 %
2002.6.07	Works B	Raw Co-settled	3.9%	99.1%	1 + 4	0.32 %
2002.7.12	Works C	1:1 Raw Prim/SAS	3.6%	96.9%	1 + 4	**
2002.7.16	Works C	Raw Primary	4.8%	91.8%	1 + 4	0.23%
2002.7.16	Works C	1:1 Raw Prim/SAS	3.8%	95.7%	1 + 4	***

* In the early trials the press was opened up after 1 hour to check the level of dewatering achieved through pressure filtration alone. The feed cycle resulted in dryness in the 20 to 30% solids range before the heat and vacuum were applied. The second stage of drying plus press emptying was consistently achieved within a further four hours.

** Polymer was added to the SAS prior to the test unit in the works' thickener at 3-4 kg/dry tonne of solids.

*** Polymer was added to the SAS prior to the test unit in the works' thickener at 3-4 kg/dry tonne of solids; the sludge was also treated with $FeCl_3$ to lower the pH from 6.6 to 4.9.

Table 3 - Product & Cake Quality

Date	Sample	% Solids	% Organic & Volatile	Faecal Coliform	E. Coli	Salmonella spp.
2002.5.27	Feed	4.1	73	5,350,000	3,400,000	
2002.5.27	Dry Prod.	94.3	74	1	< 1	0
2002.5.28	Feed	3.8	74	9,800,000	7,940,000	
2002.5.28	Dry Prod.	96.6	77	2	< 1	0
2002.7.12	Feed	3.6	80	76,100,000	40,600,000	
2002.7.12	Dry Prod.	96.9	81	2	< 1	0
2002.7.16	Feed	4.8	79	50,500,000	28,600,000	
2002.7.16	Dry Prod.	91.8	79	188	33	0
2002.7.16	Feed	3.8	80	49,000,000	27,000,000	
2002.7.16	Dry Prod.	95.7	80	50	2	0

Table 4 - Filtrate Quality

Date	Sample	pH	SS (mg/l)	Cl (mg/l)	P (mg/l)	Ammonia (mg/l)	BOD (mg/l)	COD (mg/l)
2002.5.27	Installed Press*	5.5	270	140	59	283.0	4,100	6,680
2002.5.27	Test Unit	5.2	680	81	36	73.4	4,700	5,730
2002.5.28	Installed Press *	5.9	290	126	16	72.0	2,600	3,690
2002.5.28	Test Unit	5.3	440	76	24	31.4	3,000	4,760
2002.7.16	Test Unit**	5.8	400	246	71	87.9	1,300	2,290
2002.7.16	Test Unit**	6.0	330	1790	12	173.0	1,200	2,340

* Works A

**The first results are for raw primary sludge conditioned with polymer; the second set are for a 50:50 mix of raw primary & SAS, conditioned with ferric chloride/polymer as noted in Table 2.

SLUDGE DRYING IN THE EAST MIDLANDS REGION OF THE UK

R J Wild, S D Clay, W Lilly
Severn Trent Water
Tel: 0121 7224000 Fax: 01926 403251 E-mail: robert.wild@severntrent.co.uk

ABSTRACT

The introduction of the ADAS Safe Sludge Matrix and the proposed changes to the Sludge (Use in Agriculture) Regulations 1989 has prompted Severn Trent to look at ways in which it can secure recycling routes to maximise beneficial reuse of biosolids. Severn Trent has recently concluded an in-depth evaluation of a Centridry enhanced dewatering / drying process at Wanlip Sewage Treatment Works, Leicester, in order to assess operability and the capability to produce an enhanced treated product with a range of sludges in order to evaluate strategic recycling options.

Digested and crude feed sludge were processed. Microbiological analysis for faecal coliforms, E-Coli and Salmonella was carried out on feed sludge and dried product for each. In all cases the microbiological content of the dried product was below levels of detection.

A cost summary and indicative comparison with other sludge recycling options is provided in terms of £/tds processed.

KEYWORDS

Biosolids, *Centridry*, Drying, E-Coli, Enhanced Treatment, Pathogen, Recycling, Safety, Salmonella, Sludge

INTRODUCTION – SLUDGE RECYCLING & LEGISLATION

The 150000tds of mesophilic anaerobically digested sludge that Severn Trent annually recycles are split between the four main routes shown in Figure 1, ie. agriculture, dedicated land, incineration at the Roundhill and Coleshill incinerators and landfill. Severn Trent is committed to maximising beneficial re-use of biosolids and so clearly the agricultural route is key to the current and future sludge strategy, and maintenance of that route for suitable sludge is of considerable importance. Therefore the introductions of the ADAS Safe Sludge Matrix and the proposed changes to the Sludge (Use in Agriculture) Regulations 1989 are likely to have a large impact upon Severn Trent's sludge recycling activities. It is anticipated that the following two microbiological standards will be incorporated into the revision of the Sludge (Use in Agriculture) Regulations by the end of 2003:

CONVENTIONALLY TREATED SLUDGE
- A reduction in the amount of Escherichia Coli (E-Coli) present in the sludge by not less than 99%.
- The sludge contains no more than 10^5 colony-forming units of E-Coli per gram dry weight.

Enhanced Treated Sludge:
- A reduction of E-Coli present in the sludge by not less than 99.9999%.
- The sludge contains no Salmonella.
- The sludge contains no more than 10^3 colony-forming units of E-Coli per gram dry weight.

It is anticipated that the regulations are also likely to call for the sludge to be monitored and recorded at critical control points using a HACCP (Hazard Analysis and Critical Control Point) quality control methodology to ensure that the product meets the required standards.

Sludge produced by conventional mesophilic anaerobic digestion, as found at Severn Trent's regional sludge treatment centres is "conventionally treated" as it typically involves a log2 to log3 pathogen reduction as opposed to the log6 described above for enhanced treatment. In order to meet the enhanced treated standard, alternative processes such as thermal drying need to be explored to secure the future of the agricultural recycling route in addition to short to medium term solutions such as lime addition[1].

BACKGROUND – CONTEXTING THE CENTRIDRY PROCESS

Conventional drying technology is split into two main types, direct and indirectly fired, depending upon whether or not the heat source comes into direct contact with the sludge or not. Existing UK drying installations have traditionally been associated with perceived high operating costs, long sludge residence times, health and safety issues and market uncertainty. These factors, together with apparently sustainable, lower cost recycling routes in compliance with existing sludge regulations have meant that Severn Trent's practical experience of sludge drying has mainly been as a preliminary process in sludge incineration in order to achieve an autothermic feedstock over 33%ds.

The *Centridry* process as used in the trial work at Wanlip is manufactured by Baker Hughes Inc (Baker Process) and marketed in the UK by Euroby Ltd. It is based around a conventional centrifuge as shown in Figure 2. An oil or gas fired burner (either natural gas or biogas, if available, can be used) heats incoming air and feeds this into a hot air jacket surrounding the centrifuge. The sludge, having passed through the centrifuge, is already drier than conventionally dewatered sludge cake and having passed through the centrifuge stage is then conveyed into a cyclone separator. Further direct flash drying in the stream of heated air occurs here, as well as separation of the now dried sludge from the off gas. Dried sludge from the bottom of the cyclone drops through a valve into a screw conveyor, which carries it into the pelletiser to produce a hard, glassy pellet. One of the main observations during operation of the *Centridry* process was that problems in handling the difficult "sticky" phase of dried sludge around 40-50%ds were not experienced due to the way in which the sludge is moved around the dryer, ie. no mechanical contact is made. Hence there is no need to recycle a proportion of the dried sludge into the inlet stream as with other drying technologies.

The majority of the off gas is recycled with any surplus passing through a scrubber and condenser before exhaust to atmosphere. Recycling the air in this way allows the oxygen concentration in the system to be controlled during operation.

INVESTIGATION - TRIALS & AIMS

Wanlip Sewage Treatment Works (STW) is the main works for the Leicester area. There are two process streams on site, each consists of preliminary treatment and primary settlement followed by conventional nitrifying Activated Sludge Process (ASP) with discharge to the river Soar. The works receives a high proportion of soluble trade waste leading to a higher than usual proportion of surplus activated sludge (SAS). Indigenous sludge is mixed with imported crude sludge and digested on site before being mixed with imported digested sludge from elsewhere in Leicestershire and the high proportion of indigenous SAS can lead to problems with sludge cake handling and storage. Recycling to agriculture is in either liquid or dewatered cake form depending upon the season and local recycling options. In total Wanlip handles 24820tds digested sludge per annum, typically at 3%ds before dewatering.

Given Severn Trent's previously limited experience of drying, the aims of the *Centridry* trial work were split into two main areas. These were operability and product quality. In addition to this it was also necessary to consider the regulatory position with regard to controlling atmospheric emissions from a full-scale installation.

Therefore the aims of the trial work were to:
- Confirm the operability of the process – in terms of reliability, availability, maintainability and operational resource implications.
- Investigate product quality in terms of:
  - ➤ microbiological quality
  - ➤ physical properties
  - ➤ nutrient value.
- Establish emissions - and context them in terms of the regulatory position.
- Confirm OPEX – in order to establish the total cost of ownership of a full-scale drying installation and be able to compare with existing options.

Trials were run from early November 2001 to early March 2003 at Wanlip STW. Over the course of the trial several feed sludges were used in order to evaluate the operability and performance of the process under a range of different conditions. Between November and March the following sludges were processed:
- ***Wanlip STW digested & crude.***
- Finham STW (Coventry) digested.
- Worksop STW crude.

RESULTS – OPERABILITY & PRODUCT QUALITY

Over the course of the trial work it was found that the operability of the trial plant was very similar to a conventional centrifuge in terms of operator attendance. The tasks that were involved in running the unit were largely comparable with running a centrifuge and the issues that affected plant availability were mainly in terms of the ancillaries. In particular problems were experienced with the polymer make-up system due to freezing temperatures and the low throughput in the product screw conveyor caused occasional blockages. Maintenance requirements appear to be comparable with a conventional centrifuge installation.

One of the main aspects of the process optimisation of the trial unit was logging cyclone temperature against product dryness in order to produce a calibration chart for each particular sludge, shown in Figure 3. This generally shows a linear relationship that allows specific product dryness to be achieved by selecting the appropriate temperature. Whilst this held for digested sludge, the same was true to an apparently lesser extent for crude sludge in that the product dryness was less predictable for a given temperature. This has been attributed to the method of determining dry solids using an oven at 105°C.

With dried product derived from all the digested and the Wanlip crude feed sludge, the measured %ds at higher expected dryness was in line with expectations although for the Worksop crude feed the ultimate dryness appeared to drop off with higher cyclone temperatures. This phenomenon is thought to be explained by the high level of organic and volatile material present in dried Worksop sludge (in excess of 85% by weight) compared with the other feeds used. It is thought that during analysis at 105°C some of this matter combusts, leading to a lower indicated dry solids than is actually the case.

Overall the level of dryness achieved ranged from 47%ds with Worksop crude sludge at a low cyclone temperature to in excess of 90%ds with Finham digested sludge at higher temperatures. Again this reflected historical centrifuge performance in terms of relative dewaterability. Feedback from the Severn Trent farms at Stoke Bardolph, Nottingham, indicated that the optimum product for recycling is 80%ds pelletised.

Analysis for E-Coli and Salmonella was carried out on dried sludge produced over the typical cyclone operating range of 110°C to 180°C. The results of these analyses are summarised in Table 1. It can be seen that with the Wanlip and Finham digested feed sludge, an enhanced treated product is achieved in all cases. Crude feed sludge produced more variable results with Worksop sludge producing an enhanced product but this is not the case with Wanlip. At higher temperatures the level of E-Coli present in the dried Wanlip product is below the 10^3 threshold, but a log6 reduction from the levels of E-Coli in the feed sludge has not been achieved. In addition there is a detectable level of Salmonella in the product. It should however be borne in mind that the trial unit was designed for use with digested sludge and in a full scale dedicated crude sludge installation the residence time would be increased, thus improving the quality of the treated product.

Analysis of the nutrients in the feed and dried products (outlined in Table 3) showed no significant degradation of nitrogen, phosphorous or potassium during the drying process. However it is worth noting that the dried product, in pelletised form, should prove easier than dewatered cake to apply thinly to land if a low application rate per hectare is required for recycling to agricultural land.

Analysis of the centrate and condensate streams showed nothing unexpected. The condenser was fed with final effluent from Wanlip STW and analysis of the condensate showed that this was unaffected by the process. Equally the centrate strength was entirely comparable with that produced by a conventional centrifuge, being relatively high in organic BOD load from the crude feed sludge and relatively high in ammonia from the digested feed sludge. As with conventional centrifuge installations return and treatment of this centrate must be borne in mind when designing a full-scale installation.

Investigations have been carried out into off gas quality levels and standards as these will affect any full-scale installation of the *Centridry* process. As the drying process is a sludge recovery process, as opposed to a sludge destruction process, initial regulatory advice has been that regulations such as the EU Incineration of Waste directive do not apply, and in addition the process does not fall under the Waste Management Licensing Regulations. Measurements taken at the stack on the trial unit indicate that the level of emissions will not involve excessive gas cleaning as CO measurements were found to be typically less than 50ppm. In addition SO_x and NO_x emissions are significantly below TALuft and Bimsch90 standards as the only combustion products present in the off gas are from combustion of the heating fuel. As stack emissions on a full-scale installation are likely to require odour control, these are likely to be the most onerous measures necessary in terms of additional off gas cleaning. It is anticipated that dry chemical scrubbing would be the most suitable odour control method for the load type from a *Centridry* installation due to the likelihood of intermittent operation, although ionised air systems could also be considered. The effect of ionised air on any dust particles present would have first to be considered fully before implementing an odour control solution of this type.

Sludge drying has been historically linked with Health and Safety concerns due mainly to the explosion risks associated with a dusty product. The HSE information document[2] outlines these in some detail, but there exist two main potential sources of dust explosion that are relevant to the *Centridry*:

- As heat is applied during start-up if the drier has not been purged of dried product during the previous shut-down.
- Spark generation from metal debris or stone in the drier.

During shut-down the circulating fan continues to run after the product feed has stopped in order to purge the system of dust prior to the next start-up, and similarly the fan is started before heat is applied during start-up to ensure that no residue dust is left. Internal oxygen levels are monitored and controlled using the off gas recycle proportion and feed sludge is not introduced until these concentrations are sufficiently low to limit the potential for explosion. Typical oxygen concentrations were in the order of less than 6%. During five months of daily operation, no heating or explosive events occurred.

The potential for spark propagation is limited as gross debris must be removed before entering the machinery to prevent damage to the centrifuge although, if present, debris is unlikely to pass through the centrifuge stage of the process.

INDICATIVE COST COMPARISONS & CONCLUSIONS

Figure 4 shows the impact upon volume reduction of dewatering and drying. Transport costs are a major constituent of sludge recycling costs so increased volume reduction can result in significant operational savings. Compared with the feed sludge at a typical value of 3%ds, a dried product gives a volume reduction in excess of 95%.

Table 3 shows the relative operational cost of various sludge recycling routes compared with the existing solution of cake to agriculture in terms of £/tds recycled. It can be seen that under existing legislation *Centridry* is probably not a cost effective route due to higher operational costs. However, when compared with the cost of recycling cake to agriculture under likely future legislation concerning enhanced treatment and nutrient application rates, drying begins to become competitive, particularly when digester gas can be used as a fuel source for the drying process. If the agricultural route cannot be secured then the volume reduction offered by drying compared with landfilling sludge cake offers a solution at considerably lower revenue costs. The trial work has shown that under the right circumstances sludge drying can be a robust and operable process, which can produce an enhanced treated product at a comparable whole life cost with other recycling options.

REFERENCES

1. Smith R & Foreman A: Lime Treatment of Digested Sludge – Good Idea? *Proceedings of the Joint CIWEM / Aqua Enviro Consultancy Services 6th European Biosolids & Organic Residuals Conference 2001.*

2. HSE 847/9 Control of Health and Safety Risks at Sewage Sludge Drying Plants

TABLES

Table 1: Product Microbiological Quality

Centridry Feed	Level of E-Coli in Product (CFU/gds)	Log Reduction in E-Coli	Level of Salmonella in Product (MPN/g)
Wanlip Digested	< detect	Log6	< detect
Finham Digested	< detect	Log6	< detect
Wanlip Crude	10^2	Log4	1
Worksop Crude	< detect	Log6	< detect

Table 2: Determinands Analysed During Product Analysis

Basic Parameters	Potentially Toxic Elements	Microbiology	Nutrients
%ds %Volatile Matter pH	Chromium, dry weight Copper, dry weight Nickel, dry weight Zinc, dry weight Lead, dry weight Cadmium, dry weight Arsenic, dry weight Selenium, dry weight Molybdenum, dry weight Mercury, dry weight Fluoride, dry weight	Faecal Coliforms E-Coli Salmonella	Phosphorous as P Sulphur as SO_3 Sulphur as S Magnesium as MgO Magnesium as Mg Potassium as K_2O Phosphorous as P_2O_5 Potassium as K

Table 3: Normalised Cost Indices for a Range of Recycling Strategies

Cake to Agriculture	Cake to Agriculture with NVZs	Cake to Agriculture with Lime	Cake to Agriculture with Lime & NVZs	Cake to Landfill	Dried Sludge to Agriculture		Dried Sludge to Landfill	
1.00	1.20	1.27	1.46	2.61	1.31 [1]	1.04 [2]	1.71 [1]	1.45 [2]

1 – Using 100% natural gas for drying
2 – Using 50% natural gas, 50% digester gas for drying

FIGURES

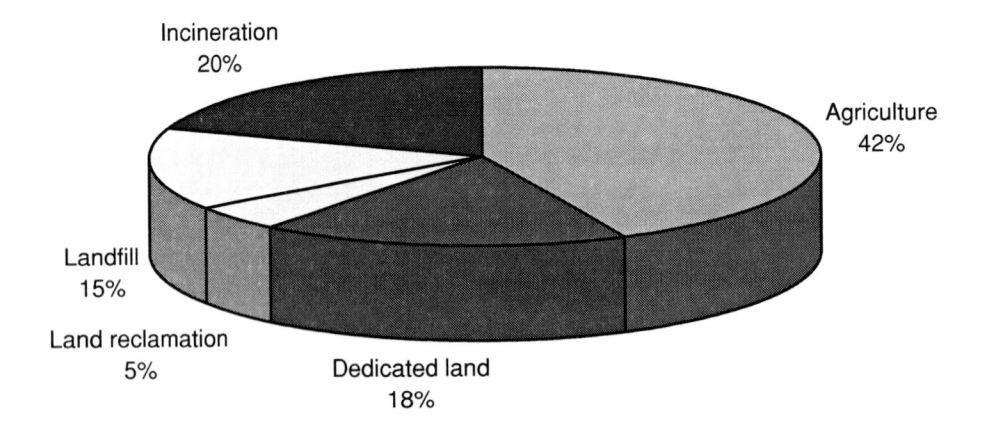

Figure 1: Biosolids Recycling in Severn Trent

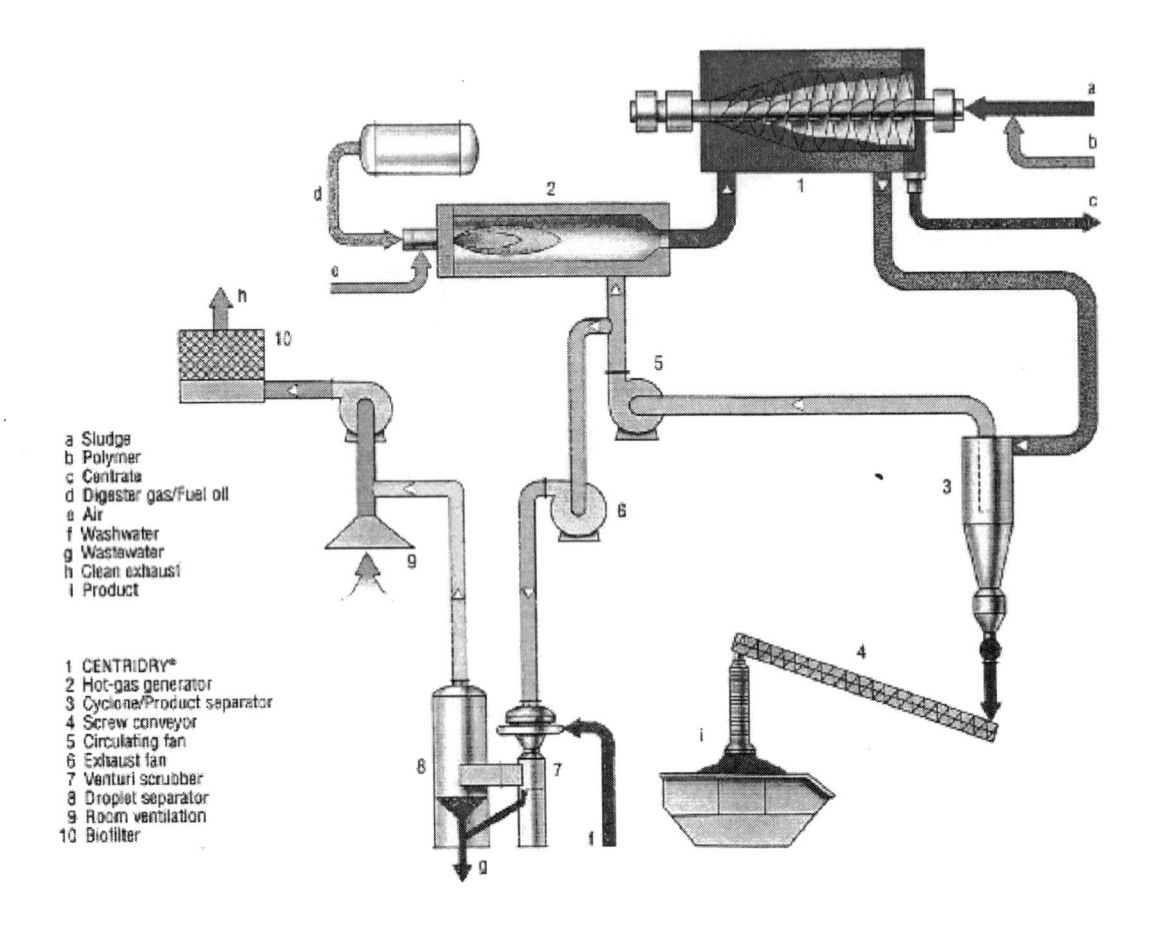

- Incineration 20%
- Agriculture 42%
- Landfill 15%
- Land reclamation 5%
- Dedicated land 18%

a Sludge
b Polymer
c Centrate
d Digester gas/Fuel oil
e Air
f Washwater
g Wastewater
h Clean exhaust
i Product

1 CENTRIDRY®
2 Hot-gas generator
3 Cyclone/Product separator
4 Screw conveyor
5 Circulating fan
6 Exhaust fan
7 Venturi scrubber
8 Droplet separator
9 Room ventilation
10 Biofilter

Figure 2: The Centridry Process

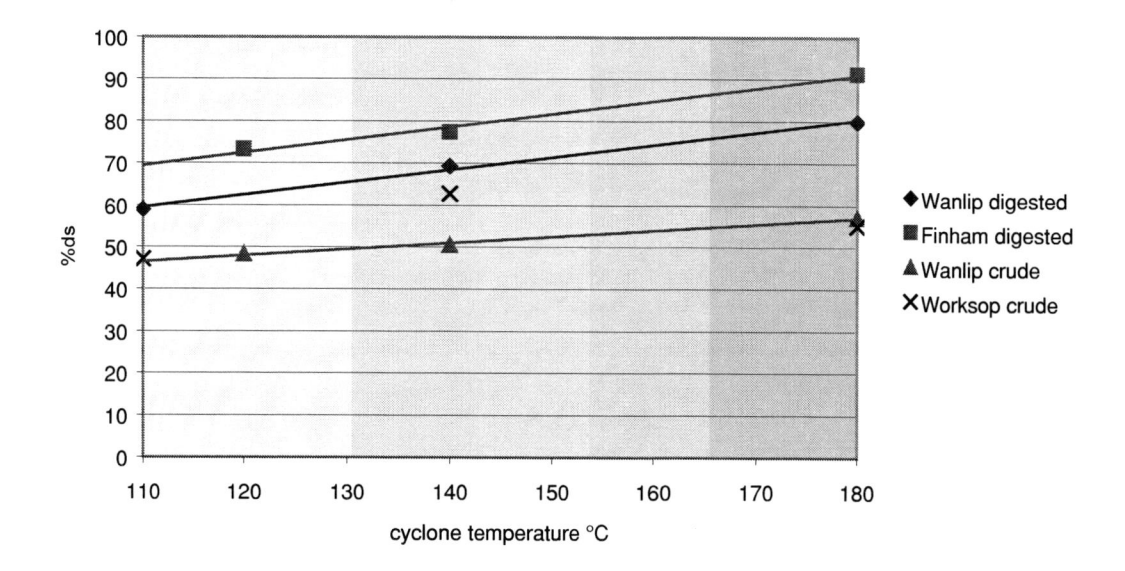

Figure 3: Product Calibration Chart

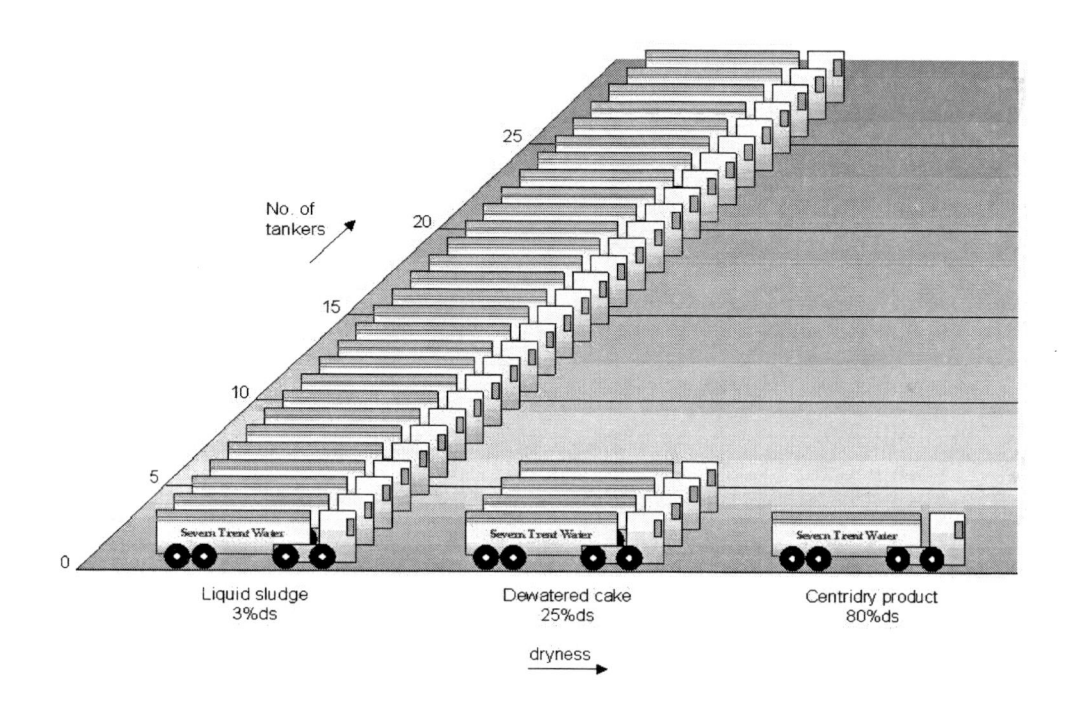

Figure 4: Graphical Representation of Volume Reduction Experienced

THERMAL SEWAGE SLUDGE DRYING – THE NEW ZEALAND EXPERIENCE

[1] Groenewegen PJ, [2] Wilson A, [3] Callander C, [4] Christison M,
[1] Flo-Dry Engineering Limited, [2] New Plymouth District Council
[3] Beca Carter Hollings and Ferner Ltd [4] Hutt Valley Water Services Ltd
Tel:+ 64 9 415 2330 Fax:+ 64 9 415 2331 E-mail: info@flow-dry.com

ABSTRACT

New Zealand and Australia are relative newcomers to Thermal Sludge Drying and the current drying facilities have made extensive use of design information and operating experiences available from the Northern Hemisphere. There are now two operating thermal-drying plants(TDP) in New Zealand and one under construction in Australia that have been designed and supplied by Flo-Dry Engineering Ltd.

New Plymouth with almost 3 years operating experience is able to offer a great deal of experience. Hutt has been operating since December 2001 and is already providing valuable experience and will continue to do so. Of interest is that both plants process completely different sludge's and therefore experiences should be different. New Plymouth is an Extended Aeration Activated Sludge plant and Hutt is a secondary treatment plant based on Primary (undigested) Sedimentation followed by High Rate Activated Sludge.

These two represent both ends of the sludge quality spectrum and gives us the opportunity to gain a great deal of additional operating experience over and above that available from the Northern Hemisphere. Thermal Sludge Drying must be considered as one of the major preferred options available for sludge processing and producing Biosolids. It offers a product that is future proofed, safe and easily handled, providing a wide variety of disposal or reuse options.

KEYWORDS

Biosolids, Class A, Drying, Thermal, New Zealand, Operating Experience, Sludge, USEPA Rule 503.

INTRODUCTION

New Zealand and Australia have only recently embraced the concept of Thermal Sludge Drying as one of the options in the ongoing quest to find the ultimate solution for the disposal or beneficial reuse of municipal sludge. Traditionally these sludge or biosolids products have been disposed of to landfill, selected and managed land applications or even given away to local residences as a fertiliser. Historically the Northern Hemisphere has made far greater use of sludge onto agricultural land then New Zealand and Australia.

As treatment plants become more sophisticated and public awareness has increased and become more sensitised these options described above are becoming less acceptable. In this way New Zealand and also Australia is no different from the evolutionary trends that the Northern Hemisphere has and is still experiencing.

In New Zealand there are two recently installed Sludge Drying Facilities. New Plymouth has now been in operation almost 3 years and Hutt Valley, which was successfully commissioned in December 2001. Both of these drying plants have been designed and built by Flo-Dry Engineering Ltd. There is currently a third plant under construction in Australia that is due for completion Jan 2003.

TECHNOLOGY EXPEREINCE

Flo-Dry Engineering Ltd is a New Zealand based company that has traditionally designed and built Low Temperature Rendering Plants. The Flo-Dry Dryer was originally developed by the company for drying protein material from meat and bone by-products processing and has been further developed for Thermal Sludge Drying. The company has over 15 years experience in drying and have installed over 40 dryers in New Zealand and Australia plus India, and the UK.

TYPICAL PROCESS FLOW

The Flo-Dry Dryer is a single pass direct gas fired rotary dryer that dries by continuously lifting the material and cascading it through the hot air stream. This dryer is one of the most efficient types available achieving actual operating efficiencies of around 3.3 GJ/tonne of water evaporated.

The Flo-Dry system is based on back mixing dried product with the wet dewatered sludge to produce a suitable mixed product prior to being feed to the Dryer. (see Figure 1 – Typical Process Flow - NPDC). This is fundamentally the same as most other rotary dryer processes although the Flo-Dry process does several unique points of difference.

1. The single pass dryer barrel that allows easy access for cleaning and maintenance and operates at low air speeds to minimise pressure drop and wear in the process loop.

2. The use of cyclones to remove dust from the process air in place of bag filters for lower housekeeping and maintenance.
3. The use of a shell and tube condenser to cool the process air prior to recycling and exhausting. The cooling water can be a single pass using plant effluent or process water. Because it is not in contact with the process airflow it remains clean and can be discharged directly to the WWTP final effluent without further treatment.
4. The process maintains a constant temperature profile by controlling process airflow and pressure; this in turn allows excellent oxygen level control in the process and high thermal efficiency.

NEW PLYMOUTH DISTRICT COUNCIL

The New Plymouth Wastewater Treatment Plant (NPWWTP) is an extended aeration waste activated sludge carrousel plant serving a community of 50,000, plus an additional 20,000-population equivalent from industrial loads. Disposal of the sewage sludge from this plant had been one of the major operational challenges during the life of the plant, and in 1997 New Plymouth District Council embarked on a project to identify a viable long-term disposal strategy for the sludge.

The dewatered sludge (approximately 14% dry solids) had historically been disposed of to land by rotary hoeing into the soil at managed sites. This practice had not been considered as a long-term solution. Primary reasons were the lack of suitable disposal sites within economic distances to the treatment plant, the presence of pathogens in the sludge necessitating quarantining of disposal sites for 18 months, and heavy metal contaminants limiting the life of disposal sites. In addition, it was a requirement of the sludge disposal resource consent from the regulatory authority (the Taranaki Regional Council TRC) that a pathogen reduction facility be installed.

The New Plymouth District Council therefore embarked on a project to implement a sustainable, beneficial and cost effective solution to the significant sludge disposal problems faced by Council. Following extensive evaluation of a wide range of sludge treatment processes, thermal drying was selected as the preferred process and Flo-Dry selected as the successful contractor.

The contract required a Design Report detailing the scope of the contract based on the Design Parameters (Table 1 New Plymouth Design Parameters). The Design needed to satisfy existing Resource Consents and because New Zealand has no formal guidelines for the treatment standard and disposal of Biosolids, USEPA Part 503 Rule for pathogen reduction and stabilisation to Class A standard was used. The final product is a pathogen free (Table 3 – New Plymouth Pathogen Analysis) comprising spherical granules between 2-4mm in diameter with a density of 810 kg/m3, that is easily handled and applied to the land as fertiliser (Figure 2 – New Plymouth Biosolids Characteristics).

A 90-day Monitored Operation period was required by contract to demonstrate compliance for biosolids quality, capacity, gas and power usage (Table 2 – New Plymouth Operating Performance). Following the successful commissioning and Monitored Operation Period, the product was initially disposed of at the managed sites. The number of truckloads reduced from about 7 per day of wet dewatered sludge to 1 per day of dried biosolids. In addition there was none of the odour problems associated with wet sludge, and it was also significantly easier to apply with no need to rotary hoe into the soil.

However, in order to maximise the benefits from the TDF, and achieve the Council's aim of delivering an environmentally favourable outcome, an application was made to the appropriate regulatory authorities to have the dried product registered as a fertiliser. This objective was achieved, with application to land of the thermally dried biosolids becoming a permitted activity under the Taranaki Regional Council's Fresh Water Plan.

New Plymouth District Council then sought tenders from horticultural organisations to accept the product and market the same in a way that would achieve a beneficial end use. The result has been the formation of a contract between Council and a private distributor to market the dried product as a biosolids fertiliser under the brand name Taranaki BIOBOOST™ 6-3-0. This is now gaining public and commercial acceptance, and it is expected to become well established in the market place, providing a long-term sustainable solution with a positive financial return for the Council and the community.

Of interest is that the majority if not all of the biosolids fertiliser is being selectively marketed for non-sensitive horticultural applications. Typically these include golf courses, instant or ready grass growing farms and ornamental nurseries. This has been a deliberate approach due to the perception that biosolids are best kept removed from any direct food chain applications even though historically sludge has been widely land applied in the Northern Hemisphere. Traditionally New Zealand has not land-applied sludge except for limited or selected applications such as forestry or marginal land that has little agricultural value.

HUTT VALLEY

Hutt Valley is a city with a population of 180,000 in the Wellington Region and in fact, is on the same harbour as Wellington City, the capital of New Zealand. For the past 20 years the Hutt Valley wastewater was fine screened to 0.5mm and discharged into the Cook Strait, which is a very turbulent stretch of water between the North and South Islands.

It became evident that this level of treatment was not sustainable so the City Council elected to undertake a full review, with public consultation of the treatment and contracting options. The Council then applied for and obtained Resource Consent to continue discharging into the Cook Strait but with a much higher level of treatment. They then tendered for a Design Build and Operate Contract over 20 years. The tender was awarded to a Joint Venture Company, Hutt Valley Water Services Ltd. (HVWS).

HVWS offered a secondary treatment process based on primary sedimentation (undigested) followed by high rate activated sludge treatment. The Hutt Valley has restricted access to long-term landfill sites, so sludge handling and disposal was always going to be an issue. For this reason the tender included an incentive for the contractor to offer alternative and innovative solutions.

The solution offered by HVWS was to dry the sludge in accordance with USEPA Part 503 Rule for pathogen reduction and stabilisation to Class A, which would then reclassify the hazardous wet sludge to a non-hazardous Biosolids as well as significantly reduce the volume. Currently HVWS are disposing of the dried Biosolids to landfill, and while this is not anticipated to be a long-term activity, this option does allow HVWS to investigate alternative options for potential beneficial reuse that could not feasibly be considered during the tendering process.

The dryer design was based on processing a blended 60:40 ratio of primary sludge (undigested) to activated sludge and generally the WWTP will produce this with minor variations (Table 4 – Hutt Valley Design Parameters). There is however periods, particularly in the first few months after start up, where the 60% primary ratio was exceeded by a significant margin (up to 90-95%). Although the difference can be seen in the process and final product there have not had any problems with drying.

Because it was not possible to develop a beneficial reuse option for the tender or before sludge was available, the Biosolids continue to be disposed of to a regional landfill as non-hazardous. However some of it is blended with clay as cover material, so in this respect has a beneficial use. HVWS are investigating other beneficial reuse options and because it has similar nutrient characteristics to BIOBOOST they are considering making use of the NPDC BIOBOOST distributor or as a fuel.

SAFETY SYSTEMS

Both plants were built in accordance with UK regulations and taking into account the latest available information from Northern Hemisphere experiences. This was necessary due to the hazardous nature of the product and process. A HAZOP was undertaken to consider the process and the design with particular attention to the potential of a dust explosion.

SUMMARY OF MAJOR PREVENTATIVE MEASURES

- Identification of hazardous zoned areas. Both dust and gas hazard zoning was identified.
- Electrical equipment rated for hazardous area operation. This included DIP rated electric motors, the use of barriers, intrinsically safe instruments, and non-invasive instruments wherever possible, SWA or shielded cabling inside zoned areas.
- Monitoring and control of key parameters such as oxygen level in the process air, flows for water, air and product, temperatures of the process and product at different parts of the process. Many of these are duplicated and in some cases hard wired back to the PLC as the ultimate safety.
- Oxygen is maintained at levels below 10% during operations and there is a staged start up procedure that will not allow the process to start or continue unless certain oxygen levels are reached with defined time limits.
- Nitrogen suppression system for the recycle and storage silos plus the dryer. This is a rapid response system that will be activated on certain alarm criteria. High temperatures or high CO levels may activate the system, depending on the application.
- All dried product leaving the dryer is cooled using both water cooled screws and air extraction. The process monitors the temperature of material before and after it is stored. If temperatures exceed 70^0C the process will shut down or in extreme cases activate the nitrogen suppression system.

- Dust extraction from all dust creating processes such as screening, crushing, hoppers and silos. It has been noted that Hutt has a higher level of dust then NPDC.
- Explosion venting for hoppers, silos, cyclones etc as required.
- Recycle silo cooling using mixing screw.

OPERATING EXPERIENCE

OPERATION - NEW PLYMOUTH DISTRICT COUNCIL

This plant was commissioned in February 2000 so has been operating for almost 3 years. The process is highly monitored and can operate totally unmanned or with minimal operator attendance. The plant is operated continuously 24h, 4-5 days per week even though there are no after hours operators on site.

The facility is started up either on a Monday or Tuesday and then operated until the weekly sludge has been dried. This normally occurs in the early hours on Saturday morning when the dryer automatically shuts down. In the event of a fault the plant will notify the duty operator, who can, via a dial up interface, fix the fault, go to site to fix the fault or shut down the plant. In all cases the plant will automatically shutdown as part of its own internal monitoring and control.

Mondays are used for housekeeping and preventative maintenance and depending on the level required for that week this could be a few hours or a whole day. Maintenance has been contracted out to local engineering company that has all the facilities required to carry out repairs. This has been very successful as it allows the operators to concentrate on the process.

OPERATION - HUTT VALLEY WATER SERVICES

Hutt was commissioned December 2001 so has nearly 12 months operational experience. The plant operates in a slightly different mode from NPDC. While still drying 24hrs 5 days per week, they split the week 2 days on, 1 day off, / 3 days on, 1 day off basis to fit in with the sludge production as they have no ability to store within the process. They are considering modifications to allow 5-7 day continuous drying, as this would make the process more efficient.

The WWTP has a single after hour's duty operator on site who also monitors the dryer operation. They use the off days for preventative maintenance.

GENERAL EXPERIENCE

The New Plymouth drying plant has now been operating for long enough to provide us with significant valuable operating experience. Flo-Dry are constantly monitoring and assessing how these plants are performing with a view to continually improve the process technology.

Because NPDC was the first drying plant in New Zealand, Flo-Dry was required to operate the plant for 90 working days in order to prove the plants operation. This was done successfully and we were able to use some of this time to train the operators. During this period the plant was to produce Class A Biosolids and to operate within certain operating performance criteria, these being throughput, Biosolids quality, power and gas usage.

The Hutt plant is still relatively new so we have limited experience but based on current information we have already noted a higher rate of wear in some equipment although at this stage it is restricted to parts that can easily and economically be refurbished.

GENERAL OBSERVATIONS

- Operator Training - The drying process is different to traditional wastewater and requires a different level of skills. Dryers are more mechanical with a higher level of process control then would be normally encountered on a WWTP. There is more wear and different safety issues that need to be considered.

For these reasons the selection and training of operators is important. Flow-dry have always recommended that the operators selected assist with and are involved with the commissioning, as it is during this period they can get a greater appreciation of all aspects of the process. Ongoing training is also important and would suggest that the first two years operation include at least two refresher-training courses to review operation.

- Wear and Tear – This has not been a major problem at New Plymouth to date although the industry has long recognised this is a problem that needs to be accepted as part of the process. It is not possible to totally eliminate wear but there are ways to minimise the effects of wear. For example for some equipment we have sacrificial and or wearing surfaces to allow refurbishment of parts.

- Instrument Reliability - For a process that is designed to operate unmanned in a hazardous environment, it is critical to select instrumentation to be robust and reliable. Critical areas of the process should be backed up using alternative monitoring equipment wherever possible. The Plant also needs to be designed to allow easy access for cleaning and maintenance.

- Sludge quality – Any process is easier to operate if the infeed is of a consistent quality. The Flo-Dryer is no different but we have found that the automatic process control systems cope well with most variations that occur over a normal day and it has not been an issue. If there is a significant product variation the Operator needs to be aware and the process monitored and changes made to suit. Some of these changes are seasonal so can slowly change and this may be more difficult to monitor.

CONCLUSION

Both these plants are proving to be successful and it can be concluded that drying of sludge to produce a pathogen free or Class A Biosolids has a great deal of attraction where there is a will on the part of Regional, Local Government and facility owners to appreciate the long term benefits.

One of today's catch cries is "future proofing". It is not certain what this mean in practical terms and how sludge drying meets this criterion.

The main advantage of Thermal Drying is that no matter how restricted Biosolids reuse becomes, this technology offers the best stage 1 process for the various options. The product is pathogen free, easily handled and stored, not readily recognisable as a sewage sludge by-product. In most cases the product has an energy value equivalent to a poor to medium grade coal. Already there are plants where the product is utilised as a source of energy and the authors perceive that ultimately this may be one of the best solutions.

World regulatory trends are what will control where and how biosolids are utilised in the future. New Zealand has not yet developed a national Biosolids Policy although a Guideline Policy is in the process of being finalised. Like most countries it is anticipated that the Regulators will be taking a conservative approach while recognising that whole question of biosolids and the long-term effects is a new science and one that will always have its detractors. To this end we continue to look at reuse alternatives.

REFERENCES

Department of Health, Public Health Services (1992) 'Public Health Guidelines for the Safe Use of Sewage Effluent and Sewage Sludge on Land'

U.S. Environmental Protection Agency (1994) 'A Plain English Guide to the EPA Part 503 Biosolids Rule'

TABLE 1: New Plymouth Design Parameters

Process Rate	3,000kg/hr dewatered sludge
Sludge Type	Extended Aeration Activated Sludge
Evaporation Rate	2,500kg/hr
Dry Tonnes / Day	4.5 Dry Tonne/day
Dewatered Sludge	Av 14% DS
Biosolids	90 –92% DS
Biosolids density	800 kg/m3
Operating hours	5 days/week @ 16-18hrs per day
Biosolids produced	35 Tonnes/wk

TABLE 2: New Plymouth Operating Performance – (May 2000 to August 2000)

	May	June	July	Aug
Wet Dewatered Sludge (tonnes/month)	1030	609	718.7	1202
Biosolids Produced (tonnes/month)	160	65.9	90	157
Avg. running hours/day	17.6	13.7	13.5	21.2
Avg. gas use (GJ/ tonne wet)	2.9	2.9	2.9	2.9
Avg. electricity use (kW/ tonne wet)	56.2	61.9	71.4	58.8

TABLE 3: New Plymouth Pathogen Analysis

Analysis	Result	Unit of Measurement	TRC* Limit	USEPA Limit
Faecal coliforms	<3	MPN/g	<200	<1,000
Salmonella	0	per 25g	<1	<3 per 4g
Campylobacter	0	per 25g	<1 per 50g	
Giardia	<1	per 50g	<1	
Cryptosporidium	<1	per 50g	<1	
Enteric Viruses	<1	pfu/4.8g	<1 pfu/4g	<1 pfu/4g
Helminth Ova	0	per 50g	<1	<1 per 4g

- Taranaki Regional Council

TABLE 4: Hutt Valley Design Parameters

Process Rate	5,000 kg/hr Dewatered sludge
Sludge Type	Undigested Primary (60%) / Activated Sludge (40%)
Evaporation Rate	3,800 kg/hr
Dry Tonnes / Day	14 Dry Tonne/day
Dewatered Sludge	Av 22% DS
Biosolids	90-92% DS
Biosolids density	550 –600 kg/m3
Operating hours	6 days/week @ 16 hrs/day
Biosolids produced	90 Tonnes/wk

FIGURE 1: Typical Process Flow Diagram - NPDC

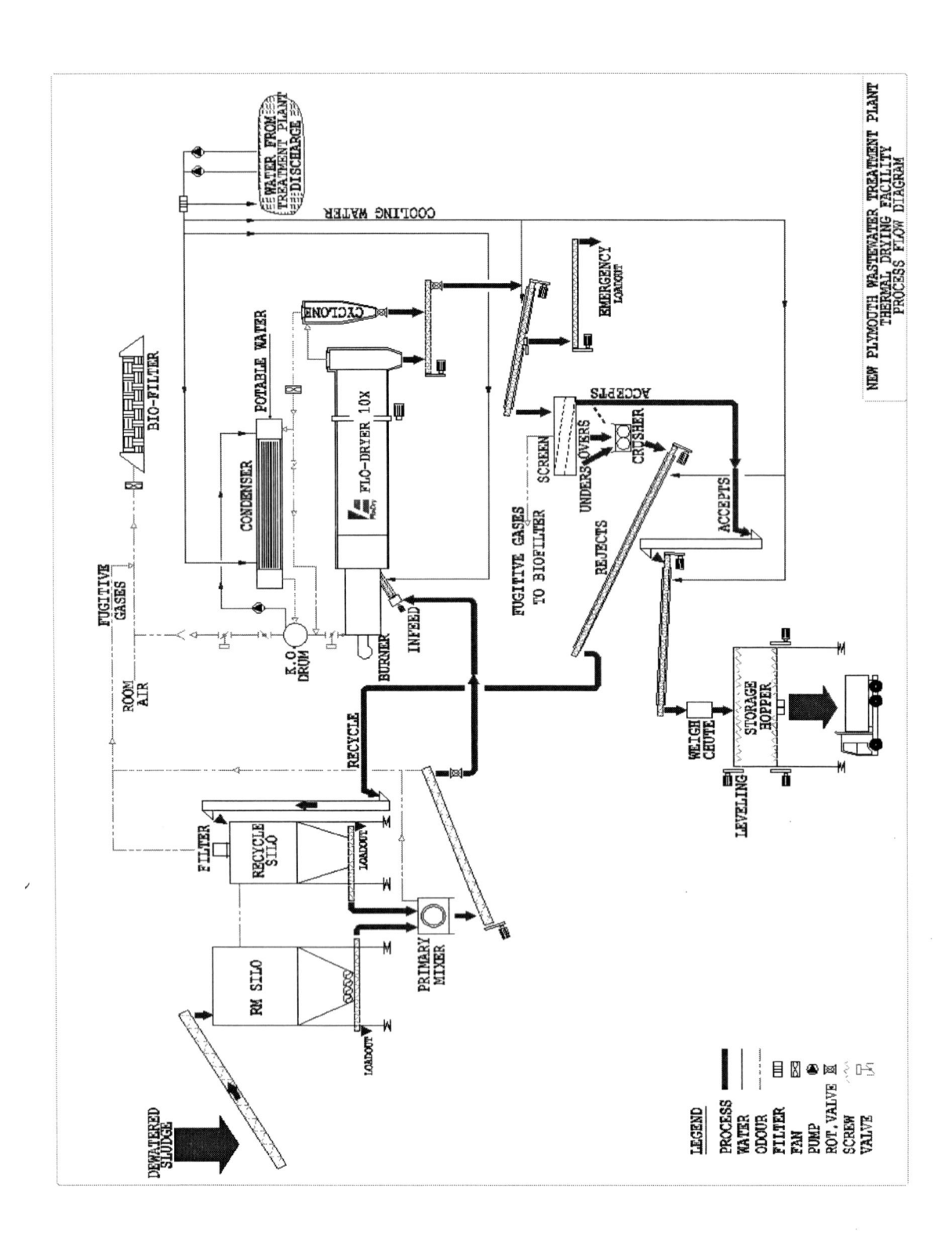

FIGURE 2: New Plymouth Biosolids Characteristics

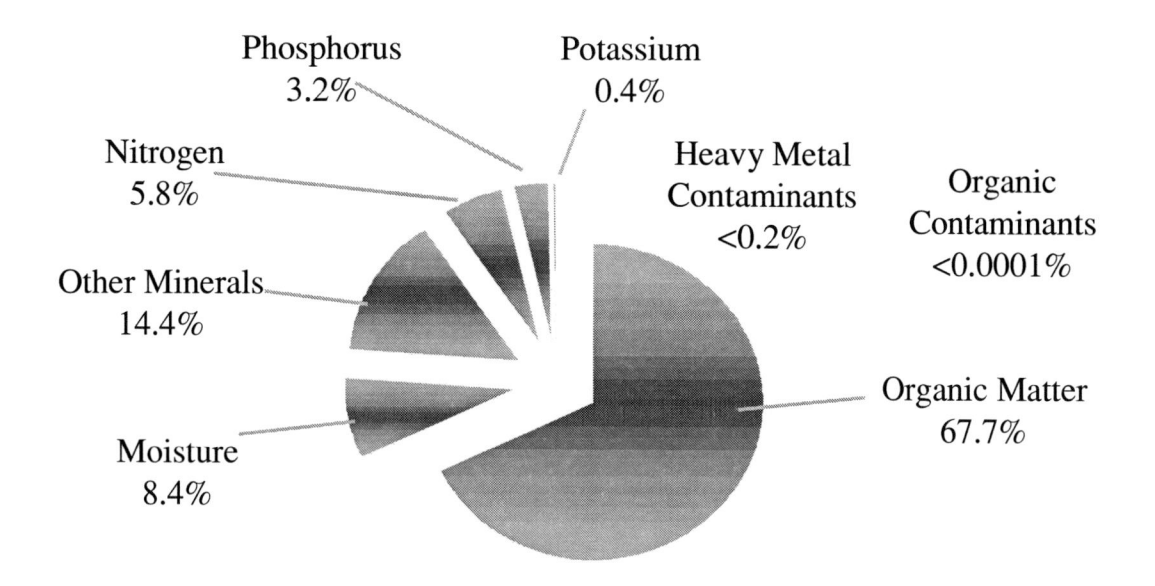

- Phosphorus 3.2%
- Potassium 0.4%
- Nitrogen 5.8%
- Heavy Metal Contaminants <0.2%
- Organic Contaminants <0.0001%
- Other Minerals 14.4%
- Moisture 8.4%
- Organic Matter 67.7%

FAST TRACK CONSTRUCTION OF A SLUDGE DRYER

[1] Gary Meades, [2] Marcel Andrews
[1] Anglian Water [2] Halcrow
e-mail: AndrewsMJ@halcrow.com

ABSTRACT
The paper is a description of the experiences of designing, procuring and constructing a sludge dryer on a fast track contract to achieve HACCPcompliance for a treated sludge to agriculture in anticipation of the proposed UK use of sewage sludge in agriculture regulations.

KEY WORDS
Advanced treatment, dryer, construction, commissioning, Health and Safety, risk, training, team, sewage sludge

OBJECTIVE
Anglian Water (AW) took a decision, in the light of the draft UK Use of Sewage Sludge in Agriculture regulations, to review the current sludge treatment technologies deployed at sewage works sites. Of the original 33 Sludge Treatment Centres investigated, 27 sites were identified as requiring process additions or modifications. A number required additional storage, lime stabilisation facilities were installed at some of the plants. 5 Pasteurisation units, one 'Hot Tin Can' hydrolytic thermophilic conditioning plant and a sludge dryer plant were also required. This paper deals with the sludge drying project.

The final decision to install a dryer was taken May 2001 resulting in a design and construction deadline of 7 months in order to meet an anticipated compliance date of the 1st Jan 2002. To add further pressure to a very tight time scale the AW sought earlier compliance to give time for their reporting systems to become fully operational.

BUILDING THE TEAM
There is no question that without a team approach it would not have been possible to fast track this project, especially a project as complex as an Advanced Sludge Treatment installation. Therefore, the team selection process was fundamental in the delivery of the scheme and probably forms the most difficult and challenging decisions for the Project Manager. Other problems tend to be engineering issues, which can be accommodated. However, personnel issues are not easily resolved. AW assembled the project team. Managing the project, contract administration was undertaken by AMEC with Halcrow offering all of the design support and Waverny pumps carrying out the installation.

ORDER PLACEMENT
DRYER
The close relationship developed during pilot studies of a SEVAR dryer at pilot plant site led to consideration being given to moving the pilot dryer to the Colchester site. It became apparent that the size of the pilot dryer was in fact too small for the Colchester duty and whilst consideration was given to modifying the strategic treatment locations to match the dryer capacity it was concluded that a larger dryer was needed for this application. The supplier looked at manufacturing availability and as a result an order was duly placed in July 2001.

PANEL
Whilst the dryer itself required the greatest number of motor drives a large number of ancillary plant items were required to interface with the dryer and the motor control centre. The design of an MCC, and the interfaces with utilities, telemetry and existing plant was outlined as far as possible within a very short time frame. However, many additional ancillary plant items including conveyors, pelletisers, pellet coolers, storage silos and bagging units, along with instrument and safety plant items that would inevitably to be added during design stage as well as those items identified by putting the plant through HAZOP revue, indicated that additional capacity would be require. The problems was that such additions radically alter the design of the MCC panel and control systems for which there are normally long-lead times. This meant an early design had to be put forward to allow the MCC panel construction to proceed in the full knowledge that the design would have to be adapted during the construction phase. The key to a successful installation was achieved by allowing sufficient capacity in the system to facilitate the add-on items. When such add-on items include a cooling tower for heat dissipation, a pelletiser and numerous other drives the generosity at the earliest stage in the end seamed a little meagre.

LAYOUT
Moving sludge from the original source via a skip filling system outside the centrifuge building, to the dryer feed hopper and subsequent movement of product from the

dryer through the post dryer process plant required development of layout of drawings which would allow civil works to proceed. CTM were selected as the conveyor suppliers who, together with Halcrow, developed the conveyor routes and layouts to define the base specifications to support the plant. CTM also supplied the final product silo and the interfaces with all the post dryer plant items. This allowed the civil design to proceed in earnest. The plant was located within the boundaries of an existing operational sludge drying bed directly adjacent to the digestion facility and two of the five site digesters. The drying bed, being some 1.5 –2.0m below ground level and adjacent to the centrifuge building enabled the top line of the dryer plant to be well below the roofline of the adjacent building.

The conveying of material was restricted by a layout that retains most plant items near the centrifuge building, which is also adjacent to the product silo. The fast track approach for the project and the shape of the old drying bed resulted in a total of 6 conveyors being installed for the transport of material through the process. All conveyors used were screw type with shafts; the longest being some 19m carrying drier sludge from the pellet cooler to the final product silo, uses an intermediate bearing.

SILO

The silo was ordered from CTM with a request that it had facilities for a bagging unit and an arch breaker screw to be added. As the safe long-term storage temperature of the product could not at this time be determined, the size of the product silo was kept to a minimum to limit the amount of un-bagged product held in a confined space.

COOLING TOWER

The SEVAR Dryer is a belt dryer with a series of tube heat exchangers and condensers in series, which are used to recover heat and condensate. However, there is a significant amount of heat that has to be removed from the system. In the short term the design and installation to facilitate the reuse of this energy was not practicable without prolonging the project. Consequently heat dump mechanisms had to be investigated. The first was the use of final effluent in a single pass system but this had potential difficulties for use due to capacity requirements and effluent temperature.

The second option was to fill a storm tank with final effluent thus acting as a buffer and to recycle through this converted storm tank final effluent make-up water. However, calculations indicated that the tank would progressively heat up resulting in an insufficient heat loss and a warm effluent that would be difficult to keep in good condition. The answer to the dilemma was to provide a cooling tower with all its accoutrements. An order was placed on Marley for the provision of a cooling tower and water treatment plant.

PELLETISER

It was always anticipated that the SEVAR product would need pelletising nevertheless the market potential for SEVAR strands was investigated. Pellets are thought to be more user-friendly for the agricultural recycling route. Combined with this there is the added advantage of having a much higher bulk density product and as a result better utilisation of the storage and transport capacity. As the pelletiser is an integral function of the SEVAR drying system it made sense for the dryer supplier to be given this responsibility for the provision of the pelletiser.

PRODUCT COOLER

One has to consider that the order for the dryer was placed before the first risk assessment could be carried out. However, the risk assessment did take place very soon after the dryer was ordered. The UK Health and Safety Executive (HSE), a government agency, was invited to discuss the safety issues involved with the plant with the aim of designing out, during development stage, as much of the potential hazards as possible. The approach was taken to reduce the risk of having the plant closed down by HSE a short period after commissioning because of the lack of provision of safety measures. Consequently a product cooler was needed to ensure thermal runaway was not a realistic possibility. As was the case with the pelletiser, this item was placed within the scope of the drying contractor. Air-cooling was selected as the means of heat dumping, as an inert atmosphere was not considered to form the basis of a sound safe installation.

BAGGING UNIT

A single-use bagging unit was selected as the method for bagging the product. Bags were sourced from Webster Griffin and the interface engineering between this unit and the product silo engineered by Halcrow.

UTILITIES

With the ordering of the main components under way it was time to turn attention to the services.

Power was available from an HV distribution board already on site. Following the power demand assessment a cable was selected ready for when the MCC kiosk and panel were in place. This cable was sized, based on the early assessment of need, before the pelletiser, product cooler and water cooling tower had been sized or selected. With a current draw of around 500A adding a 56kW pelletiser had a potential to use up surplus capacity. The original assessment for the plant suggested around 400A current load for the total installation was needed. The cable drawn had a 520A capacity but after all modifications the plant load was getting rather close to the limit. A suspected Floating Site Earth added somewhat to the drama. While all these issues were resolved the plant was operated on a generator to keep pressure on project during the finishing of installation and commissioning stage.

The next problem was the provision of energy for evaporation. With no gas production figures available for the site and no time to engineer the use of biogas a burner using fuel oil or biogas was selected as the heat source for the dryer. The sighting of the fuel tank and a double-wall pipe and tank were had to be located within a relatively small footprint.

The provision of water proved to be an even complex problem as the site pipe plans were not up to date. Water was required for a number of uses including:
- A water deluge on the dryer for over temperature/ fire suppression,
- cooling hydraulic oil in the sludge extruder system,
- the scrubbing tower and to maximise the use of the entire cooling tower.

As in many fast track projects a fundamental issue was early placement of orders for the long lead items, the dryer itself being the single largest item.

HEALTH AND SAFETY

During the period of the design development and construction, compliance with HSE 847/9 was being applied to the project. Albeit the HSE guideline document had not been written with belt-driven-drier technology in mind. An early start to a programme of HAZOPs allowed inclusion of key factors designed to ensure HSE would find it easy to accept start up of the plant.

For any dryer installation determining a suitable basis of safety within the system is mandatory within HSE guidelines. For the Colchester plant these issues were addressed at the earliest possible moment. For current Anglian Water projects, a pre-tender HAZOP and a tender HAZOP is required. This ensures the safety issues are address at an early stage in the project.

Fundamental to the provision a safe drying plant are the following:
- Installation outside to avoid many of the dust accumulation issues found on other sites. A fast track dryer construction in a building would inevitably lead to extensive additional costs.
- Use of a low dust generating dryer.
- Operation at low temperatures.
- Low product storage volumes.
- Low product storage temperature, and in the case of the SEVAR system
- No product recycle.

Other safety features common to most drying plant installations included key features such as a CO_2 extinguisher system in addition to a deluge water system. Other safety items added to the plant included: a CO monitor, and
- multipoint temperature probe to sense potential thermal runaway,
- discharge chutes to safely empty silos under fire risk,
- dual zone temperature monitoring for extinguisher discharge,
- and many others interlocks and controls.

Following the initial HAZOP study the plant design was modified to incorporate all necessary changes to deliver a plant that would meet modern safety requirements. The results of the HAZOP study and the outcome were advised to the HSE. They also attended an early design phase HAZOP meeting held on-site in advance of plant installation and commissioning. Their suggestions were incorporated into the design queries and subsequently answered by the dryer supplier. In particular the issue of dust accumulation in certain key areas had to be address.

Following the construction and commissioning of the dryer a comprehensive suite of analyses were conducted to determine the potential risks and to fix the Control Set Points for the dryer operation in light of the safety criteria. Having tested the ignition temperature, layer ignition temperature, minimum ignition energy and minimum layer ignition temperature, it was found that the initial designs selected during initial HAZOP study were found to be fully validated. The only major modification to the plant design was the inclusion of an explosion vent on the product silo. The reason for this was a rather long conveyor, which was deemed to increase the anticipated dust loading in the silo.

The basis of safety for the dryer is component specific. In general the SEVAR unit has very low dust generation potential due to its gentle handling of the product, which is helped by the fact that the plant does not have a back mix system. Both these features contribute significantly to a plant design that has a lower risk than other installations. A third contributing factor is the relatively low temperatures at which the unit operates, well below the ignition temperatures of the product.

INSTALLATION AND COMMISSIONING

Like many plant installations this system had to be constructed on a live site. Interface with operations was critical to ensure the disruption to the operational area was kept to a minimum. Fortunately the interface between the existing sludge treatment facility and the new plant was restricted to a limited number of items. For example the conveyor diverting sludge to the new facility, the water supply and control interface. Of course there were many other issues on site such as connection to the main power supplies and maintaining a favourable working environment for construction and operating staff. This was not helped by Anglian Water having to manage the Foot and Mouth sewage sludge issues during the construction period.

With design development occurring through the construction phase the drawings, especially the P&I D were subject to frequent update. A great effort was made to reconcile the nomenclature of the panel drawings to that on the P&I D to produce a single drawing that could be used for general reference. This was marked up as a master copy on site and redistributed to all parties for interface control.

The ancillary plant engineering was difficult to achieve in such a short time as information relating to either existing infrastructure or new plant was not as complete as it needed to be to resolve all the issues. A classic example was a dust control system on a silo. Only late in the design period did it become apparent that it was an active rather than a passive system and that it needed a compressed air supply. Two spare cubicles in the MCC were swallowed up in one item of plant along with all the associated cabling.

The installation company did show remarkable initiative, such as installing extra cores in many areas when pulling cables, but these were quickly used up as the site developed. Decisions had to be made on limited available information. Of course such initiatives do come at a cost but the gain in time were of such significance that the additional cable costs were not an issue. In reality the number of spare cores and cables is no greater than on a traditional construction site.

Inevitably there were a few snags during the initial start up, compounded by a number of plant item breakdowns. These included recirculation pump motors that were faulty when supplied, a fuel oil booster pump with insufficient pressure generation capability, a burner that needed a gas supply to start up and therefore requiring an independent cylinder of propane with a special regulator for start up (until the biogas connection could be made). Nevertheless, it was not long before sludge was passing through the drying unit.

At the end of construction and during the commissioning stage informal training of the operators was achieved by allowing site staff to shadow the commissioning team. This was followed up with formal certified training sessions and an extended period where the commissioning staff worked along side operators during the initial operating period.

PLANT PERFORMANCE

The dryer was purchased with the aim of meeting the enhanced treated sludge status for agricultural recycling route. In addition a pelletised and bagged product was produced to deliver the product to the marketplace.

The dryer performance was rather remarkable. For a plant installed in such a short period of time the unit was able to deliver a dried product with a concentration of 85-90% dry solids. This was achieved in part not solely by controlling the dryer performance but in maintaining the performance of the pre-dryer pelletiser performance. The product was tested for pathogens and it was found that provided HACCP parameters were met the pathogen reduction and Salmonella levels were capable of meeting the Advanced Treated status.

CONCLUSIONS

A fast track construction for an advanced treated sludge project is viable provided all parties adopt a very flexible approach. Design development during the construction period will lead to problems that could ruin the fast track timescale. Major problems arise but with the right team, the right management approach and a lot of effort the results can be startling.

ACKNOWLEDGEMENTS

The author is grateful to Anglian Water for permission to publish this paper/article. All views and opinions expressed are those of the author and they do not necessarily reflect those of the board of Anglian Water.

FIGURES

Figure 1: Colchester dryer installation

DRYING-PELLETISING OF DIFFERENT TYPES OF SLUDGE AND BIOSOLIDS

Dehing, F., Englebert, B.,
SEGHERSBetter technology , Belgium

ABSTRACT

Paper describes the sludge parameters that have significant influence on the drying process. It compares the influence of sludge origin in different types of sludge drying installations on process capacity, efficiency, safety and product quality. Case study illustrates the above.

Owning to circumstances beyond their control the authors were unable to meet the printers deadline. This paper should be available on the conference CD

SAFE STORAGE OF SEWAGE SLUDGE
Ian Birkinshaw, James Duggleby
Portasilo Ltd, New Lane, Huntington, York, YO32 9PR
Tel: 01904 624872 Fax: 01904 611760
Email: Ian.Birkinshaw@Portasilo.Co.Uk; James.Duggleby@Portasilo.Co.Uk

ABSTRACT

Recent changes in both UK and European legislation have led to the installation of sewage sludge drying plants at many Coastal Treatment Works. Sludge that was previously disposed of at sea is now being dried and granulated for subsequent use as fuel or fertiliser. In addition to the coastal treatment works many inland sites are being updated with new technologies and processes.

This has led to a growing need to store the dried solids in bulk, which presents a number of risks which are new to the sewage Industry. Failure to appreciate that the dried sludge is a highly bio-degradable material, which can present a risk of explosion or spontaneous combustion, has already led to a number of accidents and the involvement of the Health and Safety Executive. These accidents have happened despite the fact that prevention and protection controls are well established in other industries.

This paper looks at the methods for the safe storage and handling of this hazardous product, and reviews the dangers and possible consequences of explosion, spontaneous combustion, exothermic reaction, and composting. These hazards can be significantly reduced and the dangers eliminated though the use of monitoring, protection and prevention systems. This paper proposes solutions to ensure that this potentially hazardous product can be stored safely while at the same time maximising the benefits of bulk storage.

INTRODUCTION

With the rapid change in the methods of disposal of sewage sludge within both the UK and Europe to meet the ever-changing environmental legislation, storage of the dried sewage was inevitable. The storage of this product had previously been carried out in its slurry stage and not as a dried product. The need for the product to be easily handled to enable transport off site in either big bags or tankers led the drive to store the material in steel silos. In taking this decision several primary objectives are required to ensure safe storage;
- Prevention of the main storage hazards
- Constant monitoring and maintenance of the silo contents during storage
- Protection against dust explosion
- Protection against self heating
- Controlled and guaranteed silo discharge
- Dust free big bag and vehicle

These areas are explored below highlighting the potential problems along with their solutions.

SAFE AND RELIABLE STORAGE
PREVENTION OF THE MAIN STORAGE HAZARDS

These hazards fall into the following categories:
- self heating,
- composting,
- risk of dust explosion and
- bridging.

The two areas of self heating and composting lead to fires starting in the main body of the material. Self heating can be started by the material reaching the critical storage temperature. The critical temperature is the point at which the material can spontaneously combust, leading to the whole of the interior of the silo catching fire. This has happened in a number of installations and was one of the main reasons why the HSE has been taking a keen interest in these plants.

The critical storage temperature is dependant on the size and shape of the storage vessel and a tall thin vessel will have a higher critical temperature than a short fat silo (refer to figure 1 which shows the effect of vessel proportions on critical temperature). These tests, which are the "standard basket tests" were commissioned by Portasilo, and were carried out to find the critical temperatures associated with these products. The results were then extrapolated to find the critical temperature for differing sizes and volumes of storage vessels.

Unfortunately due to planning constraints the overall height of most installations is too low to achieve the ideal vessel proportions. These constraints have led to installations where the critical storage temperature has been as low as 55°C. In these situations we would recommend that the temperature of the material entering the silos should be cooled to a maximum temperature of 40° C. This then allows a small rise to take place through radiant heating or composting.

As organic dust combusts, heat is produced. In addition, oxidation of organic products can occur. This is classed as an exothermic reaction, also adding to heat production. This heat rapidly increases the volume and pressure of the combustion products. This process is correctly known as a deflagration, rather than an explosion, but the effects are no less damaging.

Dried sewage sludge is capable of self-heating and subsequent ignition, if the rate of heat production exceeds the rate of heat loss. A number of variables need to be considered when predicting the critical temperature, above which ignition can be expected. This figure can be predicted based upon the vessel proportions and test data of the sludge. Figure 1 shows the effect of silo proportions on the critical temperature.

There is a further risk associated with self-heating, which is caused by a quite different mechanism. Excessive moisture caused by condensation in the storage vessel can cause fermentation or composting of the organic solids. This biological action can raise the temperature sufficiently to cause the exothermic reaction described previously.

The risk of dust explosion with dried sewage sludge is very high and several instances have been recorded where drying equipment has been damaged due to a dust explosion. Unfortunately the product does differ across the UK in its characteristics especially in relation to its Kst rating. The Kst value of a product is the rate of pressure rise or the speed of the flame front measured in bar m/s.

According to literature published by VDI the main Engineering association in Europe, the Kst value of this product could be in the range of 36 – 104 bar m/s. Tests in the UK relating to dried sewage sludge have recorded results in the region of 80 – 195 bar m/s. As these results have a large part to play in calculating the required vessel strength and the vent area required to vent an active explosion, tests must be carried out on the actual sewage sludge in the final location and must not rely on a generic sample.

The last main item relating to prevention of storage hazards relates to ensuring that the storage silo discharges in mass flow. Figure 2 shows the main differences between mass flow and core flow. In mass flow silos all the material is moving during discharge ensuring first in first out discharge, whereas core flow leaves large areas of the silo stagnant and promotes first in last out discharge. With this product having the ability to self combust over time, it is essential that the whole of the silo is discharged without leaving pockets of material at the hip level of the vessel as would be the case if the silo was discharged conventionally using core flow techniques.

CONSTANT MONITORING AND MAINTENANCE OF THE SILO CONTENTS DURING STORAGE

In storing this potentially hazardous material it is essential that the contents be constantly monitored for both temperature and carbon monoxide. If the trend of temperature rise can be caught quickly enough then preventive action can be taken to avert a fire in the silo contents.

Monitoring of the temperature in storage can be achieved with a multi-point temperature detector suspended from the vessel roof. This will monitor the temperature at the different layers, both during storage and silo filling.

As the temperature in storage builds up, carbon monoxide gas begins to evolve. Although production rates are low at first and hence, difficult to detect, Figure 3 shows the marked increase as the temperature of the solids exceeds the critical temperature. By continuous sampling of the silo atmosphere and temperature, the onset of self-heating can be detected and alarmed.

In the event that self-heating has been detected early enough, the silo should be emptied. Flexible big bags have a limited temperature resistance, so outloading to bulk vehicles or skips is preferable. If silo discharge cannot be affected quickly or an advanced exotherm was indicated from the temperature monitor, the silo should be flooded with an inert gas to prevent ignition. Sufficient gas should be provided to fill the empty silo totally. Carbon dioxide is typically used.

This is heavier than air and is therefore best introduced through a plenum built into the silo outlet. Fig 4 shows the immediate effect on temperature and CO evolution rates after activation of the inert gas system.

PROTECTION AGAINST DUST EXPLOSION

Protection of the storage vessel against the damaging pressures associated with a dust explosion is usually best achieved by explosion relief venting, provided these can discharge to a safe area. Figure 5 shows a silo roof complete with explosion vents.

Inerting of the silo can also be considered, but this requires the silo full of inert gas at all times, to exclude oxygen, and running costs can be high. The explosion can also be suppressed by detecting the incipient pressure rise and injecting the silo with a finely divided powder to 'smother' the explosion. This protection is often best suited to smaller vessels, which are difficult to vent to a safe place.

A final option is containment, which in the case of dried sludge requires the vessel and all connected plant designed for 8-9 bar gauge. This is a very expensive option not only because the silo itself has to be strengthened to take the pressure but also the ancillary equipment, such as filters, level probes, discharge aids, conveying systems and down stream equipment. They all need to be rated to contain the pressure of 8-9 bar gauge.

PROTECTION AGAINST SELF HEATING

As one of the primary causes of self heating is the possibility that the material may compost it is therefore vital that moisture is kept away from the inside of the vessel. The main reason for moisture is condensation caused when the warm material is in contact with the cold surface of the silo body. There are two main ways in which the risk of condensation can be reduced to an acceptable level.

The first method is to insulate and clad the outside of the silo which ensures that the vessel wall does not become cold. This is quite an expensive option and the costs have prohibited its implementation in the waste water industry.

The second method is to maintain a positive airflow in the void above the material, using ambient air at a rate of approximately two full air changes per hour. This method is relatively inexpensive to implement but the additional airflow needs to be taken into account for any filtration and odour control systems fitted to the system. If the rate of air flow cannot be accommodated by the filter or odour control system it is possible to introduce smaller quantities of dehumidified air which will have the same effect.

CONTROLLED AND GUARANTEED SILO DISCHARGE

As previously mentioned dried sewage sludge is an extremely variable product. This is emphasised by all the different dryer manufacturers using a slightly different drying process to produce a range of finished product. This varies from fibrous granules, which break down easily with mechanical handling to hard pellets.

In order to overcome many of these problems associated with ensuring accurate and controlled discharge the *Rotoflo* discharge valve has been successfully designed and developed by Portasilo. The *Rotoflo* unit can be interfaced with a mass flow hopper design, allowing controlled feed rates as low as 500kg/h to downstream processes.

The *Rotoflo* system has been designed to tackle the problem of efficient silo discharge at a controlled rate. This rate can be either continuous or variable to suit the individual process requirements. Although many different dischargers are available to agitate the product and initiate flow, few will continue to control the flow throughout the discharge cycle.

Controlled discharge by the *Rotoflo* discharge valve is achieved by rotating an inverted cone on a vertical axis within the silo. At start-up, the smooth cone rotates relative to the product, thereby reducing the starting torque on the *Rotoflo* drive unit. Once the cone has started to rotate friction builds up between the rotating cone and the product. As a result of this action the product begins to rotate with the cone, and as a consequence of the rotating action of the *Rotoflo* unit the product rotates at a greater diameter than the discharger. Thus the *Rotoflo* discharge valve is capable of preventing bridging at a diameter far in excess of its own size.

The rotating cone is fitted with one or two open fronted scoops. These scoops plough product off the flat table, above which the inverted cone rotates. The product feeds through the machine body and free falls into downstream equipment at a controlled rate proportional to the *Rotoflo* speed (see cut away drawing Fig. 6).

The *Rotoflo* unit discharge valve incorporates a positive shut-off gate behind each of the scoops, to stop the flow of product when required. Reversing the *Rotoflo* unit by approximately 90° activates these shut-off gates.

In addition to shutting off the product flow, the positive shut-off gates allow complete maintenance of the *Rotoflo* unit with the silo full of material. Routine maintenance is restricted to lubrication checks, all of which are carried out external to the machine. It is however, possible to carry out a complete overhaul with the silo full.

In many instances the discharge from the silo needs to be able to be directed into two applications. In the first instance directly into a road tanker or bulker at a rate of around 50 m³/hr and secondly in to a big bag filling machine which typically requires an infeed rate of 12- 15 m³/hr. The *Rotoflo* discharge system ensures that the silo will be discharged in mass flow over the total range of discharge rates in a reliable and controlled manor.

DUST FREE BIG BAG AND VEHICLE FILLING

In order to ensure that the product discharge is dust free and to ensure that the equipment is fed at the rate which is compatible with the down stream equipment, this rate needs to be controlled. Secondly the areas around the tanker filling point and bagging machine need to be placed under negative pressure with a return duct back up in to the silo at high level.

Figure 7 and figure 8 show a typical tanker loading system used for the discharge of sewage sludge. In figure 8 the discharge of dried sewage sludge is in progress and by extracting the dust back up into the storage vessel dust emissions are eliminated.

CONCLUSIONS

The hazards of storing this product are now well known within the wastewater industry and recent instances of fires and explosions have focused the attention of those working in the industry and of the various associated bodies. A scientific approach to the system design, based on measurable parameters can go some way to eliminate the hazards.

The techniques for assessing and reducing the residual risks are now well established, but must be rigorously applied by suitably qualified and experienced Engineers. These hazards can be engineered out of the system as shown in the flowsheet Figure 9. This shows two silos fitted with both the temperature and CO monitoring equipment, inerting equipment, accurate and controlled discharge and dust free loading into either tankers or big bags via the loading hood. Figures 10 and 11 show actual installations of safe storage solutions in operation.

The magnitude of the loss associated with a runaway exotherm or dust explosion must not be under-estimated. Loss of life is a distinct possibility and serious damage to the plant is inevitable.

NB *Rotoflo* is a Registered Trade Mark

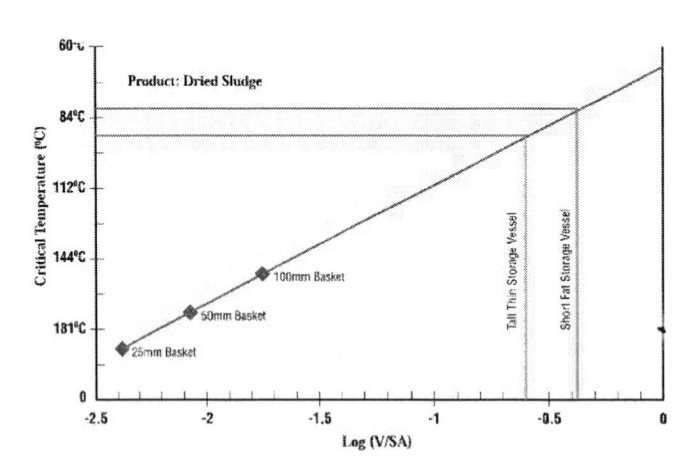

Figure 1 : Effect of vessel proportions on critical temperature

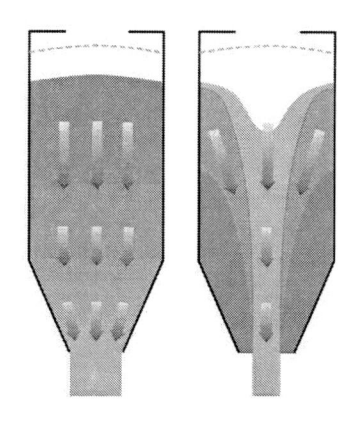

Mass Flow   Core Flow

Figure 2 Mass Flow Core Flow Discharge

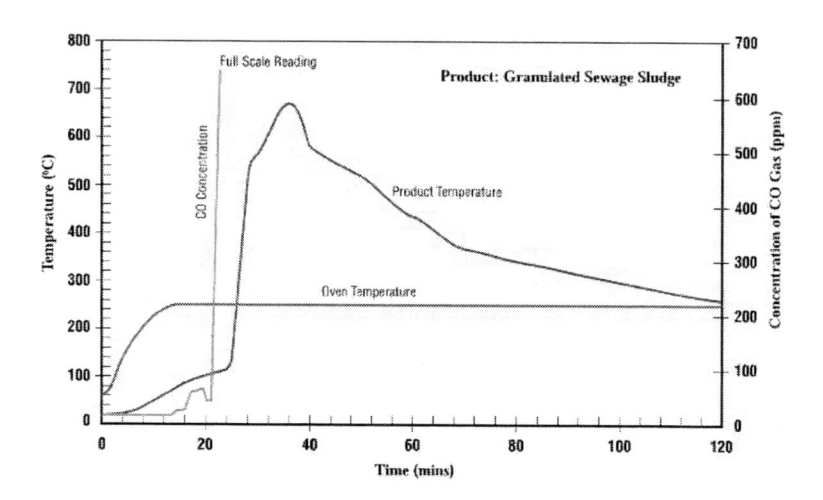

Figure 3: Temperature Rise and Carbon Monoxide Evolution Rates

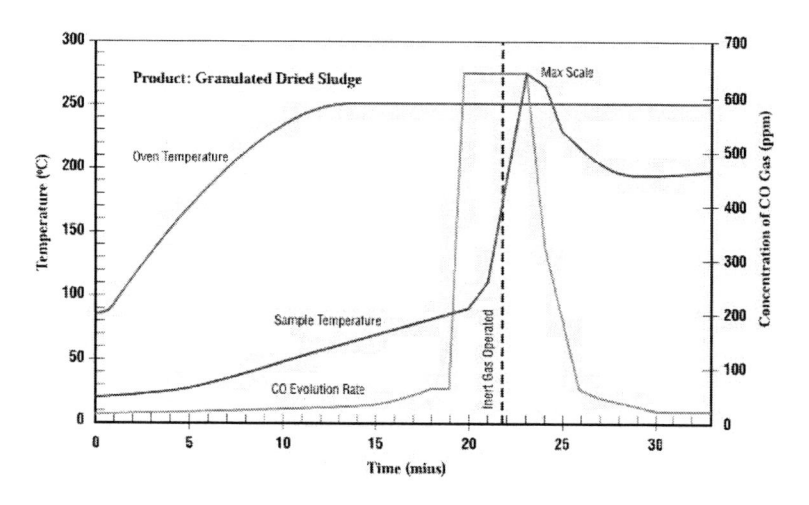

Figure 4: Effect of Inert Gas Purging

Figure 5: Explosion vents

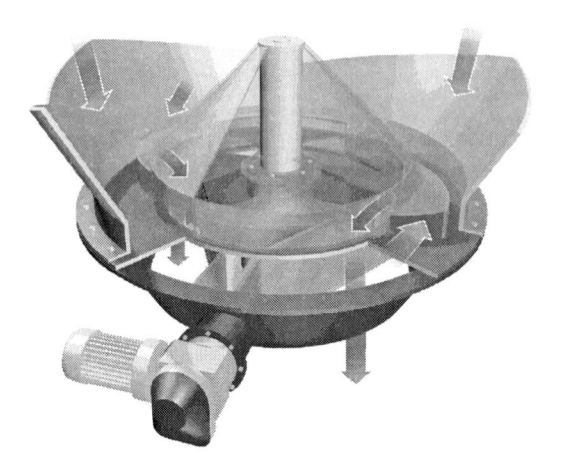

Figure 6: Rotoflo Discharge Valve

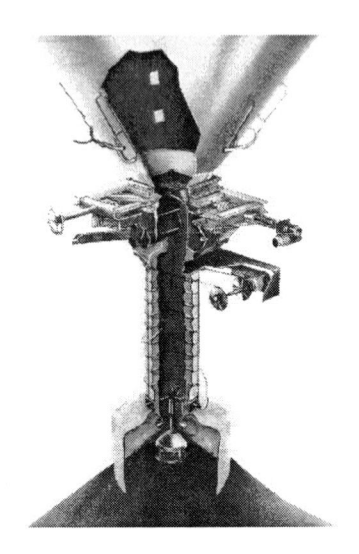

Figure 7 Tanker Loading Hood

Figure 8 Loading Hood During Discharge

Typical Flowsheet

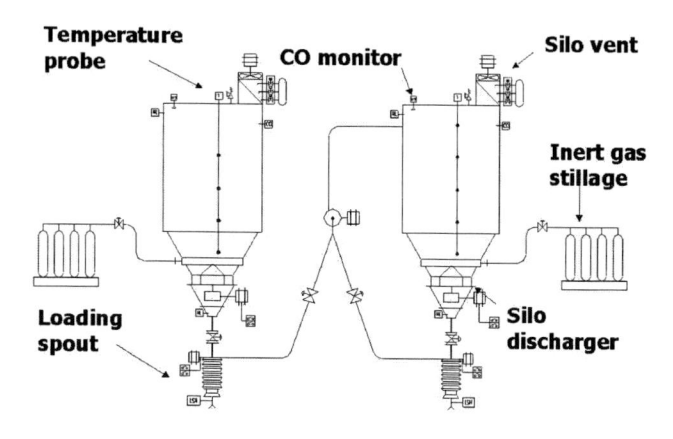

Figure 9   Flowsheet of Sewage Sludge Silos

Figure 10 Installation of a safe storage solution.

Figure 11 Tanker loading installation

CASE STUDY: FIRE IN A THERMAL SEWAGE SLUDGE DRYER

Steven J. Manchester
Fire and Risk Sciences Division (FRS), BRE Ltd Bucknalls Lane Garston Watford WD25 9XX.
Tel. 01923 664944 Fax. 01923 664910 E-mail: manchesters@bre.co.uk

ABSTRACT

This paper describes the initial dust explosion and self-heating tests that were undertaken on dried sewage sludge samples taken from a thermal sewage sludge dryer. Following a fire incident in the drying air handling unit of the plant, an investigation into the fire was undertaken by BRE in conjunction with the dryer operating personnel. A detailed HAZOP study of the process was undertaken at which BRE's findings, following the investigation, were discussed. It was recognised that the standard dust explosibility and self-heating tests used for assessing process safety hazards, were not sufficient to fully characterise the particular conditions within the dryer. Hence, a programme of research work was identified as being required in order to provide the necessary data for the safe operation of the dryer.

KEYWORDS

Dryer, dust, explosion, fire, HAZOP, investigation, self-heating, sewage, sludge.

BACKGROUND

There are approximately 110 drying plants in Europe, using a number of different processes to dry the sewage sludge. All have the following in common.

- The production of a combustible solid as granules/pellets and dust.
- In the presence of sufficient oxygen and with numerous potential sources of ignition present in the process, there is a considerable risk of an explosion or fire occurring.

Since 1997 a number of significant fire and explosion incidents have occurred at sites in the UK and throughout Europe resulting in damage to process equipment. Many of these incidents can be prevented if the hazards present during the drying process are fully identified and the risks quantified. Only then can adequate preventative and protective measures been taken.

In de-watering and drying sewage sludge a large quantity of dust and final dried product is produced, both of which are readily combustible. They will form the fuel in what is known as the "fire triangle". This refers to the necessary requirements for a fire or explosion to occur in any system. The first requirement is the presence of a fuel, and in order for the fuel to burn a sufficient quantity of oxygen is required. Finally, there must be an ignition source present of sufficient energy to ignite the fuel air mixture.

Although dust explosions have been known for over two centuries in industries such as flour milling and coal mining, they were a new phenomenon to the water industry when they started producing large amounts of dust as a consequence of their drying operations on sewage sludge. The fire hazards arising from layers of dust and bulk storage of the final product were also relatively unknown to the water industry. Many drying plants were supplied, installed and commissioned without sufficient thought to the possible fire and explosion risks posed by the production of fine dry dust and final granular/pellet product. As a result a number of fires and explosion incidents occurred, as the conditions required for an ignition of dried sewage sludge material are present during the drying operations of a de-watering plant.

INTRODUCTION

Soon after commissioning of a new sewage sludge drying plant in 1999, BRE undertook a site visit to identify the potential fire and explosion hazards, and recommend prevention and protection measures. As a consequence a range of dust explosion tests and isothermal self-heating tests on two samples of dust and the final granular product respectively were undertaken. The tests undertaken followed existing and forthcoming British, European and International standards, or where no standards existed, tests followed the descriptions given in published guidance[1,2]. About 2 years later a fire incident occurred within the drying air filter unit of the plant. BRE was asked to investigate the cause of the fire.

OVERVIEW OF THERMAL SLUDGE DRYER OPERATION

Digested sewage sludge is held in two large storage tanks and then pumped to two centrifuges where it is mixed with a polyelectrolyte thickener to produce a slurry of about 25% dry solids and 75% water. This material is then taken by screw conveyor to a twin shaft mixer, where it is mixed with fine dust from the recycle silo to increase the quantity of dry solids allowing the material to flow more easily. Material from the twin shaft mixer is then taken, via the drying drum feed screw conveyor, to the rotating drying drum operating at a nominal drying air inlet temperature of 400°C and a nominal drying air outlet temperature of 125°C.

As the sewage sludge passes along the dryer drum it is dried producing water vapour that is used to keep the oxygen level in the plant at a low enough concentration to protect against a dust explosion. At the outlet of the drum dryer the dried granular product, along with dust, then enters the drying air filter unit where a cyclone, separates the bulk solids from the fines. A reverse jet filter unit is located directly above the cyclone and this separates the fines from the drying air. The dry product is transferred via a rotary valve at the bottom of the drying air filter/cyclone to a cooled screw conveyor and into a bucket elevator. Fines collected on the reverse jet filter are added back into the product by pulsing the filter bags with compressed air to shake off the product.

The bucket elevator takes the product to a sizer unit, which separates the product into three streams; undersize, oversize and final product. The undersize and oversize material falls into the recycle silo, it is then carried by a dosing screw conveyor to the twin shaft mixer.

The final product is transferred, via a water-cooled screw conveyor, to a storage silo located outside the main building. The final product enters the storage silo at approximately 30 to 40°C and is stored here until it is taken to the packaged bagging plant, via the silo outlet bucket conveyor.

The drying process is operated under an effectively closed circuit system, which is used to maintain an inert atmosphere in the rotating drum dryer and the cyclone/filter unit during normal operation. The heated air from the combustion chamber enters the drier at a maximum temperature of 400°C and the water evaporated in the drying process reduces the oxygen concentration to between 8% and 9%. This reduces the air temperature at the outlet of the dryer to a maximum of 125°C. Excess moisture and particulates are removed before the water vapour/air mixture is reheated and recycled back to the dryer.

FIRE INCIDENT

In order to carry out routine maintenance an automatic shutdown sequence was started in the sewage sludge dryer. The plant manager and an operator were in the drying plant building when about ten minutes into the automatic shutdown procedure an alarm was heard signalling that a rotary valve was jammed and the dryer was placed into an emergency shutdown procedure. Smoke or steam was seen to come from the twin-shaft mixer along with some black dust. Water was seen to be pouring from the drum dryer outlet.

The oxygen concentration level inside the plant was seen to be dropping and, assuming a fire had started inside the drying plant, the Fire Service was called. The fire was found to be located in the top section of the drying air filter unit. The top plates of the unit were each removed in turn and the fire inside doused with water until extinguished.

Inspection of the damage after the fire showed extensive burning to the filter bags, which were located in the top of the air filter unit. The top sections of the air filter unit also showed some signs of buckling. The right-hand section of the filter unit seemed to suffer more damage than the other side.

The BRE investigation of the incident revealed that in the few days prior to the auto shutdown of the plant for maintenance, all the operating parameters were recorded as being well within the operating ranges. The drying air drum inlet temperatures were all below 400°C, the drying air drum outlet temperatures were all below 126°C and the oxygen concentration levels were below 9.5%. None of these values would have been high enough to trigger an alarm condition in the operating system and would not have been regarded as abnormal.

During the auto-shut down, two different features occurred compared to a normal shutdown:

a) The rotary valve on the air filter unit became jammed.

b) The air outlet temperature in the drum dryer reached a peak temperature of 135°C, compared to a typical temperature of about 120°C, although in at least one previous auto shutdown a peak temperature of 135 °C had been obtained. Possible reasons for this increased temperature may have been dryer than normal residual sludge in the dryer drum or the heat generated from the early stages of the fire in the air-filter unit.

Unfortunately there was no conclusive evidence to definitively identify the cause of the ignition. However, from the nature of the drying system, the uncertainty of the level of oxygen required to produce a smouldering combustion in a sewage sludge deposit and experience by the author of other similar incidents in sludge dryers, it is most likely the source was the self-heating of a dust deposit. This may have been produced in the drum dryer or in the air filter unit, in areas where dust could accumulate.

The specific points to note are:

- The most likely source of ignition of the fire was glowing/smouldering particles of dried sewage.
- The smouldering particles were most likely formed from the self-heating of a deposit of dust, either in the drum or air filter unit.
- The time to ignition of a deposit of dust, based on the previous self-heating test work undertaken, usually takes hours rather than minutes, depending on the size of the deposit. This would tend to suggest that the self-heating reaction was initiated before the start of the auto shutdown procedure.
- The oxygen concentration and temperature in the dryer plant prior to the start of the auto shutdown must have been sufficiently high enough to initiate a self-heating reaction in a deposit.
- The oxygen concentration may have been high due to the plant running at slightly higher levels of oxygen than normal, even though this may be within the plant design operating parameters, or from a leakage into the plant of air creating a localised oxygen rich environment.
- During the auto shutdown the oxygen concentration starts to climb. This would have increased the self-heating reaction leading to combustion involving glowing/smouldering particles being formed.

In the report of the fire incident investigation a number of recommendations were given for plant operation, these included:

- Maintain the oxygen concentration to a level closer to 7% until more information is available on the behaviour of smouldering deposits in a steam
- Ensure the oxygen concentration monitor is regularly calibrated and checked to ensure it is working accurately.
- As the oxygen concentration is the basis of safety for the plant, ideally a back-up oxygen concentration sensor should be installed.
- Check the plant for structural defects and weak joints to try to ensure that air cannot leak into the parts of the plant where it is necessary to maintain an inert atmosphere.
- Modify the dryer drum and air filter unit to try and remove areas where dust can accumulate.
- Maintain an inert atmosphere in the plant during an auto shutdown and an emergency shutdown, until the plant has reached a safe pre-determined temperature. This could possibly be achieved by injecting nitrogen or carbon dioxide into the plant to act as the inerting agent.
- Assess the operation and maintenance of the whole plant to ascertain potential risk from the ignition of deposits. Particularly at risk are plant items that may be liable to contain deposits of dried sewage sludge subjected to heat and oxygen concentration levels high enough to initiate self-heating.
- Following a risk assessment of the fire hazards of the plant, investigate the most appropriate fire detection and extinguishing system

Testing/research work:

- Undertake air-over layer test to determine what temperature thin dust layers in the dryer/air filter can ignite in air.
- Determine the level of oxygen required to produce smouldering combustion in the sewage sludge material, under steam inerting conditions.
- Determine the limiting oxygen concentration to prevent an ignition of a dust cloud of sewage sludge, under steam inerting conditions.

HAZOP STUDY

Following the BRE investigation of the fire incident a HAZOP study was convened to look in detail at the air re-circulation loop, which included the dryer drum, the re-circulation air filter and the heat exchanger. The purpose was to review the implications of the incident and determine what remedial action would be necessary in order to minimise the risks of such an incident occurring again.

Deviations of temperature, pressure and oxygen concentration were examined during normal operation, start-up and shut-down. BRE also presented the results of its investigations to the assembled HAZOP team.

A series of actions were agreed following discussions within the meeting. One of the actions was to undertake tests to determine the self-heating behaviour of dried sewage sludge dust deposits under steam inerting conditions and to undertake a study of the oxygen level at which dust clouds could be ignited under steam inerting conditions. A detailed programme of work was then compiled by BRE to look at these two potential ignition hazards.

RESEARCH WORK ON SEWAGE SLUDGE DUST IGNITION UNDER INERT STEAM CONDITIONS

The programme of research work was compiled in order to provide ignition data for the sewage sludge dust in two key areas. These were:
1. the maximum permissible oxygen concentration to prevent a dust cloud explosion using steam as the inert media.
2. the maximum permissible concentration of oxygen required to initiate self-heating of a deposit of dust using steam as the inerting media.

DETERMINATION OF THE LIMITING OXYGEN CONCENTRATION (LOC) TO PREVENT A DUST CLOUD EXPLOSION USING STEAM

INTRODUCTION

The aim if this part of the project was to provide data on the effect of steam on the Limiting Oxygen Concentration (LOC) of sewage sludge dust taken from the thermal drying plant. The results would be used to determine the safe oxygen concentration operating level to prevent ignition of a dust cloud. The sample used was obtained from grinding the granular final product in a ball mill and then sieving through a 125 micron sieve. A particle size analysis was then undertaken using a Malvern Instruments Laser Particle Sizer, with a dry powder feed (table 1).

APPARATUS

Routine testing of LOC's are undertaken in the 20-litre sphere pressure vessel. This apparatus was used for this work item, and is described in detail by the manufacturers Kuhner AG in their latest operating manual[3]. The apparatus (Figures 1 and 2) consists of a spherical chamber with a volume of 20 litres and surrounded by a water jacket. Dust enters the sphere from a 0.6 litre pressurised storage chamber via a pneumatically operated outlet valve. The sample is injected by compressed air and a perforated deflector plate inside the chamber ensures uniform dispersion.

However, a number of modifications to the test vessel were required in order to obtain a steam inerting atmosphere inside the vessel at the moment of ignition of the dust cloud.

Temperatures of above 100°C were obtained inside the 20-litre sphere pressure vessel by attaching three 1m lengths of heating tape around the sphere body, one length around the lid and one length around the dust storage chamber. These were connected to a Eurotherm temperature controller, set at the required temperature. The temperature inside the sphere was measured using a K-type 1mm diameter thermocouple, which was connected to the Eurotherm temperature controller.

In order to maximise the heat from the heating tapes, the sphere (including the storage chamber) was wrapped in commercially available loft insulation. The water jacket surrounding the sphere, which is normally used to provide cooling water, was filled with vegetable oil to improve the heat conductance through the wall of the sphere from the heating tape on the outside to the internal surface.

The steam used in the test as the inerting medium was generated using distilled water from a 2000ml measuring cylinder, which was pumped, through a 10m length of coiled copper tubing inside an oven. The oven was set at a temperature of 150- 200°C, depending on the temperature required inside the sphere. Air from a compressed air line via a flowmeter was mixed with the water prior to being fed into the top of the oven. The water pump was calibrated so that the exact water feed rate was known during a test. By knowing the flowrate of the water/steam and the air, a known steam/air mixture could be generated, which then flowed along a further 2m length of copper pipe before entering the sphere inlet valve. This pipework was also heated at the end nearest to the sphere by heating tape, and was lagged with loft insulation along its entire length from the oven to the sphere inlet valve.

Immediately prior to testing, the sphere is partially evacuated so that the injection of the dust to be tested, using either air or nitrogen, results in the pressure inside the sphere at ignition of 1 bar a (atmospheric). To achieve this, a vacuum pump is used to suck the inert gas/air out of the sphere. When using steam as the inerting media, a pre-vacuum liquid catchpot was installed, along with a moisture absorbing silica gel trap, to ensure excessive moisture was not sucked into the vacuum pump.

EXPERIMENTAL PROCEDURE

The 20-litre sphere was set-up as described above and work was undertaken to ensure that temperatures of at least 100°C could be achieved in the test apparatus. Once this had been established, steam was piped into the sphere with air and checks made to ensure that the steam/air inert atmosphere inside the sphere could be generated and maintained. The next stage was to ensure the steam atmosphere inside the sphere could be evacuated without impairing the vacuum pump by ensuring the catchpot and drying media removed excessive water. Once these trials had been completed and modifications made as required, tests using samples of sewage sludge dust were started.

Initially, the sample was tested at room temperature using nitrogen as the inert gas, i.e. a normal LOC test. These results were then used as a basis for undertaking tests using steam as the inert gas. Tests at temperatures of 100°C and 130°C were undertaken. The test procedure used follows the accepted method described in the draft European CEN standard (prEN14034-4:2002)[4] and in the equipment manufacturer's manual[3].

RESULTS

The test results indicated that clouds of dried sewage sludge dust from this plant will ignite at lower oxygen levels, when tested using steam as the inert gas at temperatures of 100°C and 130°C, compared to using nitrogen as the inert gas at room temperature. When tested in nitrogen at room temperature the LOC measured was 12 - 13%. Using steam as the inerting agent gave a LOC of 10 - 11%.

DETERMINATION OF THE LIMITING OXYGEN CONCENTRATION REQUIRED TO PREVENT SELF-HEATING OF SLUDGE AT ELEVATED TEMPERATURE.

INTRODUCTION

The aim of this part of the research work was to provide data on the level of oxygen required to prevent the self-heating of sludge at elevated temperature in an inert atmosphere of steam in order to assess the safe operating regime of the dryer. In order to determine the self-heating characteristics of a dust sample, isothermal tests were undertaken using a commercially available fan-assisted oven. A sample was placed inside an oven, set at a particular temperature, and monitored using thermocouples for exothermic behaviour, i.e. a rise in temperature above the oven temperature. If the temperature of the sample rises by more than 50°C above the oven temperature, ignition is deemed to have occurred. Further tests are then undertaken at successively lower oven temperatures until no ignition occurs. This procedure is undertaken using four different sample volumes as the exact quantities of material that may form a deposit inside the dryer are not known. It also enables the extrapolation of the results to plant-scale size deposits to be undertaken if required. This method is described in detail by Bowes[5] and Beever[6].

From the results of the tests it is possible to calculate the critical ignition temperature of a deposit of known size and shape, the time to ignition of a deposit and hence specific safe storage/operating temperatures, volumes and residence times.

SAMPLE PREPARATION

The sample supplied for testing was the granular final product material, as the fine dust from the drying air filter was not available in sufficient quantities. In order to test the dust, the granular material was ground in a ball mill and then sieved through a 500 micron sieve.

APPARATUS

One of the BRE large fan-assisted ovens was used for these tests. The thermocouples used were 0.5mm diameter K-type thermocouples connected to a Linseis datalogger and PC. The test samples were held in purpose built sample cells fabricated from 24 gauge stainless steel and formed into cylinders closed at the base and fitted with a detachable lid. Four sizes of sample cell were used: 115mm x 75mm diameter, 140mm x 100mm diameter, 165mm x 125mm diameter and 190mm x 150mm diameter. These hold sample volumes of 508cm^3, 1100cm^3, 2025cm^3 and 3358cm^3 respectively.

A short length of 6mm OD pipe is fitted into both the top and the base to allow the air or reduced oxygen atmospheres to pass through the sample cell. A steel mesh is inserted into the bottom of the sample cell to support the sample material, 20mm from the base and 10mm above a gas diffuser. The gas diffuser is fabricated from 95mm x 6mm OD stainless steel pipe drilled with 1mm holes at 10mm intervals along two sides. Three 0.6mm diameter holes are located in the wall and base of the sample cell to allow thermocouples to be inserted to monitor gas inlet and bulk sample temperatures.

The steam used in the tests as the inerting medium was generated using distilled water from a 10 litre container, which was pumped using a peristaltic pump through a 10m length of coiled copper tubing inside the self-heating oven. The oven was set at a temperature of between 120 - 150°C, depending on the temperature required for each test. Air from a compressed air line via a flowmeter was mixed with the water prior to being fed into the top of the oven. The water pump was calibrated so that the exact water feed rate was known during a test. By knowing the flowrate of the water/steam and the air a known steam/air mixture could be generated, which then flowed into the sample cell inside the oven containing the test sample.

TEST PROCEDURE

The dust sample under test was placed in the test cells of different sizes (75 mm, 100 mm, 125 mm, 150 mm). The cells were filled with the sample and levelled with a straight edge. The material was not compacted into the cell. The filled cell was then suspended in an oven, pre-heated and thereafter maintained at a known temperature to within ± 1°C.

The centre and side temperatures of the sample were monitored using 0.5 mm, stainless steel sheathed, chromel/alumel thermocouples. These were connected to a chart recorder and personal computer so that the self-heating process could be observed and recorded.

Two types of behaviour were observed for the tests in air:

1   the central sample temperature rising by a relatively small amount, < 50°C, above the pre-set oven temperature and then gradually falling back to the oven temperature (sub-critical behaviour, see Figure 3), or

2   the central sample temperature rising to a high value, between 500°C and 700°C, associated with combustion and indicating ignition (super-critical behaviour, see Figure 4).

Isothermal self-heating tests were initially undertaken with the sample in air. If the sample ignited, then the oven was reset to a lower temperature and a fresh sample tested. Conversely, if a sample failed to ignite, the test was repeated with a fresh sample at a higher temperature. In this way, the minimum ambient temperature for ignition was determined for each cell size used. This temperature is called the critical ignition temperature and is specific to the volume of the sample tested. The 100mm diameter test cell was used to start with and then the remaining three sample volumes were tested.

Once the critical ignition temperatures in air had been determined for each sample size, the test was repeated at this temperature using steam to reduce the oxygen concentration to the required level. It became evident when undertaking tests with the 75mm and 100mm sample cells, that the oxygen levels required for ignition were as high as 18%. Hence, the temperature at which the tests were undertaken was increased to a level at least 5°C above the critical ignition temperature.

RESULTS

The results of the tests undertaken clearly show that the temperature at which ignition occurs reduces as the sample size increases. For the range of sample sizes tested, it was found that oxygen concentration levels above 7 - 8% will undergo self-heating leading to an ignition.

The time taken to ignite the sample tested varies with the volume of material. The smaller the volume the shorter the time to ignition. For the smallest sized samples tested (508cm^3), the time taken for ignition to start was in the range 3.0 - 4.5 hours.

MODIFICATIONS TO THE PLANT

During normal operation, steam is produced from the drying process of the wet sludge, which is used to reduce the oxygen level within the drying drum and drying air filter unit. The level of oxygen maintained inside the plant is now based on the values of the LOC determined using steam. Shutdown of the plant was recognised as being a potentially hazardous operation due to the rapid rise of the oxygen level inside the plant, while still at relatively high temperatures. Nitrogen is now injected into the plant to replace the steam in order to keep the oxygen level at a safe concentration while the temperature inside the dryer decreases. The LOC tests in nitrogen and in steam have provided the data on which to base the required quantity of nitrogen required to maintain a safe oxygen concentration.

Ideally, in order to prevent fires from occurring inside the plant as a consequence of self-heating, areas inside the plant where dust can accumulate over periods of time should be either eliminated or regularly cleaned. In practice it is difficult to ensure the complete absence of deposits. However, keeping the volume of material of accumulated dust deposits to as low a level as possible is advantageous, as the smaller the volume of material the higher the ignition temperature that will lead to self-heating. Maintenance procedures during shut-down that encompass the removal of deposits will help to reduce the risk of self-heating.

Running the plant at as low a temperature as practically possible will also be beneficial in that it means that small deposits present may not reach their critical ignition temperature and only larger volumes can be ignited.

The level of oxygen present in the plant also has a bearing on the ignition potential of a deposit. During normal operation, oxygen levels should ideally be kept below 8%, based on a deposit volume of less than 3358cm^3 (the maximum size deposit tested in this study). If, either as a result of operating experience with the plant or in theoretical calculations, it is determined that deposits greater in volume than 3358cm^3 may be present in the plant, then isothermal calculations can be undertaken to determine the critical ignition temperature for this volume.

CONCLUSIONS

- Dried sewage sludge can self-heat to ignition and the dust cloud can pose an explosion hazard.

- The limiting oxygen concentration, under steam inerting conditions needs to be known to enable safe operational parameters to be set and to ensure safety measures are adequate.

- The fire investigation and subsequent studies identified plant shut-down as a potential high risk operation, if the oxygen concentration is allowed to climb to a sufficiently high enough concentration while there is a dust cloud present in the drying air filter/drying drum. An engineering solution was provided by injecting nitrogen into the plant during shut-down to maintain the inert atmosphere inside the plant.

REFERENCES

1. Field, P. Dust explosions. Handbook of Powder Technology Vol. 4, Elsevier, Amsterdam 1982.

2. Dust Explosion Prevention and Protection - A Practical Guide. Edited by John Barton. IChemE. 2002.

3. CESANA, C SIWEK, R. Operating instructions for the 20-litre sphere. 5.2. 1999.

4. prEN 14034-4 Determination of the explosion characteristics of dust clouds - *Part 4 Determining the limiting oxygen concentration of dust/air atmospheres*. 2001

5. BOWES, P.C. Self-heating: Evaluating and Controlling the Hazards. Building Research Establishment. HMSO. 1984

5. BEEVER, P. F., Spontaneous Combustion - Isothermal Test Methods, Building Research Establishment Information Paper IP23/82 (1982).

TABLES
Table 1. Particle size distribution using 100mm lens

Sample	<500µm	<125µm	<74µm % volume	<32µm	<20µm	Median (µm)
Sewage sludge	100	78	49	23	16	77

Figure 1: 20-litre sphere

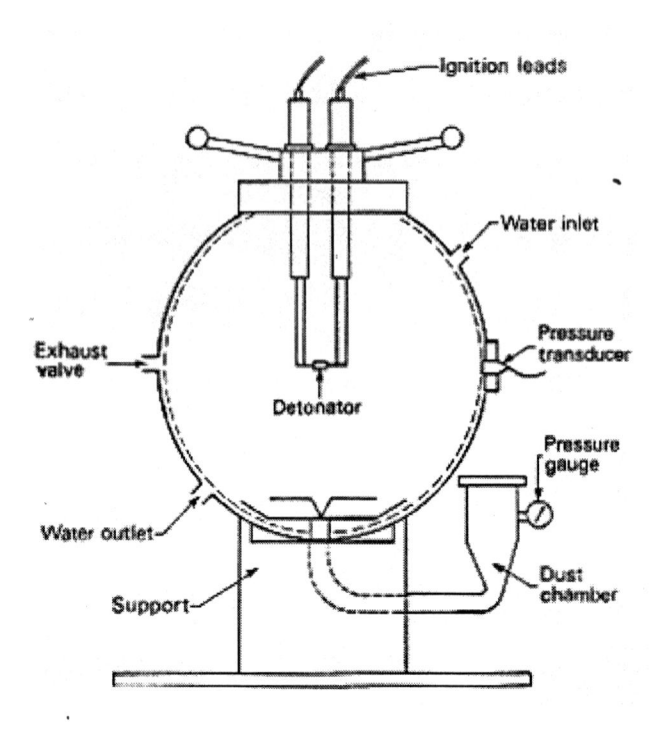

Figure 2   : 20-litre sphere (schematic)

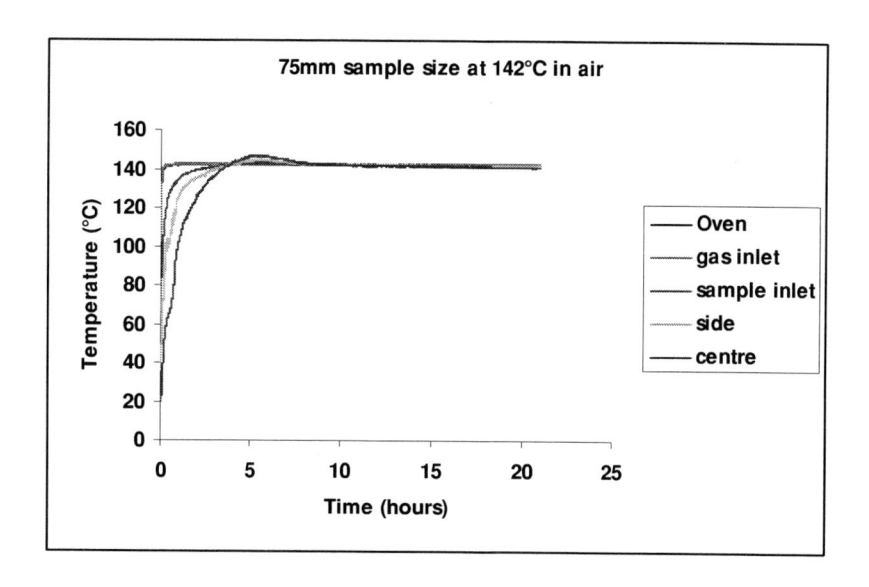

Figure 3: Sub-critical self-heating

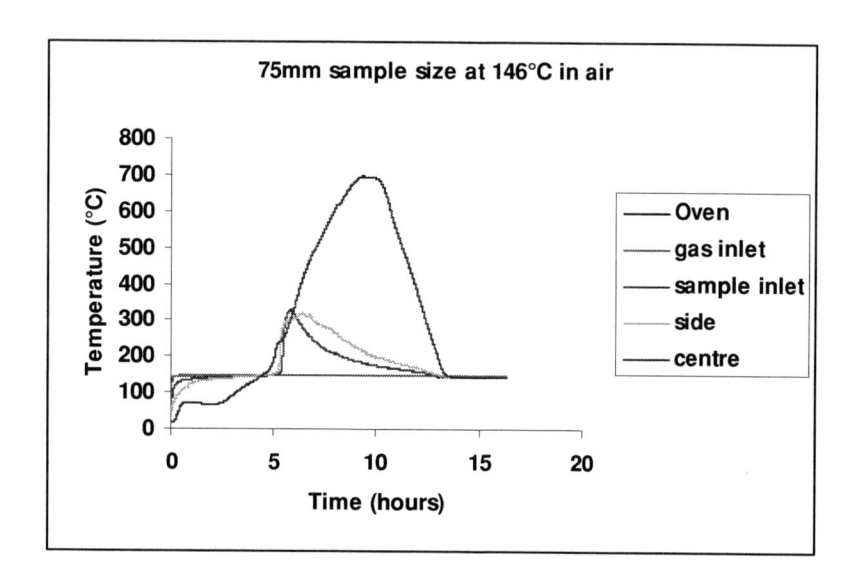

Figure 4: Super-critical self-heating

THE IMPACT OF THE HSE AND DSEAR ON THE UK WATER INDUSTRY

[1]Alan Whipps, [2]Robin Lloyd
[1]Pell Frischmann Consultants, Exeter. [2]South West Water Limited, Exeter.
Corresponding Author Tel: +44 1 392 444345   Fax: +441392 444880
e mail:awhippps@pellfrischmann.com

ABSTRACT

The paper presents the findings of HSE Inspectors visit to a sewage sludge thermal drying plant in the context of the HSE Information Document (847/9). The consequential impact of the visit and further remedial work is discussed. The impact of the Dangerous Substances and Explosive Atmosphere Regulations (DSEAR), that are released this autumn, are reviewed in the light of the HSE Information Document (847/9). DSEAR introduce into UK legislation the requirements of a number of EU ATEX Directives dealing with potentially explosive and hazardous situations. The impact of DSEAR on thermal drying installations is relatively clear. However, there are greater and far-reaching implications for the UK Water Industry and these are considered in detail.

KEY WORDS

Explosion, Dangerous Substances and Explosive Atmosphere Regulations. , hazards, HSE, risk assessment, sewage sludge, thermal drying .

BACKGROUND TO THE HSE INFORMATION DOCUMENT

In the UK, considerable investment has been made in thermal drying facilities to accommodate the significant increase in sewage sludge production. Within the UK, almost 40 thermal drying facilities have been installed from Scotland, Ireland and nine out of the ten Water Service Companies recognising the advantages the process can provide. The Health and Safety Executive (HSE) became involved with sewage sludge drying processes in the UK whilst investigating safety arrangements following a number of explosion/combustion incidents. During the investigation of explosion incidents at Nash STW, Newport (1998) and Wigan STW (1999), it became apparent to the HSE that other sewage sludge thermal drying facilities had experienced minor thermal events that had gone unreported.

Following considerable investigation and consultation with the Water Industry and suppliers of thermal drying systems in the UK, the HSE issued Information Document (HSE 847/9) "Control of Health and Safety Risks at Sewage Sludge Drying Plants" in July 2001. The Information Document seeks to identify good practice, formalises safety aspects and seeks to ensure compliance with legal requirements of drying facilities. The Information Document does not restrict consideration to the thermal drying plant but considers the complete drying process as a whole. The document includes the properties of sewage sludge, risk assessments, pre-dryer storage, the thermal drying process, dry product handling, dry product storage, bagging and final transportation and storage systems.

Discussions with the HSE have confirmed the principal objective of producing the Information Document to be raising awareness of risks associated with thermal drying processes and protecting the health, safety and welfare of personnel, in particular, plant operators. The HSE is not concerned with the protection and security of the process plant. However, this is of key interest to the Water Industry and additional explosion suppression and other protective systems have been introduced at a number of installations.

The HSE have undertaken a programme of inspections at operational thermal drying facilities throughout the UK. The following section describes the outcome of inspections to facilities in South West Water Ltd' s area.

THERMAL DRYING AND SOUTH WEST WATER

South West Water Ltd (SWWL) have two sewage sludge thermal drying facilities, located at Barnstaple (Ashford) STW and Plymouth Central STW. The plants were supplied by OSC Process Engineering Ltd based on the indirect two-stage drying process developed by Buss AG of Switzerland. The sites are strategic regional sludge facilities and the drying plant are designed to process raw (ie undigested) sewage sludge from a number of sewage treatment works.

No serious thermal events have been recorded for the SWWL plants, although high ferric concentrations were linked with high operating temperatures in the second stage (Rovactor) dryer at the Ashford site (1997/98). Also, smouldering dry product was removed from the storage silo at Plymouth Central (December 1998).

Safety systems were reviewed by OSC Process Engineering Ltd following the explosion incident at Nash STW, Newport. Additional equipment, including explosion suppression, explosion vent panels and temperature monitoring were provided to both SWWL sites. Following a review of SWWL installations and the Information Document (847/9), the following key activities were identified:-

- review HAZOP studies and hazard assessments;

- review plant documentation to ensure consistency;
- separately document the Basis of Safety of the process;
- review plant start-up and shutdown procedures;
- review potential for contamination and place greater scrutiny on the import of wastes to the site;
- improve operational records, in particular atypical operating conditions;
- revise procedures for calibration and maintenance of safety critical systems

In March 2002, an Operational HAZOP was conducted to identify variances with the initial Hazop studies carried out at the design stage and to gain the operational experiences of SWWL staff.

Also in March 2002, dry sludge samples were transported to Chilworth Technology Ltd to determine the properties of the sewage sludge and safety critical parameters. The following key parameters were determined at that time:-

Minimum Ignition Energy:	> 100mJ
Minimum Ignition Temperature:	440°C
Layer Ignition Temperature:	> 260°C
Explosion Severity (K_{st}):	150 barg/m/s
Dust Classification:	St 1
Maximum Pressure (P_{max}):	7.1 bar
Limiting Oxygen Concentration:	10%

HSE INSPECTIONS

HSE Inspectors visited the SWWL sites of Ashford and Plymouth Central STW in April and May 2002 respectively. The formal inspection was undertaken by two HSE Inspectors with experience of industrial applications and explosion/combustion incidents. The inspection commenced with a review of the P&I diagrams of the process with specific reference to the Information Document (847/9). The inspection systematically followed the Information Document through each stage of the process. A review of the operational plant was also undertaken. The following items were discussed in detail:-

- at the time of the inspection, fully certified results had not been produced by Chilworth Technology Ltd for the properties of the Ashford sludge. However, from the initial data, the K_{st} value of 150 barg/m/s was significantly higher than the original design provision of 100 barg/m/s. The HSE sought confirmation of adequacy of the explosion suppression system and what changes occurred to safety parameters during the holiday season when sludge properties may alter:
- develop/enhance explosion relief panel to the dry product bucket elevator;
- develop a specific safety file containing details of site safety provisions;
- produce summary statements for hazard assessments and the various Hazop studies;
- investigate improvements to ventilation within the Dryer Building to lower temperatures for process and personnel safety;
- develop a specific calibration file to record and monitor instrument calibration.

The HSE Inspectors advised that a return visit would be arranged within 12 months to assess the progress being made on the items identified above. Additional sludge analysis is currently being undertaken by Chilworth Technology Ltd to determine the impact of seasonal changes on sludge properties.

BACKGROUND TO EU LEGISLATION

A number of EU Directives have been introduced over the past 10 years that have both a direct and indirect impact on the Water Industry. The following are summarised:-

Machinery Directive: 98/37/EC[2] is aimed at removing trade barriers throughout the EU and requires protection systems to be provided for fire and explosion associated with machinery.

ATEX Directive: 94/9/EC[3] identifies requirements for equipment and protective systems to be used in potentially explosive atmospheres. The definition of "equipment" is broader than "machinery" in 98/37/EC and is principally aimed at suppliers of equipment. In the UK, these are implemented by the Equipment and Protective Systems for Use in Potentially Explosive Atmospheres Regulations 1996.

Chemical Agents Directive: 98/24/EC[4] requires an assessment of hazards to be undertaken and measures introduced to prevent hazardous concentrations, avoid ignition sources or mitigate the effects of fires and explosions.

ATEX Directive (137): 1999/92/EC[5] introduces minimum requirements for improving the safety and health protection of workers potentially at risk from explosive atmospheres. Similar to the hazard requirements of the Chemical Agents Directive, this ATEX Directive has a broader influence across processes and applications. The Directive requires employers to assess the specific risks from explosive atmospheres and record these in an "explosive protection document". This document should be prepared prior to commencement of work and revised when changes occur during installation, commissioning and throughout plant operation. The document therefore remains a "live document" for ongoing protection throughout the life of the plant. Employers are to classify hazardous areas, provide appropriate equipment and signs to designate the areas.

EU member states are required to implement the ATEX Directive (137) with local legislation for new or modified facilities after 30 June 2003. Existing operational facilities are to comply fully by 30 June 2006. In the UK, however, fire and explosion hazards are covered by a large number of existing pieces of legislation.

Some of the existing UK legislation is old and has limited relevance to modern industrial applications. The HSE considered the option of simply adding the requirements of ATEX Directive (137) as a new stand-alone regulation would confuse issues. As a consequence, the decision was taken to review and completely overhaul UK legislation, including repealing unnecessary or conflicting legislation.

THE DANGEROUS SUBSTANCES AND EXPLOSIVE ATMOSPHERES REGULATIONS 2002 (DSEAR 2002)

The HSE developed draft regulations to implement the requirements of ATEX Directive (137) and the Chemical Agents Directive, seeking consultation from all industry sectors. It is anticipated that the final Regulations become law in November 2002 and aim to provide a consistent and coherent focus on hazard assessment and minimising risks for employees.

APPLICATION OF DSEAR

DSEAR 2002 applies to all areas where "dangerous substances" may be present irrespective of the quantity of the substances, the size of the process or a minimum number of employees affected. A "dangerous substance" is defined as a substance (or mixture of substances) that can give rise to fire, explosion or similar energetic event which affects the safety of employees or others. As a consequence, DSEAR 2002 will apply to most commercial and industrial sectors and all associated work activities.

Of significance, DSEAR 2002 places joint responsibility on employers, where employees of one employer works at another employer's premises, to collaborate and co-operate to ensure compliance with the Regulations. Also, the Regulations apply to self-employed persons as if they were both the employer and the employee.

The HSE have stressed that DSEAR 2002 does not introduce new requirements for the process industry but provides a clear focus on the requirements and clarifies the situation for those less experienced with UK process industry procedures and risk assessments.

KEY REQUIREMENTS OF DSEAR

The following summarises the key requirements of DSEAR 2002. Employers, which include contractors, subcontractors and self-employed persons, are to ensure:

Risk Assessments: These are to be undertaken for risks arising from dangerous substances. Risks shall be assessed for people not directly employed but remain at risk from the dangerous substance.

Documentation: Significant findings of risk assessments are to be recorded at the time the assessment is made. The "explosion protection document" as required by ATEX 137 is not specifically required by DSEAR 2002 provided details are formally recorded.

Elimination or Reduction of Risks: Where practical, elimination of risks may be possible to avoid the presence or use of dangerous substances. However, for many process applications, the following hierarchy should be adopted:-

- Reduction - reduce the quantity to a minimum, control or minimise the release of the dangerous substance and prevent the formation of an explosive atmosphere.
- Avoidance - avoid an ignition source, or adverse conditions which could cause dangerous substances to give rise to harmful effects. Ensure secure segregation of incompatible dangerous substances.
- Mitigation - reduce number of employees exposed, avoid propagation of fires and explosions, provide explosion relief or suppression systems, provide suitable containment, provide personal protective equipment and safe handling, storage and transport facilities.

Hazardous Area Classification: Where explosive atmospheres may occur, the areas are to be classified as hazardous or non-hazardous. Hazardous classification means special precautions are required to protect the health and safety of workers.

Hazardous Areas: Work places that are classified as hazardous must, prior to their use for the first time, have overall explosion safety independently verified. Employers must ensure that explosion protection devices are maintained and appropriate work clothing provided.

Equipment and protective systems must be selected in accordance with the zoning classification and be suitable for gases, vapours, mists and dusts as appropriate. DSEAR 2002 also require hazardous areas to be designated with signs at points of entry to aid the safety of employees.

Accidents, Incidents and Emergencies: Accident procedures for first aid and safety drills must be provided.

For incidents, warning or communication systems must be available for response, remedial action and rescue operations. Also, warnings must be available before an explosive condition is reached for safe evacuation of employees, together with escape facilities.

For the emergency services, information on site emergency procedures must be made available so that their own response and precautionary measures can be prepared.

Information, Instruction and Training: When a dangerous substance has been identified, the employer shall name the substance and the risk it poses, linking to Material Safety Data Sheets where appropriate. Legislative requirements and significant findings of the risk assessment shall be identified, together with appropriate instruction and training to safeguard employees.

The HSE consider the principal objective of the DSEA Regulations is to ensure that risks to workers from explosions are as low as reasonably practicable. This is consistent with the existing risk assessment requirements of the Management of Health and Safety at Work Regulations 1999[6]. As a consequence, HSE's view is that the majority of DSEA Regulations requirements apply with immediate effect.

LEGISLATION REPEALED BY DSEAR 2002

The following key pieces of legislation will be replaced by DSEAR 2002:-

- The Factories Act 1961, Section 31
- The Highly Flammable Liquids and Liquefied Petroleum Gases Regulations 1972
- The Workplace (Health, Safety and Welfare) Regulations 1992, Regulation 6(3)(b)
- The Carriage of Dangerous Goods by Road Regulations 1996, Schedule 12
- The Carriage of Dangerous Goods (Classification, Packaging and Labelling) and Use of Transportable Pressure Receptacles Regulations 1996, Regulation 22(b)

APPROVED CODE OF PRACTICE

As DSEAR repeals existing pieces of legislation, the specific guidance associated with this legislation will be lost. To overcome this issue, the HSE have developed Approved Codes of Practice and Guidance. The following activity specific codes and guidance are being prepared to complement the DSEA Regulations:-

- assessing and controlling the risks from hot work;
- the storage of dangerous substances;
- material separation;
- using ventilation to control risks;
- identification and control of ignition sources;
- design of plant, equipment and process areas;
- permits to work;
- procedures and equipment for dealing with emergencies, leaks, spills, etc;
- safe maintenance, repair and cleaning procedures;
- the safe disposal of waste materials and redundant equipment.

IMPACT OF DSEAR 2002

The DSEA Regulations focus the attention of issues relating to employee safety, explosion hazards and clarify the requirements of existing UK legislation in conjunction with the ATEX Directive (137).

It can be seen from the structure and approach that has been adopted in the formulation of HSE Information Document (847/9) the clear objectives that have now been laid down within the DSEA Regulations: what the original Information Document lacked in terms of mandatory requirements have now been verified and clarified. As a consequence, the HSE Information Document (847/9) is supported and strengthened by DSEA Regulations 2002 and this will provide renewed focus of attention on certain compliance issues. In particular, the documentation of hazard and risk assessments, together with the hazardous zoning classification, are seen as key requirements that need to be addressed.

However, for the UK Water Industry, the DSEA Regulations 2002 are not limited to thermal drying systems with potential fire and explosion hazards. A wide range of plant, equipment and processes will fall under DSEAR 2002 and employers will need to take account of that position from November 2002. For the Water Industry, relatively simple equipment such as heating plant and gas-fired boilers fall within the scope of DSEAR 2002 and this will impact on all sludge digestion systems installed throughout the UK.

Of significance, with greater interest of outsourcing in the UK Water Industry, employers' responsibilities will need to be co-ordinated between the various parties involved on a site to comply with DSEAR 2002.

With reference to the HSE Information Document (847/9), the involvement of the HSE has brought the Water Industry together, particularly the users of the thermal drying processes. Regular meetings between individuals operating thermal dryers in the UK have continued in the guise of the Dryer User Group. This Group is a proactive forum that continues to co-operate on health and safety issues associated with thermal drying plants. This has included a recent meeting with the HSE where a number of issues relating to the Information Document were discussed. In particular, the outcome of recent thermal testing of sludge and the impact of the classification and packaging requirements of the dried product as dangerous goods (ISBN 0-7176-1221-X)[7] and the ability of dried sludge to self-heat.

It is understood that the HSE will update Information Document (847/9) in 2003 once the full results of the thermal testing programme have been assessed and to link specifically with the DSEA Regulations.

REFERENCES

1. HEALTH & SAFETY EXECUTIVE (2001). Control of Health and Safety Risks at Sewage Sludge. Drying plants - Information Document (HSE 847/9)

2. COUNCIL OF EUROPEAN COMMUNITIES. The Machinery Directive (98/37/EC)

3. COUNCIL OF EUROPEAN COMMUNITIES. The Explosive Atmospheres Directive (94/9/EC)

4. COUNCIL OF EUROPEAN COMMUNITIES. The Chemical Agents Directive (98/24/EC)

5. COUNCIL OF EUROPEAN COMMUNITIES. The Explosive Atmospheres Directive (1999/92/EC)

6. STATUTORY INSTRUMENT 3242. Management of Health and Safety at Work Regulations, 1999.

7. HEALTH & SAFETY EXECUTIVE (1999). ISBN 0-71 76-1221-X. Approved Requirements and

   Test Methods for the Classification and Packaging of Dangerous Goods for Carriage

POTENTIAL IMPACT OF THE EU SOIL STRATEGY ON THE REVISION OF THE SLUDGE DIRECTIVE.

McDonnell, E.,
DEFRA, UK

ABSTRACT

This paper looks at the recent draft publication by the EU relating to the development of a soil strategy, which can be adopted by Member States. This is expected to lead to a new soil directive, which will form the basis for subsequent directives and member State regulations for the recycling of organic materials to land.

This paper failed to make the printers deadline. It should be available on the conference CD

THE IMPACT OF THE REVISED SLUDGE TO LAND REGULATIONS UPON PRACTICES IN ENGLAND AND WALES

Rowlands, C.,
Severn Trent Water, UK

ABSTRACT

During the extended period of the development of the revised UK regulations, there has been much concern that the heightened focus upon biosolids might adversely affect stakeholder's perceptions making it increasingly difficult to obtain sufficient land and maintain the goodwill of the farmer customer. In the run up to the new regulations, the a UK water Industry and the Environment Agency have been in discussion to develop a common understanding of how HACCP will be introduced. As well as the above consideration of the above aspects, the paper puts biosolids use into perspective when considered alongside other materials used in agriculture in terms of relative quantities, qualities and regulatory control. The impact of the redesigninations of Nitrogen Vulnerable Zones is also discussed.

This paper failed to make the printers deadline. It should be available on the conference CD

IMPLEMENTATION OF HACCP CONTROLS UNDER THE NEW SLUDGE (USE IN AGRICULTURE) REGULATIONS.

[1]Brian Crathorne, [2]Chris Rowlands [3]Paul Bryson, [3]Janet Cochrane, [3]Nina Sweet.
[1]Thames Water and UK Water, [2]Severn Trent and UK Water, [30]Environment Agency

ABSTRACT

The revised UK Sludge (Use in Agriculture) Regulations will require sewage sludge treatment operators to develop and operate a Hazard Analysis & Critical Control Point (HACCP) system for those parts of the treatment process involved in achieving obligations for pathogen reduction. HACCP procedures were developed in the food industry and are now widely used for the control of processes in this sector. Water Companies were therefore able to draw on experience from the food industryin setting out the framework for a HACCP control system for sludge treatment. However the development of a robust HACCP system capable of controlling treatment processes such that specific pathogen reduction requirements are met routinely, presented a considerable challenge. This was a challenge, not only for the operators in developing and applying suitable systems but also for the Environment Agency, responsible for enforcing the regulations. Appropriate enforcement is of particular importance in providing stakeholders with the confidence that the regulations are being implemented. In November 2000, Water UK set up a HACCP Group to discuss and develop a set of common procedures within which HACCP systems should operate. Regular meetings have also taken place with a group set up by the Environment Agency to cover this issue and over the last 2 years a common framework has been developed and agreed. This paper will set out some of the issues tackled by the Group, present details of the operating framework agreed and discuss the way in which the regulatory controls will be enforced.

This paper failed to make the printers deadline. It should be available on the conference CD